APPLETON & LANGE'S
OUTLINE REVIEW
CLINICAL CHEMISTRY

Robert H. Christenson, Ph.D., DABCC, FACB
Professor of Pathology
Professor of Medical and Research Technology
Department of Pathology
University of Maryland School of Medicine
Baltimore, Maryland

Linda C. Gregory, Ph.D., MT(ASCP), DABCC
Consultant
Rochester, Minnesota

Lisa J. Johnson, M.H.S., MT(ASCP)SC
Assistant Professor
Department of Medical and Research Technology
University of Maryland School of Medicine
Baltimore, Maryland

Appleton & Lange Reviews/McGraw-Hill
Medical Publishing Division

New York   Chicago   San Francisco   Lisbon   London
Madrid   Mexico City   Milan   New Delhi
San Juan   Seoul   Singapore   Sydney   Toronto

McGraw-Hill

*A Division of The **McGraw·Hill** Companies*

Appleton & Lange's Outline Review: Clinical Chemistry

Notice

Medicine is an ever-changing science. As new research and clinical experience broaden our knowledge, changes in treatment and drug therapy are required. The authors and the publisher of this work have checked with sources believed to be reliable in their efforts to provide information that is complete and generally in accord with standards accepted at the time of publication. However, in view of the possibility of human error or changes in medical sciences, neither the authors nor the publisher nor any other party who has been involved in the preparation or publication of this work warrants that the information contained herein is in every respect accurate or complete, and they disclaim all responsibility for any errors or omissions or for the results obtained from use of the information contained in this work. Readers are encouraged to confirm the information contained herein with other sources. For example and in particular, readers are advised to check the product information sheet included in the package of each drug they plan to administer to be certain that the information contained in this work is accurate and that changes have not been made in the recommended dose or in the contraindications for administration. This recommendation is of particular importance in connection with new or infrequently used drugs.

This book was set in Garamond by Rainbow Graphics.
The editor was Patricia Casey.
The production supervisor was Phil Galea.
Project management was provided by Rainbow Graphics.
The cover designer was Elizabeth Pisacreta.
Courier Printing was the printer and binder.

This book is printed on acid free-paper.

Library of Congress-in-Publication Data

Appleton & Lange's outline review of clinical chemistry / [edited by] Robert H. Christenson, Linda C. Gregory, Lisa J. Johnson.
    p. cm.
    ISBN 0-07-031847-6
    1. Clinical chemistry.   I. Christenson, Robert H. II. Gregory, Linda C. III. Johnson, Lisa J. IV. Title: Clinical chemistry.
    [DNLM: 1. Chemistry, Clinical—Examination Questions.   Qy 18.2 A649 2001]
    RB40.A67 2001
    616.07´56´076—dc21

                                                                00-062469

To my mother:
   Deirdre H. Driscoll
and my family:
   Vicki, Mark, Eric, and Paul

R.H.C.

To my son:
   Michael Federspiel
and my husband:
   Mark Federspiel

L.C.G.

To my parents:
   Talliette and Leon
for their love and support

L.J.J.

Contents

Contributors

F. Philip Anderson, Ph.D., DABCC
Assistant Professor
Co-Director, Clinical Chemistry Laboratory
Department of Pathology
Medical College of Virginia/Virginia Commonwealth
  University
Richmond, Virginia
Separation Techniques

**Hassan M.E. Azzazy, Ph.D., SC(ASCP), DABCC,
  FACB**
Assistant Professor
Department of Pathology
University of Maryland School of Medicine
Baltimore, Maryland
*Molecular Diagnostics, Disease-Specific Analytes:
  Biochemical Markers of Cardiac Injury, Biochemical
  Markers of Bone Metabolism*

Larry A. Broussard, Ph.D., DABCC
Associate Professor
Department of Clinical Laboratory Sciences
School of Allied Health Professions
Louisiana State University Health Sciences Center
New Orleans, Louisiana
Therapeutic Drug Monitoring, Toxicology

Robert H. Christenson, Ph.D., DABCC, FACB
Professor of Pathology
Professor of Medical and Research Technology
Department of Pathology
University of Maryland School of Medicine
Baltimore, Maryland
Separation Techniques

Janine Denis Cook, Ph.D., DABCC
Assistant Professor
Department of Medical and Research Technology
University of Maryland School of Medicine
Baltimore, Maryland
*Endocrine Function, Method Evaluation, Management
  and Supervision*

Patrick J. Cummings, Sc.D., MT(ASCP)
Associate Program Chair
Master of Science in Biotechnology Program
Johns Hopkins University
Baltimore, Maryland
*Immunoassay; Disease-Specific Analytes: Markers of
  Viral Disease, Tumor Markers*

John S. Davis, M.B.A., MT(ASCP)SC, DLM
Program Director & Assistant Professor
Department of Clinical Laboratory Sciences
School of Allied Health Professions
Louisiana State University Health Sciences Center
Shreveport, Louisiana
Photometric and Electrochemical Measurements

Glenn C. Flodstrom, M.S., MT(ASCP)
Visiting Lecturer
Department of Clinical Laboratory Sciences
Community College of Philadelphia
Philadelphia, Pennsylvania
Calculations & Statistical Analyses

Linda C. Gregory, Ph.D., MT(ASCP), DABCC
Consultant
Rochester, Minnesota
Lipids and Lipoproteins, Metabolic Analytes

Paul A. Griffey, M.B.A., MT(ASCP)
Assistant Professor
Department of Medical and Research Technology
University of Maryland School of Medicine
Baltimore, Maryland
Management and Supervision

Denise M. Harmening, Ph.D., MT(ASCP), CLS(NCA)
Professor and Chair
Department of Medical and Research Technology
University of Maryland School of Medicine
Baltimore, Maryland
Carbohydrates

Audrey E. Hentzen, Ph.D., MT(ASCP)
Program Director & Assistant Professor
Clinical Laboratory Science Program
Illinois State University
Normal, Illinois
Regulatory Issues for the Clinical Laboratory

Debra R. Johnson, M.A.Ed., MT(ASCP)SC
Assistant Manager, Pathology/Chemistry
Pitt County Memorial Hospital
Greenville, North Carolina
Clinical Enzymology

Lisa J. Johnson, M.H.S., MT(ASCP)SC
Assistant Professor
Department of Medical and Research Technology
University of Maryland School of Medicine
Baltimore, Maryland
Carbohydrates

Steven C. Kazmierczak, Ph.D., DABCC, FACB
Professor
Department of Pathology and Laboratory Medicine
East Carolina University School of Medicine
Greenville, North Carolina
Clinical Enzymology

Mary Kay Boehmer O'Connor, MT(ASCP)
Laboratory Services Manager
Group Practice Services Corporation
Randolph, New Jersey
Quality Assurance and Quality Control

Christine Papadea, Ph.D., MT(ASCP)SC
Associate Professor
Department of Pathology and Laboratory Medicine
Division of Laboratory Medicine
Medical University of South Carolina
Charleston, South Carolina
Amino Acids and Proteins

Wendy R. Sanhai, Ph.D.
Postdoctoral Fellow in Clinical Chemistry
Department of Pathology
University of Maryland School of Medicine
Baltimore, Maryland
Automation and Computerization

John G. Toffaletti, Ph.D., DABCC
Associate Professor
Department of Pathology
Duke University Medical Center
Durham, North Carolina
Critical Care Analytes

Preface

Like most areas in clinical science and medicine, the discipline of clinical chemistry has experienced great evolution over the years, bringing to mind the old adage *"the only thing constant is change."* There are several critical companions of this constant change: One is the need for professionals involved in science, technology, and medicine to keep up and remain competent in the practice of their chosen profession. Indeed, regulatory bodies mandate documentation of competency. A second critical companion is the need for professionals who have completed their training to demonstrate proficiency through successful performance on national certification examinations. The central purpose of *Appleton & Lange's Outline Review: Clinical Chemistry* is to provide a resource that focuses on these needs. The audience of this work is intended to be broad, including a diverse group of scientific and medical laboratory professions. The overall method has involved the recruitment of contributions from individuals who have dedicated their careers to the training, mentoring, and evaluation of laboratory professionals.

The chapters in this professional review are divided into three parts. The first is "Classic Analytes," which discusses biochemical species within the purview of clinical chemistry. The second part is "Instrumentation and Analytical Techniques," which deals with measurement principles central to performance of clinical laboratory testing. The third part is "Laboratory Operations"; this part encompasses material critical for regulatory compliance, technical operation, and management of a clinical laboratory.

A practical approach was taken in the design of this work. Each chapter begins with a list of key terms, the learning objectives, a list of analytes or methods (where applicable), followed by multiple choice questions intended for self-assessment. The text is intended to follow the general flow from theory, physiology, and medical overview, through technical issues and testing procedures, and the interpretation of results and medical relevance. At the end of each chapter, the authors extend some "Items for Further Consideration." The format is intended to facilitate establishing personal expectations, quantitative self-assessment, and focused review of material, all toward the end point of competency in clinical chemistry.

Our goal with *Appleton & Lange's Outline Review: Clinical Chemistry* was not to produce another detailed and comprehensive clinical chemistry reference; rather, the intention for this work is to provide laboratory professionals with a useful resource to periodically refresh and polish their skills.

Robert H. Christenson
Linda C. Gregory
Lisa J. Johnson

Acknowledgments

The editors wish to express deep gratitude to Gershon J. Shugar, Ph.D., and John E. Hutton III, B.S., for their essential work in initiating *Appleton & Lange's Outline Review: Clinical Chemistry.*

part I

Classic Analytes

Carbohydrates

Lisa J. Johnson and Denise M. Harmening

Key Terms

AMADORI REARRANGEMENT	*GLYCOSYLATED HEMOGLOBIN*	*MICROALBUMINURIA*
CARBOHYDRATES	*HEMOGLOBIN A_{1c}*	*POLYDIPSIA*
DIABETES MELLITUS	*HYPERGLYCEMIA*	*POLYPHAGIA*
GESTATIONAL DIABETES MELLITUS	*HYPOGLYCEMIA*	*POLYURIA*
GLUCAGON	*INSULIN*	*TYPE 1 DIABETES MELLITUS*
GLUCOSE	*KETOACIDOSIS*	*TYPE 2 DIABETES MELLITUS*
GLYCATED HEMOGLOBIN	*KETONES*	

Chapter Objectives

Upon completion of this chapter, the reader will be able to:

1. Identify the primary biological function of carbohydrates in humans.
2. Describe, in general terms, the metabolism of carbohydrates in humans.
3. Identify and explain the role of the following hormones in glucose metabolism: insulin, glucagon, epinephrine, thyroxine, adrenocorticotropic hormone (ACTH), cortisol, and growth hormone.
4. Discuss the symptomatology of hypo- and hyperglycemia in humans.
5. Characterize the following types of diabetes mellitus: type 1, type 2, and gestational diabetes mellitus.
6. Discuss the major acute and chronic complications associated with diabetes mellitus.
7. Describe how the following laboratory tests are used in the evaluation of hypo- or hyperglycemia: blood glucose, glycated hemoglobin (hemoglobin A_{1c}), and ketones.
8. Discuss the role of self-monitoring devices for diabetics in the measurement of blood glucose.

9. State the criteria recommended by the American Diabetes Association for the diagnosis of diabetes mellitus.
10. Compare methods for glycated hemoglobin analysis in regard to product measured and frequency of use in clinical laboratories.
11. Discuss laboratory tests used to evaluate the presence of ketoacidosis and microalbuminuria.

Analytes Included ◀

ALBUMIN

CREATININE

GLUCOSE

GLYCATED HEMOGLOBIN/HBA$_{1c}$

KETONES

Pretest

Directions: Choose the single best answer for each of the following items.

1. The hormone that facilitates the entry of glucose into the cell is
   a. epinephrine
   b. cortisol
   c. insulin
   d. glucagon

2. In cases of hypoglycemia, the body responds by releasing
   a. insulin and thyroxine
   b. ketoacids and proteins
   c. sorbitol and fructose
   d. glucagon and epinephrine

3. One of the acute complications of diabetes mellitus is
   a. ketoacidosis
   b. blindness
   c. coronary artery disease
   d. microalbuminuria

4. In a type 1 diabetic, insulin concentration is
   a. elevated
   b. low
   c. normal
   d. correlated to age of onset

5. Autoimmune disorders are associated with which of the following types of diabetes?
   a. type 1
   b. type 2
   c. non–insulin-dependent
   d. gestational

6. A 47-year-old woman with a history of diabetes mellitus in her family is most likely to develop which of the following types of diabetes?
   a. type 1
   b. type 2
   c. gestational
   d. secondary

7. The type of diabetes that develops in pregnancy is
   a. hereditary
   b. associated with certain human lymphocyte antigens (HLA)
   c. permanent
   d. gestational

8. Diagnostic criteria for diabetes mellitus, as defined by the American Diabetes Association, is based on which of the following tests?
   a. albumin:creatinine ratio
   b. glycated hemoglobin
   c. oral glucose tolerance
   d. plasma glucose

9. Which of the following blood collection additives inhibits continued glycolysis?
   a. potassium citrate
   b. sodium fluoride
   c. lithium heparin
   d. ethylenediaminetetraacetic acid (EDTA)

10. In regard to carbohydrate analysis, enzymatic methods measure
    a. glucose
    b. all reducing sugars
    c. mono- and disaccharides
    d. total glycated proteins

11. Which of the following tests is used to monitor the glycemic status of diabetics during the previous 3 to 4 months?
    a. urine dipstick for glucose
    b. fructosamine
    c. ketostix
    d. glycated hemoglobin

12. One of the most common errors associated with the self-monitoring blood glucose (SMBG) devices is
    a. insertion of the test strip in the device
    b. application of sample on the test strip
    c. interference from tissue juices in the sample
    d. improper calibration

13. How does whole blood glucose concentration, measured in SMBG devices, compare to plasma glucose measured in the clinical laboratory?
    a. whole blood glucose is typically lower than plasma glucose
    b. whole blood glucose is typically higher than plasma glucose
    c. plasma glucose represents a fraction of whole blood values
    d. there is no difference in whole blood and plasma values

14. The retrospective time period that glycated hemoglobin concentration represents is dependent on the
    a. hemoglobin concentration in the red blood cell
    b. amount of insulin administered
    c. life span of the red blood cell
    d. amount of protein ingested

15. Which of the following conditions might cause an increase in glycated hemoglobin?
    a. hemolytic anemia
    b. sickle cell disease
    c. iron deficiency anemia
    d. ABO blood type discrepancy

16. The term that describes the stable ketoamine structure when glucose is irreversibly bound to hemoglobin is
    a. glycosylated
    b. glycated
    c. aldimine
    d. Amadori rearrangement

17. The fraction of adult hemoglobin that is often measured to assess glycemic control for the past 3 to 4 months is
    a. HbA_{1a}
    b. HbA_{1b}
    c. HbA_{1c}
    d. HbA_2

18. Glycated hemoglobin analysis is currently utilized for
    a. diagnosis of diabetes mellitus
    b. screening high-risk individuals
    c. prognosis of acute complications
    d. monitoring of glycemic status

19. In regard to diabetics, the presence of microalbuminuria is indicative of
    a. developing nephropathy
    b. low-molecular-weight proteins that attack the kidney
    c. advanced retinopathy
    d. the fractionation of albumin due to microvascular disease

20. Which of the following laboratory assessments is used to screen for the presence of microalbuminuria in diabetics?
    a. protein dipstick on a random spot urine
    b. quantitative turbidimetric analysis on a 24-hour urine
    c. blood urea nitrogen (BUN) on a 24-hour urine
    d. albumin:creatinine ratio on a random spot urine

I. INTRODUCTION

Carbohydrates serve as a major energy source for the sustainment of life. Hence, the role and importance of the regulation of carbohydrate metabolism for living organisms cannot be overestimated. In addition to their main role in energy storage and generation, carbohydrates also serve as structural materials, such as in the cell walls of bacteria, and aid in molecular recognition and communication, such as the blood group sugars on cell surfaces. Since disorders of carbohydrate or glucose metabolism have a wide range of effects, it is critical to maintain control of the blood concentration. The principal disease resulting from a disturbance of glucose metabolism is diabetes mellitus, which is characterized by high levels of glucose in the plasma and deficient levels in the cells where energy is needed. Diabetes is a common disorder, affecting 3% of the U.S. population and approximately 10% of the population over the age of 65. The cost of care for diabetes ranges from 12 to 14% of the total U.S. health care expenditures. In 1997, the American Diabetes Association (ADA) estimated the cost of diabetes care to be $98.1 billion. Moreover, a recent report from the Diabetes Treatment Centers of America stated that a commercial health plan member of a managed care organization who has diabetes costs approximately $368 per member per month, whereas members without the disease can be managed for about $90 per member per month.

II. CARBOHYDRATE METABOLISM

A. BIOCHEMISTRY

▶ **Carbohydrates** serve as a major source of energy for the body and are classified into three main groups: monosaccharides, oligosaccharides, and polysaccharides. Monosaccharides are simple sugars, such as glucose and fructose, which have four to eight carbon atoms and contain an aldehyde or ketone group. The monosaccharide in highest concentration is glucose, a six-carbon aldose with the formula ($C_6H_{12}O_6$). Oligosaccharides have two to six molecules of simple sugars linked together. Disaccharides, the most frequently encountered oligosaccharides in nature, are composed of two monosaccharides linked together. Large numbers of monosaccharides linked together form a group of compounds called polysaccharides, as exemplified by glycogen and starch.

Carbohydrates must be broken down to monosaccharides before they can be absorbed by the intestinal mucosal cells. In the digestive process, salivary amylase mixes with food, hydrolyzing starch. As food is swallowed, this enzyme continues digestion until it is inactivated by gastric hydrochloric acid. In the duodenum, the action of pancreatic amylase hydrolyzes these partially digested polymers further into intermediate-sized glucosans. Other intestinal enzymes, such as maltase, lactase, and sucrase, continue the degradation until monosaccharide units are formed. Following absorption through the intestinal mucosa, the monosaccharides are transported to the liver, where the liver con-
▶ verts fructose and galactose into **glucose.** Glucose is either released into the circulation to be used for energy in the tissues, converted to

Table 1–1. Major Biochemical Pathways Associated
           with Carbohydrate Metabolism

• Glycolysis	Glucose → pyruvate ↔ lactate
• Glycogenesis	Glucose → glycogen
• Tricarboxylic Acid Cycle	Glucose → pyruvate → CO_2 + H_2O + ATP
• Pentose Phosphate Pathway/ Hexose Monophosphate Shunt	Glucose → ribose + CO_2 + NADPH
• Uronic Acid Pathway	Glucose → glucuronic acid
• Glycogenolysis	Glycogen → glucose
• Gluconeogenesis	Noncarbohydrate sources → glucose

glycogen in the liver for storage, utilized in skeletal muscle for protein synthesis, or used in adipose tissue for triglyceride synthesis. All energy production in the cells begins with glycolysis. However, there are multiple metabolic pathways that are involved. Table 1–1 lists some of the major metabolic pathways.

B. HORMONAL CONTROL

Glucose is under tight hormonal control. Though a number of different hormones affect plasma glucose concentration, insulin and glucagon are the primary regulators. Under conditions of normal dietary intake and nutrition, the adult blood glucose level during fasting is usually between 70 and 105 mg/dL. After a meal high in carbohydrates, it may rise to 130 to 160 mg/dL. **Insulin,** secreted by the ◄ beta cells of the pancreas, facilitates the entry of glucose into hepatocytes, adipocytes, red cells, and monocytes by making the cell membrane permeable to glucose. **Glucagon,** secreted by the alpha ◄ cells of the pancreas, stimulates glycogenolysis (breakdown of glycogen into glucose) and gluconeogenesis (production of glucose from noncarbohydrate sources such as proteins and fats). Thus, insulin secretion ultimately reduces plasma glucose while glucagon secretion results in increased plasma concentration. Other chemical regulators that cause an increase in blood glucose include epinephrine, human growth hormone (hGH), cortisol, adrenocorticotropic hormone (ACTH), and thyroxine (see Table 1–2).

III. CLINICAL OVERVIEW

A. HYPOGLYCEMIA

Hypoglycemia, characterized by low blood glucose concentration, ◄ can result from many different causes including overadministration of insulin or oral hypoglycemic agents, severe hepatic dysfunction, insulinoma, and inborn errors of metabolism; hypoglycemia may range in severity from insignificant to life threatening. When plasma glucose levels reach approximately 65 mg/dL, the body responds by secreting glucagon and epinephrine. The symptoms of hypoglycemia, however, are typically not manifested until blood glucose concentrations reach 50 to 55 mg/dL. Most symptoms are related to the release of epinephrine and include anxiety, dizziness, chills, tachycardia, and increased perspiration in the early stages. If the body is unable to correct the blood glucose concentration through

Symptoms of Hypoglycemia

Early stages:
   Anxiety
   Dizziness
   Chills
   Tachycardia
   Increased perspiration
Later stages:
   Slurred speech
   Loss of motor coordination
   Loss of consciousness/coma
   Irreversible nerve damage

Table 1–2. Hormones that Regulate Glucose Metabolism

HORMONE	SECRETED BY	PRIMARY FUNCTION	ACTIONS
Insulin	Pancreas (β cells of islets of Langerhans)	Decreases blood glucose levels Movement of glucose from blood into cells	• Increases uptake of glucose by muscle and fat cells through increased cell membrane permeability to glucose • Increases uptake of glucose by the liver (promotes glycogenesis and lipogenesis) • Inhibits hepatic output of glucose into circulation • Increases synthesis of protein in liver, muscle, and fat cells
Glucagon	Pancreas (α cells of islets)	Increases blood glucose levels and free fatty acids Mobilization of energy stores	• Stimulates the breakdown of liver glycogen • Promotes liver gluconeogenesis • Promotes hepatic lipolysis
Epinephrine	Adrenal medulla	Increases blood glucose levels Mobilization of energy stores	• Stimulates breakdown of glycogen • Decreases insulin release from the pancreas • Promotes breakdown of triglycerides in fat tissue
Growth hormone	Anterior pituitary	Increases blood glucose levels Mobilization of energy stores	• Inhibits glucose uptake by the tissues • Increases hepatic glucose availability • Inhibits lipogenesis from carbohydrate • Antagonist to insulin
Adrenocorticotropic hormone (ACTH)	Anterior pituitary	Increases blood glucose levels Mobilization of energy stores	
Glucocorticoids (cortisol)	Adrenal cortex	Increases blood glucose levels Mobilization of energy stores	• Promotes protein catabolism • Promotes deamination of amino acids • Promotes gluconeogenesis (diversion of amino acids) • Inhibits glucose metabolism in the peripheral tissues • Antagonist of insulin
Thyroid hormones Thyroxine (T_4) Triiodothyronine (T_3)	Thyroid gland	Increases blood glucose levels Mobilization of energy stores	• Increases absorption of glucose from the gastrointestinal tract • Stimulates glycogenolysis • Accelerates the degradation of insulin

hormonal action, other, more severe symptoms such as slurred speech, loss of motor coordination, loss of consciousness or coma, and irreversible nerve damage may develop.

B. HYPERGLYCEMIA—DIABETES MELLITUS

▶ **1. Etiology and Symptomatology.** The most common cause of **hyperglycemia,** or elevated blood glucose, is **diabetes mellitus.** Diabetes mellitus is a chronic disease that represents a heterogeneous group of metabolic disorders. The development of diabetes involves several etiologies, including the absolute decrease or absence of insulin, the relative decrease in insulin (insulin resistance), an abnormality in the control of insulin secretion, and a defect in the action of insulin on the target tissue cell. Defective action of insulin on target tissues results in abnormalities in carbohydrate, fat, and protein metabolism. Both defects in insulin action and deficiencies of insulin secretion frequently occur in the same patient.

The symptoms of diabetes are related to the pathophysiology of hyperglycemia and are commonly characterized by the "3 Ps"—polyuria, polydipsia, and polyphagia. Increased glucose excretion in the urine causes an osmotic diuresis, which results in **polyuria,** or increased urine output. This, in turn, causes an increase in thirst, referred to as **polydipsia.** The depletion of cellular energy and utilization of stored energy results in **polyphagia,** or an increase in appetite. Other symptoms include weight loss and an increased susceptibility to infection. Table 1–3 summarizes the symptoms of diabetes mellitus.

> **The 3 Ps of Diabetes Symptoms**
>
> Polyuria—increased urine output
> Polydipsia—increased thirst
> Polyphagia—increased appetite

2. Acute and Chronic Complications of Diabetes. The complications of diabetes, which result from hyperglycemia, are multifaceted and affect almost every organ system in the body. Both macrovascular and microvascular disease develop. Table 1–4 lists the major acute and chronic complications of diabetes mellitus.

Hypoglycemia may develop due to overadministration of insulin or oral hypoglycemic agents, as well as impaired glucagon response to hypoglycemia. The opposite pattern occurs in diabetic **ketoacidosis,** in which an insulin deficiency or defect in insulin action results in a low concentration of intracellular glucose accompanied by a high concentration in the blood. The cellular response to the lack of glucose is the mobilization of fats for energy. The by-products of this process are the ketoacids acetone, acetoacetate, and β-hydroxybutyrate. The accumulation of these ketoacids produces a metabolic acidosis. The acute complications of hypoglycemia and ketoacidosis are of particular importance and may develop into life-threatening situations.

In chronic complications, hyperglycemia results in glucose binding to proteins, creating excessive protein glycosylation. This buildup of excessive glycated end products causes abnormalities in vascular permeability, enhanced lipoprotein deposits, cross-linking of collagen, and adverse effects on capillary endothelial cells. The walls of the capillaries thicken due to deposits of the glycated proteins. This may eventually cause the vessels to become occluded and is the underlying pathogenesis of renal complications (nephropathy) and vision complications (retinopathy) associated with diabetes mellitus. The retinopathy associated with diabetes is the number one cause of blindness in

Table 1–3. Symptoms of Diabetes Mellitus

- Polyuria, polydipsia, and polyphagia
- Weight loss
- Increase in bacterial and yeast infections
- Fatigue
- Blurred vision
- Frequent vaginal infections and cessation of menstruation (women)
- Impotence (men)

Table 1–4. Complications of Diabetes Mellitus

Acute
- Hypoglycemia
- Ketoacidosis

Chronic
- Neuropathy
- Retinopathy
- Nephropathy
- Angiopathy (coronary artery disease, stroke)
- Infection
- Tissue necrosis (gangrene)

the United States. Diabetes is also the leading cause of end-stage renal disease. Moreover, it is one of the major causes of coronary artery disease, myocardial infarction, and stroke.

Hyperglycemia also causes another type of sugar, sorbitol, to be produced in high concentrations in tissue cells, such as nerve cells, that do not rely on insulin for the uptake of glucose. High levels of intracellular sorbitol increase the intracellular osmotic pressure, causing water to enter. The nerve damage (neuropathy) caused by diabetes has been attributed to this type of swelling in nerve cells. This swelling often results in a loss of feeling in the extremities so that injury is often not detected. This injury, coupled with poor circulation and reduced healing, leads to gangrene, which results in amputation. For this reason, diabetes is the number one cause of nontraumatic amputations.

Changes in blood flow characteristics are also a result of hyperglycemia, which may result in high blood pressure and the development of macrovascular disease. The term *macrovascular* refers collectively to the risks of developing coronary artery disease (two to three times greater than that of a nondiabetic population) and cerebral vascular disease manifested by stroke and peripheral vascular disease (fivefold increase).

One of the major questions that plagued clinicians and diabetics alike was whether the risk of any of these complications could be reduced by tightly regulating glycemic status. Though early information clearly demonstrated a correlation between complications and glycemia, it was not until 1993, when the Diabetes Control and Complications Trial (DCCT) was completed, that unequivocal evidence demonstrated that normalization of plasma glucose reduces the risk of complications associated with diabetes mellitus. A major disadvantage of the intensive therapy used in the DCCT required to maintain such control is hypoglycemia, particularly in the elderly and young children. Despite the increased risk of hypoglycemia, the overall results support intensive treatment to maintain glycemic control and improve patient outcome.

C. CLASSIFICATION OF DIABETES MELLITUS

The first classification of diabetes mellitus and disorders of glucose regulation was published in 1979 by the National Diabetes Data Group and endorsed in 1980 by the World Health Organization (WHO) Expert Committee on Diabetes. In 1998, the International Expert Committee, sponsored by the ADA, modified this classification based on the research findings of the last 18 years to a system based on disease etiology. In this classification, there are three major types of diabetes: type 1, type 2, and gestational diabetes.

1. Type 1 Diabetes Mellitus. Type 1 diabetes, previously classified ◀ as insulin-dependent diabetes mellitus (IDDM), represents approximately 10% of all diabetes mellitus cases. Although the onset can occur at any age, it develops more frequently during childhood or young adulthood with an abrupt onset. Type 1 diabetes is the result of beta cell destruction, which usually leads to absolute insulin deficiency. Therefore, these individuals are dependent on exogenous insulin for survival. Consequently, they are at high risk for developing acute and/or chronic complications. Moreover, they are frequently lean due to recent weight loss and are also prone to developing certain autoimmune disorders such as Graves' disease, Hashimoto's thyroiditis, Addison's disease, myasthenia gravis, and pernicious anemia. Autoantibodies are detected in 85 to 90% of individuals with type 1 diabetes. Table 1–5 lists the autoantibodies that serve as markers of the immune-mediated destruction of beta cells of the pancreas. There also seems to be a correlation between the development of type 1 diabetes and the incidence of viral infections such as mumps and measles, and also between the presence of certain HLA types such as B8, BW15, DR3, and DR4.

2. Type 2 Diabetes Mellitus. Type 2 diabetes, previously classi- ◀ fied as non–insulin-dependent diabetes mellitus (NIDDM), represents approximately 90% of all cases of diabetes. Type 2 is a classification used for individuals who have an insulin resistance with relative (rather than absolute) insulin deficiency or an insulin secretory defect with insulin resistance. Consequently, insulin levels are normal to elevated. Type 2 diabetes mellitus, and the risk of developing type 2 diabetes, increases with age and a lack of physical activity. Approximately 80% of type 2 diabetics are obese at the time of diagnosis, and the onset is insidious rather than acute. There is also an increased incidence in those with a family history of type 2 diabetes. Type 2 diabetes occurs more frequently in women with previous diabetes mellitus diagnosed during pregnancy and in individuals with hypertension or dyslipidemia. Its frequency varies in different racial/ethnic groups with a higher incidence in Hispanic Americans, African Americans, Native Americans, Asian Americans, and Pacific Islanders. The hyperglycemia in type 2 diabetes can be treated with diet, drugs, or insulin. Although these patients usually do not need insulin treatment for survival, they are not spared the complications of type 1 diabetes. Table 1–6 gives a comparison of the characteristics associated with type 1 and type 2 diabetes.

Table 1–5. Autoantibodies Detected in Type 1 Diabetes

- Islet cell autoantibodies (ICAs)
- Insulin autoantibodies (IAAs)
- Glutamic acid decarboxylase (GAD) autoantibodies
- Tyrosine phosphatase autoantibodies (TPA)

Table 1–6. Comparison of Type 1 and Type 2 Diabetes

FEATURE	TYPE 1	TYPE 2
Usual age of onset	Children, young adults	Middle-aged, elderly
Type of onset	Acute	Insidious
Weight	Lean, usually weight loss	Often obese, weight loss uncommon
Occurrence of acute complications (ketosis, hypoglycemia)	Often seen with poor glycemic control; may be frequent	Not as frequent as type 1
Development of chronic complications (retinopathy, nephropathy, angiopathy, neuropathy)	Common	Common
Serum insulin concentration	Low or absent	Often normal; may be increased
Family history of diabetes	Uncommon	Common
Autoimmune association	Common	Uncommon

Complications of GDM

- Perinatal morbidity and mortality
- Increased rate of cesarean delivery
- Chronic hypertension

Risk Factors for GDM

- Age 25 and over
- Less than age 25 and obese
- History of diabetes in first-degree relative
- Belonging to ethnic/racial group with a high prevalence of diabetes

▶ **3. Gestational Diabetes Mellitus (GDM). Gestational diabetes mellitus** develops in approximately 4% of all pregnancies in the United States and represents 90% of the diabetes classified during pregnancy. Any degree of glucose intolerance detected during pregnancy is classified as GDM. Complications of GDM can occur in both the baby and the mother, including the well-documented GDM-associated perinatal morbidity and mortality, increased rate of cesarean delivery, and chronic hypertension. Although GDM occurs only temporarily during pregnancy, with glucose regulation returning to normal after delivery, up to 40% of these women develop type 2 diabetes later in life. The ADA recommends screening for GDM between the 24th and 28th weeks of pregnancy for all women who fall into a high-risk group. This group is defined as those: (1) age 25 and older, (2) less than age 25 and obese, (3) having a family history of diabetes in first-degree relatives, or (4) belonging to an ethnic/racial group with a high prevalence of diabetes.

IV. TEST PROCEDURES

One of the clinical laboratory's most important roles is based in the screening, diagnosis, and monitoring of disturbances in carbohydrate metabolism. The measurement of blood glucose, one of the most common assays performed in the laboratory, provides critical information to physicians regarding the glycemic status of their patient. Crucial diagnostic and therapeutic decisions are made based on the blood concentration of this single biochemical analyte. In fact, the diagnostic criteria for diabetes mellitus, as published by the ADA, is based on blood glucose concentration with accompanying symptomatology. Use of blood glucose measurement in this capacity underscores the importance of the accuracy, precision, and reliability of the assay. Furthermore, since screening and periodic follow-up is now recommended by the ADA for apparently healthy subjects age 45 and older, as well as for younger people at defined high risk, the frequency of testing for blood glucose is expected to increase. This recommendation represents an important change, as screening in apparently healthy subjects was previously discouraged by the ADA.

Once a diagnosis is made, other biochemical markers are used in conjunction with blood glucose for monitoring glycemic status, screening for renal and cardiovascular complications, and detection of acute crisis situations such as ketoacidosis. Some of the laboratory tests in these categories include glycated hemoglobin and other glycated serum proteins, urinary albumin, urinary creatinine, and ketones.

A. Screening and Diagnostic Procedures

1. Blood Glucose. As indicated, the measurement of blood glucose is one of the most important analytes measured in regard to the assessment of glycemic status. Though no defined criteria exist for the diagnosis of hypoglycemia, the determination of blood glucose concentration allows the caregiver to make appropriate therapeutic decisions that could be critical in the prevention of crisis situations. This is of particular importance because the symptoms for hyper- and hypoglycemia may be similar on patient presentation. Moreover, the detection of asymptomatic hypoglycemia is of particular importance in diabetics treated with insulin.

Diagnostic criteria for diabetes mellitus, as defined by the ADA in 1998, are given in Table 1–7. These values relate to plasma or serum glucose, which is the body fluid typically analyzed by laboratory methods. Whole blood glucose, measured by home monitors, is approximately 12 to 15% lower than the corresponding plasma or serum concentration in individuals with a normal hematocrit. Because plasma glucose determination is the primary diagnostic parameter, proper sample handling procedures to minimize glycolysis before analysis are imperative. At room temperature, serum or plasma glucose decreases by approximately 5 to 7% per hour due to continued glycolysis. This rate is actually higher in individuals with leukocytosis or in cases of bacterial contamination. Glycolysis is inhibited by the addition of sodium iodoacetate or sodium fluoride to the blood collection tube. If no additive is present, blood samples must be processed very soon after collection by separation of plasma or serum from the cells in order to minimize the effects of continued metabolism.

2. Blood Glucose Methodologies. Practically all methods employed in the laboratory today to determine blood glucose are enzymatic and measure glucose specifically, as opposed to inclusion of all reducing sugars. The precision of these assays is typically ≤ 3%. Enzymatic assays are based on the glucose oxidase or hexokinase

---

Table 1–7. Diagnostic Criteria for Diabetes Mellitus[a][b]

- Symptoms of diabetes plus glucose concentration ≥ 200 mg/dL; concentration not dependent on time of collection or time since last meal

OR

- Fasting plasma glucose ≥ 126 mg/dL

OR

- 2-hour postload glucose ≥ 200 mg/dL during an oral glucose tolerance test (OGTT). Test should be performed according to WHO protocol using a glucose load containing the equivalent of 75 g anhydrous glucose dissolved in water. (*The OGTT is not recommended for routine clinical use.*)

---

[a]Criteria recommended by the ADA.

[b]Diagnostic if one or any combination of the listed criteria is present on two or more occasions.

methodology, with hexokinase being recognized as the reference method. Reference ranges do not differ significantly between the two methods.

3. Oral Glucose Tolerance Test. As indicated by the diagnostic criteria, the oral glucose tolerance test (OGTT) is no longer recommended for routine screening except in the case of pregnancy. The OGTT may still be useful in the screening and diagnosis of some cases of suspected GDM. Abandonment of the OGTT is an important practical change in light of the inherent problems and infrequent use of the test. There are major concerns with the OGTT: poor reproducibility, expense, time, patient preparation requirements, lack of standardization in glucose load administration, multiple venipunctures, and general unpalatability among individuals being tested.

4. Reducing Sugars. Though not of significance in adults, screening for reducing sugars in the urine of neonates and infants is still important in the detection of inborn errors of metabolism. Depending on the biochemical defect, reducing sugars such as galactose and fructose may appear in the urine and are most commonly detected by the copper reduction methodology.

B. *Tests for Monitoring Diabetic Patients*

Advances in technology have enabled clinicians and their diabetic patients to more closely monitor and maintain appropriate glycemic status. With the availability and use of home monitors, diabetics can now pursue a realistic goal of maintaining stable blood glucose concentrations. Although diabetics relied on the urine dipstick to monitor their glucose level in the past, the information provided by urine testing is limited in terms of its practical use because the average renal threshold for glucose is approximately 180 mg/dL. A negative result from a urine specimen would indicate only that the blood glucose concentration was less than 180 mg/dL at the time of urine collection, given a normal glomerular filtration rate. The ADA now recommends that urine glucose be used only in those patients who are unwilling or unable to use the home monitors to measure blood glucose concentration.

1. Self-Monitoring of Blood Glucose. Self-monitoring of blood glucose (SMBG) is now recommended by the ADA for all treatment programs. For type 1 diabetics in particular, SMBG is recognized as being a crucial factor in the achievement of reasonable glycemic goals and the minimization of adverse effects caused by acute complications. Though overly ambitious efforts to reach euglycemia are not recommended, SMBG allows subjects to strive toward normal or near normal glycemia. SMBG is encouraged for all diabetics, but the frequency of monitoring is determined individually. In general, monitoring is recommended three to four times per day for type 1 diabetics. The optimum frequency of SMBG is not known for type 2 diabetics but should facilitate the achievement of glycemic goals.

Although significant advantages of home monitoring are known and realized, a minority of diabetics actually perform SMBG on a regular basis. Some of the barriers associated with the current limited practice include cost, inadequate understanding by patients and clin-

---

Problems Associated with SMBG

- Cost
- Lack of understanding by patients and clinicians as to the benefits and proper use of test results
- Discomfort associated with finger-stick
- Inconvenience in relation to time and physical setting
- Complexity of the technique
- Accuracy is user and instrument dependent

icians regarding the benefits and proper use of test results, subject discomfort associated with the fingerstick, inconvenience of testing in relation to time and physical setting, and complexity of the technique. Moreover, the accuracy of the data obtained by SMBG is user and instrument dependent. It is therefore critical that health care providers evaluate each patient's monitoring technique, initially and at regular time intervals thereafter. Calibrators and controls should be used on a regular basis. Though the ADA recommends that clinicians use SMBG instead of a laboratory assessment of blood glucose to determine glycemic status, comparisons between the lab and SMBG are important in evaluating the accuracy of patient results. Moreover, the laboratory must also be involved in proficiency testing of home monitoring devices since the ADA now recommends that such programs be implemented for SMBG.

The most common errors in SMBG are associated with the application, timing, and removal of specimen from the monitor. Newer monitors have addressed such errors by eliminating the need for timing, blotting, and wiping. Other factors affecting the accuracy and reproducibility of test results include user variability and hematocrit.

Most SMBG monitors employ the glucose oxidase methodology; however, hexokinase systems do exist. As monitors measure whole blood glucose, concentrations are typically 12 to 15% lower than those of plasma determinations. During fasting, capillary blood glucose is approximately 2 to 5 mg/dL greater than the venous concentration. However, after a glucose load, capillary concentration rises 20 to 70 mg/dL greater than that of concurrently drawn venous samples.

2. Glycated Hemoglobin. Another important laboratory test used to monitor the glycemic status of diabetics is **glycated hemoglobin,** or **hemoglobin A$_{1c}$ (HbA$_{1c}$).** As opposed to documenting a current blood glucose concentration that fluctuates greatly, glycated hemoglobin analysis provides a retrospective time-averaged assessment of the previous 2 to 3 months of glycemic control. Moreover, values are unaffected by the day-to-day changes in glucose concentration, degree of exercise, or recent food ingestion. It is also now accepted that glycated hemoglobin concentration is predictive for the risk of developing many of the chronic complications associated with diabetes mellitus. Consequently, the ADA recommends that the measurement of HbA$_{1c}$ be performed in all diabetics, initially to document the patient's glycemic status at the time of diagnosis, and then periodically as part of continuing care. It is suggested that testing be performed approximately every 3 months to determine whether a patient's metabolic control has remained within his or her target range. Testing frequency, of course, depends on the individual's treatment regimen and clinician's judgment.

As the concentration of circulating glucose increases, more of the carbohydrate enters the cell and more protein is glycated. Figure 1–1 diagrammatically represents the glycation process. This glycation process is a slow, three-step, nonenzymatic reaction that takes place in the red cell. First, glucose reversibly binds to accessible amino groups on the hemoglobin molecule to form a labile aldimine. Second, this aldimine intermediate undergoes a rearrangement that is irreversible and forms a stable ketoamine through a process known as the **Amadori rearrangement.** Once this rearrangement has taken place, the linkage between glucose and hemoglobin is permanent and lasts

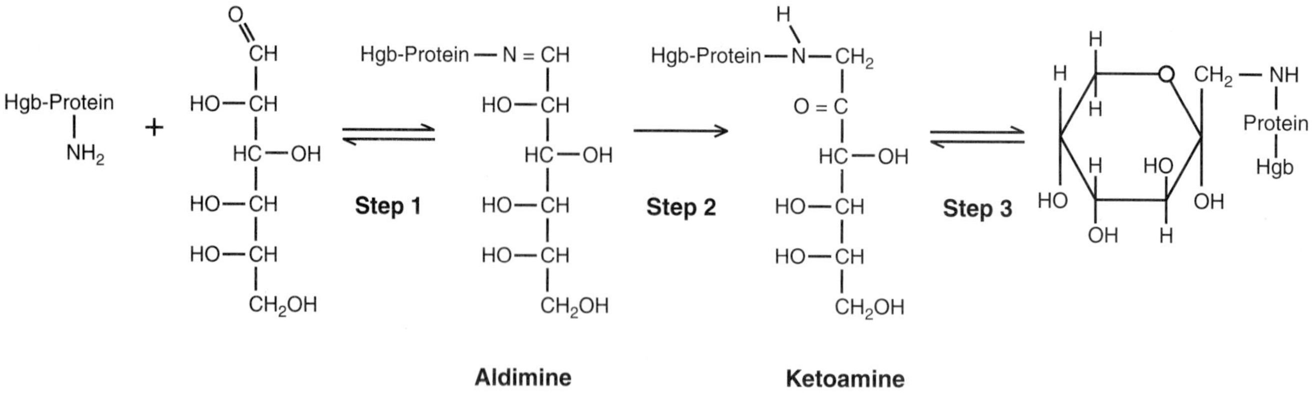

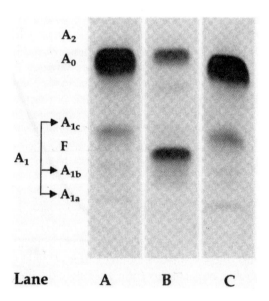

Figure 1–1. Glycation reaction of hemoglobin.

the remainder of the red cell's life span. The final step in the glycation reaction is a conformational change of glucose to a cyclic structure that is favored at the body's pH of 7.4. The terminology used to describe this process as well as the product measured in the laboratory is often confusing. The term *glycated hemoglobin* refers to the ketoamine structure where glucose is irreversibly bound to hemoglobin. **Glycosylated hemoglobin** refers to the reversible stage where glucose may equilibrate (attach and detach) with the hemoglobin molecule.

Many of the current laboratory methods measure HbA_{1c}. By age 1, the predominant normal form of hemoglobin present is HbA, which is further divided into the primary fractions A_o and A_1. HbA_1, which is post-translationally modified, can further be separated into the subfractions A_{1a}, A_{1b}, and A_{1c}. Figure 1–2 shows the electrophoretic separation of these hemoglobin fractions. HbA_{1c} comprises 60 to 80% of HbA_1 and thus represents the major subfraction of HbA_1. The primary site of glycation on the HbA_{1c} molecule is the N-terminal valine on the beta chain. Laboratory methods that mea-

Figure 1–2. Electrophoretic separation of hemoglobin fractions. Lane A, normal, nondiabetic adult; Lane B, normal infant; Lane C, diabetic in poor glycemic control.

sure "total glycated hemoglobin" include glycation at other sites as well. Because methods that determine glycated hemoglobin differ greatly in regard to what is actually measured, interferences in measurement, and nondiabetic reference range, it is very important for the laboratory to properly characterize and define the method employed. Furthermore, reference intervals must be appropriate for the chosen method and area population.

The recommended 3-month testing frequency is based on the average life span of red blood cells. Because glucose is freely permeable across the cell membrane and virtually all hemoglobin in circulation is found within the red cell, the level of glycated hemoglobin in the blood correlates with the red cell's 120-day average life span. Figure 1–3 demonstrates the correlation between red cell survival and glycated hemoglobin concentration. Consequently, diabetics having conditions that affect the life span of the red cell may have different glycation kinetics as well as variations in physical properties that are not fully understood. For example, patients with hemolytic disease or other conditions with shortened red cell lifetime demonstrate significantly lower values for glycated hemoglobin. If the hemolytic process is due to the presence of an abnormal hemoglobin, the effect on glycated hemoglobin will be constant. In such cases, values should be compared with previous levels from the patient and not with the laboratory's published reference intervals. At the other end of the spectrum, iron deficiency anemia may result in the delayed replenishment of expired red cells. In this case, insufficient iron is available for normal hemoglobin assembly. This results in a shift of the red blood cell life span curve toward older cells. Hence, the normal distribution of red cells in regard to age is dis-

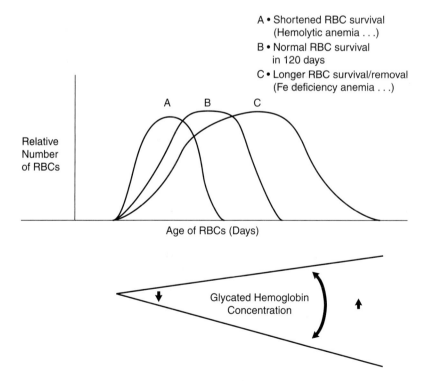

Figure 1–3. Relationship between red cell survival and glycated hemoglobin.

rupted and glycated hemoglobin levels are increased. Increased levels may also be seen following splenectomy due to increased red cell survival. Blood transfusion affects glycated hemoglobin concentration as levels are biased toward that of the transfused hemoglobin.

3. Glycated Hemoglobin/HbA$_{1c}$ Methodologies.

Current methods for the determination of glycated hemoglobin can be divided among three categories based on differences in charge, structure, and chemical makeup. Techniques used to measure differences in charge include ion exchange chromatography, high-performance liquid chromatography (HPLC), and electrophoresis. Structural differences are determined by affinity chromatography and immunoassay, while spectrophotometry is used to perform chemical analysis. The most popular methods currently in use are immunoassay and chromatography, primarily HPLC. An advantage of HPLC is that the chromatogram provides a visual view of the separation pattern of the hemoglobin fractions. This may serve as a quality assurance indicator and also allows for the detection and identification of hemoglobin variants. The identification of hemoglobin variants may be an important feature, especially in areas where known variants are in high prevalence. The most common hemoglobin variants encountered in the United States are S, C, and E, in decreasing order of prevalence. In immunoassay methodologies, since the antibodies used are highly specific, hemoglobin variants may not be measured or accounted for. Moreover, the high specificity of the immunologic methods typically produces a lower reference range than that of other techniques. One of the most desirable features of immunoassay is the ease of adaptability to automation. All methods require a whole blood sample, collected in ethylenediaminetetraacetic acid (EDTA), for analysis.

Some of the major problems associated with glycated hemoglobin analysis include the lack of standardization between methods and poor precision. In regard to standardization, results between laboratories using the same method have shown poor correlation. Thus, in 1996, the National Glycohemoglobin Standardization Program, sponsored by the ADA, began an initiative to standardize values to correlate with those from the DCCT. Manufacturers are awarded a certificate of traceability to the DCCT reference method if their assays pass the criteria set for precision and accuracy. Hence, the ADA recommends that laboratories use methods that have passed the certification testing. The ADA also encourages participation in proficiency testing programs.

The ADA recommends that the goal of therapy should be to maintain HbA$_{1c}$ concentration to less than 7%, as measured by a method certified as traceable to the DCCT reference assay, which is HPLC. As standardization of methods improves, health care providers will be better able to use glycated hemoglobin results to monitor their patients. Though numerous studies have documented the clinical utility of glycated hemoglobin testing, the lack of correlation among assays has prevented it from being accepted for use as a diagnostic procedure. However, though not currently appropriate for diagnostic purposes, glycated hemoglobin remains a primary indicator for monitoring the glycemic control of diabetics.

4. Detection of Microalbuminuria.

In addition to monitoring glycemic control, physicians also seek to identify impending chronic complications such as nephropathy. Diabetes is the most common cause of end-stage renal disease in the United States and Europe. **Microalbuminuria,** defined as low but abnormal levels of albumin ◀ in the urine (> 30 mg/day), is the earliest indicator of nephropathy in diabetics. Moreover, the detection of microalbuminuria indicates the need to also screen for other vascular disease.

The ADA recommends that a routine urinalysis be performed yearly in adult diabetics. If the routinely used urinary dipstick test for protein is negative, further testing for the presence of microalbuminuria should be done because the sensitivity of the dipstick method does not detect the low concentration of protein present in microalbuminuria. Positive reagent strip tests for protein should be confirmed by more specific assays. Screening for microalbuminuria is currently performed by one of the following methods:

1. The albumin:creatinine ratio is determined on a random spot urine.
2. Serum and urine creatinine is measured and a creatinine clearance performed on a 24-hour urine collection.
3. Urine creatinine is measured in a timed urine collection.
4. Albumin is measured in a timed urine collection.

Because of the wide day-to-day variability in albumin excretion, at least two of three urine collections done in a 3- to 6-month period should be positive before a patient is designated as having microalbuminuria. Microalbuminuria is present if urinary albumin excretion is greater than 30 mg in a 24-hour collection or 30 mg/g creatinine in a random urine sample.

C. DETECTION OF ACUTE CRISIS SITUATIONS— HYPOGLYCEMIA AND KETOACIDOSIS

As previously described, the two major acute crisis situations that occur in diabetics are ketoacidosis and hypoglycemia. Though more commonly associated with type 1 diabetes, these complications may be found in type 2 diabetics as well. Home monitoring systems have been an important tool in differentiating hyper- and hypoglycemia. Once a diabetic is identified as being hyperglycemic, it is then important to screen for the presence of **ketones**, since ketoacidosis ◀ can occur with only moderate elevations in blood glucose. The ADA recommends that diabetics test their urine for ketones during acute illness or stress, when blood glucose is consistently high, during pregnancy, or when symptoms such as nausea, vomiting, or abdominal pain are present. Ketones are normally present in the urine during fasting and also in approximately 30% of first morning urine samples from pregnant women.

None of the commonly used screening tests for ketones in the blood or urine detects all three ketone bodies (i.e., acetone, acetoacetate, β-hydroxybutyrate). All of the commercial tests are based on the nitroprusside reaction. However, nitroprusside is most sensitive to acetoacetate, followed by acetone. No reaction takes place between nitroprusside and β-hydroxybutyrate, which is the predominant ketone body found in the blood. Consequently, it is possible to have a negative test for ketones but the patient still be in ketoacidosis. A commercial method for β-hydroxybutyrate in blood, serum, or plasma is now available, but it requires considerable technical expertise to perform and is not practical for home monitoring. Though the current technology is limited, it is recommended that home testing be performed on blood rather than urine, since semiquantitation of ketones in the blood is more accurate than in urine. However, urine testing remains popular due to the ease of sample collection.

Other laboratory parameters that may be helpful in diagnosing and monitoring acute crisis situations include blood gas analysis, electrolyte determination plus the calculation of anion gap, and osmolality. These tests are discussed in more detail in other chapters of this text.

SUGGESTED READINGS

American Diabetes Association. Diabetic nephropathy (position statement). *Diabetes Care* 21(Suppl. 1):S50–S53, 1998.

American Diabetes Association. Gestational diabetes mellitus (position statement). *Diabetes Care* 21(Suppl. 1):S60–S61, 1998.

American Diabetes Association. Management of dyslipidemia in adults with diabetes (position statement). *Diabetes Care* 21(Suppl. 1):S36–S39, 1998.

American Diabetes Association. Report of the Expert Committee on the Diagnosis and Classification of Diabetes Mellitus. *Diabetes Care* 21(Suppl. 1):S5–S19, 1998.

American Diabetes Association. Standards of medical care for patients with diabetes mellitus. *Diabetes Care* 21(Suppl. 1):S23–S31, 1998.

American Diabetes Association. Tests of glycemia in diabetes (position statement). *Diabetes Care* 21(Suppl. 1):S69–S71, 1998.

Goldstein DE, Little RR. Monitoring glycemia in diabetes—short-term assessment. *Endocrin Metab Clin North Am* 26:475-86, 1997.

John WG. Glycated haemoglobin analysis. *Ann Clin Biochem* 34:17–31, 1997.

Lapuz MHS. Diabetic nephropathy. *Med Clin North Am* 81:679–88, 1997.

Skyler JS. Diabetic complications—the importance of glucose control. *Endocrin Metab Clin North Am* 25:243–54, 1996.

Zimmet PZ. The pathogenesis and prevention of diabetes in adults—genes, autoimmunity, and demography. *Diabetes Care* 18:1050–64, 1995.

Items for Further Consideration

1. What factors should be considered in choosing or changing a method for the determination of glycated hemoglobin?

2. How should diabetics with dysfunctional hemoglobins (S, C, E, etc.) be monitored in regard to glycemic status?

3. Should glycated hemoglobin analysis be used as a diagnostic test for diabetes mellitus?

4. What is the advantage for diabetics in striving to achieve and maintain euglycemia?

5. What is the significance of detecting microalbuminuria at the earliest possible stage?

6. How has the use of SMBG devices affected the clinical laboratory's role in diagnosing and monitoring diabetics?

Lipids and Lipoproteins

Linda C. Gregory

Key Terms

AMPHIPATHIC

APOLIPOPROTEINS

ATHEROSCLEROSIS

CHOLESTEROL

CHOLESTEROL ESTERS

CHYLOMICRONS

CORONARY HEART DISEASE

DYSLIPIDEMIAS

ENDOGENOUS PATHWAY

EXOGENOUS PATHWAY

FAT-SOLUBLE VITAMINS

FATTY ACIDS

HIGH-DENSITY LIPOPROTEINS

INTERMEDIATE-DENSITY LIPOPROTEINS

LDL RECEPTOR–MEDIATED UPTAKE

LECITHIN CHOLESTEROL ACYLTRANSFERASE

LIPASES

LIPIDS

LIPOPROTEIN(a)

LIPOPROTEIN LIPASE

LIPOPROTEINS

LOW-DENSITY LIPOPROTEINS

NONRECEPTOR-MEDIATED ENDOCYTOSIS

PHOSPHOLIPIDS

REVERSE CHOLESTEROL TRANSPORT

SPHINGOLIPIDS

TRIGLYCERIDES

VERY LOW-DENSITY LIPOPROTEINS

Chapter Objectives

Upon completion of this chapter, the reader will be able to:

1. Give a general definition for (a) lipid and (b) lipoprotein.
2. Describe the general chemical structure, biological function(s), and clinical significance of the following lipids: fatty acids, triglycerides, phospholipids, and cholesterol.
3. For the major lipoprotein classes: (a) relate physical characteristics and functions to the composition of the lipoprotein particle, and (b) identify the characteristics used to distinguish each class.

4. Discuss the exogenous and endogenous pathways of lipoprotein metabolism. Predict the effects of specific metabolic defects on plasma lipid/lipoprotein concentrations.
5. In general terms, discuss the pathogenesis of atherosclerosis and coronary heart disease.
6. Describe reverse cholesterol transport and discuss its possible role in atherosclerosis.
7. Describe the major primary lipid disorders and the associated changes in plasma lipids/lipoproteins.
8. Identify factors or clinical conditions that may be secondary causes of hyperlipidemias.
9. Create protocols to minimize effects of preanalytical variation on lipid/lipoprotein testing.
10. Describe the common routine methods for measuring total cholesterol, triglycerides, high-density lipoprotein (HDL) cholesterol, low-density lipoprotein (LDL) cholesterol, and apolipoproteins. Identify any technical issues that may affect the results of these assays.
11. Given the necessary data, use the Friedewald formula to calculate LDL cholesterol.
12. Predict consistent lipid/lipoprotein abnormalities given the appearance of a plasma or serum sample.
13. Given concentrations of the major serum lipids, lipoproteins, or lipoprotein constituents and/or risk factor status of an individual, determine whether a patient is likely to be at increased or decreased risk for coronary heart disease.
14. Compare and contrast the value of population-based reference ranges versus medical decision limits in evaluating lipid/lipoprotein test results.
15. Summarize the current efforts to improve identification of individuals at risk for atherosclerotic disease and to improve lipid/lipoprotein measurements.

Analytes Included

APOLIPOPROTEINS:	*HDL CHOLESTEROL (HDL-C)*	*TOTAL CHOLESTEROL (TC)*
APOLIPOPROTEIN AI (APO AI)	*LDL CHOLESTEROL (LDL-C)*	*TRIGLYCERIDES (TG)*
APOLIPOPROTEIN B (APO B)		
APOLIPOPROTEIN E (APO E) ISOFORMS		
LIPOPROTEIN(a)		

Pretest

Directions: Choose the single best answer for each of the following items.

1. Lipids are
   a. polar
   b. hydrophobic
   c. readily soluble in aqueous solutions
   d. hypertonic

2. Lipid malabsorption will likely cause a deficiency of
   a. folate
   b. vitamin B_6
   c. iron
   d. vitamin E

3. Complete hydrolysis of a triglyceride molecule will yield
   a. glyceraldehyde and a fatty acid ester
   b. a monoglyceride and three fatty acids
   c. glycerol and three fatty acids
   d. fatty acid and three glycerols

4. The essential fatty acids include
   a. linoleic and linolenic acids
   b. myristic and linoleic acids
   c. oleic and linolenic acids
   d. stearic and oleic acids

5. Cholesterol is a precursor in the synthesis of
   a. fatty acids
   b. prostaglandins
   c. steroid hormones
   d. vitamin A

6. Phospholipids are
   a. poor detergents
   b. important components of biological membranes
   c. phosphate esters of cholesterol
   d. transported in the core of lipoprotein particles

7. Hyperalphalipoproteinemia refers to a condition with an elevation in lipoprotein fraction
   a. chylomicrons
   b. very low-density lipoprotein (VLDL)
   c. LDL
   d. HDL

8. The lipoprotein fraction with the greatest proportion of cholesterol is
   a. chylomicrons
   b. VLDL
   c. LDL
   d. HDL

9. Apolipoprotein B-100 is a constituent of
   a. chylomicrons and VLDL
   b. VLDL, intermediate-density lipoprotein (IDL), and LDL
   c. IDL and HDL
   d. VLDL, LDL, and HDL

10. The enzyme that catalyzes the hydrolysis of triglycerides in chylomicrons and VLDL is
    a. lipoprotein lipase
    b. lecithin cholesterol acyltransferase (LCAT)
    c. apolipoprotein CI
    d. lipohydrolase

11. A genetic variant of the LDL receptor with decreased binding capacity would be expected to cause an elevation in plasma
    a. HDL cholesterol
    b. total cholesterol
    c. triglycerides
    d. apolipoprotein AI

12. A 49-year-old woman stood for 45 minutes before having blood drawn at a health fair in the mall. It is likely that her
    a. triglycerides will be falsely elevated due to hemodilution
    b. triglycerides will be falsely decreased due to hemoconcentration
    c. total cholesterol will be falsely decreased due to hemodilution
    d. total cholesterol will be falsely elevated due to hemoconcentration

13. A patient was instructed to fast for 12 hours prior to her 8 A.M. doctor's appointment. However, she ate her usual breakfast of orange juice, coffee, and a bagel with cream cheese. It would be expected of the lipid test results that
    a. both the total cholesterol and triglycerides results will be invalid
    b. chylomicrons will be increased but total cholesterol and HDL cholesterol will be decreased
    c. both the triglycerides and the calculated LDL cholesterol will be affected by food intake
    d. only the triglycerides result will not be affected by food intake

14. A plasma sample appeared turbid 1 hour after venipuncture. Sample appearance was unchanged after standing undisturbed overnight. Upon lipid testing, the technologist would expect an increased
    a. total cholesterol
    b. LDL cholesterol
    c. triglycerides due to VLDL
    d. triglycerides due to chylomicrons

15. Choose the test analyte that is correctly matched with its reagent and the correct purpose of the reagent.

Analyte	*Reagent/Purpose*
a. total cholesterol	polyethylene glycol to precipitate HDL
b. LDL cholesterol	lipase to hydrolyze endogenous glycerol
c. triglycerides	cholesterol oxidase to hydrolyze cholesterol esters
d. HDL cholesterol	dextran sulfate/$MgCl_2$ to precipitate LDL and VLDL

16. Calculate the LDL cholesterol given that:
    Total cholesterol = 448 mg/dL
    HDL cholesterol = 51 mg/dL
    Triglycerides = 323 mg/dL
    a. 332 mg/dL
    b. 182 mg/dL
    c. 74 mg/dL
    d. the equation for calculating LDL cholesterol is not valid under these circumstances; it should be measured directly or by a reference laboratory

17. A fasting sample from a 33-year-old type 1 diabetic was submitted for lipid testing. Given the following laboratory results, predict the most likely lipoprotein abnormality.
    Total cholesterol = 235 mg/dL
    HDL cholesterol = 28 mg/dL
    Triglycerides = 194 mg/dL
    Plasma appearance clear with slight orange discoloration
    a. elevated chylomicrons
    b. elevated VLDL
    c. elevated LDL
    d. elevated HDL

18. Choose the lipoprotein abnormality that is most consistent with the following laboratory results for a fasting plasma sample.
    Total cholesterol = 351 mg/dL
    HDL cholesterol = 32 mg/dL
    Triglycerides = 537 mg/dL
    Plasma appearance (after standing) moderately turbid throughout
    a. familial (monogenic) hypercholesterolemia
    b. familial combined hyperlipidemia
    c. familial chylomicronemia
    d. familial hypertriglyceridemia

19. On physical examination, a child demonstrated motor weakness and enlarged tonsils and adenoids. Given the following results for a fasting sample, predict the most likely lipoprotein abnormality.
    Total cholesterol = 134 mg/dL
    HDL cholesterol = 2 mg/dL
    Triglycerides = 115 mg/dL
    Plasma appearance clear
    a. abetalipoproteinemia
    b. familial dysbetalipoproteinemia
    c. analphalipoproteinemia
    d. hypobetalipoproteinemia

20. Choose the laboratory finding associated with increased risk for coronary heart disease.
    a. increased lipoprotein(a)
    b. increased apolipoprotein AI
    c. decreased apolipoprotein B
    d. increased HDL cholesterol

21. Based on the patient's history and results of lipoprotein testing, a physician has determined that a 58-year-old man is at high risk for coronary heart disease. The goals for reducing the patient's risk include:
    (1) decreasing HDL cholesterol
    (2) quitting cigarette smoking
    (3) increasing LDL cholesterol
    (4) decreasing apolipoprotein B
    (5) increasing apolipoprotein AI
    a. (1), (2), and (4)
    b. (3), (4), and (5)
    c. (2), (4), and (5)
    d. (1), (2), and (3)

22. During a cholesterol screening, a 34-year-old man was found to have a total cholesterol of 225 mg/dL and an HDL cholesterol of 83 mg/dL. Based on these results, this individual
    a. is at borderline high risk for coronary heart disease
    b. should be counseled to modify his diet to reduce his total cholesterol
    c. should be counseled to see his physician immediately for follow-up testing
    d. is probably not at borderline high risk for coronary heart disease

I. INTRODUCTION

▶ Lipids are essential for normal cell structure, function, and metabolism. **Dyslipidemias** (disorders of lipid metabolism) are associated with various pathogenic states. In particular, certain hyperlipidemias have shown a relationship to the development of atherosclerotic vascular disease, including coronary heart disease (CHD). Laboratory measurements of lipids and lipoproteins are used to detect, diagnose, guide, and monitor treatment of lipid disorders, most often for the purpose of managing vascular disease. Lipid analysis is commonly used in wellness testing to assess risk for cardiovascular disease.

II. BIOCHEMISTRY

Lipids are hydrophobic (nonpolar), organic molecules—insoluble in water, but readily soluble in organic solvents. As a group, lipids are diverse in structure and biological function. **Lipoproteins,** complexes of lipids and proteins, act as vehicles for transporting lipids in the aqueous milieu of the circulatory system.

The major lipids found in plasma (in order of descending concentration) are triglycerides, cholesterol, phospholipids, and fatty acids.

A. FATTY ACIDS

Fatty acids are carboxylic acids of straight-chain hydrocarbons; they are represented as R-COOH, where R is the alkyl chain. Long-chain fatty acids containing even numbers of carbon atoms (from 12 to 26) are the most clinically significant. Although most fatty acids can be synthesized by the body (see exceptions below), fatty acids are largely provided by the diet.

Fatty acids provide energy through β-oxidation. They are components of other lipids as esters of alcohols (e.g., to form triglycerides, phospholipids, and cholesterol esters), and they are precursors of prostaglandins. Fatty acids may be *saturated* (all single bonds between carbons), *monounsaturated* (one double bond between two carbons in the chain), or *polyunsaturated* (more than one double bond in the carbon chain). Two polyunsaturated fatty acids, linoleic and linolenic acids, are *essential* fatty acids—they are required precursors of other molecules but cannot be synthesized by the human body and must be obtained from dietary sources. Free fatty acids (i.e., unesterified) are present in plasma at low concentrations and are bound to albumin for transport.

B. TRIGLYCERIDES

Triglycerides are formed from three fatty acids in ester linkage at the three alcohol groups of glycerol. In a single triglyceride molecule, the three fatty acids (represented as R_1, R_2, and R_3 in the chemical formula below) may be the same or different.

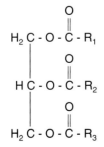

Triglycerides are the most prevalent fat in the diet; triglycerides from animal sources contain mostly saturated fatty acids and are solid

---

Major Lipids Found in Plasma
• Triglycerides • Cholesterol • Phospholipids • Fatty acids

Fatty Acid Classification
Saturated—all single bonds between carbons Monounsaturated—one double bond in the chain Polyunsaturated—more than one double bond in the chain

at room temperature, while those from plant sources contain unsaturated fatty acids and are liquid even upon refrigeration. Saturated fats from meat and dairy products are believed to be more likely to contribute to vascular disease than unsaturated fats from vegetable sources. Triglycerides (TG) can also be synthesized in the liver.

Triglycerides yield chemical energy upon hydrolysis and subsequent β-oxidation of the fatty acids. Triglycerides are unique as a metabolic fuel because any excess can be stored as reserve in the tissues (i.e., as fat in adipose tissue). Hydrolysis of triglycerides yields three fatty acids and one glycerol molecule. Enzymatic hydrolysis of
▶ triglycerides is catalyzed by **lipases;** hydrolysis also occurs nonenzymatically under acidic and alkaline conditions. Triglycerides are the most prevalent lipid in plasma (i.e., nonfasting); they are transported in the core of lipoprotein particles.

C. CHOLESTEROL

▶ **Cholesterol** is a 27-carbon alcohol derivative of the four-ring sterane carbon skeleton.

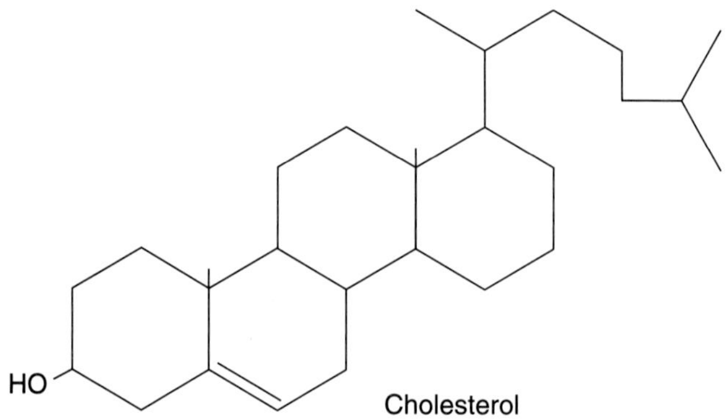

Cholesterol

Most body cholesterol is synthesized in the liver and to a lesser extent in other tissues; a small amount of cholesterol is absorbed from meat and dairy products in the diet. Cholesterol is a component of cell membranes and is the precursor for steroid hormones, vitamin D, and
▶ bile acids. **Cholesterol esters** have one fatty acid linked at the alcohol group at carbon three (see hydroxyl group in structure above) and are more hydrophobic than free (i.e., unesterified) cholesterol. Cholesterol is secreted unchanged in bile or is metabolized to bile acids. About 70% of cholesterol in circulation is in the ester form. Cholesterol esters and free cholesterol are transported in lipoproteins.

D. PHOSPHOLIPIDS

▶ **Phospholipids** are similar to triglycerides except that a polar phosphoryl group is substituted for a fatty acid at one of the terminal carbons of glycerol. The phosphoryl groups include a phosphate and one of several additional groups (represented by **-X** in the chemical formula below), such as hydrogen, choline, serine, ethanolamine, glycerol, and inositol to yield phosphatidic acid, phosphatidylcholine, phosphatidylserine, phosphatidylethanolamine, phosphatidylglycerol, and phosphatidylinositol, respectively.

$$
\begin{array}{c}
\overset{\displaystyle O}{\displaystyle \|} \\
H_2C-O-C-R_1 \\
|\overset{\displaystyle O}{\displaystyle \|} \\
HC-O-C-R_2 \\
|\overset{\displaystyle O}{\displaystyle \|} \\
H_2C-O-P-O-\mathbf{X} \\
|\\
O^-
\end{array}
$$

Phospholipids are mostly synthesized in the liver. Phospholipids are structural and functional components in cell membranes and act as lung surfactants. Phospholipids have **amphipathic** (both hydrophilic and hydrophobic) properties; the end of the molecule composed of two fatty acid tails participates in hydrophobic interactions, while the polar end with the phosphoryl group associates with hydrophilic species. This amphipathic nature is crucial to the role of phospholipids in the structure of bilayer membranes and allows them to function as detergents and components of micelles. Phospholipids are the major component of the lipoprotein outer shell (discussed below).

E. OTHER LIPIDS

This chapter focuses on lipids implicated in the pathogenesis of atherosclerotic vascular disease. Other lipids of clinical importance will be mentioned only briefly. *Steroid hormones* and *bile acids* are both derivatives of cholesterol. Steroid hormones have wide-ranging effects on metabolic processes, water and electrolyte balance, and gonadal function. Bile acids aid digestion by solubilizing fats and promoting their hydrolysis. **Sphingolipids** are found in cell membranes and the central nervous system. Genetic disorders in sphingolipid catabolism result in their accumulation in lysosomes as seen in Gaucher's disease, Niemann–Pick disease, and Tay–Sachs disease. *Prostaglandins* are derivatives of fatty acids that have potent physiological effects and are produced at the site of action by the affected tissue(s). The **fat-soluble vitamins** are themselves lipids and include vitamins A, D, E, and K. Their physiological roles are varied, but their absorption, transport, and storage are similar to those of other lipids.

F. LIPOPROTEINS

Lipoproteins are roughly spherical complexes of lipids and proteins (Figure 2–1) with a design uniquely suited to transport lipids and facilitate the metabolism of lipids in the hydrophilic environment of the circulatory system. Triglycerides and cholesterol esters, the most hydrophobic components of the lipoprotein, occupy the central core of the complex. The core is surrounded by an outer shell of phospholipids, protein, and free cholesterol. The molecules in the outer shell are amphipathic, oriented with their hydrophilic ends toward the outer surface and the hydrophobic ends toward the inner core. The protein components of the lipoproteins are called **apolipoproteins** (apo).

There are five major classes of lipoproteins categorized according to their relative density: **chylomicrons, very low-density**

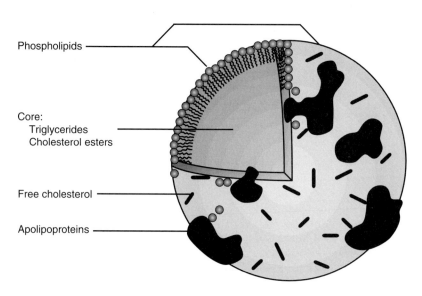

Phospholipids

Core:
 Triglycerides
 Cholesterol esters

Free cholesterol

Apolipoproteins

Figure 2–1. Schematic of lipoprotein particle. Exterior: polar (hydrophilic) environment; interior: nonpolar (hydrophobic) environment. Phospholipid head groups cover the surface of the lipoprotein not occupied by cholesterol and apolipoproteins.

The Five Major Classes of Lipoproteins

- Chylomicrons
- Very low-density lipoproteins (VLDL)
- Intermediate-density lipoproteins (IDL)
- Low-density lipoproteins (LDL)
- High-density lipoproteins (HDL)

▶ lipoproteins (VLDL), **intermediate-density lipoproteins** (IDL),
▶ **low-density lipoproteins** (LDL), and **high-density lipoproteins** (HDL). Each class performs particular functions and, while there is variation in particles within any one class, each class has characteristic physical and chemical properties (Table 2–1). In general terms, lower density corresponds to larger particle size and lower protein-to-lipid ratios, while higher-density particles are smaller and have higher proportions of protein. Chylomicrons have the lowest density (i.e., they float at the density of normal plasma) and are largely com-

Table 2–1. Lipoproteins: General Characteristics

	CHYLOMICRONS (CM)	VERY LOW-DENSITY LIPOPROTEINS (VLDL)	INTERMEDIATE-DENSITY LIPOPROTEINS (IDL)	LOW-DENSITY LIPOPROTEINS (LDL)	HIGH-DENSITY LIPOPROTEINS (HDL)
Density, g/mL	< 0.95	0.95–1.006	1.006–1.019	1.019–1.063	1.063–1.21
Particle size, nm	80–1,200	30–80	23–35	18–25	5–12
Major lipids[a]	90% TG	65% TG 15% CHOL	35% PL 25% CHOL	50% CHOL	25% PL 20% CHOL
Protein content[a]	1–2%	8–10%	20%	20%	50–55%
Major apolipoproteins	AI, B-48, CI, CII, CIII	B-100, CI, CII, CIII	B-100, E	B-100	AI, AII
Electrophoretic mobility	Remains at origin[b]	Prebeta	Broad beta[b]	Beta	Alpha
Source	Intestine	Liver	Product of VLDL catabolism	Product of VLDL (IDL) catabolism	Liver
Principal function	Transport of exogenous triglycerides	Transport of endogenous triglycerides	Precursor of LDL	Cholesterol transport	Reverse cholesterol transport

TG, triglycerides; CHOL, cholesterol; PL, phospholipids.

[a]Approximate percentage of lipoprotein composition by weight.

[b]The chylomicron and broad beta bands are not seen in normal fasting sera.

posed of exogenous triglycerides (from diet). VLDL mainly transports triglycerides synthesized in the liver (endogenous). Cholesterol is the major component of LDL, while HDL has a high protein content with phospholipids and cholesterol as the major lipid components. The lipoprotein classes also differ in their electrophoretic mobility on agarose at pH 8.6, and bands are named in relation to the migration of major serum proteins. The lipoprotein electrophoretic nomenclature is often used to name various dyslipoproteinemias.

Each lipoprotein class has a characteristic set of associated apolipoproteins: apolipoprotein AI (apo AI) and apo AII are found in HDL; apo B-48 is a component of chylomicrons; apo B-100 is a constituent of VLDL, IDL, and LDL; apo CI, CII, and CIII are primarily found in chylomicrons and VLDL; and apo E is a component of VLDL and IDL. Apolipoproteins are functional as well as structural components, participating in lipoprotein metabolism as enzyme activators, enzyme inhibitors, or binding sites for cellular receptors (Table 2–2).

There is a sixth category of lipoprotein known as **lipoprotein(a)** ◄ [Lp(a)]. Lp(a) is a variant of LDL in which a molecule of the protein apo(a) is bound to apo B-100. The function of Lp(a) is unknown, although the homology of apo(a) with plasminogen has led to investigations of a possible role in promoting thrombosis by inhibiting fibrinolysis.

III. PHYSIOLOGY

Lipoprotein metabolism in the plasma is a dynamic process in which several events may occur simultaneously. Lipoproteins (1) transfer and accept lipid and protein components to and from other lipopro-

Table 2–2. Apolipoproteins: Functions

APOLIPOPROTEIN	LIPOPROTEIN CLASS(ES)[a]	FUNCTION
AI	HDL, CM	Activator or cofactor for LCAT
		Structural component of HDL
AII	HDL	Role uncertain (may activate hepatic lipase and/or inhibit LCAT)
		Structural component of HDL
B-48	CM	Structural component of chylomicrons required for transport of lipids from small intestine
B-100	VLDL, IDL, LDL	Ligand for LDL receptor
		Structural component of VLDL, IDL, and LDL
CI	CM, VLDL	Activator of LCAT
CII	CM, VLDL	Activator of lipoprotein lipase
CIII	CM, VLDL	Inhibitor of lipoprotein lipase
		Activator of LCAT
		May inhibit uptake of chylomicron and VLDL remnants
E	VLDL, IDL	Ligand for hepatic apo E receptors
		Ligand for cellular LDL receptors
		(Several apo E isoforms exist differing in binding affinity)

CM, chylomicrons; LCAT, lecithin cholesterol acyltransferase.

[a] The apolipoprotein is most likely a constituent of the lipoprotein class(es) listed.

Lipoprotein Metabolism

Exogenous pathway—the processing
  of dietary lipids
Endogenous pathway—the processing of lipids synthesized in the
  body
Reverse cholesterol transport—a
  pathway involving HDL metabolism

teins and tissues, (2) are subjected to actions of plasma and tissue enzymes, and (3) are removed from circulation through receptor-mediated uptake. For the purpose of discussion, lipoprotein metabolism can be separated into: (1) an **exogenous pathway,** describing the processing of dietary lipids (Figure 2–2); (2) an **endogenous pathway,** describing the processing of lipids synthesized in the body (Figure 2–3); and (3) **reverse cholesterol transport,** a pathway involving HDL metabolism.

A. EXOGENOUS PATHWAY

Following digestion and absorption of fats, the intestinal epithelium assembles dietary TG and cholesterol, along with apo B-48, apo A, and phospholipids, into chylomicron particles. Chylomicrons, initially released into the lymph, enter the blood through the thoracic duct. In the plasma, there is an exchange of components between chylomicrons and HDL, in which apo CII, apo CIII, apo E, phospholipids, and cholesterol are initially transferred from HDL to chylomicrons, while apo A (AI, AII, and AIV) is transferred from chylomicrons to

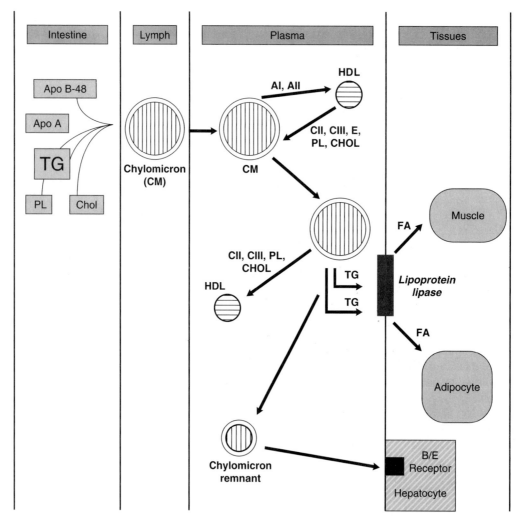

Figure 2–2. Simplified schematic of exogenous pathway of lipoprotein metabolism. (Chol, cholesterol; CM, chylomicrons; FA, fatty acids; PL, phospholipids; TG, triglycerides; AI, AII, CII, CIII, and E are apolipoproteins; B/E, Apo B/Apo E).

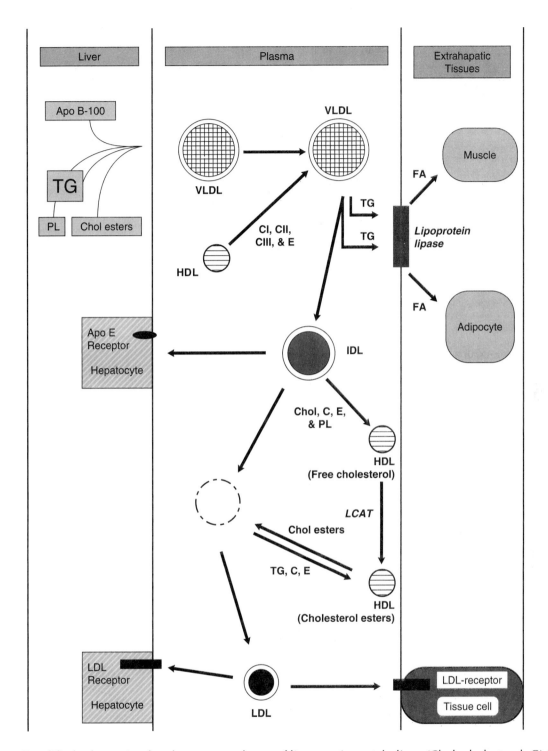

Figure 2–3. Simplified schematic of endogenous pathway of lipoprotein metabolism. (Chol, cholesterol; CM, chylomicrons; FA, fatty acids; LCAT, lecithin cholesterol acyltransferase; PL, phospholipids; TG, triglycerides; C, CI, CII, CIII, and E are apolipoproteins; dashed circle, lipoprotein in transition from IDL to LDL).

HDL. Apo CII in the chylomicrons activates **lipoprotein lipase** ◀
found in the capillary epithelium of muscle and adipose tissue.
Triglycerides in the chylomicron core are hydrolyzed by lipoprotein
lipase, and liberated fatty acids enter either muscle cells, where they
are metabolized for energy via β-oxidation, or adipocytes, where

triglycerides are resynthesized for storage. As core triglycerides are depleted, the chylomicrons shrink and apo CII, apo CIII, phospholipids, and free cholesterol are transferred back to HDL. The chylomicron remnant, containing apo B-48, apo E, and cholesterol esters, binds to the hepatic apo B/E receptor and then is internalized and catabolized.

B. ENDOGENOUS PATHWAY

Endogenous lipids, including those synthesized *de novo* in the liver and those previously delivered to the liver, are transported to extrahepatic sites by VLDL particles. The VLDL, containing endogenous triglycerides and cholesterol, phospholipids, and apo B-100, are assembled and secreted by the hepatocytes. In the plasma, apo C (CI, CII, and CIII) and apo E are transferred to VLDL from HDL. Apo CII–activated lipoprotein lipase hydrolyzes triglycerides from the VLDL core to yield a smaller, more dense IDL particle (or VLDL remnant). Some IDL are removed from circulation and catabolized by the liver via an apo E–dependent receptor pathway. Remaining IDL particles transfer free cholesterol, apo C, apo E, and phospholipids to HDL. The free cholesterol in HDL is converted to cholesterol es-

▶ ters by **lecithin cholesterol acyltransferase** (LCAT), a plasma enzyme activated by apo AI. Conversion of IDL is complete after the cholesterol esters are transferred from HDL to IDL and triglycerides and apoproteins (except apo B-100) are removed from IDL. The final product is the LDL particle composed primarily of cholesterol esters and apo B-100. LDL delivers cholesterol to hepatocytes and other tis-

▶ sues (e.g., adrenal gland) by **LDL receptor–mediated uptake.** Scavenger cells in the vasculature may become overloaded with choles-

▶ terol esters due to **nonreceptor-mediated endocytosis** of LDL, especially when an excess is present, contributing to atherosclerotic plaque formation.

C. REVERSE CHOLESTEROL TRANSPORT

In addition to roles in chylomicron and VLDL metabolism, HDL mediates the return of extrahepatic cholesterol to the liver in a process called reverse cholesterol transport. Nascent HDL particles containing apo A and apo E are assembled and secreted by the liver. Unesterified cholesterol from other lipoproteins or extrahepatic cells is transferred to HDL where LCAT converts it to cholesterol esters. The cholesterol esters are transferred to other lipoproteins such as LDL for delivery to hepatocytes or are transported by HDL to the liver.

IV. MEDICAL RELEVANCE

▶ **Coronary heart disease** (CHD) due to **atherosclerosis** is the leading cause of death and disability in the United States. Atherosclerosis is the formation of arterial lesions containing lipid material, including cholesterol. Lesions start as fatty streaks, progress to fibrous plaques, and then to complicated lesions that partially occlude the vessel. Disruption of the complicated lesion promotes formation of an occlusive thrombus, causing acute myocardial infarction, stroke, gangrene, or aneurysm (depending on the vascular bed affected). While many

Sequelae of Atherosclerosis

- Acute myocardial infarction
- Stroke
- Gangrene
- Aneurysm

factors have been implicated in promoting CHD, certain dyslipidemias (or dyslipoproteinemias) have been identified as major risk factors.

The primary purpose of clinical lipid and lipoprotein testing is for CHD risk assessment and management of atherosclerotic disease by detecting elevations in *atherogenic* lipoprotein components or decreases in *antiatherogenic* lipoprotein components (Table 2–3). Higher amounts of LDL (especially the smaller, denser LDL particles), IDL, Lp(a), and the apo B-100 associated with these lipoproteins have been related to progressive atherosclerotic disease. In contrast, decreased concentrations of HDL and its associated apo AI are predictors of CHD. Markedly elevated triglycerides in chylomicronemia is associated with increased risk for pancreatitis due to triglyceride deposition in the pancreas. The association of CHD with elevations in triglycerides is unclear; triglycerides are often increased in CHD patients, but this may or may not be an independent indicator of disease.

Dyslipoproteinemias result from defects or disturbances in lipoprotein metabolism. These disorders are increasingly classified according to the metabolic defect and the observed lipoprotein abnormality, extending the older Fredrickson classification of the hyperlipidemias, which is based solely on the serum lipoprotein phenotype.

A disorder may be *primary* (i.e., congenital) or *secondary* (i.e., caused by other conditions). Some primary disorders occur with an inherited mutation in a single gene; others are influenced by multiple genes. In certain cases a genetic predisposition may exist, but expression depends on the presence of environmental factors (i.e., acquired). The more common primary *hyper-* and *hypolipidemias* of clinical importance are listed and described in Tables 2–4 and 2–5, respectively. Some of these dyslipidemias represent a group of disor-

Table 2–3. Lipoprotein Changes Associated with Atherosclerotic Disease

LIPOPROTEIN	CHANGES IN ANALYTES ASSOCIATED WITH INCREASED RISK FOR ATHEROSCLEROTIC DISEASE
Atherogenic Lipoproteins	
LDL	↑ LDL-C (with ↑-N TC)
	↑ Apo B
	↑ Proportion of small, dense LDL
	↑ Susceptibility to LDL oxidation
IDL	↑ TC (↑LDL-C[a])
	Abnormal apo E isoform
	LP Electrophoresis: Broad beta band
Lp(a)	↑ Lp(a)
	Small isoform of Lp(a)
Chylomicron and VLDL	↑ Triglycerides
Chylomicron/VLDL remnants (?)[b]	
Antiatherogenic Lipoprotein	
HDL	↓ HDL-C
	↓ Apo AI

TC, total cholesterol; N, normal.
[a]LDL-C will be elevated if the IDL-C is measured with LDL-C in the assay used.
[b]Recent studies have associated these particles with increased risk for CHD.

Table 2–4. Primary Hyperlipidemias

DISORDER	LIPID/LP CHANGES	CHD RISK	DESCRIPTION
Familial combined hyperlipidemia	↑ VLDL only OR ↑ LDL only OR ↑ VLDL *and* LDL	↑	Autosomal dominant; defect may be hepatic overproduction of apo B; often accompanied by ↓ HDL-C; there may be a variant in which apo B is increased without lipid abnormalities
Familial hypercholesterolemia	↑ LDL *Heterozygote:* TC = 250–500 mg/dL *Homozygote:* TC > 500 mg/dL	 ↑ ↑↑	Autosomal codominant; various mutations in LDL receptor or LDL binding or apo B-100 causing a defect in LDL clearance; TG usually normal; the homozygote has no receptors, ↓ HDL-C, and exhibits eruptive xanthomas
Polygenic familial hypercholesterolemia	↑ LDL (TC > 190 mg/dL)	↑	Expression of disease depends on several genes and is influenced by environmental factors; likely to respond to diet modification; defect unknown; has been associated with apo E4 isoform
Familial dysbetalipoproteinemia	↑ IDL and ↑ VLDL	↑	Remnant removal disease; autosomal recessive; defect in removal of IDL; associated with apo E2/2 phenotype (defective binding) or apo E deficiency; expression influenced by other conditions (e.g., obesity, excess ethanol, hypothyroidism, diabetes); xanthomas on palms of hands typical; broad beta band on electrophoresis
Familial chylomicronemia	↑ chylomicrons and ↑ VLDL	↑ (?)	Autosomal recessive; lipoprotein lipase deficiency or apo CII deficiency causes decreased clearance of chylomicrons and VLDL; eruptive xanthomas typical; risk of pancreatitis due to ↑↑ TGs
Familial hypertriglyceridemia	↑ VLDL	↑ (?)	May be several types; defects may be ↑ VLDL-TG synthesis in the liver or ↓ VLDL clearance; often associated with ↓ HDL (with ↑ CHD risk) due to ↑ HDL catabolism; exacerbated by obesity and diabetes; risk of pancreatitis
Familial lipoprotein(a) excess	↑ Lp(a)	↑	Highly heritable trait; Lp(a) is LDL with one apo(a) bound to the apo B-100; ↑ Lp(a) associated with CHD; apo(a) isoforms may play a role; no xanthomas
Familial hyperalphalipoproteinemia	↑ HDL	↓	No treatment required; associated with increased longevity; demonstrates the need for measurement of both TC and HDL-C in patient screenings

TC, total cholesterol.

ders exhibiting similar lipid abnormalities caused by one of several specific molecular defects (e.g., familial hypercholesterolemia).

Secondary causes of hyperlipidemia usually fall into one of four categories: diet, drugs, disorders of metabolism, and diseases (Table 2–6). Altered lipid metabolism due to secondary causes may lead to elevations in cholesterol or triglycerides or both. Some secondary factors may cause a concomitant decrease in HDL. Secondary hyperlipidemias are treated by controlling the underlying condition, and therefore must always be considered before diagnosis of a primary lipid disorder and before treating any lipid disorder. While secondary factors may cause secondary or acquired hyperlipidemias, they may also exacerbate a preexisting primary lipoprotein disorder.

Table 2–5. Primary Hypolipidemias

DISORDER	LIPID/LP CHANGES	CHD RISK	DESCRIPTION
Familial hypoalphalipoproteinemia	↓ HDL	↑	Various possible defects (e.g., LCAT deficiency, apo AI deficiency, apo AI variants) causing ↓ apo AI production or ↑ apo AI degradation; may have increases in the apo B lipoproteins
Analphalipoproteinemia	Absence of HDL	↑↑	Tangier disease; rare autosomal recessive; may be due to ↑↑ catabolism of HDL or apo AI; accumulation of cholesterol esters in tissues; progressive motor weakness
Familial hypobetalipoproteinemia	↓ LDL and ↓ VLDL	↓	Autosomal dominant; treatment unnecessary; associated with increased longevity
Abetalipoproteinemia	Absence of chylomicrons, VLDL, and LDL	?	Rare autosomal recessive; may be defect in synthesis or secretion of apo B–containing lipoproteins; clinical manifestations include fat malabsorption, growth retardation, eventual degeneration of the central nervous system

Table 2–6. Secondary Causes of Hyperlipidemia

	SECONDARY CAUSES OF INCREASED CHOLESTEROL	SECONDARY CAUSES OF INCREASED TRIGLYCERIDES
Diet	High saturated fats and cholesterol	Alcohol excess Alcoholism Excess caloric intake
Drugs	Anabolic steroids[a] Beta blockers[a] Carbamazepine Cyclosporine Estrogens Progestins[a] Thiazide diuretics[a]	Beta blockers[a] Corticosteroids Estrogens Thiazide diuretics[a]
Disorders of metabolism	Diabetes mellitus[a] Hypothyroidism Pregnancy	Diabetes mellitus[a] (under poor control) Hypothyroidism Obesity[a] Pregnancy
Diseases	Hepatoma Multiple myeloma Nephrotic syndrome[a] Obstructive liver disease (Lp-X[b]) Renal failure (chronic)[a]	Nephrotic syndrome[a] Renal failure[a]

[a] Also associated with decreases in HDL-C.

[b] An abnormal lipoprotein found in obstructive liver disease.

V. TEST PROCEDURES

A. PREANALYTICAL FACTORS

Lipid/lipoprotein analyses may be affected by a number of preanalytical factors. Patients must be advised to maintain their customary diet and lifestyle for 3 days prior to testing. Evaluation of lipid/lipoprotein status should not be performed at the time of an acute or recent illness. It is necessary for patients to fast for at least 12 hours, and abstain from alcohol for 24 hours, before samples are drawn for analysis of triglycerides (TG), apo B, or calculated LDL-C; fasting is not required, but is usually preferred, for measurement of total cholesterol (TC), HDL-C, LDL-C by a direct method, apo AI, apo AII, or Lp(a). To avoid hemoconcentration of blood samples resulting in falsely elevated lipid/lipoprotein results, patients should be seated for at least 15 minutes prior to sample collection, and phlebotomists should minimize tourniquet use. Testing is usually performed on serum or EDTA plasma; concentrations in EDTA plasma are 4.7% lower than serum due to osmotic shifts. For long-term sample stability, EDTA may reduce oxidation and enzymatic cleavage of lipoproteins by chelating metal ions. Heparinized samples may be acceptable; however, heparin activates lipoprotein lipase. Due to intraindividual and analytical variation, it is recommended that lipid/lipoprotein testing be performed on more than one occasion when evaluating a patient.

Patient Preparation for Lipid/ Lipoprotein Testing

- Maintain customary lifestyle for preceding 3 days
- Abstain from alcohol for preceding 24 hours
- Fasting for at least 12 hours before sampling recommended (required for TG, apo B, and calculated LDL-C)
- Remain sitting for at least 15 minutes prior to sampling

B. PLASMA (SERUM) APPEARANCE

A plasma or serum sample that appears milky is *lipemic*. The physical appearance of a fasting plasma or serum sample after standing undisturbed at 4°C for 16 hours may be informative in lipid/lipoprotein testing. A creamy layer floating at the top signals the presence of chylomicrons, an abnormal finding in a fasting sample. A sample that remains turbid or milky throughout is likely to contain a high level of VLDL (or IDL). Samples with visible evidence of chylomicrons or VLDL would be expected to have high TG concentrations. While elevated levels of LDL do not cause turbidity, they may cause an orange discoloration of the sample due to carotenoids bound to LDL particles.

Appearance of Plasma (Serum) Samples with Elevations in Lipoprotein Fractions

- Increased chylomicrons—creamy layer on top
- Increased VLDL and/or IDL—remains turbid or milky throughout
- Increased LDL—no turbidity, may have orange discoloration
- Increased HDL—normal, clear

C. TOTAL CHOLESTEROL

TC is routinely measured by enzymatic end-point assays. Analysis typically begins with the following two coupled reactions:

$$\text{Cholesterol esters} \rightarrow \text{Cholesterol + fatty acids} \quad \text{(Enz = } \textit{cholesterol esterase}\text{)}$$

$$\text{Cholesterol + O}_2 \rightarrow \text{Cholest-4-ene-3-one + H}_2\text{O}_2 \quad \text{(Enz = } \textit{cholesterol oxidase}\text{)}$$

Most commonly, the hydrogen peroxide formed in the second reaction becomes a reactant in subsequent coupled reactions. Some of these assays use a third coupled reaction catalyzed by peroxidase where H_2O_2 reacts with a chromogen to form a chromophore that can be measured spectrophotometrically. In other assays, there is a third reaction catalyzed by catalase, plus an additional reaction to form a chromophore. Another type of assay measures the consumption of O_2 in the second reaction electrochemically with an oxygen electrode.

The Centers for Disease Control and Prevention (CDC)-modified Abell–Kendall reference method is an accurate chemical assay, but is too cumbersome for routine use. In this method, cholesterol esters are hydrolyzed and then cholesterol (all in free form) is extracted. Cholesterol is then oxidized by Liebermann–Burchard reagent (containing acetic acid, acetic anhydride, and sulfuric acid) to form colored compounds that can be measured spectrophotometrically. The definitive method for measuring TC using isotope-dilution mass spectrometry is beyond the technical and financial limits for routine or even reference testing.

D. HDL Cholesterol

Prior to HDL-C measurement, VLDL and LDL are precipitated from the sample using either polyanions in the presence of divalent cations (e.g., heparin/$MnCl_2$, heparin/$CaCl_2$, dextran sulfate/$MgCl_2$, or Na phosphotungstate/$MgCl_2$) or neutral polymers (e.g., polyethylene glycol). After centrifugation, the cholesterol remaining in the supernatant is assumed to belong to the HDL fraction and is analyzed by an appropriate cholesterol method. The method used for precipitation may be verified by analyzing the supernatant and precipitate by lipoprotein electrophoresis or by measuring apo AI and apo B content.

Newer methods using precipitant bound to magnetic particles obviates the need for centrifugation. Methods in which physical separations are unnecessary are under development.

No validated HDL-C reference method is established; however, the CDC HDL-C procedure is widely used to gauge accuracy. This CDC HDL-C reference method first uses ultracentrifugation to separate chylomicrons and VLDL from HDL and LDL. In the second step, LDL is precipitated using a heparin/$MnCl_2$ reagent. The remaining cholesterol (i.e., HDL-C) is measured using the CDC-modified Abell–Kendall procedure.

E. LDL Cholesterol

For common analysis purposes, the LDL-C includes cholesterol from LDL, IDL, and Lp(a). Currently, LDL-C concentrations are routinely estimated using the empirically derived Friedewald equation:

$$[LDL\text{-}C] = [TC] - [HDL\text{-}C] - ([TG]/5)$$

where [TC], [HDL-C], and [TG] are measured on a fasting sample. The equation is not valid if the TG concentration exceeds 400 mg/dL or if chylomicrons are present. If triglycerides are elevated, LDL-C should be measured by the CDC reference method or a direct LDL-C method (described below).

Direct methods for LDL-C developed more recently are under evaluation and are increasingly being used for routine testing. One direct LDL-C assay uses immunoseparation where HDL and VLDL are bound by latex beads coated with anti-HDL and anti-VLDL antibodies and removed by centrifugation. The remaining LDL-C is measured by an appropriate cholesterol method.

As with HDL-C, there is no validated LDL-C reference method and the CDC LDL-C procedure is most commonly used as a target for accuracy. The CDC procedure (sometimes known as beta quantification):

1. Separates chylomicrons and VLDL from the LDL and HDL by ultracentrifugation
2. Precipitates LDL using heparin and $MnCl_2$ to leave HDL
3. Measures cholesterol in the infranate from step 1 to determine the sum of LDL-C and HDL-C and in the supernate from step 2 for HDL-C using the cholesterol reference method
4. Calculates LDL-C as:

$$LDL\text{-}C = (LDL\text{-}C + HDL\text{-}C) - (HDL\text{-}C)$$

F. TRIGLYCERIDES

Triglycerides are routinely measured enzymatically beginning with the following two reactions:

Triglycerides $\rightarrow$ Glycerol + 3 fatty acids   (Enz = *lipase*)

Glycerol + ATP $\rightarrow$ Glycerol-3-phosphate (G-3-P) + ADP (Enz = *glycerol kinase*)

Assays use one of several available enzyme systems coupled to a product of the second reaction to generate a quantifiable indicator; coupled systems include: (1) glycerol phosphate oxidase and peroxidase; (2) glycerol phosphate dehydrogenase; or (3) pyruvate kinase and lactate dehydrogenase. Since all of the enzymatic methods include the reaction converting glycerol to G-3-P, any preexisting or *endogenous glycerol* in the sample will cause a falsely elevated triglyceride result. Endogenous glycerol is typically very low in normal serum; however, levels may be increased depending on certain medical conditions or sample handling procedures (e.g., diabetes, infusion of glycerol-containing intravenous fluids, contamination of blood collection devices, prolonged storage of whole blood). Correction for endogenous glycerol is recommended using a glycerol-blanking method, where preexisting glycerol is measured by omitting the TG hydrolysis step using a reagent without lipase. The result for endogenous glycerol is subtracted from the triglyceride result to give the corrected triglyceride value. Due to the low frequency of problems with endogenous glycerol, many laboratories do not blank for glycerol but concentrate on proper sample handling instead.

There is no validated reference method for triglycerides measurement, but the CDC reference method is most often used to assess accuracy. The CDC uses a chemical method in which triglycerides are extracted with methylene chloride and then hydrolyzed to glycerol and fatty acids under alkaline conditions; the liberated glycerol is measured spectrophotometrically after color development with a chromotropic acid reagent.

G. APOLIPOPROTEINS

Measurement of apo B-100 (apo B), the major protein in LDL, as well as a component of the atherogenic lipoproteins VLDL, IDL, and Lp(a), may be the most clinically useful of the apolipoprotein determinations. Increased apo B concentrations have been highly associated with CHD. Decreased apo AI values, indicating low amounts of HDL, have also shown a relationship to CHD. Lp(a) concentration in an individual is largely genetically determined with significant variation between ethnic groups. Increased Lp(a) is associated with expression of a smaller size isoform of apo(a) and is an independent risk factor for premature atherosclerotic disease. Quantification of apos AII, AIV, CI, CII, CIII, and E may be available in specialized laboratories for the purposes of research or for characterizing familial lipoprotein abnormalities.

Apolipoproteins are measured by any one of a number of immunoassay methods, including automated immunonephelometric or immunoturbidimetric systems (most common for apo AI and apo B), radioimmunoassay, enzyme-linked immunosorbent assay (ELISA), fluorescence immunoassay, and radial immunodiffusion. Apolipoprotein determinations have been notorious for intermethod variability of results, leading to a lack of confidence in their value. For the most part, variations have been attributed to the lack of standardization in calibrators between methods. Through considerable effort by the International Federation of Clinical Chemistry, reference materials have been identified for the calibration of apo B and apo AI. Apo B and apo AI values may become more clinically useful after medical decision limits are established from epidemiological and clinical studies.

There are three common *isoforms* of apo E, designated E2, E3, and E4, each differing in binding affinity for the LDL receptor. Apo E3 is the wild-type isoform binding normally to the receptor (i.e., 100% of normal); apo E2 demonstrates only 2% of normal binding affinity; and apo E4 binding is greater than 100%. Individuals homozygous for apo E2 (E2/2) have increased serum triglyceride levels due to inability to clear or catabolize IDL. Total cholesterol is increased in those with apo E4/3 or E4/4 phenotypes, presumably due to increased LDL-C. Apo E *phenotypes* can be determined by isoelectric focusing with immunoblotting; apo E *genotypes* can be determined by nucleotide sequencing of allele-specific oligonucleotides or by restriction fragment length polymorphisms (RFLPs) following polymerase chain reaction (PCR) amplification.

Immunoassay Methods for Measuring Apolipoproteins

- Automated immunoephelometric or immunoturbidimetric systems
- Radioimmunoassay
- Enzyme-linked immunosorbent assay (ELISA)
- Fluorescence immunoassay
- Radial immunodiffusion

Apo E Isoforms

E2—demonstrates only 2% of normal binding affinity
E3—wild-type isoform; binding defined as 100% of normal
E4—binding is greater than 100%

H. OTHER TESTS

Less common lipid/lipoprotein tests are available in specialized laboratories for particular purposes. Lipoprotein electrophoresis is a qualitative test now primarily useful only in identifying abnormal amounts of IDL found in familial dysbetalipoproteinemia (also known as broad beta disease). An increased proportion of small LDL particles or small HDL particles (determined by lipoprotein subfractionation), susceptibility to LDL oxidation, and increased chylomicron or VLDL remnants have all been related to increased risk for CHD; however, testing is typically only available on a research basis. Some familial dyslipidemias may be characterized by evaluation of lipoprotein lipase activity, LCAT activity, Lp(a) isoforms, and genes for familial hypercholesterolemia.

I. STANDARDIZATION

The CDC has established a Cholesterol Reference Method Laboratory Network (CRMLN) to facilitate a national accuracy base in lipid testing for clinical laboratories and manufacturers of *in vitro* diagnostics.

Table 2–7. NCEP Analytical Goals for Routine Lipid Measurements

LIPID ANALYTE	TOTAL ERROR	ACCURACY (BIAS)	PRECISION (COEFFICIENT OF VARIATION)
Total cholesterol	≤ 8.9%	≤ ±3%	≤ 3%
Triglycerides	≤ 15%	≤ ±5%	≤ 5%
HDL-C	≤ 14%	≤ ±5%	≤ 4% for HDL-C ≥ 42 mg/dL SD ≤ 1.7 mg/dL for HDL-C < 42 mg/dL
LDL-C	≤ 12%	≤ ±4%	≤ 4%

SD, standard deviation.

The CRMLN currently provides testing for total cholesterol, HDL-C, and triglycerides using reference or designated comparison methods traceable to the CDC reference methods. The National Cholesterol Education Program (NCEP) has recommended analytical goals for routine lipid testing in an effort to minimize interlaboratory variation and misclassification of patients (Table 2–7).

VI. TEST RESULTS AND INTERPRETATION

The typical lipid/lipoprotein results expected with the major lipoprotein abnormalities are listed in Table 2–8.

Overall, the reference ranges for total cholesterol, LDL-C, and triglycerides are higher with increasing age. However, because a significant proportion of the "normal," apparently healthy population has preclinical atherosclerotic lesions, population-based reference ranges for total cholesterol, HDL-C, LDL-C, and triglycerides have been abandoned in favor of medical decision limits with the object of identifying those at risk for developing clinical disease. The NCEP has established guidelines for the detection, evaluation, and management of individuals at increased risk for CHD. The current NCEP-recommended lipid/lipoprotein decision levels are given in Table 2–9; some suggested cut-off values and reference ranges for important apolipoproteins are given in Table 2–10.

Initial measurement of total cholesterol and HDL-C is the primary strategy used for case finding. Individuals are also assessed for major risk factors (listed in Table 2–11) by interview or questionnaire. Further testing is done when indicated (e.g., when HDL-C is less than 35 mg/dL). Generally, treatment is initiated in individuals with: (1) LDL-C ≥ 100 mg/dL when there is evidence of CHD; (2) LDL-C ≥ 130 mg/dL and two risk factors in the absence of disease; or (3) LDL-C ≥ 160 mg/dL and less than two risk factors in the absence of disease. Suspicion of a primary dyslipidemia in an individual may lead to recommendations for testing in other family members.

Guidelines for CHD risk assessment, disease prevention, and disease management will need to be adjusted as the pathogenic processes of atherosclerosis become better characterized and understood. Testing of nonlipid factors such as homocysteine, hypercoagulability, or markers of inflammation, such as high-sensitivity C-reactive protein, may become more important in the future.

Table 2–8. Summary of Expected Lipid/Lipoprotein Results

ABNORMALITY	APPEARANCE OF STANDING PLASMA	TOTAL CHOLESTEROL <200 mg/dL	HDL CHOLESTEROL >60 mg/dL	LDL CHOLESTEROL <130 mg/dL	TRIGLYCERIDES <200 mg/dL	APO AI	APO B	LP ELECTROPHORESIS	FREDRICKSON CLASSIFICATION[a]	CHD RISK
None (optimal)	Clear	<200 mg/dL	>60 mg/dL	<130 mg/dL	<200 mg/dL	N	N	Normal alpha, prebeta, and beta bands	—	Desirable
↑Chylomicrons	Top creamy layer; clear to slightly turbid below	N - ↑	N - ↓	N	↑↑↑	N - ↓	N	Prominent band at origin	Type I	↑ (?) (↑ Risk of pancreatitis)
↑LDL	Clear; possible orange discoloration	↑ (sometimes N)	N - ↓	↑	N	N - ↓	↑	Prominent beta band	Type IIA	↑↑
↑LDL and VLDL	Clear to slightly turbid throughout	↑	N - ↓	↑	↑	N - ↓	↑	Prominent prebeta and beta bands	Type IIB	↑
↑IDL and remnants	Slightly turbid to opaque	↑	N - ↓	(↑)	↑	N - ↓	N - ↑	Broad beta band	Type III	↑
↑VLDL	Turbid to opaque	N - ↑	N - ↓	N	↑↑	N - ↓	↑	Prominent prebeta band	Type IV	↑
↑VLDL and chylomicrons	Top creamy layer, turbid to opaque below	↑	N - ↓	N	↑↑↑	N - ↓	↑	Prominent prebeta band and band at origin	Type V	↑ (↑ Risk of pancreatitis)
↑HDL	Clear	N - ↑	↑	N	N	↑	N	Prominent alpha band	—	↓
↓LDL	Clear	↓ - ↓↓↓	N	↓ - absent	N - ↓	N	↓ - ↓↓	Depressed beta band or absent prebeta and beta bands	—	↓
↓HDL	Clear	N - ↓	↓ - absent	N - ↑	N - ↑	↓	N	Depressed alpha band or absent alpha band	—	↑

N, normal.
[a] Of historical interest only.

Table 2–9. NCEP Lipid/Lipoprotein Medical Decision Levels for Adults Without Evidence of CHD

ANALYTE	MEDICAL DECISION LEVEL	
Total cholesterol	Desirable	< 200 mg/dL
	Borderline high risk	200–239 mg/dL
	High risk	≥ 240 mg/dL
LDL cholesterol	Desirable	< 130 mg/dL
	Borderline high risk	130–159 mg/dL
	High risk	≥ 160 mg/dL
HDL cholesterol	Desirable	> 60 mg/dL
	High risk	< 35 mg/dL
Triglycerides	Normal	< 200 mg/dL
	Borderline high triglycerides	200–400 mg/dL
	High triglycerides	400–1,000 mg/dL
	Very high triglycerides[a]	> 1,000 mg/dL

[a] Associated with risk of pancreatitis.

Table 2–10. Suggested Medical Decision Levels for Apolipoproteins AI and B[a]

ANALYTE	MEDICAL DECISION LEVEL	
Apolipoprotein AI	Reference ranges:	
	Male	0.88–1.8 g/L
	Female	0.98–2.1 g/L
	Desirable	> 1.2 g/L
Apolipoprotein B	Reference ranges:	
	Male	0.55–1.51 g/L
	Female	0.44–1.48 g/L
	Desirable	< 1.2 g/L

Sources: Contois JH, McNamara JR, Lammi-Keefe CJ, et al. Reference intervals for plasma apolipoprotein A-I determined with a standardized commercial immunoturbidimetric assay: results from the Framingham Offspring Study. *Clin Chem* 42:507–14, 1996; and Contois JH, McNamara JR, Lammi-Keefe CJ, et al. Reference intervals for plasma apolipoprotein B determined with a standardized commercial immunoturbidimetric assay: results from the Framingham Offspring Study. *Clin Chem* 42:515–23, 1996.

Table 2–11. NCEP Major Risk Factors for CHD

Gender and age	Males: ≥ 45 years
	Females: ≥ 55 years, or premature menopause without estrogen replacement
Family history of premature coronary heart disease	Definite myocardial infarction or sudden death:
	Before 55 years of age in father or other first-degree male relative, or
	Before 65 years of age in mother or other first-degree female relative
Current cigarette smoking	
Hypertension	≥ 140/90 mm Hg or on antihypertensive therapy
HDL cholesterol	< 35 mg/dL
Diabetes mellitus	

SUGGESTED READINGS

Bachorik PS, Ross JW. National Cholesterol Education Program recommendations for measurement of low-density lipoprotein cholesterol: executive summary. *Clin Chem* 41:1414–20, 1995.

Contois JH, McNamara JR, Lammi-Keefe CJ, et al. Reference intervals for plasma apolipoprotein A-I determined with a standardized commercial immunoturbidimetric assay: results from the Framingham Offspring Study. *Clin Chem* 42:507–14, 1996.

Contois JH, McNamara JR, Lammi-Keefe CJ, et al. Reference intervals for plasma apolipoprotein B determined with a standardized commercial immunoturbidimetric assay: results from the Framingham Offspring Study. *Clin Chem* 42:515–23, 1996.

Jialal I. A practical approach to the laboratory diagnosis of dyslipidemia. *Am J Clin Path* 106:128–38, 1996.

National Cholesterol Education Program Expert Panel. Summary of the second report of the National Cholesterol Education Program expert panel in detection, evaluation, and treatment of high blood cholesterol in adults (Adult Treatment Panel II). *JAMA* 269:3015–23, 1993.

Rifai N, Warnick GR, Dominiczak MH (eds.). *Handbook of lipoprotein testing.* Washington, DC: AACC Press, 1997.

Stein EA, Myers GL. National Cholesterol Education Program recommendations for triglyceride measurement: executive summary. *Clin Chem* 41:1421–26, 1995.

Warnick GR, Wood PD. National Cholesterol Education Program recommendations for measurement of high-density lipoprotein cholesterol: executive summary. *Clin Chem* 41:1427–33, 1995.

Items for Further Consideration

1. Explain how the structure of a lipoprotein particle allows the transport of lipids in the blood.

2. What is the role of apolipoprotein CII in the metabolism of VLDL and chylomicrons?

3. By what mechanism(s) does excess plasma LDL apparently promote atherosclerosis?

4. Devise a protocol for obtaining samples for lipoprotein evaluation minimizing preanalytical effects.

5. Predict the effect of the following situations on the results of lipid measurements: (a) incomplete hydrolysis of triglycerides by lipase in a triglycerides assay; (b) endogenous glycerol in a sample for triglycerides assay; and (c) incomplete precipitation of LDL in an HDL cholesterol assay. Propose a plan to evaluate or correct for these problems.

6. If a patient is homozygous for an abnormal apolipoprotein E variant that is defective in binding to its receptor, predict the disease present and the expected laboratory findings.

7. List the changes in lipid/lipoprotein analytes associated with increased risk for coronary heart disease. List the major risk factors associated with increased risk for coronary heart disease.

Amino Acids and Proteins

Christine Papadea

Key Terms

ACUTE PHASE RESPONSE

ALBUMIN

α_1-*ANTITRYPSIN*

α_1-*ACID GLYCOPROTEIN*

α_1-*FETOPROTEIN*

α_2-*MACROGLOBULIN*

AMINO ACIDS

AMINOACIDOPATHIES

AMINOACIDURIAS

β_2-*MICROGLOBULIN*

CERULOPLASMIN

COMPLEMENT

C-REACTIVE PROTEIN

CRYOGLOBULINS

FIBRINOGEN

GLOBULINS

HAPTOGLOBIN

HOMOCYSTEINE

IMMUNOGLOBULINS (G, A, M, D, E)

MONOCLONAL GAMMOPATHY

MULTIPLE MYELOMA

MULTIPLE SCLEROSIS

OLIGOCLONAL BANDING

PARAPROTEIN

PEPTIDE BONDS

POLYPEPTIDES

PREALBUMIN

TRANSFERRIN

TRANSTHYRETIN

WALDENSTRÖM'S MACROGLOBULINEMIA

Chapter Objectives

Upon completion of this chapter, the reader will be able to:

1. Describe the biochemical structure and metabolism of proteins.
2. List the major physiological functions of proteins.
3. Distinguish between primary, renal, overflow, and secondary disorders of amino acid metabolism.
4. Discuss the underlying metabolic defect and the clinical consequence of the defect for three aminoacidopathies.
5. Describe the major function(s) and/or associated clinical abnormalities of the following proteins: transthyretin; albumin; α_1-antitrypsin; α-fetoprotein; α_1-acid glycoprotein; α_2-macroglobulin; haptoglobin; ceruloplasmin;

transferrin; hemopexin; β_2-microglobulin; fibrinogen; C-reactive protein; C3 component of complement; immunoglobulins G, A, M, and E; and oligoclonal immunoglobulin bands.

6. Describe the structure and identify important components of an immunoglobulin molecule (i.e., Fab, Fc, hinge region, constant/variable region, amino-/carboxy-terminal ends, heavy/light chains, and J chain).
7. Distinguish between positive and negative acute phase reactants in serum and give an example of each type.
8. Discuss the use of the following analytical techniques in regard to the assessment of proteins: refractive index, immunofixation electrophoresis, capillary zone electrophoresis, Biuret reaction, dye-binding methods, immunochemical analysis, and reagent strip testing.
9. Differentiate the cerebrospinal fluid (CSF)/serum albumin index from CSF/serum IgG index and explain the significance of each determination.
10. Distinguish between a transudate and an exudate in regard to body fluid protein concentration.
11. Correlate clinical findings and the laboratory assessment of certain amino acids or proteins in regard to the following conditions: hyperhomocysteinemia, nephrotic syndrome, inflammation, infection, monoclonal gammopathies, and multiple sclerosis.

Analytes Included

AMINO ACIDS:	PROTEINS:	C-REACTIVE PROTEIN
HISTIDINE	ALBUMIN	FIBRINOGEN
HOMOCYSTEINE	α_1-ACID GLYCOPROTEIN	HAPTOGLOBIN
HOMOGENTISIC ACID	α_1-ANTITRYPSIN	HEMOPEXIN
ISOLEUCINE	α_1-FETOPROTEIN	IMMUNOGLOBULINS A, G, M, D, AND E
LEUCINE	α_2-MACROGLOBULIN	OLIGOCLONAL BANDS
PHENYLALANINE	β_2-MICROGLOBULIN	TRANSFERRIN
TYROSINE	CERULOPLASMIN	TRANSTHYRETIN
VALINE	COMPLEMENT C3	

Pretest

Directions: Choose the single best answer for each of the following items.

1. The primary function of albumin in the circulation is to
   a. enhance the ionization of calcium
   b. regulate colloidal oncotic pressure
   c. promote the antigen activity of plasma proteins
   d. provide a neutral binding site for cell receptors

2. Complement is responsible *in vivo* for
   a. enhanced phagocytosis of infectious agents
   b. attraction of polymorphonuclear cells to the site of injury
   c. agglutination of red cells
   d. cross-linking of immunoglobulins

3. In testing a healthy 6-month-old child for serum immunoglobulin G (IgG) concentration, which of the following would be expected?
   a. lower concentration relative to the normal adult concentration of IgG
   b. undetectable IgG concentation
   c. higher concentration relative to normal concentration of adult of IgG
   d. immunoglobulin A (IgA) concentration higher than IgG concentration

4. Which of the following proteins transports thyroxine and vitamin A in the blood and serves as a marker of nutritional balance?
   a. α_1-acid glycoprotein
   b. hemopexin
   c. ceruloplasmin
   d. transthyretin

5. Transferrin is a protein typically associated with the transport of
   a. ferritin
   b. ferric ions
   c. ferrous ions
   d. heavy metals

6. Which of the following proteins is decreased in hemolytic disorders?
   a. transferrin
   b. β_2-microglobulin
   c. fibrinogen
   d. haptoglobin

7. Maple syrup urine disease is caused by a deficiency of
   a. isoleucine synthetase
   b. branched chain keto-acid decarboxylase
   c. fumarylacetoacetate hydroxylase
   d. glucose oxidase

8. A heavy distinct band located in the beta–gamma region on serum protein electrophoresis preliminarily indicates the presence of
   a. nephrotic syndrome
   b. a polyclonal gammopathy
   c. a monoclonal protein
   d. an acute inflammatory response

9. Homocysteinemia, as related to cardiovascular disease, is associated with
   a. myocarditis
   b. a deficiency of folate and vitamin B_{12}
   c. an excess of methionine and vitamin B_6
   d. an excess of methionine and cysteine

10. Which of the following analytical techniques is used to identify the immunoglobulin class that is elevated in monoclonal gammopathies?

    a. radial immunodiffusion
    b. biuret reaction
    c. dye binding assay
    d. immunofixation electrophoresis

11. Plasma proteins that decrease in the acute phase response include
    a. albumin, transthyretin, and IgG
    b. ceruloplasmin, haptoglobin, and fibrinogen
    c. albumin, transthyretin, and transferrin
    d. C-reactive protein, fibrinogen, and haptoglobin

12. α_1-Antitrypsin exhibits extensive polymorphism due to its numerous
    a. protease inhibitors
    b. alleles
    c. polymerases
    d. phenotypes

13. α_1-Fetoprotein may be included in a testing panel for
    a. prenatal screening for neural tube defects
    b. prenatal screening for pulmonary disease
    c. the evaluation of protein-losing renal disease
    d. the identification of an acute-phase reaction

14. Abnormal CSF protein bands located in the gamma region on electrophoresis are referred to as _____, and are indicative of _____.
    a. myoglobins; meningitis
    b. prealbumins; Guillain–Barré syndrome
    c. oligoclonal bands; multiple sclerosis
    d. transarachnoid bands; neurosyphilis

15. Using zone electrophoresis at alkaline pH, which of the following lists of plasma components (from anode to cathode) indicates a normal pattern?
    a. albumin, transthyretin, transferrin, orosomucoid
    b. transthyretin, haptoglobin, transferrin, complement C3
    c. albumin, lipoprotein, transferrin, fibrinogen
    d. transthyretin, albumin, C-reactive protein, ceruloplasmin

I. INTRODUCTION

The word *protein* is derived from the Greek word *proteios*, meaning "first or primary." Proteins are indeed of primary importance, comprising a major proportion (50 to 70%) of the dry weight of cells. They are components of the structural framework of cells and tissues, and have key roles in the transport, synthesis, storage, and clearance of many biological substances.

Although hundreds of different proteins have been identified in plasma, this chapter focuses on the major plasma proteins (about 25) that have established clinical correlation with health and disease. Protein analyses are important in a number of clinical situations:

1. Diagnosis and monitoring of paraproteinemias
2. Evaluation of renal disease
3. Assessment of nutritional status
4. Evaluation of inflammatory conditions
5. Evaluation of patients with unexpected findings

II. BIOCHEMISTRY

▶ Proteins are polymers of building block units called **amino acids.** The structural backbone of an amino acid includes an amino ($-NH_2$) group and a carboxyl ($-COOH$) group linked to an α-carbon atom; the backbone has the basic formula $RCH(NH_2)COOH$. The side chain (R) on the α-carbon confers the amino acid's identity and is responsible for its unique properties; R groups may be classified by their chemical characteristics (e.g., acidic, basic, hydrophilic, hydrophobic, or aromatic).

The *primary structure* of a protein is the sequence of amino acids comprising the polypeptide chain that is genetically encoded by the deoxyribonucleic acid (DNA). Messenger ribonucleic acid (RNA) transcribed from DNA interacts with the cell's protein assembly machinery to translate the genetic code into a primary amino acid sequence. During protein assembly, amino acids are joined by ▶ peptide bonds. **Peptide bonds** are covalent linkages between the α-carboxyl group and the α-amino group of adjacent amino acids; a series of peptide bonds forms a chain of amino acids that is the primary structure of proteins. *Secondary protein structure* describes recurring formations along the polypeptide chain (e.g., α-helix, β-pleated sheet, or random coil) caused by intramolecular forces such as hydrogen bonding. The *tertiary structure* is the folded, three-dimensional conformation of the protein that is stabilized by both covalent bonds and noncovalent forces mainly involving amino acid R groups. The protein's environment influences tertiary structure; in the polar aqueous environment, hydrophobic side chains tend to turn toward the interior of a folded protein, while hydrophilic regions form the outer surface of the protein. *Quaternary protein structure* is the association of two or more polypeptide chains, or subunits, to form a functional protein complex.

Proteins are structurally and functionally diverse due to the heterogeneity that is possible in the sequences and arrangements of amino acids. Although the molecular size of biologically active protein macromolecules spans a range from about 5,000 daltons to several million daltons for some structural proteins, many hormones are much smaller. Chains of five amino acids, having molecular weights of about 500 daltons, to peptide chains up to about 3,000 daltons, ▶ are termed **polypeptides.**

Proteins carry a net charge determined by the protein's constituent amino acids and the pH environment; the isoelectric point (pI) of a protein is the pH at which its net charge is zero. If the pH of the protein's environment is higher than its pI, the molecule will be negatively charged; conversely, if the pH is lower than its pI, the protein will be positively charged. Differences in the net charge

---

Protein Structure

Primary structure—the sequence of amino acids comprising the polypeptide chain that is genetically encoded by the DNA

Secondary structure—recurring formations along the polypeptide chain caused by intramolecular forces such as hydrogen bonding

Tertiary structure—the folded, three-dimensional conformation of the protein that is stabilized by both covalent bonds and noncovalent forces, mainly involving amino acid R groups

Quaternary structure—the association of two or more polypeptide chains, or subunits, to form a functional protein complex

are used as a basis for analytical and preparative separation of proteins.

Proteins of clinical importance are associated with numerous physiologic and biological activities. Examples of major protein functions include:

- Oncotic pressure: albumin
- Transport: transferrin, albumin, lipoproteins
- Catalytic (enzymes): creatine kinase, alkaline phosphatase
- Coagulation: fibrinogen
- Hormones: thyroid-stimulating hormone
- Structural and connective tissue: collagen, elastin
- Receptors: hormone receptors
- Defense: antibodies
- Buffer mechanisms: hemoglobin (in addition to O_2 transport)

Proteins ingested in food are hydrolyzed in digestive processes in the stomach and small intestine. The action of endopeptidases, exopeptidases, and other hydrolytic enzymes converts peptides into their constituent amino acids, which are absorbed by the small intestine through various active transport systems. The absorbed amino acids are then carried to the liver, the main organ for protein synthesis. While many plasma proteins are produced in the liver, the immunoglobulins are products of plasma cells.

III. MEDICAL RELEVANCE

A. AMINO ACIDS

There are approximately 40 major inherited **aminoacidopathies** ◄ (i.e., diseases of amino acid metabolism), most of which are autosomal recessive disorders. Persistent elevations in the concentrations of plasma amino acids (aminoacidemias) are clinically significant, especially when found in newborns and children; concentrations may be transiently elevated or decreased during the first few days of life in premature or low-birth-weight babies. Decreased plasma concentrations of amino acids may occur in the first half of pregnancy but are otherwise rare, because the amino acid pool is relatively constant even in starvation due to catabolism of endogenous body proteins. Increased concentrations of plasma amino acids are either primary (inherited) defects or secondary (acquired) disorders. Some primary aminoacidemias are lethal if not treated within the early days of life, while others are apparently benign. Secondary increases are usually a result of severe liver disease.

The primary aminoacidopathies are inborn errors of metabolism caused by congenital defects in the metabolic pathway of a specific amino acid (see Table 3–1 for some examples). The result is accumulation of the specific amino acid (or substrate) such that normal products are not formed. Also, substrate may be diverted to other metabolic pathways, leading to an increase in products of alternate pathways; in either case, the defect may give rise to an excess of toxic metabolites.

Primary **aminoacidurias,** the increased urinary excretion of ◄ amino acids, may be categorized as either overflow or renal. Overflow aminoacidurias arise from increased plasma amino acid concentrations. Aminoaciduria results because the normal resorptive mecha-

Table 3–1. Characteristics of Selected Primary Aminoacidopathies

| DISORDER | FREQUENCY | SUBSTANCES INCREASED IN | | ENZYME DEFECT | CLINICAL MANIFESTATIONS[a] |
		BLOOD	URINE		
Phenylketonuria (PKU), classic form, type I	1:10,000	Phenylalanine	Phenylalanine and metabolites	Phenylalanine hydroxylase, absent	Mental retardation, severe
PKU, variant form, type II	1:14,000	Phenylalanine	Phenylalanine	Phenylalanine hydroxylase, deficient	Mental retardation, mild
Transient neonatal hyperphenylalanemia	1:30,000	Phenylalanine	Phenylalanine not usually found	Phenylalanine hydroxylase, deficient	Normal
Histidinemia	1:20,000	Histidine	Imidazole	Histidinase, absent	Neurological deficits
Tyrosinemia, type I	1:100,000	Tyrosine; methionine	Tyrosine and its metabolites	Fumarylacetoacetate hydroxylase, deficient	Liver disease
Homocystinuria	1:200,000	Homocysteine; methionine	As in blood	Cystathionine β-synthase, absent or deficient	Mental retardation; ocular, skeletal, and vascular damage
Maple syrup urine disease (branched-chain ketoaciduria)	1:250,000	Leucine; isoleucine; valine	As in blood	Branched-chain ketoacid decarboxylase, deficient	Central nervous system; mental retardation; acidosis
Carbamyl phosphate synthetase deficiency	Rare	Ammonia increased; citrulline absent		Carbamyl phosphate synthetase, absent or reduced	Hyperammonemia; severe brain damage
Alkaptonuria	1:250,000	Homogentisic acid	As in blood	Homogentisic acid oxidase, absent	Cartilage pigmentation

[a] Consequences without timely medical intervention.

nisms in the kidney tubules are saturated and unable to handle the increased amino acids present in the glomerular filtrate. Renal aminoacidurias are associated with defects in the renal tubular reabsorption mechanisms when plasma amino acids are within normal blood concentrations. Secondary aminoacidurias may be associated with liver disease, malabsorption, or malnutrition.

One amino acid that has received particular attention recently is **homocysteine.** Elevations in plasma concentrations of homocysteine are most often caused by either (1) genetic defects in enzymes involved in homocysteine metabolism or (2) nutritional deficiencies of folate, vitamin B_{12}, and vitamin B_6, the cofactors required for homocysteine metabolism. Two- to tenfold higher concentrations are seen in intermediate and severe hyperhomocysteinemia, respectively. The latter form is rare; the classical form of homocystinuria results from a genetic deficiency of cystathionine β-synthase. Recently, the association of homocysteinemia and cardiovascular disease has received widespread attention. Increased homocysteine is often included with other well-known cardiovascular risk factors such as smoking, increased cholesterol, and hypertension.

Primary Aminoacidurias
Overflow—arise from increased plasma amino acid concentrations Renal—associated with defects in the renal tubular reabsorption mechanisms

B. Plasma Proteins

About 40 to 60% of the total protein in plasma is **albumin;** the remaining plasma proteins are collectively termed the **globulins.** In contrast to the albumin fraction, which is highly homogeneous, the globulins consist of many structurally and functionally diverse proteins. As shown in Figure 3–1, serum protein electrophoresis (SPE)

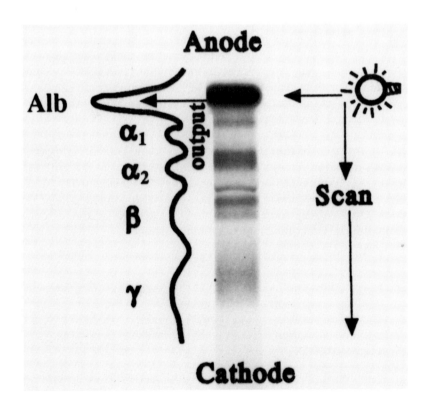

Figure 3–1. Normal serum protein electrophoresis pattern superimposed on a densitometric tracing of the pattern. Albumin migrates toward the anode (+). (Reprinted with permission from AACC.)

▶ separates proteins into five fractions. A sixth fraction, **prealbumin,** which is faster than albumin, is faintly visible in some patients. After the albumin region, the globulins are designated: alpha$_1$ (α_1), alpha$_2$ (α_2), beta (β), and gamma (γ); each globulin fraction is composed of multiple proteins that migrate similarly on agarose SPE (see Table 3–2).

Total protein concentration reflects both albumin and globulin. However, total protein may remain within normal limits (6.0 to 8.0 g/dL) when there are inverse changes in protein fractions, such as a decrease in albumin and an increase in one or more of the globulins.

▶ This inverse change may occur as a result of the **acute phase response** (APR), in which levels of several proteins increase or decrease in response to physiological stresses such as tissue necrosis, inflammatory conditions, or trauma. Increased components are termed *positive APR proteins;* some examples include C-reactive protein, fibrinogen, and α_1-antitrypsin. Conversely, *negative APR proteins* decrease after physiological stress; examples include albumin and transferrin. In general, the APR proteins respond within hours to a few days after injury.

The properties, clinical significance, and reference ranges of selected plasma proteins, grouped by their electrophoretic protein fraction, are summarized in Table 3–2 and are briefly reviewed in the following section.

▶ **1. Prealbumin. Transthyretin,** or prealbumin, is a glycoprotein that is typically a faint band on SPE. Prealbumin is, however, a prominent band on CSF electrophoresis since it is synthesized in the choroid plexus as well as in the liver. Transthyretin transports thyroxine, triiodothyronine, and, in a complex with retinol-binding protein, vitamin A. The half-life of transthyretin in the circulation is relatively short at ~2 days. Transthyretin measurement is used clinically to assess nutritional status. A rise in transthyretin indicates a positive nitrogen balance in patients treated with total parenteral nutrition. A decrease reflects a negative balance. Transthyretin is increased in salicylate therapy; it is a negative APR protein.

▶ **2. Albumin. Albumin** is the most abundant protein in normal plasma; the adult reference interval in plasma is 3.5 to 5.2 g/dL. Albumin has numerous functions including responsibility for about 80% of the intravascular oncotic pressure, also termed *colloid osmotic pressure,* the force that confines fluids within the vascular compartment. Other functions of albumin include:

1. Serving as a source or reserve of amino acids
2. Serving as a plasma buffer
3. Providing antioxidant activity (by binding free fatty acids, copper ions, and bilirubin)
4. Altering capillary permeability by binding specific membrane-associated glycoproteins to increase permeability toward small proteins

Albumin is characterized by a low pI that confers a strong overall negative charge at physiologic pH. At the pH of 8.6 generally used for SPE, albumin migrates strongly toward the positively charged anode as shown in Figure 3–1. Albumin's highly negatively charged surface at physiological pH 7.4 is a characteristic feature that

---

Albumin Functions

- Responsible for 80% of intravascular oncotic pressure
- Serves as a source or reserve of amino acids
- Serves as a plasma buffer
- Provides antioxidant activity
- Alters capillary permeability

Table 3–2. Properties and Features of Selected Plasma Proteins

PROTEIN	ELECTRO-PHORETIC MOBILITY[a]	MOLECULAR WEIGHT[b]	REFERENCE, ADULT,[c] G/L	PRIMARY FUNCTION	CLINICAL ASSOCIATION[d]	CLINICAL LABORATORY METHODS[e]
Prealbumin	Precedes albumin	55,000	0.1–0.4	Transports retinol and thyroxine	↓, Malnutrition; NAPR(+)	Nephelometry
Albumin	Albumin	66,000	35–52	Plasma oncotic pressure; transport	↓, Malnutrition, liver, and kidney disease; NAPR(+)	Dye-binding; nephelometry
α₁-Antitrypsin	α-1	51,000	0.9–2.0	Protease inhibitor	↓, Pulmonary and liver diseases; PAPR(++)	Nephelometry
α₁-Acid glycoprotein (orosomucoid)	α-1	40,000	0.5–1.2	Inhibits platelet aggregation; binds progesterone	↓, Nephrotic syndrome; PAPR(++)	Nephelometry
α₁-Fetoprotein	α-1	69,000	< 2 µg/L	Fetal albumin	Prenatal screening; tumor marker	Enzyme immunoassay
α₂-Macroglobulin	α-2	750,000	1.3–3.0	Protease inhibitor	↑, Nephrotic syndrome; ↓, acute pancreatitis	Nephelometry
Haptoglobin	α-2	100,000	0.3–2.0	Binds free hemoglobin	↓, Hemolysis; PAPR(++)	Nephelometry
Ceruloplasmin	α-2	160,000	0.2–0.6	Ferroxidase; binds copper	↓, Wilson's disease; PAPR(+)	Nephelometry
Hemopexin	β-1	59,000	0.5–1.15	Binds heme	↓, Hemolytic anemia	Nephelometry; radial immunodiffusion
Transferrin	β-1	80,000	2.0–3.6	Iron transport (Ferric)	Iron metabolism; NAPR(+)	Nephelometry
β₂-Microglobulin	β-2	11,800	1–2.4 mg/L	Leukocyte antigen	Renal tubular function, cancer monitoring	Enzyme immunoassay
Complement C3	β-2	185,000	0.9–1.8	Complement factor	↓, Complement activation; PAPR(+)	Nephelometry
Fibrinogen	Between β-2 and γ	340,000	2.0–4.0	Precursor of fibrin	↓, Fibrinolysis; PAPR(++)	Optical detection of fibrin
C-Reactive protein	Between β and γ, or as slow γ	~120,000	< 10 mg/L	Interactions with humoral and cellular effectors of inflammation	PAPR(+++); cardiovascular risk (high-sensitivity assay)	Nephelometry
Immunoglobulin G	Broad γ	150,000	7–16	Humoral immunity	↑, Liver disease; infections; multiple myeloma	Nephelometry
Immunoglobulin A, monomer	Broad γ	160,000	0.7–4.0	Humoral immunity	Same as IgG	Nephelometry
Immunoglobulin M, pentamer	Broad γ	970,000	0.4–2.3	Humoral immunity	Same as IgG; ↑, Waldenström's macroglobulinemia	Nephelometry

[a] Electrophoresis at alkaline pH on agarose gel.

[b] Approximate, given as daltons.

[c] Units except as indicated. Intervals may vary with method and standardization, age, and gender. Reference: Tietz Textbook of Clinical Chemistry, 3rd ed. Philadelphia: WB Saunders, 1998.

[d] ↓, decreased in; ↑, increased in. NAPR, negative acute phase reactant; PAPR, positive acute phase reactant. APR graded according to the percent change in concentration relative to normal plasma concentration: (+), 50–100%; (++), 200–400%; (+++) greater than 400%.

[e] Primary source: 1998 College of American Pathologists Proficiency Testing Program, Participants Summary Reports. Nephelometry refers to immunochemical methods using commercially available antisera.

promotes solubility as well as the binding with many different endogenous molecules such as long-chain fatty acids, unconjugated bilirubin, hormones, amino acid metabolites, and plasma cations including calcium, magnesium, and heavy metals such as copper. Plasma calcium binds to albumin in an equilibrium that functions to maintain the physiologic levels of free (ionized) calcium. Albumin also binds exogenous substances such as drugs, drug metabolites, and dyes.

Synthesis of albumin occurs in hepatic parenchymal cells, except during early fetal development when it is synthesized in the yolk sac. The rate of synthesis is controlled by changes in colloidal osmotic pressure, the APR, and dietary factors. Albumin is catabolized by various tissues and has a half-life of 15 to 19 days. Because of its molecular weight at about 66,000 daltons and the fact that albumin and proteins comprising the kidney's glomerulus both have strong repelling negative charges at physiologic pH 7.4, less than 0.1% of plasma albumin passes into the glomerular filtrate. Normally, most of the albumin filtered is reabsorbed and catabolized by the proximal renal tubular cells, resulting in very little urine albumin excretion.

Elevated albumin is seldom seen, except in dehydration. Decreased serum albumin (hypoalbuminemia) may occur in both renal and liver diseases and indicates significant pathology. Albumin is a negative APR protein, so levels are decreased in acute inflammation and in chronic infections. This is probably related to the production of interleukin-6, which decreases the production of albumin. Because the liver's reserve capacity is substantial, even a large reduction of liver function does not adversely affect serum albumin levels; albumin excretion in renal disease is the most common cause of hypoalbuminemia. Other causes of hypoalbuminemia include reduced synthesis due to liver disease, malnutrition, impaired absorption or digestion, loss during bleeding, albumin excretion with burns, and gastrointestinal disorders. Artifactually low concentrations are frequently seen in pregnancy due to increased plasma volume.

Urinary albumin up to 30 mg/24 hours is normal. Urine albumin is an indicator of glomerular permeability and in excess of 30 mg/24 hours suggests either excessive glomerular filtration or impaired reabsorption of albumin by the proximal tubule. *Microalbuminuria* refers to mildly increased excretion, 30 to 300 mg/24 hours, of albumin. Assessment of microalbuminuria in patients with diabetes mellitus is critically important for monitoring progression of their disease. Mechanistically, it is thought that glycation of kidney proteins results from poor glucose control, impairing the reabsorption of albumin. Thus, normal albumin values in urine indicate that the diabetic patient's disease is stable, whereas microalbuminuria indicates disease progression and is predictive of worsening renal disease and an indicator of the progression of other vascular complications. Nephrotic syndrome, defined as edema with hypoalbuminemia and proteinuria, is a glomerular disease that is associated with protein (mostly albumin) loss exceeding ~3 g/24 hours. Altered glomerular function allows passage of proteins that are normally restricted by the healthy glomerulus.

Causes of Hypoalbuminemia

- Albumin excretion in renal disease (most common cause)
- Reduced synthesis due to liver disease
- Malnutrition
- Impaired absorption or digestion
- Loss during bleeding
- Excretion with burns
- Gastrointestinal disorders

▶ **3. α_1-Globulins.** α_1-**Antitrypsin** (AAT) is a glycoprotein that comprises approximately 90% of the alpha$_1$ region of the SPE. The main function of AAT is to inhibit lysosomal elastase released from polymorphonuclear leukocytes during their response to particles and inhaled bacteria. Also, AAT inhibits chymotrypsin and other enzymes with trypsin-like activity.

Due to its genetic polymorphism, at least 25 alleles and 75 genetic variants have been identified by isoelectric focusing electrophoresis (IEF). The majority of the U.S. population has the genotype MM (allele Pi^M), indicating homozygosity for the M protein and fully functional protease inhibition. While IEF reveals most common AAT phenotypes, specific DNA probes used in polymerase chain reaction (PCR) techniques are now available to detect the clinically important Z protein. Individuals who are homozygous for Pi ZZ develop obstructive pulmonary disease and/or emphysema. In Pi ZZ–affected individuals, AAT protein accumulates in hepatic cells without release; these patients have deficient AAT protease inhibition in their blood. Heterozygous (Pi MZ) individuals who are cigarette smokers are at increased risk for pulmonary emphysema or liver disease. AAT is a positive APR protein.

α_1-**Fetoprotein** (AFP) is a glycoprotein normally present in ◄ very low concentrations (< 2 μg/L) in the serum of healthy men and nonpregnant women. AFP is the major protein in fetal serum and is found in high amounts through the first year of life. During pregnancy, AFP is found in maternal serum; any abnormal fetal condition that results in increased passage of fetal plasma into the amniotic fluid will increase AFP in maternal circulation. The normal concentration of AFP in maternal serum at 20 weeks' gestation is variable within the interval of 20 to 100 μg/L and reaches the highest concentrations at approximately 30 weeks of gestation. The measurement of AFP in maternal serum is included in prenatal screening tests, and increased levels are found in conditions such as neural tube defects, atresia of the gastrointestinal tract, multiple gestations, and fetal blood loss; these are situations with increased risk of premature delivery or fetal death. The risk of Down syndrome is associated with decreased AFP. AFP is increased in hepatitis and is also a tumor marker associated with primary hepatomas, yolk sac tumors of the ovary, and testicular carcinomas.

α_1-**Acid glycoprotein** (AGP), or orosomucoid, is a glycoprotein ◄ synthesized both by the liver and by granulocytes and monocytes. AGP's biological function remains unclear, but it has been shown to inhibit the phagocytic activity of neutrophils and also to inhibit platelet aggregation. AGP binds drugs such as propranolol, quinidine, and chlorpromazine, as well as some lipophilic hormones, particularly progesterone.

AGP is a positive APR protein, with levels increasing three- to fourfold above the normal reference interval after inflammation or necrosis. Increased concentrations are found in ulcerative colitis and in association with elevated glucocorticoids (endogenous or exogenous). Low serum concentrations are found with protein-losing enteropathies and in nephrotic syndrome.

4. α_2-Globulins. α_2-**Macroglobulin** (A_2M) has a molecular weight ◄ of about 750 kilodaltons and is one of the largest plasma proteins. A_2M consists of four identical polypeptide chains arranged in two pairs joined by disulfide bonds. A_2M is synthesized by hepatocytes, monocytes, and macrophages. An important function of A_2M is protease inhibition by blocking the active sites of enzymes such as plasmin, thrombin, and kallikrein. Also, a significant increase in A_2M is seen in the nephrotic syndrome, which is explained, in part, both by selective retention by the damaged glomeruli and increased synthesis. These increases are thought to represent a compensatory mechanism to maintain oncotic pressure when there is increased loss of albumin. In carcinoma of the prostate, A_2M forms a complex with prostate-specific antigen that is cleared rapidly.

▶ **Haptoglobin** (Hp) is a glycoprotein synthesized in the liver. The primary function of Hp is related to its ability to bind free hemoglobin (Hb) released during intravascular hemolysis. The half-life of Hp in the circulation is ~5 days. Formation of the Hp–Hb complex promotes clearance within a few hours and helps prevent renal glomerular loss of Hb and consequent kidney injury. The Hp–Hb complex is cleared by the reticuloendothelial system, and its heme iron and protein recycled by the liver. Both Hp and the Hp–Hb complex are thought to be important in the control of local inflammatory processes by inhibiting peroxides and cathepsin B released during inflammation and phagocytosis.

Hp is a positive APR protein; thus, elevated concentrations occur with infection and other inflammatory conditions, neoplasia, trauma, and tissue injury. Decreased Hp concentrations are found in hemolytic disorders.

▶ **Ceruloplasmin** (Cer) is a copper oxidase enzyme synthesized in the hepatocytes, consisting of a single polypeptide chain that carries six to eight copper atoms. Bound copper ions impart a blue color to the protein, hence its name from the Latin *caeruleus* meaning blue.

The primary functions of Cer are related to its ferroxidase activity in the regulation of iron transport and utilization. In the presence of Cer, iron in the ferrous state (Fe^{+2}) is oxidized to the ferric form (Fe^{+3}), which can then be bound by transferrin. Cer is also important as an antioxidant to prevent lipid peroxidation and cellular damage. Cer together with albumin is involved in the transport of copper to tissues. Approximately 95% of total serum copper is bound to Cer.

Cer is a positive APR protein, and increased concentrations also occur in late pregnancy and with estrogen therapy. Cer is most often measured to screen for Wilson's disease, also termed *hepatolenticular degeneration,* a disorder in which serum copper is decreased. In this condition, copper is not incorporated into Cer, but is instead deposited in the skin, liver, brain, and periphery of the iris (Kayser–Fleischer rings). Although Cer plasma concentrations are greatly decreased in Wilson's disease, tissue biopsy and staining for copper is necessary for diagnosis.

▶ **5. β-Globulins. Transferrin** (Trn) is a glycoprotein synthesized in the liver whose major function is the transport of oxidized iron (Fe^{+3}). Trn has a half-life of approximately 8 to 10 days and contains two iron-binding sites per molecule; the iron–protein complex is stable, and release of iron is regulated by pH. At pH 4.5, the iron–Trn complex dissociates into apotransferrin and free iron. Normally, free iron concentration in plasma is low and there is excess Trn capacity for binding iron; the protein is only about 35% iron saturated. On SPE, Trn accounts for a major proportion of the β-globulin region. Another common laboratory test, *total iron-binding capacity* (TIBC), is an indirect measure of Trn; Trn values may be estimated by a simple equation: Trn (mg/dL) = TIBC (μg/dL) × 0.70.

The primary clinical use for Trn involves evaluation of iron status. Increased serum Trn is seen in iron deficiency anemia, supposedly as a compensatory mechanism to enhance iron absorption from the gastrointestinal tract. Increased Trn saturation with iron (> 70%) is indicative of iron overload, a serious clinical condition that is frequently associated with the hereditary disease hemochromatosis, in which excess iron is absorbed from the gastrointestinal tract and deposited in the liver and other tissues, leading to cell injury and sclerosis. Trn is also increased in pregnancy, with estrogen therapy, and with anabolic steroid use.

Trn is a negative APR protein and is decreased in inflammatory conditions, malignancies, and after acute thermal burns. Trn is a relatively small protein, ~80,000 daltons, that is excreted into urine in renal glomerular disease. Decreased serum Trn due to either increased renal loss or decreased synthesis is seen in acquired and chronic conditions such as liver disease, malnutrition, infections, renal disease, and protein-losing gastroenteropathies.

β_2-**Microglobulin** (B_2M) is a small protein expressed as the light or β chain of the human leukocyte antigen on the surface of most nucleated cells. Because of its low molecular weight of 11,800 daltons, B_2M is easily filtered through the glomerulus. However, essentially 100% is reabsorbed by functional proximal tubular cells where the protein is metabolized. As a result of its metabolism, serum B_2M determinations are clinically useful for assessing renal tubular function in kidney transplant recipients. A separate clinical use is derived from the strong association between increased serum B_2M levels and the presence or recurrence of certain malignancies. B_2M measurements are frequently used for monitoring multiple myeloma and B-cell tumors.

Fibrinogen is a glycoprotein synthesized in the liver, having a half-life of approximately 3 days. Thrombin activates fibrinogen to form fibrin by cleaving the fibrinopeptides A and B from the fibrinogen polypeptide chains. These fibrin monomers are released and polymerize to form the fibrin clot.

Fibrinogen is a positive APR protein, and is increased in inflammation and tissue injury; increased concentrations also occur in pregnancy and with estrogen therapy. Like other large proteins such as A_2M, fibrinogen is increased in renal disease. Decreased concentrations indicate consumption of fibrinogen typically associated with intravascular coagulation.

Because of loss during serum formation (i.e., the clotting process), fibrinogen does not appear on SPE. In normal plasma, however, fibrinogen will be visualized as a distinct band between the β- and γ-globulins. Thus, it is important to be aware of the sample type used for electrophoresis to distinguish fibrinogen from an abnormal immunoglobulin (paraprotein) with mobility in the beta–gamma region.

Complement refers to the group of approximately 25 proteins that interact in a sequential manner with each other and with cell membranes in host defense mechanisms. Activation of each component results in a cleavage product that participates in the subsequent reaction to continue the cascade that balances activation and inhibition products. The function of the system is threefold:

1. To cause lysis of most foreign agents, such as bacteria, viruses, and yeast
2. To opsonize cellular debris prior to phagocytosis

Causes of Increased Serum Transferrin

- Iron deficiency anemia
- Pregnancy
- Estrogen therapy
- Anabolic steroid use

3. To mediate the production of effector peptides that regulate the inflammatory response

C3 is the complement protein in highest serum concentration. Activation of C3 results in the release of C3b, which leads to increased vascular permeability, histamine release from mast cells, chemotaxis of phagocytes, and opsonization that enhances phagocytic ingestion of foreign cellular debris.

Complement-associated disorders are either inborn defects or acquired abnormalities. The most common congenital error is hereditary angioedema, a disease manifested by subcutaneous episodes of bronchial or gastrointestinal edema that results from a deficiency of C1 esterase inhibitor. During attacks of hereditary angioedema C4 is commonly decreased. The most common acquired alterations in serum complement components occur in patients with infections and acute and chronic inflammatory diseases. Increased serum complement occurs in inflammatory conditions such as acute rheumatic fever, rheumatoid vasculitis, bacteremia, dermatomyositis, and ulcerative colitis. C3 and other members of the complement systems are positive APR proteins. Despite being a positive APR protein, C3 levels may be decreased due to consumption in autoimmune diseases such as systemic lupus erythematosus as well as in certain inflammatory processes.

▶   **C-reactive protein** (CRP) was so named because of its ability to bind the C-polysaccharide on the cell wall of *Streptococcus pneumoniae*. Normally, CRP levels are too low to allow visualization on SPE. With physiologic stress, CRP rises rapidly within a few hours to concentrations 1,000-fold (or more) greater than normal serum levels (< 10 mg/L) within 2 days, allowing visualization on SPE. The migration of CRP varies from β- to γ-globulin regions. The main functions of CRP are (1) forming complexes with microorganisms to facilitate activation of the classical complement pathway; (2) initiating opsonization, phagocytosis, and lysis of foreign cells; and (3) binding and clearing toxic autogenous substances from necrotic tissue.

The clinical significance of CRP is mainly due to its characteristic brisk, early response as a positive APR after infection, inflammation, surgery, tissue injury, or necrosis. However, CRP increase is nonspecific and must be used with supporting clinical information. CRP is commonly used to screen for infection and for monitoring antibiotic therapy in newborns. CRP is also used to monitor inflammatory conditions such as rheumatoid arthritis, and to assess intrauterine infections and rejection of transplanted organs. Large epidemiologic studies have shown that increased CRP concentrations are predictive of future coronary events in both men and women. Because CRP increases predictive of coronary events tend to be subtle and within or near the normal reference interval, high-sensitivity CRP methods that have high precision in the normal range must be used for this purpose. There has also been much recent interest in the use of CRP measurements for predicting outcome in acute heart disease, including unstable angina and acute myocardial infarction.

▶   **6. γ-Globulins. Immunoglobulins** (Igs) are a group of proteins secreted at different stages of development of plasma cells (see Table 3–3). The diversity of these proteins is reflected in the wide range of biologic functions and antigen-binding capabilities, which characterize the humoral component of the immune system. The broad, diffusely stained γ-region on the SPE pattern in Figure 3–1 is

Table 3–3. Stages and Characteristics of Development and Proliferation of Plasma Cells

STAGE OF PROLIFERATION	SITE	SURFACE RECEPTOR IMMUNOGLOBULIN	IMMUOGLOBULIN SECRETED	ASSOCIATED NEOPLASMS
Stem cell	Bone marrow	None	None	Acute lymphoblastic leukemia
Early B lymphocyte Surface immunoglobulins develop; encounters antigen to begin the primary immune response and development	Lymph nodes	IgM, IgD	None	Lymphoma, chronic lymphocytic leukemia
Late B lymphocyte Antigen encountered to begin secondary immune response; M heavy chain changes to other types during development to plasma cell	Lymph nodes	IgM	IgM	Lymphoma, chronic lymphocytic leukemia, Waldenström's macroglobulinemia
Plasma cell Secretion of highly specific antibody into blood	Lymphatic tissue	IgG, IgA	IgG, IgA	Multiple myeloma

due to the Igs. This pattern reflects the heterogeneous (polyclonal) Ig population with varied structural properties.

The basic structural unit of an Ig consists of two identical heavy chains (50,000 to 70,000 daltons each) and two smaller (23,000 daltons each) identical light chains. Heavy and light chains are joined by disulfide bridges and noncovalent forces. The Ig molecule can be visualized as four components that fit together in a "Y" shape (see Figure 3–2). The amino (NH$_2$)-terminal portion of the light

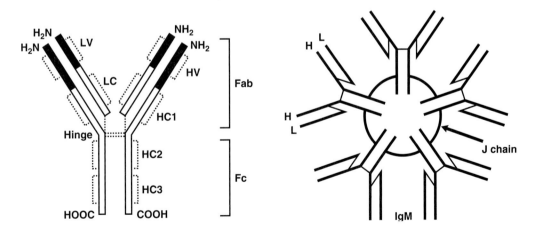

Figure 3–2. Models of immunoglobulin molecules represent a monomer such as IgG (left) and the pentamer IgM (right). The short inside arms of the basic four-chain structure represent the light (L) chains with variable (V) and constant (C) domains. The long outside pieces represent the heavy (H) chains with variable (V) and three constant (C1, C2, and C3) domains. Amino (NH$_2$) groups terminate the antigen binding (Fab) sites; carboxyl (COOH) groups terminate the Fc unit. The segments of the pentamer are joined by a polypeptide J chain. (Adapted from Parslow TG. Immunoglobulins and immunoglobulin genes. In Stites DP, Terr AI, Parslow TG (eds.). *Medical Immunology*, 9th ed. Stamford, CT: Appleton & Lange, 1997.)

chain and the heavy chain form the antigen-binding (Fab) region of each arm of the Y, and the carboxy (C)-terminal ends of the two heavy chains comprise the leg (Fc region). A short region of the heavy chains called the hinge region forms the diverging section of the Y.

The five classes of heavy chains are represented by the Greek letters γ, α, μ, δ, and ϵ, corresponding to G, A, M, D, and E, respectively. The variable regions of the heavy and light chains of the Igs are so named due to the highly variable sequences of the amino acids at the NH_2-terminal regions within the same class. This hypervariability is responsible for recognition and binding of a myriad of antigenic substances encountered by the host. The constant portion of the heavy chain is very similar for all molecules in a given Ig class. This region is recognized by Fc receptors on many cells and is responsible for secondary biologic or effector properties. Igs have two major functions related to the critical role of host defense: (1) binding to antigen, a function that depends on the structural variability of the antibody combining site formed by the NH_2-terminal portions of the heavy and light chains, and (2) various biological and effector activities such as complement fixation, placental transfer, and others. Properties of the five classes of Igs are summarized in Table 3–4.

The five major classes—IgG, IgA, IgM, IgD, and IgE—are included in one group of proteins because they have the following common properties:

Table 3–4. Properties of Human Immunoglobulins (Igs)

	IgG	IgA	IgM	IgD	IgE
Heavy-chain class	γ	α	μ	δ	ϵ
Molecular forms	Monomer	Monomer; dimer; trimer; secretory	Pentamer	Monomer	Monomer
Subclasses	IgG_1, IgG_2, IgG_3, IgG_4	IgA_1, IgA_2	None	None	None
Molecular weight[a], daltons	150,000	160,000[b]	970,000	180,000	190,000
Carbohydrate content[a] (%)	3	7	10	9	13
Electrophoretic zone	γ	β to γ	β to γ	γ	γ
Concentration[a] in adult serum, mg/dL	700–1,600	70–400	40–230	< 10	~ 0.05
Percentage[a] of total Igs	75	15	10	< 1	< 0.01
Half-life in serum, days	21 (IgG_1, IgG_2, IgG_4); 7 (IgG_3)	9 (IgA_1); 5 (IgA_2)	5	3	2
Antibody binding sites	2	2 or 4	10	2	2
Placental transfer	Yes	No	No	No	No
Binding to mast cells and basophils	±	0	0	0	4+
Antibacterial activity	1+	1+	3+	?	?
Antiviral activity	0	3+	1+	?	?
Antibody activity	Secondary response	Primary response in secretions	Primary response	B-cell surface marker	In allergic and parasitic responses
Complement fixation	1+ (IgG_1, IgG_2, IgG_3); 0 (IgG_4)	0	4+	0	0

[a] Approximate.

[b] Monomer.

1. Synthesis in reticuloendothelial tissues, including plasma cells and other lymphoid tissue cells
2. Antibody activity
3. Similar structure
4. Polyclonality within each class

Each class has been characterized according to the structural similarities of their heavy chains and their biologic and physical properties. In the circulation IgG, most of IgA, IgD, and IgE occur as monomers, while IgM occurs as a pentamer as shown in Figure 3–2. Approximately 15% of IgA occurs as dimers or trimers.

Light chains are one of two types, kappa or lambda. Identical kappa or lambda chains are found on an Ig molecule; the light chains are never mixed on the same Ig. A slight excess of light chains is produced in healthy individuals. The serum concentration ratio of kappa to lambda production is 2:1 in the healthy individual. The functional differences between types of light chains are unclear. Patients with B-cell lymphoproliferative disorders or renal tubular damage may have excess light chains in serum or urine specimens. The presence of free (unbound) kappa or free lambda light chains in urine on electrophoresis is termed *Bence Jones proteinuria* and generally indicates a poor prognosis for patients with plasma cell malignancy.

The rate at which concentrations of the Igs mature from early childhood to adult levels are dependent on genetic and environmental factors such as antigenic exposure. Infants have a high concentration of maternal IgG, which declines rapidly in the first 3 months of life. Soon after birth, synthesis of IgM begins and is followed by production of IgG. IgM reaches adult levels at about 9 months of age, IgG at 3 years, and IgA, IgD, and IgE about 10 years later.

Immunoglobulin G (IgG) is the most abundant serum immunoglobulin, comprising approximately 75% of total. IgG is responsible for long-term physiologic protection and the majority of neutralization of bacteria and virus antigens. Total IgG is composed of four subtypes: IgG_1, IgG_2, IgG_3, and IgG_4. IgG_1 and IgG_3 activate killer monocytes by binding their Fc receptors on phagocytic cells. Response of many IgG clones to various antigens gives rise to the polyclonal pattern and homogeneous staining normally observed in the gamma electrophoretic region.

Immunoglobulin A (IgA) comprises about 10 to 15% of total plasma immunoglobulins. IgA exists as two subtypes: a dimeric subtype, IgA_1, and a secretory form, IgA_2. IgA is found in both polymeric (15%) and monomeric (85%) forms in plasma.

IgA's major function may be in secretions since IgA is found in tears, saliva, sweat, milk, and gastrointestinal and bronchial secretions. Secretory IgA has a molecular weight of 380 kilodaltons, and consists of two IgA molecules plus a secretory component of 70 kilodaltons, and a J chain of 16 kilodaltons. The secretory component is believed to enhance solubility in secretions and nonplasma fluids.

Immunoglobulin M (IgM) comprises about 10% of total Igs in plasma. IgM is the largest immunoglobulin, existing in plasma mainly as a pentamer having a molecular weight of > 900 kilodaltons. Along with the five IgM monomeric units, the pentamer contains a J chain; disulfide bonds connect the J chain and monomers. IgM is the most primitive immunoglobulin, and is found on the surface of early B lymphocytes; IgM is excreted in blood by late B lymphocytes (see

The Five Major Classes of Immunoglobulins

IgG—the most abundant serum immunoglobulin, comprising approximately 75% of total

IgA—comprises 10 to 15% of total plasma immunoglobulins; has a secretory form

IgM—comprises 10% of total plasma immunoglobulins

IgD—a monomeric plasma immunoglobulin, not routinely measured

IgE—the lowest-concentration immunoglobulin, monomeric in plasma

Table 3–3). IgM is the first or primary response immuoglobulin to be produced following antigen assault.

Immunoglobulin D (IgD) is a monomeric immunoglobulin in plasma. IgD is active on the cell surface of B lymphocytes (Table 3–3), but its function is largely unknown. IgD is not measured routinely in clinical laboratories. It is labile upon storage, so sample handling and measurement must be completed expeditiously.

Immunoglobulin E (IgE) is the lowest-concentration Ig and is monomeric in plasma. IgE attaches rapidly to the membranes of mast cells by its Fc region. Cross-linking of two IgE molecules by antigen causes release of histamine, heparin, and vasoactive amines by mast cells. IgE's activity plays a key role in the allergic reactions associated with urticaria, hay fever, and asthma. IgE's plasma concentration is about 0.05 mg/dL so sensitive immunoassays are necessary for quantification.

Hypogammaglobulinemia refers to decreased serum concentrations of one or more classes of normal Igs and is caused by primary (congenital) or secondary immune deficiencies. The latter may result from chronic disorders, including malignancies, or by immunosuppressive agents that interfere with the expression of humoral and/or cellular immunity.

Hypergammaglobulinemia refers to polyclonal immunoglobulin increases that are *secondary* to inflammatory conditions in response to an antigenic stimulus. Chronic inflammatory conditions including infections, hepatitis, and rheumatoid arthritis are usually associated with polyclonal increases in one or more Ig classes. Production of IgM by the fetus increases with uterine infections, which can be detected by measuring IgM in cord blood at birth. In liver cirrhosis, there is a generalized increase of IgA and other immunoglobulins, presumably because the increased liver scarring decreases clearance of antigens from the circulation by the reticuloendothelial system. In addition, decreased clearance of immune complexes frequently results in prominent beta and gamma regions that merge on the SPE pattern. This merge is commonly described as beta–gamma bridging.

▶ **Monoclonal gammopathy** is a disease process in which an Ig from a single B cell clone is produced in excess as a result of neoplastic transformation. Monoclonal gammopahy can be conceptualized as a disease continuum, ranging from a relatively benign condition, which may be stable over many years, to the malignancy of

▶ **multiple myeloma.** Plasma cells produce Igs from a specific clone; Table 3–3 shows the proliferation of various cell types and the conditions associated with each. When a neoplastic clone multiplies beyond normal control (i.e., when a monoclonal gammopathy occurs),

▶ the **paraprotein** (also known as the *M protein* or *M spike*) can be identified on SPE as a dark area or spike having restricted electrophoretic mobility as shown in Figure 3–3. Detection, characterization, and temporal monitoring of the paraprotein are important for clinical care of patients having monoclonal gammopathy. Monoclonal gammopathy will be uncovered in approximately 5% of patients older than 50 years, and 8% of patients over 70 years.

When suspected clinically or suggested by laboratory findings such as a high globulin fraction on total protein and albumin measurement, clinicians should strongly consider requesting SPE with interpretation. Suspicious or abnormal findings should be followed up with serum *immunofixation electrophoresis* (IFE; see description in

Monoclonal Protein Pattern

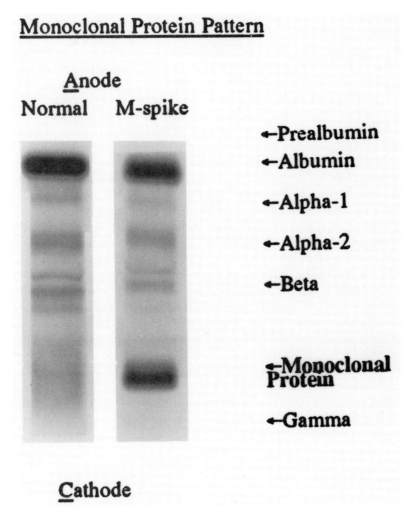

Figure 3–3. Normal serum protein electrophoresis pattern (left) and pattern showing a monoclonal gammopathy (right). (Reprinted with permission from AACC.)

later section) and timed urine collection for total protein and urine protein electrophoresis, with IFE as appropriate. Quantitative measurement and monitoring of the paraprotein and other immunoglobulin classes are also important for assessing disease progression. The major conditions associated with paraproteins are discussed below.

Monoclonal gammopathy of undetermined significance (MGUS) (also termed *benign monoclonal gammopathy* [BMG]) is defined as follows:

1. The M protein concentration does not exceed 2 g/dL if an IgG, or 1 g/dL if an IgA or IgM
2. Polyclonal immunoglobulin concentrations are within the normal reference interval and stable
3. No Bence Jones proteinuria
4. No osteolytic bone lesions on radiographic studies
5. No increase in plasma cells in the bone marrow
6. Stable concentrations of the paraprotein(s) over time

MGUS accounts for many of the M proteins found. Patients demonstrating MGUS require periodic monitoring with serum and urine studies. Treatment of patients with MGUS is not recommended unless laboratory abnormalities progress or symptoms of multiple myeloma develop, because these patients may remain stable for years. Figure 3–4 shows the findings of follow-up over a 20- to 35-year period for a group of MGUS patients.

Smoldering multiple myeloma (smoldering MM) is characterized by a serum M protein > 3 g/dL and > 10% plasma cells in bone marrow. Also, serum Igs produced by non-neoplastic clones are reduced, and often small amounts of light chains are excreted in urine. This designation implies that there is no anemia, no skeletal lesions, and no renal insufficiency. As with MGUS, these patients may remain stable for years, so treatment is not recommended unless laboratory abnormalities or symptoms of multiple myeloma progress.

Multiple myeloma (MM) results from neoplastic proliferation of a single clone of plasma cells and is characterized by paraprotein(s) in serum (80% of cases) and urine (50% of cases), bone pain, anemia, fatigue, lytic bone lesions, hypercalcemia, renal insufficiency, and > 10% plasma cells in the bone marrow. Reliable tests for discriminating MGUS and smoldering MM from multiple myeloma include serial measurements of the M protein in serum and urine as well as periodic evaluation of clinical and other laboratory features. The incidence of MM is about 4 per 100,000 per year. MM comprises about 1% of all malignant disease. Mean age at diagnosis is 62 years; MM is slightly more common in men than women. Seventy percent of MM cases survive longer than 30 months with treatment. This prognosis may improve in the future because encouraging results have been found with bone marrow transplantation, which is fast becoming an important treatment option.

Characteristics of Multiple Myeloma

- Paraprotein(s) in serum and urine
- Bone pain
- Anemia
- Fatigue
- Lytic bone lesions
- Hypercalcemia
- Renal insufficiency
- > 10% plasma cells in bone marrow
- Rouleaux formation on peripheral blood smear

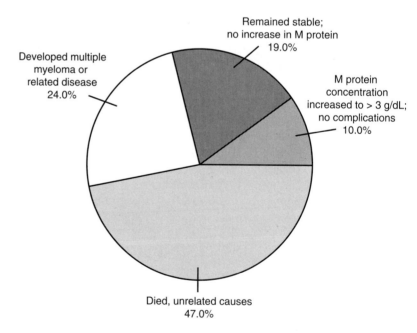

Figure 3–4. Follow-up findings of 20 to 35 years (median 22 years) in 241 patients initially presenting with monoclonal gammopathy of unknown significance (MGUS). (Data from Kyle RA. "Benign" monoclonal gammopathy—after 20 to 35 years of follow-up. *Mayo Clin Proc* 68:26–36, 1993.)

Major complications in cases of MM result from infection, renal failure, hypercalcemia, spinal cord compression, and hyperviscosity. Overall, IgG MM cases appear to have a less severe disease course than those associated with IgA. However, all MM patients who have Bence Jones proteinuria have a worse overall survival prognosis. Also, many patients later develop acute leukemia. Table 3–5 shows the distribution of Ig classes associated with MM.

As indicated in an earlier section, β_2-microglobulin is a prognostic indicator and important for monitoring these diseases.

Waldenström's macroglobulinemia is associated with mono- ◄ clonal paraprotein of IgM type. Waldenström's macroglobulinemia differs from MM in several important ways:

- Less bone pain and fewer lytic bone lesions
- Fewer bacterial infections
- Greater incidence of serum hyperviscosity
- Higher incidence of visual changes and less hepatosplenomegaly and lymphadenopathy
- Some textbooks state that virtually all patients have Bence Jones proteinuria; however, others have reported that about 50% of patients excrete free light chains.

Cryoglobulinemia is characterized as the presence of proteins that precipitate at lower temperatures (some as high as 35°C). **Cryo-** ◄ **globulins** can be divided into three types:

1. Type I features an associated multiple myeloma protein (25% of all cases)
2. Type II mixes a multiple myeloma component and other proteins (25% of cases)
3. Type III is composed of mixed polyclonal proteins

Patients with types II and III disease often have associated immune complex disease.

Technically, samples should be refrigerated at 4°C for 72 hours before assessing the presence of cryoprotein. The cryoprotein should be redissolved in saline warmed to 37°C and then studied by electrophoresis and IFE for confirmation.

C. PROTEINS IN OTHER FLUIDS

The concentrations of proteins in urine, CSF, and other body fluids are, in general, directly proportional to the individual protein con-

Table 3–5. Distribution of the Immunoglobulin Classes Associated with Multiple Myeloma

PROTEIN IN PLASMA	FREQUENCY OF ASSOCIATION WITH MULTIPLE MYELOMA	BENCE JONES PROTEIN (IN URINE)[a]	MEAN AGE OF OCCURRENCE
IgG	50%	60%	65
IgA	25%	70%	65
IgD	2%	100%	57
Free light chains only	20%	100%	56
None detected	< 1%	0%	
Biclonal	1%	Unknown	
IgM	1%	50–100%	

[a] Some authors state that the incidence of Bence Jones proteinuria is about 50%.

centrations in the intravascular compartment and inversely related to the molecular size of the respective proteins.

1. Urinary Proteins. The glomerular membrane in the healthy nephron excludes proteins > 60,000 daltons from passing into the filtrate except in trace amounts. Proteins < 60,000 daltons in the filtrate are actively reabsorbed by the proximal tubular cells. Increased urinary proteins usually indicate renal or extrarenal disease and can be caused by the following four mechanisms:

1. Increased permeability and destruction of the glomeruli
2. Decreased tubular reabsorption
3. Abnormal elevation of proteins in the circulation
4. Abnormal secretion of proteins from the urinary tract

The quantitative measurement of proteins in timed collections (usually 24-hour collections) correlates with disease better than screening tests in random specimens which may be transiently dilute or concentrated.

As described earlier, protein (albumin) excretion from 30 to 300 mg/24 hours is termed microalbuminuria. Protein excretion in excess of 300 mg/24 hours is termed clinical proteinuria. The nephrotic syndrome is characterized by gross changes in glomerular permeability with protein excretion exceeding 3 g/24 hours, hypercholesterolemia, and edema. Clinical proteinuria is associated with other conditions including chronic glomerulonephritis, malignant hypertension, toxemia of pregnancy, heavy metal poisoning, neoplasias, nephrosclerosis, multiple myeloma, chronic pyelonephritis, chronic interstitial nephritis, and polycystic kidney disease.

2. Proteins in Cerebrospinal Fluid. CSF protein is derived primarily by ultrafiltration of the plasma across the blood–brain barrier. Individual constituents, as a percentage of total protein in CSF, are prealbumin, 10%; albumin, 46 to 66%; α-globulins, 2 to 12%; β-globulins, 8 to 18%; and γ-globulins, 3 to 12%.

The following conditions, with examples, are associated with increased total protein in CSF:

- Increased blood–CSF permeability (arachnoidosis, meningitis, hemorrhage, and drug toxicity)
- CSF circulation defects (mechanical obstructions such as tumors)
- Increased IgG synthesis (neurosyphilis; multiple sclerosis)
- Increased IgG synthesis and increased blood–CSF permeability (tuberculous meningitis; brain abscess)

▶ **Multiple sclerosis** (MS), coined the crippler of young adults, is a demyelinating disease that typically affects the brain, optic nerve, and spinal cord. Although advances in radiologic technology, such as magnetic resonance imaging (MRI), have become increasingly important for diagnosis, 90 to 95% of MS patients show **oligoclonal banding** on CSF agarose electrophoresis and IFE. Oligoclonal banding is useful for diagnosis after other causes of banding, such as central nervous system (CNS) syphilis, chronic meningitis, and subacute sclerosing panencephalitis, are excluded. The CSF/serum albumin index (see Calculated Results section later in this chapter) is typically not elevated in MS patients because the disease does not often show increased permeability of the blood–CSF barrier. CSF IgG index (see

Disease Conditions Associated with Increased Urinary Protein Excretion

Nephrotic syndrome
Chronic glomerulonephritis
Malignant hypertension
Toxemia of pregnancy
Heavy metal poisoning
Neoplasias
Nephrosclerosis
Multiple myeloma
Chronic pyelonephritis
Chronic interstitial nephritis
Polycystic kidney disease

Calculated Results section later in this chapter) studies are less sensitive but are considered useful for diagnosis and monitoring.

Because paraproteins in serum may traverse the blood–CSF barrier and cause banding in CSF, performing electrophoresis on both serum and CSF is vital to avoid false positives. IFE with IgG antisera allows more precise definition of oligoclonal banding. In addition to the conditions listed above, Guillian–Barré syndrome, systemic lupus erythematosus, and bacterial meningitis may show CSF abnormalities that overlap with MS.

3. Proteins in Other Biological Fluids. The abnormal accumulation of a fluid in a body cavity is an effusion and is usually due to an imbalance between production and reabsorption. Fluids in peritoneal, pericardial, and pleural cavities are serous fluids and are either *transudates* (total protein < 3 g/dL) due to increased hydrostatic pressure or decreased oncotic pressure, or *exudates* (total protein > 3 g/dL), reflecting increased excretion by cells. The measurement of total protein in serous fluids is usually interpreted in combination with other relevant testing (e.g., cell and differential counts, cytologic studies, glucose, lipids, and enzymes) to determine the nature of the effusion (i.e., transudate versus exudate).

IV. TEST PROCEDURES

Refer to Table 3–6 for a review of the clinical methods used to evaluate proteins.

A. AMINO ACIDS

Amino acid concentrations in plasma and urine can be determined by qualitative and quantitative analytical methods. Qualitative tests are applied for screening, while quantitative tests are used for confirming a provisional diagnosis or monitoring treatment after the diagnosis is established. Screening or semiquantitative procedures are based on various techniques, including the microbiological inhibition (Guthrie) test, thin-layer chromatography, and spectrophotometry. The Guthrie test is widely used in laboratories performing large-scale population screening of neonates. Methods for quantitative analyses of specific amino acids are based on techniques such as gas–liquid chromatography and high-pressure liquid chromatography.

B. PROTEINS

1. Qualitative and Semiquantitative Methods

Refractive Index. Refractometry is a rapid semiquantitative method based on the assumption that the dissolved solids in serum are composed primarily of proteins and nonprotein substances that vary little among various serum samples. This assumption is valid only for nonpigmented samples with normal concentrations of glucose, urea, and other nonprotein solutes. The protein range measured by this method is 3.5 to 11 g/dL.

Immunofixation Electrophoresis. IFE is a technique that takes advantage of both antibody specificity for antigens and the ability of antibod-

Table 3–6. Methods for Clinical Evaluation of Proteins

COMPONENTS	METHOD	PRINCIPLE	SPECIMEN	CLINICAL UTILITY	COMMENTS
Total protein	Electrophoresis	Separation based on differences in charge	Serum; urine; CSF	Screening; diagnostic monitoring	Semiquantitative; different proteins have unequal dye-binding properties
	Biuret	Purple chromogen formed with cupric ions at alkaline pH	Serum	Diagnostic	Semiquantitative; specific for proteins with two or more peptide bonds
	Refractive index (RI)	RI of test sample compared to RI of water	Serum	Screening	Rapid; semiquantitative; reliable only if protein is > 2.5 g/dL and other solutes are not increased
	Dye-binding	Dye forms a chromogen with the protein in proportion to its concentration	Urine; CSF	Diagnostic	Quantitative; Coomassie Brilliant Blue underestimates globulins
	Turbidimetric	Chemical precipitate formed in proportion to protein concentration	Urine; CSF	Diagnostic	Quantitative; TCA, SSA, BZC are widely used
	Kjeldahl	Acid digestion; total nitrogen is measured	Serum	Reference method	Quantitative; not used routinely
Albumin	Electrophoresis	Separation based on differences in electrical charge	Serum; urine; CSF	Screening; post-diagnostic monitoring	As above (total protein); albumin is the densest staining band closest to the anode
	Dye-binding	As above for total protein; anionic dyes such as BCG and BCP	Serum	Diagnostic	Quantitative; BCG is widely used; BCP is more specific
	Immunoassay	Target protein combines with specific antibodies; nephelometric method of analysis	Serum; urine; CSF	Diagnostic	Quantitative; greater sensitivity and specificity than above methods especially with CSF
α_1 globulins: AAT; AGP; AFP	Immunoassay	Nephelometry (AAT; AGP); EIA (AFP)	Serum; urine	Diagnostic	Quantitative
α_2 globulins: AMG; Hp; Cer	Immunoassay	Nephelometry	Serum; urine	Diagnostic	Quantitative
β globulins: Transferrin; C3; C4	Immunoassay	Nephelometry (Transferrin, C3, C4); RIA (C3; C4)	Serum; urine; CSF	Diagnostic	Quantitative
γ-globulins: IgG; IgA; IgM; light chains	Immunoassay	Nephelometry	Serum; urine; CSF	Diagnostic	Quantitative; IgG primarily in CSF
IgE	Immunoassay	Nephelometry; EIA	Serum; urine	Diagnostic	Quantitative
IgD	Immunoassay	RID; RIA	Serum; urine	Diagnostic	Greater sensitivity than nephelometry for very low concentrations
C-Reactive protein	Immunoassay	Nephelometry; RID; RIA	Serum	Diagnostic	Nephelometry preferred for speed and ease
	Electrophoresis	As above for total protein	Serum	Screening	Semiquantitative; detection only at elevated concentrations

AAT, α_1-antitrypsin; AFP, alpha fetoprotein; AGP, α_1-acid glycoprotein; AMG, α_2-macroglobulin; BCG, bromcresol green; BCP, bromcresol purple; BZC, benzylthonium chloride; Cer, ceruloplasmin; CSF, cerebrospinal fluid; Hp, haptoglobin; TCA, trichloroacetic acid; SSA, sulfosalicylic acid; C3, C4 complement factors; AFP, alpha fetoprotein; AMG, α_2-macroglobulin; Cer, ceruloplasmin; C3, C4 complement factors; EIA, enzyme immunoassay; RIA, radioimmunoassay; RID, radial immunodiffusion.

ies to fix (immunoprecipitate) proteins in a gel matrix. IFE is performed after the specimen is subjected to electrophoresis as described for SPE. After electrophoresis, a template cutout is placed on the gel such that it isolates the area of sample migration. Each isolated electrophoretic lane is then overlaid with specific antisera for the protein–antigen of interest. The antiserum reacts with the specific protein, fixing it in the gel matrix. After washing away unreacted reagent, the gel is stained and the protein of interest identified visually. The ability to visualize a specific protein is enhanced severalfold with IFE compared to conventional electrophoresis, because the additional protein from the antibody–antigen reaction magnifies the intensity of antigen staining.

IFE of serum is typically performed using specialized gels in which electrophoresis is performed simultaneously in six adjacent lanes for each specimen. Following separation and placement of the template cutout around each lane, the first lane is overlaid with a general protein fixative. Each of the second through sixth lanes is overlaid with respective antisera against the IgG, IgA, and IgM heavy chains and kappa and lambda light chains. Upon washing, each specific protein remains fixed in gel matrix for staining and visual identification. An example IFE gel showing an IgG kappa paraprotein is displayed in Figure 3–5.

IFE of urine requires electrophoresis of the preconcentrated urine specimen in multiple lanes. Generally, the same six-lane specialized gels as for serum IFE are used. Because the focus of urine IFE is usually detection of Bence Jones proteinuria, a general fixative

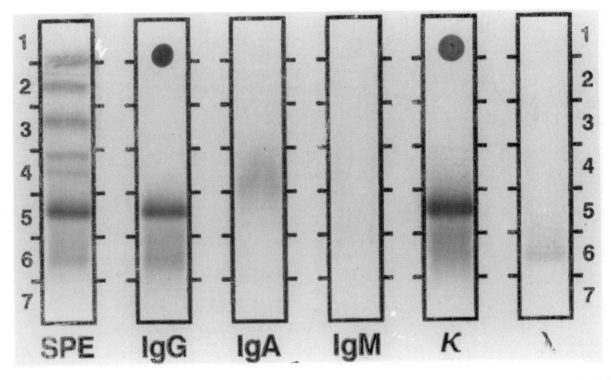

Figure 3–5. Typical 6-lane immunofixation gel in which a general fixative is applied to the leftmost lane. Each of the other lanes, 2 through 6, is overlaid with specific antisera IgG, IgA, IgM, kappa, and lambda, respectively. In lanes 2 through 6, the protein of interest is fixed for staining and visualization. Note that the identity of the monoclonal protein is IgG, with a kappa light chain. (The stained "dot" at the top of the IgG and kappa lanes is for quality control purposes only.)

is typically overlaid on the first lane; antisera to kappa, free (unbound to heavy chain) kappa, lambda, and free lambda are overlaid on in the second through fifth lanes, respectively.

IFE of CSF typically requires specimen concentration or sensitive stains, using the same specialized gels as are used for serum IFE. CSF IFE is usually performed to assess the presence of oligoclonal banding, defined as multiple bands of IgG identity migrating in the gamma region. The strategy for CSF IFE typically includes a general fixative in the first lane and IFE with IgG, kappa, and lambda antisera in separate lanes. CSF IFE allows a more precise definition of oligoclonal banding.

Capillary Zone Electrophoresis. Capillary zone electrophoresis is a relatively new technique that separates proteins on a capillary column of fused silica. The narrow diameter (~50 microns) of the lumen presents a large, negatively charged surface area to proteins moving across it toward the cathode. The absorbance of proteins is measured at 200 nm. This technique remains unproven for most clinical laboratories, but its recognized advantages of high resolution and automation offer exciting possibilities for the future.

2. Quantitative Methods

Biuret Reaction. This method is based on the reaction between copper ions in an alkaline solution and substances containing at least two peptide bonds. The intensity of the purple complex is proportional to the number of peptide bonds and thus protein concentration. The biuret reaction detects protein as low as 15 mg/dL.

Dye-binding. This method is based on the property of proteins to bind various dyes such as Coomassie Brilliant Blue and Amido Black, as used for staining electrophoresis plates. A major limitation of dye-binding is related to the unequal affinities of various proteins for the different dyes. Bromcresol green, bromphenol blue, and bromcresol purple, in particular, are bound specifically by albumin even in the presence of other proteins, and are widely applied in various automated spectrophotometric methods for the accurate quantitation of albumin.

Immunochemical. Immunochemical or immunoassay methods utilize highly specific antisera for individual proteins detected as antigen–antibody complexes in either liquid or solid media. Accurate quantitation of most clinically significant proteins in serum and other biological fluids can be achieved using automated nephelometers, which measure the rate at which immune complexes are formed in solution. Various configurations of immunoassays are described in Chapter 12.

C. Urine and Cerebrospinal Fluid

1. Qualitative Tests. Reagent strips (dipsticks) are qualitative tests for screening random urine specimens. The reaction is based on the principle known as the protein error of indicators. A buffered indicator coated on the strip develops a color change in the presence of protein < 10 mg/dL. The indicator strips are more sensitive to albumin than to globulins, which are poorly reactive; immunoglobulin free light chains (Bence Jones protein) are nonreactive.

2. Quantitative Tests. Quantitative and semiquantitative measurements for total protein and albumin in urine and CSF can be performed spectrophotometrically using dye-binding methods and turbidimetric methods, using protein precipitating agents such as trichloracetic acid or sulfosalicylic acid. The advantages of these methods include sensitivity to albumin, small sample volume (~0.5 mL), simplicity, and speed. The important disadvantage, however, is the underestimation of globulins due to the lack of sensitivity for these proteins. In urine, this limitation presents a serious problem since low-molecular-weight proteins excreted in tubular proteinuria, paraproteins in multiple myeloma, and immunoglobulin free light chains may be measured inaccurately low or missed entirely.

Quantitative assay of specific proteins in urine and CSF is best performed by nephelometric immunochemical methods calibrated against standardized reference preparations for the individual proteins.

3. Calculated Results. The measurement of total protein in CSF is useful but nonspecific for determining the concentrations of specific proteins. The degree of increased permeability of the blood–brain barrier can be estimated by the quantitation of albumin and IgG in CSF and serum specimens obtained at the same time. Increased intrathecal synthesis of IgG is clinically important in the diagnosis of a demyelinating disease such as multiple sclerosis. Albumin is a good reference protein because its synthesis occurs only in the liver and not in the central nervous system. Filtration through the damaged blood–brain barrier accounts for its presence in the CSF.

Albumin and IgG are measured by highly specific and sensitive immunochemical methods to calculate the CSF/serum albumin index and CSF/serum IgG index, shown as ratio 1 and ratio 2, respectively:

$$\textbf{Ratio 1.}\ \text{CSF/Serum Albumin Index} = \frac{\text{CSF Albumin (mg/dL)}}{\text{Serum Albumin (g/dL)}}$$

$$\textbf{Ratio 2.}\ \text{CSF/Serum IgG Index} = \frac{\text{CSF IgG (mg/dL)}}{\text{Serum Albumin (g/dL)}}$$

A CSF/serum albumin index of < 9 is compatible with a healthy blood–brain barrier; values of 14 to 30 suggest moderate impairment, while values > 30 indicate severe impairment. Increased intrathecal IgG synthesis can be estimated from the CSF/serum IgG index, which is elevated by intrathecal IgG synthesis or by IgG passing across a damaged blood–brain barrier.

IgG from plasma can be estimated by dividing ratio 2 by ratio 1 to obtain the CSF IgG Index shown as ratio 3 below:

$$\textbf{Ratio 3.}\ \text{CSF IgG Index} = \frac{\text{CSF IgG/Serum IgG}}{\text{CSF Albumin/Serum Albumin}}$$

The CSF IgG index normalizes the IgG concentration in the CSF to the IgG concentration in the serum. The reference interval for the CSF IgG index varies between 0.3 and 0.8. Values > 0.8 suggest increased intrathecal IgG synthesis. The CSF IgG index is increased in most patients with MS, and assists in the assessment of this and other inflammatory neurologic diseases.

SUGGESTED READINGS

Check IJ, Papadea C. Immunoglobulin quantitation. In Rose NR, de Macario EC, Folds JD, Lane HC, Nakamura RM (eds.). *Manual of clinical laboratory immunology,* 5th ed. Washington, DC: ASM Press, 1997; 134–46.

Christenson RH, Azzazy HME. Amino acids. In Burtis CA, Ashwood ER (eds.). *Tietz textbook of clinical chemistry,* 3rd ed. Philadelphia: WB Saunders, 1999; 444–76.

Johnson AM, Rohlfs EM, Silverman LM. Proteins. In Burtis CA, Ashwood ER (eds.). *Tietz textbook of clinical chemistry,* 3rd ed. Philadelphia: WB Saunders, 1999; 477–540.

Kahn WA, Wigfall DR, Frank MM. Complement and kinins: mediators of inflammation. In Henry JB (ed.). *Clinical diagnosis and management by laboratory methods,* 19th ed. 1996; 928–46.

Kelgenhauer K. Differentiation of the humoral immune response in inflammatory diseases of the central nervous system. *J Neurol* 228:223–37, 1982.

Keren DF. *High-resolution electrophoresis and immunofixation: Techniques and interpretaton,* 2nd ed. Boston: Butterworth-Heinemann, 1994.

Kokko JP. Approach to the patient with renal disease. In Bennett JC, Plum F (eds.). *Cecil textbook of medicine,* 20th ed. Philadelphia: WB Saunders, 1996; 511–17.

Kyle RA. "Benign" monoclonal gammopathy. *JAMA* 251:1849–54, 1984.

Kyle RA, Grafton JP. The spectrum of IgM monoclonal gammopathy in 430 cases. *Mayo Clin Proc* 62:719–31, 1987.

Merlini G, Abuzzi F, Whicher J. Monoclonal gammopathies. *J Int Fed Clin Chem* 9:171–76, 1997.

Richie RF, Navolotskaia O (eds.). *Serum proteins in clinical medicine,* Vol. 1, Laboratory Section. Portland, ME: Foundation for Blood Research, MPX Maine Printing Co., 1996.

Virella G, Wang A-C. Biosynthesis, metabolism, and biological properties of immunoglobulins. In Virella G (ed.). *Introduction to medical immunology,* 4th ed. New York: Marcel Dekker, 1998; 91–103.

Items for Further Consideration

1. What important information can be obtained by the electrophoresis of a 24-hour urine from a patient with hypogammaglobulinemia?

2. Iron studies are used in the evaluation of disorders of iron metabolism. Discuss the physiological roles of the major biochemical serum analytes of iron metabolism: iron, ferritin, apoferritin, hemosiderin, and transferrin.

3. What is the acute phase response, and what is the function of this response?

4. After performing protein electrophoresis on a specimen, a discrete band was seen in the alpha–beta region of the stained gel. What would you do as the next step(s) in each case if the specimen was (a) serum; (b) plasma; (c) urine?

5. Immunofixation results identified a monoclonal protein as being IgG κ. However, the quantitation of immunoglobulins showed IgA to be in the greatest concentration. What could cause such a discrepancy in results?

Clinical Enzymology

Steven C. Kazmierczak and Debra R. Johnson

Key Terms

ACTIVE SITE

APOENZYMES

COENZYME

COFACTOR

CONTINUOUS-MONITORING
METHODS

DENATURATION

END-POINT ASSAYS

ENZYMES

FIRST-ORDER KINETICS

FIXED-TIME METHODS

HOLOENZYME

INHIBITORS

INTERNATIONAL UNIT

ISOENZYMES

KINETIC ASSAYS

KINETIC METHODS

LAG PHASE

LINEAR PHASE

MASS ASSAYS

MICHAELIS–MENTEN
CONSTANT (K_m)

MOLECULAR ACTIVITY

PROSTHETIC GROUPS

SPECIFIC ACTIVITY

SUBSTRATE

SUBSTRATE DEPLETION

ZERO-ORDER KINETICS

Chapter Objectives

Upon completion of this chapter, the reader will be able to:

1. Differentiate between mass and activity enzyme assays, and between fixed-time and continuous-monitoring kinetic enzyme assays.
2. Explain the use of indicator reactions in enzyme assays; identify the compound commonly measured at 340 nm in an indicator reaction.
3. Identify and discuss isoenzymes of clinical importance.
4. Compare and contrast zero- and first-order reaction kinetics.

5. Give examples of substances that act as inhibitors of enzyme activity.
6. Describe the changes in absorbance during the phases of a kinetic enzyme assay (i.e., lag phase, linear phase, substrate depletion).
7. Discuss how parameters such as pH, temperature, substrate and enzyme concentration, and the presence or absence of cofactors and inhibitors affect assays designed to measure enzyme activity.
8. Differentiate between competitive, uncompetitive, and noncompetitive enzyme inhibition.
9. Discuss the clinical utility for the measurement of the following enzymes: alanine aminotransferase, aspartate aminotransferase, alkaline phosphatase, amylase, creatine kinase, gamma-glutamyltransferase, lactate dehydrogenase, lipase, serum cholinesterase.
10. Assess specimen integrity for enzyme analysis based on sample collection, processing, and storage conditions.
11. Explain the clinical use of each of the following parameters: dibucaine number, amylase:creatinine clearance ratio, fluoride number.

Analytes Included

ALANINE AMINOTRANSFERASE (ALT)

ALKALINE PHOSPHATASE (ALP)

AMYLASE (AMY)

ASPARTATE AMINOTRANSFERASE (AST)

CREATINE KINASE (CK)

GAMMA-GLUTAMYLTRANSFERASE (GGT)

LACTATE DEHYDROGENASE (LD)

LIPASE (LPS)

SERUM CHOLINESTERASE (PSEUDOCHOLINESTERASE; SChE)

Pretest

Directions: Choose the single best answer for each of the following items.

1. Which of the following compounds serves as a cofactor for AST and ALT in the transfer of amino groups from one substance to another?
   a. magnesium
   b. calcium
   c. vitamin B_6
   d. NADH

2. Clinical assays that measure enzyme activity are designed to follow
   a. first-order kinetics
   b. zero-order kinetics
   c. the Q_{10} reaction order
   d. end-point analysis

3. In a kinetic enzyme assay, the maximum rate of activity occurs in the
   a. substrate depletion phase
   b. log-linear phase
   c. first-order kinetics phase
   d. zero-order kinetics phase

4. The Michaelis–Menten constant (K_m) is expressed in terms of
   a. time
   b. substrate concentration
   c. velocity
   d. wavelength

5. The period of an enzyme reaction in which the rate of change in absorbance measured is maximal and constant is the
   a. lag phase
   b. logarithmic phase
   c. linear phase
   d. substrate depletion phase

6. Which of the following compounds is often measured at 340 nm in a secondary or indicator reaction for enzyme analysis?
   a. $FADH_2$
   b. MDH
   c. NAD^+
   d. NADH

7. Enzyme assays in which absorbance is measured at frequent intervals over a period of time are referred to as
   a. fixed-time assays
   b. continuous-monitoring methods
   c. mass measurements
   d. turnover measurements

8. One International Unit of enzyme activity catalyzes the transformation of what amount of substrate to product in one minute under defined conditions?
   a. 1 mole
   b. 1 mmole
   c. 1 μmole
   d. 1 nmole

9. Which of the following enzymes is inhibited by organophosphate compounds?
   a. amylase
   b. lipase
   c. alkaline phosphatase
   d. cholinesterase

10. The dibucaine number is determined to detect atypical forms of which of the following enzymes?
    a. creatine kinase (CK)
    b. gamma-glutamyltransferase (GGT)
    c. serum cholinesterase (SChE)
    d. aspartate aminotransferase (AST)

11. Serum "cholinesterase" is synonymous with the term
    a. true cholinesterase
    b. pseudocholinesterase
    c. acetylcholinesterase
    d. erythrocyte cholinesterase

12. Which of the following enzymes shows the greatest specificity for hepatocellular liver damage?
    a. lactate dehydrogenase (LD)
    b. alkaline phosphatase (ALP)
    c. aspartate aminotransferase (AST)
    d. alanine aminotransferase (ALT)

13. The ratio of ALT to AST is typically > 1.0 in which of the following disease states?
    a. viral hepatitis
    b. alcoholic hepatitis
    c. cirrhosis
    d. hepatocellular carcinoma

14. The most sensitive indicator of biliary obstruction is
    a. AST
    b. ALT
    c. GGT
    d. ALP

15. Alcohol may induce the synthesis of which of the following enzymes?
    a. AST
    b. ALT
    c. GGT
    d. ALP

16. A woman in her third trimester of pregnancy would be expected to demonstrate an elevated activity of which of the following enzymes?
    a. AST
    b. LD
    c. GGT
    d. ALP

17. An elevated activity of which of the following enzymes would be consistent with a diagnosis of Paget's disease?
    a. creatine kinase MM band (CK-MM)
    b. LD
    c. ALP
    d. amylase (AMY)

18. Which of the following enzymes would demonstrate an increase in activity following hemolysis?
    a. CK
    b. LD
    c. ALP
    d. GGT

19. Refrigeration or freezing of specimens for LD analysis should be avoided due to degradation of which of the following isoenzyme fractions?
    a. LD-1 and LD-2
    b. LD-2 and LD-3
    c. LD-3 and LD-4
    d. LD-4 and LD-5

20. Strenuous exercise prior to biochemical serum testing may result in an increased
    a. ALT
    b. SChE
    c. CK
    d. ALP

21. Which of the following enzymes is currently used in the assessment of myocardial damage due to infarction?
    a. AMY
    b. lipase (LPS)
    c. AST
    d. CK

22. Which of the following serum enzyme assays would be most helpful in differentiating between normal bone growth and mild liver dysfunction in a teenager?
    a. AST
    b. LD
    c. GGT
    d. CK

23. In addition to pancreatitis, an increase in serum lipase activity may be observed in
    a. acute myocardial infarction
    b. renal insufficiency

    c. cystic fibrosis
    d. viral hepatitis

24. Which of the following diagnoses would be consistent with the finding of a normal lipase but an elevated amylase?
    a. pulmonary infarction
    b. acute myocardial infarction
    c. acute hepatitis
    d. macroamylasemia

I. INTRODUCTION

The measurement of enzymes in biological fluids and tissues has become an important tool for helping to establish clinical diagnoses, for assessing patient prognosis, and for evaluating the effectiveness of different treatment regimens. New enzyme markers continue to be identified that may further our abilities to diagnose disease processes and advance our understanding of disease mechanisms. These new discoveries bring about a constant reappraisal of the clinical utility of enzyme markers resulting in an upsurge in the importance of some enzymes and a decline in the significance of others. Examples of this include the use of lactate dehydrogenase (LD) and creatine kinase MB (CK-MB) supplanting the use of aspartate aminotransferase (AST) as an indicator of myocardial damage and, more recently, the use of the troponins to supersede LD, and in some cases CK-MB, as markers of myocardial injury. Thus, the field of diagnostic enzymology is dynamic, necessitating an awareness of new discoveries in the field of diagnostic enzymology.

The utility of enzymes as diagnostic tools was first demonstrated over 80 years ago when amylase was advocated as a sensitive and reliable test for various pancreatic disorders. However, the discipline of clinical enzymology did not advance significantly over the next 40 years, until 1953 when AST was found to help diagnose cardiac and hepatic disease. Further study of enzymes and their usefulness as indicators of certain pathophysiological conditions gave rise to the discipline of diagnostic enzymology. Today, thousands of enzymes have been identified in various tissues and body fluids. However, the typical hospital clinical laboratory measures less than 15 different enzymes for diagnostic purposes. Further understanding of the roles that enzymes play in health and disease should enable more effective utilization of these cellular components as diagnostic markers.

II. BIOCHEMISTRY

▶ **Enzymes** are integral components of many cellular processes. Some enzymes are localized within specific organelles in the cell (e.g., the mitochondria), appearing within the circulation only following injury that is severe enough to disrupt the organelle. Other enzymes are cytosolic and are released from the cell when the outer membrane is weakened or made porous. Certain enzymes are contained in intracytoplasmic vesicles or granules, such as lysosomes or zymogen granules. Others are released due to increased synthesis and cellular "leakage" or excretion. Finally, some enzymes are integral components of, or are firmly attached to, membranes within the cell; the release of membrane-associated enzymes typically occurs only following massive cell necrosis. Knowing the characteristic distribution of enzymes within the cell allows the assessment of the severity of cell injury.

The diagnostic utility of many enzymes is due to differences in the amount of enzymes in various tissues, the presence of isoenzymes specific to a particular tissue or organ, as well as the localization of an enzyme in the cytoplasm, mitochondria, or membrane(s) of the cell. Table 4–1 provides examples of enzymes that are associated with specific organelles and tissues. The location of an enzyme within a cell, its tissue specificity, the size of the molecule, and its solubility all contribute to the diagnostic utility of an enzyme. Small, non–membrane-bound, cytoplasmic enzymes are the most readily released from the cell and the most likely to appear in the blood shortly after cell injury. Large enzymes located within the mitochondria or attached to a membrane or other cell component are released from the cell only upon disruption of both the cellular and mitochondrial membrane, indicating substantial cellular damage. Some enzymes become increased in the blood due to causes other than cell injury or death. Inducible enzymes become increased due to increased synthesis within the cells that contain the enzyme; for example, the synthesis of gamma-glutamyltransferase (GGT), a hepatic enzyme, may be increased by the effects of drugs (e.g., antiepileptics, antidepressants, and alcohol) on the cell. Likewise, the increase in intracanalicular pressure that occurs with biliary obstruction results in an increase in synthesis of alkaline phosphatase (ALP) and GGT by the cells lining the canalicular space.

The intracellular concentration of enzymes is often thousands of times greater than the corresponding concentration in the serum.

Table 4–1. Tissue Source of Enzymes Commonly Used for Diagnosis

ENZYME	MAJOR TISSUE SOURCES (IN ORDER OF DECREASING TISSUE ACTIVITY)
Alkaline phosphatase (ALP)	Placenta, intestinal mucosa, kidney, bone, liver
Alanine aminotransferase (ALT)	Liver, kidney
Amylase (AMY)	Pancreas, salivary glands, fallopian tube
Aspartate aminotransferase (AST)	Heart, liver, skeletal muscle, kidney, pancreas
Creatine kinase (CK)	Skeletal muscle, myocardium, brain, colon, stomach, urinary bladder
Gamma glutamyltransferase (GGT)	Kidney, biliary tract of liver
Lactate dehydrogenase (LD)	Brain, heart, erythrocytes, kidney, lung, skeletal muscle, liver, pancreas, stomach
Lipase (LPS)	Pancreas
Cholinesterase (pseudocholinesterase)	Plasma, liver

Cell membranes act as barriers and anchors, holding enzymes in their appropriate compartment and preventing the passage of enzymes out of the cell. However, serum contains measurable activities of many cellular enzymes. Their presence in the peripheral circulation of healthy individuals is due to a homeostatic balance between (a) normal cell turnover (apoptosis) with release of enzyme into the circulation, (b) normal efflux of a small amount of enzyme through the cell membrane, and (c) clearance of enzyme from the peripheral circulation by the reticuloendothelial system or by the kidneys.

A. ENZYME COMPOSITION AND STRUCTURE

Enzymes are proteins, and thus composed of amino acids. As described in Chapter 3, the sequence of amino acids in the polypeptide chain determines the primary structure of an enzyme. Secondary enzyme structure refers to recurrent structures (e.g., an α-helix), usually formed by hydrogen bonding, along the one-dimensional axis of the polypeptide chain. Tertiary structure refers to the three-dimensional folding that occurs when the polypeptide chain bends or folds back upon itself. Tertiary structure of the enzyme may be stabilized by hydrogen bonds, hydrophobic interactions, ionic interactions, and disulfide bridges between residues in the polypeptide chain. Finally, some enzymes are composed of more than one polypeptide chain. The association of two or more of these polypeptide subunits is referred to as the quaternary structure. Depending on the enzyme, the subunits that associate may have identical, slightly different, or very different amino acid sequences (and hence structures). The different multimeric forms that may exist for a particular enzyme (i.e., forms differing in their quaternary structure) are known as **isoenzymes.** The structure of an enzyme is integral to its function as a *biological catalyst*. The enzyme structure includes an **active site** (or catalytic site), a region with specific structure that binds the substrate and facilitates conversion to product. It is the active site that confers the specificity of an enzyme for its substrate(s) and the reaction it catalyzes. In some enzymes, each subunit may contain an active site.

B. FACTORS AFFECTING ENZYME ACTIVITY

The activity of enzymes can be enhanced or inhibited by a variety of conditions and factors. Because enzymes are proteins, they are susceptible to **denaturation** due to changes in their physical surroundings. Loss of catalytic activity by enzymes can occur due to changes in pH, temperature, and the concentration of certain ions in the surrounding medium. The loss of enzyme activity can be permanent (irreversible denaturation), or it can be temporary (reversible denaturation).

1. Cofactors and Coenzymes. Many enzymes require a **cofactor** or a **coenzyme** for full activity. Cofactors are small, inorganic compounds such as metal ions; coenzymes are organic compounds, including vitamins. Some cofactors or coenzymes may be only loosely associated with the protein, and thus, easily removed by dialysis or by the addition of another compound that has a higher affinity for the cofactor than does the enzyme. Cofactors or coenzymes that are tightly bound and difficult to dissociate from the enzyme are called

prosthetic groups. Enzymes lacking their prosthetic groups are referred to as **apoenzymes;** the combined enzyme and prosthetic group is called a **holoenzyme.** *Metalloenzymes* are enzymes that have a metal ion prosthetic group. Coenzymes usually function as carriers of special compounds or electrons that are transferred in the enzymatic reaction. For example, pyridoxal phosphate (vitamin B_6) is a coenzyme for AST and ALT; its function is to transfer amino groups.

2. pH. Enzyme activity is pH dependent. Enzymes have characteristic pH activity profiles for a given set of conditions (e.g., buffer type, temperature, substrate concentration, and enzyme concentration). The peak of the pH versus activity plot identifies the *pH optimum* for maximum enzyme activity (Figure 4–1). When possible, enzyme activity should be measured at the optimal pH using a buffer having a pK_a within one unit of the desired pH. The pH optimum is not necessarily the best pH for physiological control *in vivo.* In many cases, the enzyme functions at an *in vivo* pH that corresponds to a point on the slope of the pH versus activity curve; in fact, this allows efficient regulation of enzyme activity by small changes in pH.

3. Temperature. Like pH, temperature also affects the rate of enzyme-catalyzed reactions. The rate at which most enzymatic reactions proceed approximately doubles with each 10°C rise in temperature. The actual increase in activity with each 10°C increment in temperature is referred to as the temperature coefficient or Q_{10}, and is slightly different for each enzyme. Increasing the temperature at which an enzyme reaction takes place results in increased activity until a temperature optimum is reached, above which any increase in temperature results in a decrease in enzyme activity. The optimum temperature for maximum enzyme activity represents the point at which the usual increase in enzyme activity seen with a temperature

> **Terminology for Enzymes and Their Associated Cofactors and Coenzymes**
>
> Prosthetic groups—cofactors or coenzymes that are tightly bound and difficult to dissociate from the enzyme
> Apoenzymes—enzymes lacking their prosthetic groups
> Holoenzyme—combined enzyme and prosthetic group
> Metalloenzymes—enzymes that have a metal ion prosthetic group

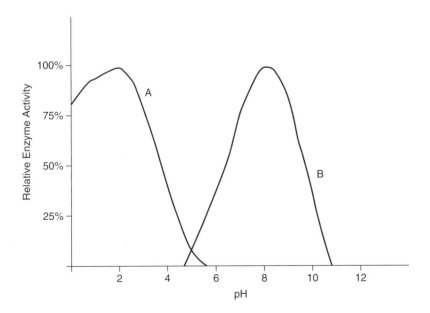

Figure 4–1. Relative activity of enzymes as a function of pH: A, pepsin (pH optimum 2.0); B, trypsin (pH optimum 8.0).

increase is counterbalanced by an increasing rate of thermal denaturation of the enzyme. Most enzymes are inactivated at temperatures above 55 to 60°C. Figure 4–2 shows an example of a temperature versus activity profile for a typical enzyme. The sudden loss of enzyme activity that occurs at temperatures above the temperature optimum is due to rapid thermal denaturation. Refrigeration or freezing of enzymes normally helps to maintain enzyme activities; there are, however, exceptions to this rule. For example, LD at refrigeration temperatures loses its enzymatic activity, especially the LD-4 and LD-5 isoenzyme fractions.

C. ENZYME KINETICS

The role of enzymes in cellular processes is to function as *biological catalysts*. Enzymes accelerate the rate at which a chemical reaction takes place, without being consumed in the reaction process. The Michaelis–Menten theory of enzyme-catalyzed reactions postulates that enzyme (E) combines with **substrate** (S) to form an enzyme–substrate (ES) complex. The ES complex then dissociates releasing the enzyme and new product (P). The sequence of reactions may be written as:

$$E + S \underset{k_{-1}}{\overset{k_1}{\rightleftharpoons}} ES \rightarrow E + P$$

These reactions are assumed to be reversible. The rates of the forward and reverse reactions are given by rate constants, k_1 and k_{-1}, respectively. By design, the concentration of substrate is much greater than the concentration of enzyme; thus, the amount of ES complex is negligible compared with the total substrate concentration. Enzyme-catalyzed reactions exhibit a feature known as *substrate saturation* (i.e., all catalytic sites are occupied with substrate) that is not observed in non–enzyme-catalyzed reactions. Figure 4–3

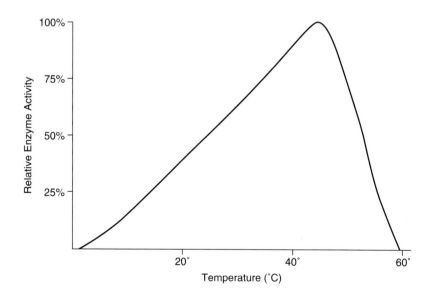

Figure 4–2. Effect of temperature on enzyme activity. Note the rapid loss of activity at temperatures greater than 45°C due to thermal inactivation of the enzyme.

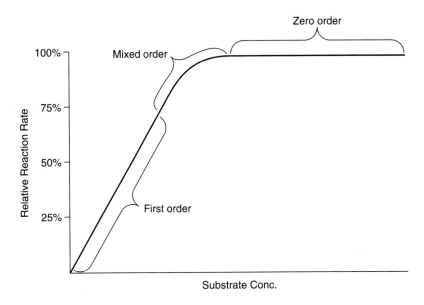

Figure 4–3. Rate of an enzyme-catalyzed reaction as a function of substrate concentration.

illustrates the effect of substrate concentration on the rate at which a given amount of enzyme can catalyze the formation of product. As shown in Figure 4–3, at low concentrations of substrate, the rate of the enzyme-catalyzed reaction is directly proportional to the concentration of substrate (i.e., one reactant); in this section of the curve, the reaction follows **first-order kinetics.** Second-order rate reactions are characterized by a rate that is proportional to the product of the concentrations of two reactants.

As the substrate concentration continues to increase (in the substrate concentration versus reaction velocity curve), the reaction rate begins to taper off and eventually plateaus until the reaction rate becomes constant and independent of substrate concentration; in this region of the curve (i.e., at high substrate concentrations), the reaction follows **zero-order kinetics.** Assays to measure enzyme activity in serum are designed for zero-order kinetics since the rate of the reaction should be dependent only on the concentration of the enzyme present in the specimen being analyzed.

The **Michaelis–Menten constant, K_m,** can be ascertained from the plot of the rate of an enzyme-catalyzed reaction versus substrate concentration (Figure 4–4). This constant is equal to the substrate concentration at which the reaction rate is one-half the maximal velocity. The K_m has units of moles per liter and is independent of enzyme concentration. For enzymes that can use different substrates, each has a characteristic K_m value. For example, for the enzyme hexokinase, the K_m of glucose (0.15 mmol/L) is 10-fold less than the K_m for fructose (1.5 mmol/L). The K_m is characteristic for a particular enzyme–substrate reaction under defined reaction conditions (e.g., pH and temperature). The Michaelis–Menten constant is important because the substrate concentrations used in assays are usually designed to be about 10-fold higher than the K_m for measurement of enzyme activity in blood and tissue specimens.

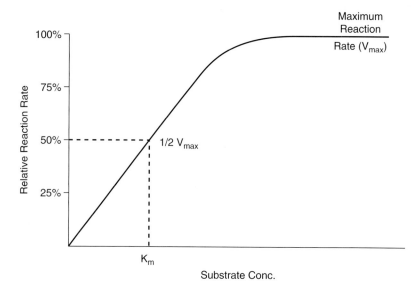

Figure 4–4. Graphical determination of the K_m for an enzyme using a specific substrate.

Types of Enzyme Inhibition
Competitive inhibition—occurs when a substance binds free enzyme at the active site, preventing binding of substrate
Noncompetitive inhibition—occurs when an inhibitor combines with either the free enzyme or the enzyme–substrate complex
Uncompetitive inhibition—occurs when the inhibitor binds to the enzyme–substrate complex only

D. *Enzyme Inhibition*

A variety of substances can reduce (inhibit) the rate at which an enzyme-catalyzed reaction proceeds. Substances that bind to particular enzymes with a resulting decrease in reaction velocity are known as **inhibitors.** Inhibition of enzyme activity by specific molecules *in vivo* is an important means of regulating cellular processes. Three types of reversible *enzyme inhibition* are generally recognized: competitive inhibition, noncompetitive inhibition, and uncompetitive inhibition.

Competitive inhibition occurs when a substance binds free enzyme at the active site, preventing binding of substrate; in this case, the inhibitor directly competes with the normal substrate for binding sites. Competitive inhibition can be overcome by increasing the substrate concentration, thereby overwhelming the amount of inhibitor. Consequently, a competitive inhibitor increases the apparent K_m of the enzyme for substrate, since a higher substrate concentration is required for the enzyme to reach a maximum rate of catalysis. Competitive inhibition can be represented schematically as follows:

$$E + S + I \rightleftharpoons ES + EI \rightarrow E + P$$

Noncompetitive inhibition occurs when an inhibitor combines with either the free enzyme or the enzyme–substrate complex, as given by:

$$E + S + I \rightleftharpoons ES + EI + ESI \rightarrow E + P$$

These inhibitors bind to the enzyme at a site other than the enzyme's active site. Unlike competitive inhibition, noncompetitive inhibition cannot be overcome by the addition of more substrate because the additional ES complexes simply bind with the inhibitor.

Uncompetitive inhibition results when the inhibitor binds to the ES complex only. This type of inhibition can be represented as:

$$E + S + I \rightleftharpoons ES + ESI \rightarrow E + P$$

Increasing the substrate concentration in this type of inhibition may actually increase the inhibitory effect.

III. CLINICAL OVERVIEW

Enzymes are measured in biological samples for diagnosis, prognosis, assessment of treatment, and monitoring the progression of the disease. As mentioned previously, most hospital laboratories perform assays for no more than 15 different enzymes. This section discusses the clinical relevance of the most commonly monitored enzymes.

A. ASPARTATE AMINOTRANSFERASE

AST is one of the two clinically important enzymes involved in the interconversion of amino acids and α-oxoacids by transfer of amino groups. While tissues contain many different aminotransferases, serum activities of AST and ALT have proven useful in the diagnosis and monitoring of various diseases. Both enzymes are sensitive indicators of necrosis in the many tissues that contain high activities of AST and/or ALT. The aminotransferases have no significant metabolic function in serum, and their presence in serum at normal levels reflects normal cellular function and turnover.

The reaction catalyzed by AST is shown below:

$$\text{L-Aspartate} + \alpha\text{-Ketoglutarate} \overset{\text{AST}}{\rightleftharpoons} \text{Oxaloacetate} + \text{L-Glutamate}$$

The coenzyme pyridoxal-5′-phosphate (P-5′-P) is required for full catalytic activity. Normally, patients' sera contain adequate amounts of P-5′-P (vitamin B_6) to sustain enzyme activity in AST activity assays. However, some individuals, such as chronic ethanol abusers, may be deficient in vitamin B_6 and show falsely decreased AST activities. Studies suggest that the inclusion of P-5′-P in the assay reagent may increase measured AST activities up to 50% in samples from persons deficient in this coenzyme.

The cytoplasmic form of AST is found in high concentration in liver, skeletal muscle, and kidney cells. Myocardial cells also contain an appreciable amount of AST; in fact, AST was the first biochemical marker utilized for the diagnosis of acute myocardial infarction. Today, however, markers such as CK, CK-MB, and the troponins are of greater value for diagnosis of myocardial infarction. Other organs containing significant amounts of AST include the pancreas, brain, spleen, and lung. Also, erythrocytes contain sufficient AST to adversely affect the accuracy of AST measurements in hemolyzed specimens.

In addition to the cytoplasmic form of AST there is a mitochondrial AST isoenzyme that is distinctly different. Although both forms are dimers with a molecular mass of approximately 90,000 daltons, mitochondrial AST exhibits a shorter half-life and greater heat stability as compared to cytoplasmic AST. As a percentage of total AST in serum, patients with liver disease have been found to exhibit greater amounts of mitochondrial AST (up to 15%) compared to patients without liver disease.

Macro-AST, a variant of AST, is an antibody–AST complex found in patients having an anti-AST autoantibody (typically im-

---

The Most Commonly Monitored Enzymes

- Aspartate aminotransferase (AST)
- Alanine aminotransferase (ALT)
- Gamma-glutamyltransferase (GGT)
- Alkaline phosphatase (ALP)
- Lactate dehydrogenase (LD)
- Creatine kinase (CK)
- Amylase (AMY)
- Lipase (LPS)
- Acetylcholinesterase (AChE) and serum cholinesterase (SChE)

Tissues Containing Significant Amounts of AST

- Liver
- Skeletal muscle
- Kidney cells
- Myocardium
- Pancreas
- Brain
- Spleen
- Lung

munoglobulin G [IgG]). Macro-AST is usually not clinically significant; however, patients with macro-AST may exhibit persistently increased serum AST activities (up to 10 to 15 times normal) due to the prolonged half-life of the IgG–AST complex. If macro-AST can be identified in cases of persistently elevated serum AST, unnecessary studies to determine the cause of the increased AST activity could be avoided.

Indications for the measurement of serum or plasma AST include the diagnosis and monitoring of diseases of the liver and biliary tract and the assessment of skeletal muscle damage. The use of AST for diagnosis of myocardial injury is now of limited clinical utility. Patients with viral hepatitis may exhibit extremely elevated AST activities, with values up to 100-fold higher than the upper reference limit. Liver hypoxia also causes extreme increases in AST. Lesser increases in AST, usually less than 10 times the upper reference limit, are seen in autoimmune hepatitis, cholestasis, alcoholic liver disease, and cirrhosis. AST may also reach exceptionally high values in patients with diseases or injuries of skeletal muscle, such as polymyositis and rhabdomyolysis. Patients with Duchenne muscular dystrophy can exhibit 100-fold increases in AST activity. Strenuous physical activity may cause increases in AST, although they rarely exceed three times the upper reference limit.

Serum or plasma collected with heparin, ethylenediaminetetraacetic acid (EDTA), oxalate, or citrate, is an acceptable specimen. AST is stable in refrigerated or frozen samples and stability can be further enhanced by adding P-5′-P to the specimen. Although rarely performed, AST can be measured in cerebrospinal fluid (CSF) and urine; AST activity in CSF has been found to correlate directly with the extent of brain injury, and AST found in urine has been used as an indicator of renal function following transplantation. Serum AST activity tends to be higher in infants and children (up to about 18 years old) than in adults. Adult males exhibit higher AST activities than adult females, probably due to differences in muscle mass. Actual reference intervals are dependent on the temperature used to measure enzyme activity and, to a lesser extent, whether the assay incorporates P-5′-P.

B. ALANINE AMINOTRANSFERASE

ALT, like AST, catalyzes the reversible transfer of an amino group and requires the coenzyme P-5′-P. ALT is primarily measured for the diagnosis and monitoring of hepatocellular disease. Similar to AST, patients with hepatocellular injury can show marked increases in serum ALT activity, often reaching 25- to 100-fold higher than the upper limit of the reference interval. Calculation of the ratio of ALT to AST, also called the *De Ritis ratio,* is sometimes useful in determining the etiology of hepatocellular disease by distinguishing among several conditions. ALT is present only in the cytosol of hepatocytes, while approximately 30% of AST is present in the cytosol and the remainder in the mitochondria; however, the relative activity of AST in liver tissue is severalfold higher than the activity of ALT. Thus, in health, or with mild liver damage, the ALT:AST ratio is typically < 1.0. In severe liver disease with necrosis (i.e., when mitochondrial AST is released), as in alcoholic hepatitis, cirrhosis, and hepatocellular carcinoma, the ALT:AST ratio will be depressed. ALT is often as high or higher than AST in infectious hepatitis and some inflammatory conditions, leading to an ALT:AST ratio of 1.0 or greater.

Unlike AST, ALT is not present in skeletal muscle or heart tissue to any appreciable extent. Thus, elevations of serum ALT are more specific for hepatocellular liver damage, while elevated AST also occurs with skeletal muscle or myocardial damage. Table 4–2 shows the increases that occur in ALT and AST in various disease states.

Serum is the preferred sample for ALT activity measurements; heparinized or EDTA plasma is acceptable. Samples stored at room temperature are stable for 3 days; refrigerated samples are stable for 1 week.

Reference intervals are method and temperature dependent. Ranges are generally higher in infants than in adults and higher in adult males than in adult females.

C. GAMMA-GLUTAMYLTRANSFERASE

GGT is an amino acid transferase; it transfers a terminal γ-glutamyl group from one compound (e.g., a peptide) to an acceptor compound (e.g., a peptide, amino acid, or water). GGT is a membrane-bound enzyme found primarily in cells that have a high secretory or absorptive capacity. GGT plays a role in the metabolism of glutathione and in the resorption of amino acids from the glomerular filtrate and the lumen of the intestine. Although the kidneys contain the highest activity of GGT per gram of tissue, GGT is primarily used as a marker for the diagnosis of certain liver-related disorders. Tissues containing significant amounts of GGT, in descending order of activity per gram of tissue, are kidney, liver, pancreas, and intestine. At least five different GGT isoenzymes have been described; however, no clear association has been established between a particular isoenzyme and a specific disease state. Most of the GGT present in normal serum is derived from the normal turnover of hepatic tissue. The half-life of GGT can vary between individuals, but is approximately 7 to 10 days in normal, healthy subjects.

GGT is an important diagnostic tool in the evaluation of hepatobiliary diseases, particularly when *cholestasis* (obstruction) is present. In cholestasis, serum GGT activity usually increases before serum bilirubin concentrations increase, making GGT a more sensitive test for early detection. Patients with advanced cholestasis typically show increased total bilirubin, ALP, and GGT. Since GGT is bound to the cell membrane, it would be expected to be present in serum only following cell necrosis; however, GGT is released even in non-necrotic hepatobiliary diseases. It is thought that cholestasis causes an increase in pressure within the hepatobiliary tract inducing increased cellular synthesis of GGT. In addition, the hepatocytes are exposed to higher concentrations of bile acids in cholestasis, promoting solubilization of GGT from the membrane and its release into the peripheral circulation.

Tissues Containing Significant Amounts of GGT
• Kidney
• Liver
• Pancreas
• Intestine

Table 4–2. Abnormalities of AST and ALT in Various Disease States

	DISEASE STATE					
	SKELETAL MUSCLE INJURY	AMI	CHF	HEPATITIS	CIRRHOSIS	BILIARY OBSTRUCTION
AST	↑ to ↑↑	↑ to ↑↑	↑ to ↑↑	↑↑↑	↑	↑
ALT	→	→	↑ to ↑↑	↑↑↑	↑	↑

AMI, acute myocardial infarction; CHF, congestive heart failure; →, no change.

The magnitude of increase in serum GGT activities is helpful in determining the extent and etiology of hepatobiliary damage. When hepatocellular damage occurs without significant cholestatic involvement, GGT activities usually range from two to five times the upper limit of normal; patients with viral hepatitis and alcoholic hepatitis generally fall into this category. Activities five- to tenfold higher than the upper reference limit suggest that cholestasis is the primary or secondary disease process. Patients who have long-term cholestasis (e.g., due to malignancy) may show GGT values up to 100-fold higher than the upper limit of the reference interval.

GGT levels can also aid in the interpretation of an elevated serum ALP. An increased ALP with a normal GGT activity is suggestive of ALP release from nonhepatic tissue; this occurs in bone diseases, such as Paget's disease, rickets, and osteomalacia. ALP is likely of hepatic origin when both ALP and GGT are increased.

Increased GGT activity in serum has long been used as a marker of alcohol intake. Alcohol, as well as many drugs (including tricyclic antidepressants, anticonvulsants, and barbiturates such as phenobarbital), can induce cellular synthesis of GGT resulting in increased serum activities of the enzyme due to leakage.

Reference intervals for GGT are dependent on the method and reaction conditions employed for measurement. Reference intervals are also age and gender dependent. Values are lowest in adolescents and rise steadily throughout life. Males tend to show higher GGT activities than females because of the additional contribution of GGT from the prostate gland.

Serum is the preferred specimen for analysis of GGT. Heparinized plasma is also acceptable in most applications; however, EDTA plasma and citrated plasma have been found to cause interference. Although erythrocytes contain little or no GGT, moderate hemolysis does interfere with some methods that employ L-γ-glutamyl-3-carboxy-4-nitroanilide as substrate. GGT is stable at 4°C for 5 to 7 days, and for 2 to 3 months when frozen.

D. ALKALINE PHOSPHATASE

Alkaline phosphatase is a general name encompassing a group of phosphohydrolases that demonstrate an alkaline pH optimum (around pH 9.0 to 10.5) for hydrolyzing phosphomonoesters. The enzyme requires Zn^{2+} and Mg^{2+} for activity. The actual *in vivo* substrates of ALP are unknown, but the enzyme appears to be involved in a number of metabolic processes, including lipid transport in the intestine and osteoblastic activity in the synthesis of bone matrix.

ALP is present in most body tissues; highest concentrations are found in organs that possess high absorptive or excretory function, including the liver, bone, small intestine, kidneys, and placenta. These organs contain certain characteristic forms of ALP. Some of the tissue-specific forms of ALP are true isoenzymes, whereas others have identical protein structures but differ in carbohydrate moieties. The various forms of ALP may be distinguished by differences in specific resistance to denaturation by heat and certain chemicals, electrophoretic migration, response to inhibitors, and reactivity with specific antibodies. Certain malignancies may excrete a tumor-derived ALP that is identical (or nearly identical) to the placental isoenzyme; the presence of the carcinoplacental form (sometimes called the *Regan isoenzyme*) has been used for diagnostic and prognostic purposes.

Tissues Containing Significant Amounts of ALP

- Liver
- Bone
- Small intestine
- Kidneys
- Placenta

The ALP normally present in the serum of healthy individuals is of bone and liver origin. Intestinal ALP is present in the serum of about 20% of healthy individuals after meals, primarily those with blood types B and O. Not all elevations in ALP activity are associated with an abnormal condition; increases may also occur as a result of normal physiological processes in healthy individuals. Serum ALP activity is high in childhood due to bone growth (hence, age-specific reference intervals) and during the healing of bone fractures, as well as in the third trimester of pregnancy due to placental release of ALP.

ALP determination is of particular importance in detecting osseous and hepatobiliary disease. ALP facilitates bone formation and is synthesized by osteoblasts. Consequently, ALP activity is elevated in bone diseases with increased osteoblastic activity. The highest serum ALP activities (10 to 25 times the upper reference limit) are found in patients with Paget's disease. Elevations of bone ALP are also found in rickets (two to four times normal), osteomalacia (up to three times normal), and with bone tumors. ALP levels are frequently within the normal reference interval, and hence of little diagnostic value, in patients with osteoporosis.

Serum ALP activity is also elevated in hepatobiliary disease. Typically, more than threefold—even up to ten- or twelvefold—increases in ALP may be observed in extrahepatic obstruction of bile flow, as may occur with cholelithiases (gallstones) or pancreatic tumors. Intrahepatic biliary obstruction more commonly causes smaller rises in ALP activity (i.e., two to three times normal). In liver disorders with hepatocellular injury but without cholestasis, ALP may be slightly to moderately increased or even normal.

The measurement of total ALP is a sensitive test for hepatobiliary disorders and osteoblastic diseases and is considered an excellent screening test; however, total ALP is relatively nonspecific due to the number of tissues that may release the enzyme. There are now specific immunoassays for measuring bone ALP to aid in identifying the enzyme form contributing to an elevated ALP.

The specimen of choice for ALP determination is freshly collected serum, free of hemolysis. Plasma is acceptable when collected in heparin; however, calcium/magnesium-chelating anticoagulants, such as oxalate, EDTA, and citrate, must be avoided. ALP activity exhibits less than 10% change over 2 to 3 days when stored at room temperature (25°C) or refrigerated (0 to 4°C). In specimens frozen at −25°C, ALP activity is stable for up to 1 month.

Reference intervals for serum total ALP are typically stratified by age and sex with respect to the methodology employed. ALP activities in sera from newborns and children range between three- and sevenfold higher than ranges seen in healthy adults due to an increased release of bone ALP during stages of active bone growth. At age 40 and above, ALP levels increase in men and women.

E. LACTATE DEHYDROGENASE

LD is a cytoplasmic enzyme found in nearly all cells. Highest activities are present in tissues that have a high capacity for utilizing glucose. LD catalyzes the interconversion of pyruvate and L-lactate, the final step of the glycolytic pathway:

$$\text{L-lactate} + \text{NAD}^+ \overset{\text{LD}}{\rightleftharpoons} \text{Pyruvate} + \text{NADH} + \text{H}^+$$

Tissues Containing Highest Levels of LD

- Brain
- Erythrocytes
- Kidneys
- Leukocytes
- Liver
- Lungs
- Myocardium
- Spleen

Tissue Specificities of the LD Isoenzymes

- Predominant isoenzymes in myocardium and erythrocytes
  —LD-1 (HHHH)
  —LD-2 (HHHM)
- Predominant isoenzymes in lungs and spleen
  —LD-3 (HHMM)
  —LD-4 (HMMM)
- Predominant isoenzyme in liver and skeletal muscle
  —LD-5 (MMMM)

Highest activities of LD, in descending order, are found in brain, erythrocytes, kidney, leukocytes, liver, lung, myocardium, and spleen. LD is increased in a wide range of abnormalities and is a sensitive, but nonspecific, marker of organ pathology. LD is also increased in a wide range of malignancies; some reports indicate that the magnitude of LD elevation correlates inversely with the prognosis for patient survival.

LD is a tetramer composed of combinations of two subunits designated H (heart) and M (muscle). There are five possible isoenzyme combinations: HHHH (LD-1), HHHM (LD-2), HHMM (LD-3), HMMM (LD-4), and MMMM (LD-5). Each of the five LD isoenzymes possesses distinct physical and biochemical properties (e.g., electrophoretic mobility), and tissues differ in their relative content of the five isoenzymes. LD-1 and LD-2 are the predominant forms in myocardium and erythrocytes; highest activities of LD-3 and LD-4 are found in lungs and spleen; LD-5 predominates in liver and skeletal muscle. Because LD is present in many tissues, analysis of isoenzymes aids in identifying the tissue source of an increased total LD activity. Assays are available for specifically measuring LD-1.

LD activity found in serum of healthy individuals is derived primarily from erythrocytes and platelets and has a half-life of approximately 2 to 3 days. The enzyme has no known function in serum and, after inactivation in the circulation, the metabolized enzyme is eliminated via the biliary tract.

Serum is the preferred specimen for LD analysis; heparinized plasma is also acceptable. Plasma collected in other anticoagulants should be avoided, especially oxalate, which has an inhibitory effect on the enzyme. Because LD is present in erythrocytes in high concentrations, hemolyzed specimens are not acceptable. Serum or plasma must be separated from the red cells as soon as possible to prevent artifactual elevations of LD due to leakage of enzyme from the cellular elements.

Storage of specimens for subsequent measurement of LD activity is a complex issue. Because the isoenzymes differ in stability at different temperatures, there is no absolute optimum storage temperature. Refrigeration or freezing results in loss of activity, predominantly in the LD-4 and LD-5 isoenzyme fractions, and an increase in the ratio of LD-1 to LD-2. The distribution of LD isoenzymes in the specimen will determine the actual decrease in total LD activity upon freezing. To avoid problems associated with sample storage, specimens should be analyzed soon after specimen collection.

Reference intervals for LD are age and gender dependent. Values in infants (up to 1 month of age) may be fivefold higher than the adult upper reference limit. Values gradually decline and reach adult levels by approximately 12 years of age. LD activities in adult males are approximately 10 to 15% higher than activities in females. The LD reference interval for individuals 60 years of age or older is slightly higher than the range for younger adults.

F. CREATINE KINASE

CK catalyzes the reversible conversion of creatine phosphate, a form of stored chemical energy, to adenosine triphosphate (ATP), the body's main source of available chemical energy:

$$\text{Creatine} + \text{ATP} \overset{CK}{\rightleftharpoons} \text{Creatine phosphate} + \text{ADP} + \text{H}^+$$

Magnesium (Mg^{2+}) is required for full enzyme activity.

CK is found in highest concentrations in skeletal muscle, but is also found (in order of decreasing activity) in cardiac muscle, brain tissue, smooth muscle of the colon, small intestine, uterus, prostate, lungs, and kidneys. While measurement of total CK is primarily used as an indicator of myocardial damage and skeletal muscle injury, the enzyme lacks specificity in distinguishing between various pathologic states. Other conditions that may cause a marked elevation in total CK activity include intramuscular injection, pregnancy, strenuous exercise, surgery, trauma, myopathies, hypothyroidism, and certain neoplastic disorders of the prostate, gastrointestinal tract, or bladder.

CK exists as a dimer composed of the two monomer subunits, designated M (muscle) and B (brain). These two subunits combine to form three isoenzymes of CK: CK-BB (CK-1), CK-MB (CK-2), and CK-MM (CK-3). The majority of CK present in normal serum is the CK-MM isoenzyme found in high concentration in skeletal muscle. The CK-MB isoenzyme is found in highest concentration in myocardial tissue and a small percentage of the CK in skeletal muscle is CK-MB. The small amount of CK-MB detected in the serum of healthy patients is derived from the skeletal muscle. The high concentration of CK-MB in cardiac muscle, relative to other tissues, makes CK-MB a useful marker of disorders that involve the myocardium, especially acute myocardial infarction. Chapter 10B discusses the use of CK-MB, as well as other markers, in the evaluation of cardiac disease. CK-BB is found primarily in brain tissue; however, high activities are also found in the prostate, bladder, gastrointestinal tract, and smooth muscle of the stomach. Historically, CK isoenzymes were evaluated by electrophoresis; now there are methods available that specifically measure CK-MB, the most clinically useful CK isoenzyme, by immunoassay.

CK may exist in two different macromolecular forms, identifiable by an alteration in electrophoretic migration. Macro CK-1 is formed by the complexing of CK to an IgG or IgA antibody. In CK electrophoresis, macro CK-1 migrates between the CK-MB and CK-BB bands. Macro CK-2, also called CK-mt, is mitochondrial CK and migrates cathodally to CK-MM. Although the presence of macromolecular forms of CK are thought to be of no clinical significance, they may be the cause of an increased total CK activity in serum due to their prolonged half-life, and result in unnecessary testing.

CK activity may be analyzed in serum, heparinized plasma, CSF, and amniotic fluid. Because CK requires magnesium ion for activity, samples cannot be collected with magnesium-chelating anticoagulants (i.e., EDTA, citrate, or oxalate). Hemolyzed samples should be avoided, although slight hemolysis has little effect on measured CK activity. Adenylate kinase, found in high concentrations in erythrocytes, catalyzes the interconversion of ATP and ADP and may interfere with assays of CK activity if not inhibited. Methods for measuring CK activity that are based on production of ATP include an inhibitor of the enzyme adenylate kinase in the reagent.

CK is relatively unstable at room temperature. Specimens should be stored at 4°C or frozen at −20°C if analysis is delayed. Since sunlight can inactivate CK, specimens should be kept in the dark, including those that are refrigerated or frozen.

Reference intervals for CK are largely dependent on the relative muscle mass of the individual. CK values are typically greater in males than in females. Age- and race-related differences in normal

Tissues Containing Highest Levels of CK

- Skeletal muscle
- Cardiac muscle
- Brain
- Smooth muscle of the colon
- Small intestine
- Uterus
- Prostate
- Lungs
- Kidneys

Conditions Causing a Marked Elevation in Total CK Activity

- Acute myocardial infarction
- Muscular dystrophy
- Intramuscular injection
- Pregnancy
- Strenuous exercise
- Surgery
- Trauma
- Myopathies
- Hypothyroidism
- Neoplastic disorders of the prostate, gastrointestinal tract, or bladder

CK activities have also been noted. Patients confined to long-term bed rest may show markedly decreased CK due to skeletal muscle inactivity. CK values in newborns may be up to 10-fold higher than those seen in normal adults due to birth-related trauma.

G. AMYLASE

Amylase is a digestive enzyme. It hydrolyzes glycosidic bonds in polymers of glucose, such as starch and glycogen, to produce maltose and other polysaccharides and dextrins. Amylase is a metalloenzyme, requiring calcium ion as a cofactor. Salivary (S-type) and pancreatic (P-type) isoenzymes have been identified, each produced by a unique gene locus. Measurement of the pancreatic isoenzyme has been found to be more useful for the diagnosis of pancreatitis as compared with measurement of total amylase only.

Amylase was one of the earliest biochemical markers to be used for the diagnosis of a particular disease. Amylase activity in serum increases within the first 12 to 24 hours following an attack of acute pancreatitis and then declines to normal values over the next 3 to 5 days. Amylase is eliminated from the circulation via renal excretion; patients with impaired renal function may exhibit longer elimination times following acute pancreatitis. In addition, patients with renal insufficiency, but without acute pancreatitis, may exhibit increased amylase activity in serum (see below).

The specificity of amylase for acute pancreatitis is poor when compared to other pancreatic enzymes such as lipase, trypsin, and carboxypeptidase A, which have all been found to provide greater diagnostic utility. However, amylase was the first marker used for this purpose and remains firmly entrenched in the diagnostic protocols for acute pancreatitis. The specificity of amylase is low due to its presence in many tissues. Amylase may be increased in conditions other than pancreatitis, including renal insufficiency, ovarian malignancy, ectopic pregnancy, salpingitis, lung carcinoma, cardiac surgery, mumps, and manipulation of the salivary glands.

Elevated amylase activity may also be due to macroamylasemia. In this condition, amylase is bound to immunoglobulin (typically IgG) in serum. The immunoglobulin–amylase complex is too large to be excreted into the glomerular filtrate, thus it remains in the circulation longer and results in increased serum amylase activities, but without signs or symptoms of acute pancreatitis. Although the condition of macroamylasemia is benign, the resulting rise in serum amylase activity may lead to unnecessary investigation of pancreatic function. Patients with macroamylasemia evaluated using the amylase:creatinine clearance ratio (ACCR; i.e., the ratio of urinary clearance of amylase to urinary clearance of creatinine, expressed as a percentage) will show decreased urine amylase activities with low ACCRs; the ACCR is typically increased in acute pancreatitis. Electrophoresis of serum amylase isoenzymes can also be used to diagnose macroamylasemia by demonstrating a broad smear instead of the normally distinct salivary and pancreatic isoenzyme bands.

Amylase is extremely stable in serum; in specimens free of bacterial contamination, it is stable for up to 1 week at room temperature and for several months at 4°C. Amylase activity requires calcium ion; therefore, plasma collected in EDTA, citrate, or oxalate is not accept-

---

Conditions Causing Increased Levels of Amylase

- Pancreatitis
- Renal insufficiency
- Ovarian malignancy
- Ectopic pregnancy
- Salpingitis
- Lung carcinoma
- Cardiac surgery
- Mumps

able for analysis. Heparinized plasma may be used for some assays; however, some investigators have found significantly higher amylase results in heparinized plasma when compared to serum specimens. Urine specimens for amylase analysis should not be acidified.

Reference intervals for amylase can vary significantly among the different commercially available assays. Activities in newborns are approximately 20% of those found in adults; however, adult values are attained by approximately age 4 years. No gender- or race-related differences in amylase activity have been noted.

H. LIPASE

Lipase is also a digestive enzyme secreted by the pancreas. Lipase catalyzes the hydrolysis of ester bonds in triglycerides to produce fatty acids and 2-acylglycerol. In the digestive process:

1. Bile acids form micelles that emulsify the substrate triglycerides.
2. Colipase, a required coenzyme, attaches to the micelle.
3. Lipase binds to the bile acid–colipase complex and hydrolyzes the substrate at the interface between water and the emulsified substrate.

Lipase, along with amylase, is utilized in the diagnosis of pancreatitis. In the past, amylase was the test of choice for the diagnosis of pancreatic disease because, until recently, the precision and accuracy of lipase assays were poor. Measurements of lipase activity have been improved by (1) incorporating colipase in the assay to ensure a sufficient supply of coenzyme and (2) developing defined substrates for the lipase reaction. Lipase is now considered to be more sensitive and specific than amylase as a marker of pancreatitis.

In patients with acute pancreatitis, lipase increases sooner and usually remains elevated for a longer period of time compared to amylase. Although lipase may offer greater diagnostic accuracy for the diagnosis of acute pancreatitis, concurrent measurement of these two enzymes has been used to help distinguish among the different causes of acute pancreatitis. For example, some studies have suggested that patients with acute pancreatitis due to ethanol abuse tend to have a higher ratio of lipase to amylase as compared with patients with acute pancreatitis due to other causes.

Pancreatic lipase has been shown to have at least three *isoforms* (forms of the same protein that differ somewhat, possibly due to post-translational modifications). In the future, lipase isoforms may provide more accurate diagnostic information than the measurement of total enzyme activity.

Lipase in the peripheral circulation is removed via glomerular filtration. However, unlike amylase, lipase is completely reabsorbed by the proximal tubule and not normally excreted in the urine. Although patients with renal tubular disorders may excrete lipase in the urine, measurement of urine lipase is not appropriate or useful.

Renal insufficiency also leads to increased lipase activity in the serum (generally at a level < 10 times the upper reference limit) that remains consistently increased. In patients with acute pancreatitis, lipase (and amylase) values > 10 times the upper limit of the reference interval have been characterized as being diagnostic of the illness.

Serum is the preferred specimen for measurement of lipase activity. Lipase activity is very stable: 1 week at room temperature, 3 weeks when refrigerated, and 5 months when frozen; repeated freezing and thawing should be avoided.

Reference intervals for lipase can vary greatly between the different assays used to measure the enzyme; patients with acute pancreatitis can show discrepancies as large as 13-fold greater than the upper reference limit between the various assays. Lipase in healthy, younger individuals tends to be somewhat lower than lipase activity in healthy, older individuals.

I. THE CHOLINESTERASES

There are two cholinesterases of clinical interest: (1) acetylcholinesterase (AChE), also referred to as true cholinesterase or RBC cholinesterase, and (2) serum cholinesterase (SChE), also referred to as pseudocholinesterase or acylcholine acylhydrolase. AChE catalyzes the hydrolysis of acetylcholine released from nerve endings and is found in the nervous system, erythrocytes, lung, and spleen. SChE also hydrolyzes choline esters; however, its actual biological role and true substrate have not yet been determined. SChE is found in the serum, heart, liver, pancreas, and white matter of the brain. AChE and SChE differ somewhat in their substrate specificities. SChE may play a role in protecting AChE from inhibition by hydrolyzing butyrylcholine, a product of fatty acid metabolism; butyrylcholine appears to be a preferred substrate for SChE.

Of the two enzymes, the clinical laboratory is involved in the measurement of SChE activity: (1) as a marker for poisoning by organic phosphorus insecticides, (2) for detection of SChE variants that are associated with susceptibility to certain muscle relaxing drugs, and (3) as an indicator of synthetic capacity of the liver.

There are multiple variants of the gene for SChE, including:

1. The normal gene
2. An atypical gene that produces a weakly active SChE that is resistant to inhibition by dibucaine
3. A variant gene that produces a weakly active SChE that is resistant to inhibition by fluoride
4. A "silent" gene associated with an inactive SChE (or complete lack of the enzyme)

Patients with weakly active variants or low SChE activity may develop apnea requiring mechanical ventilation when muscle relaxants such as succinyldicholine are used during surgery. These patients are unable to metabolize the drug at a normal rate. Individuals with phenotypes associated with susceptibility to succinylcholine and related drugs can be identified by measuring SChE activity, the percent inhibition of SChE by dibucaine (i.e., the *dibucaine number*), and the percent inhibition of SChE by fluoride (i.e., the *fluoride number*). Patients likely to experience complications with succinylcholine administration typically have low SChE activity and a low dibucaine number and/or fluoride number. Family members of susceptible individuals are screened to identify others with variant SChE phenotypes.

Some organic phosphorus compounds bind to the active site of SChE causing irreversible inhibition of activity. Workers in agriculture and chemical manufacturing exposed (acutely or chronically) to or-

---

Tissues Containing the Cholinesterases

AChE:
- Nervous system
- Erythrocytes
- Lungs
- Spleen

SChE:
- Serum
- Heart
- Liver
- Pancreas
- White matter of the brain

ganic phosphorus compounds may absorb sufficient amounts to inhibit AChE and SChE leading to neuromuscular disturbances, progressing from blurred vision, possibly to convulsions, respiratory paralysis, and death. Serum SChE can be measured in suspected cases of insecticide poisoning and the results used to assess the extent of exposure and to guide treatment.

SChE is synthesized by hepatic cells. SChE activity is decreased when hepatic synthetic capacity is diminished (e.g., in acute hepatitis, advanced cirrhosis, and carcinoma with liver metastases).

The desired specimen for analysis of SChE is serum or heparinized plasma; SChE activity is inhibited in plasma collected in oxalate, fluoride, or citrate. SChE activity is stable for several weeks in refrigerated samples. The enzyme is stable for up to 3 years in frozen serum, but only up to 1 month in frozen plasma. SChE can be analyzed for variants using dried blood specimens. Samples to assess SChE in patients experiencing complications after administration of muscle relaxants should not be drawn until paralysis has passed.

Reference intervals for SChE are dependent on both age and gender. Infants up to 6 months of age have serum activities that are 40 to 50% of adult values. Females have been found to have SChE activities that are approximately 75% of values seen in males, and pregnant women demonstrate lower activities than nonpregnant females.

IV. TEST PROCEDURES

In the clinical laboratory today, enzymes of clinical interest are measured by activity assays or by mass assays. Assessment of functional enzyme activity, the most commonly used method, is addressed thoroughly in this chapter. **Mass assays** measure protein mass or actual enzyme concentration and have been successfully applied to the measurement of certain isoenzymes such as CK-MB, LD-1, and the bone fraction of ALP. In most institutions, mass determinations have replaced electrophoretic techniques for detecting and measuring these isoenzymes. Immunoassays are used to measure enzyme concentration. A more detailed discussion of immunoassay techniques may be found in Chapter 12.

A. *DETERMINATION OF ENZYME ACTIVITY IN BLOOD AND TISSUE*

The quantity of enzyme that is present in a blood or tissue specimen can be determined from its catalytic effect on an appropriate substrate. In order to design assays to measure enzyme activity, the following criteria must be established:

1. Requirements for certain cofactors
2. Optimal concentration of substrate and cofactor needed
3. pH and temperature required for optimum activity
4. A mechanism for measuring the appearance of the product or disappearance of the substrate resulting from enzyme activity

Ideally, activity is measured at the optimal temperature and pH for the enzyme of interest. Assays are designed so that the reaction follows zero-order kinetics; the substrate must be present in high enough concentration to saturate all enzyme active sites so that only the amount of enzyme determines the rate of reaction. Measurement of the amount of product formed is the usual mechanism for quantitating enzyme activity since the substrate concentration is kept at very high levels to maintain zero-order kinetics. However, some enzymes, such as AST and ALT, are analyzed by monitoring substrate depletion. The product formed by the enzyme-catalyzed reaction may be (1) measured directly (e.g., by spectrophotometry), or (2) coupled to a secondary reaction, catalyzed by another enzyme included in the reagent system, to produce a second product that can be more easily measured. The following is a reaction scheme that utilizes a secondary, or *indicator*, reaction to quantitate AST activity:

1. L-Aspartate + α-Ketoglutarate $\rightleftharpoons$ Oxaloacetate + L-Glutamate

2. Oxaloacetate + NADH $\rightleftharpoons$ Malate + NAD + H$^+$

The first reaction is catalyzed by AST, with sufficient L-aspartate and α-ketoglutarate available to saturate the enzyme. The second, *coupled reaction* is catalyzed by the enzyme malate dehydrogenase, which is present in amounts sufficient to immediately remove oxaloacetate produced in the first reaction. When monitored by spectrophotometry, the disappearance of NADH (the reduced form of nicotinamide-adenine dinucleotide) causes a corresponding decrease in absorbance at 340 nm. Measurement of production or depletion of NADH by monitoring absorbance at 340 nm is the indicator system used in a number of clinical assays.

If the reaction rate of an enzyme is carefully monitored (e.g., by absorbance), it can be noted that there are three basic phases in the reaction between enzyme and substrate. Following the initial mixing of specimen and reagent, there is a **lag phase** in which little change in absorbance occurs. Once thermal and kinetic equilibrium is established, the reaction rate enters a **linear phase** in which zero-order kinetics prevail and the absorbance change becomes constant over time. Eventually, the substrate in the reaction mixture is depleted so that zero-order kinetics no longer hold, and no further change in absorbance occurs. The time it takes to reach **substrate depletion** is dependent on the activity of the enzyme in the specimen. Figure 4–5 shows the reaction kinetics for three different samples containing different amounts of enzyme; each shows differences in the lag phase, the slope of the linear phase, and the time it takes to reach substrate depletion. If the concentration of enzyme present in different specimens is to be measured accurately under identical assay conditions, the change in absorbance must be measured over a time interval that includes a portion of the linear phase for all three specimens.

Methods for enzyme activity measurement are based on the rate at which enzyme in the specimen catalyzes the reaction over a given time interval; these are known as **kinetic methods.** Most automated analyzers monitor absorbance continuously or take multiple readings during the course of the assay to confirm that the assay rate is constant and substrate supply has not been consumed; these are called **continuous-monitoring methods. Fixed-time methods,** rarely used today, determine enzyme activity from a single absorbance reading taken at a predetermined time after the sample and reagent are mixed to commence the reaction. Ideally, in fixed-time assays, there should be no lag phase and the reaction rate should be constant throughout the entire reaction period; in practice, these conditions may be difficult to meet. Very high enzyme activities may cause substrate depletion prior to the time that the absorbance reading is taken, resulting in an underestimation of en-

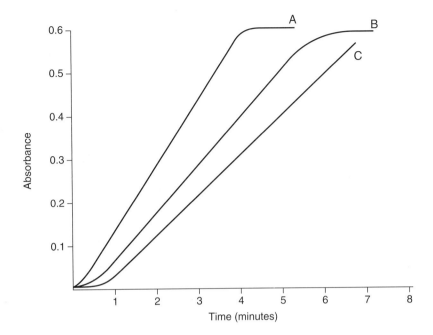

Figure 4–5. Reaction sequence showing absorbance change over time for three different specimens, each containing a different amount of enzyme activity. The time interval, which includes a portion of the linear phase for all three specimens, is between minutes 1 and 3. Note that readings taken before 1 minute would include the lag phase of sample C, while readings taken at 4 minutes would include the substrate depletion phase of sample A.

zyme activity; usually sample dilution with an appropriate matrix and reassay will resolve this problem.

Various means have been employed for describing the amount of enzyme that is present in a specimen. Early enzymologists defined activity units specific for each enzyme assay based on the ability of a specific enzyme to catalyze a reaction using a specific substrate under specific conditions of pH, temperature, substrate concentration, and so on. Use of different substrates and reaction conditions, along with the discovery of new enzymes, led to an increasing number of reporting schemes. In order to standardize the reporting of enzyme activity, regardless of substrate or reaction conditions used, the International Unit was adopted. The **International Unit** (U or IU) is defined as the amount of enzyme that catalyzes the transformation of 1.0 μmol (10^{-6} moles) of substrate per minute of reaction at 25°C using optimal measuring conditions. While the use of the IU has standardized the terminology used to report enzyme activities, it has not standardized reference intervals for any particular enzyme. The use of different substrates and reaction conditions results in differences between IU values that can be obtained for a single specimen that is analyzed using different assays. Therefore, enzyme activities are often evaluated in relative terms, for example, as multiples of the upper reference limit. Table 4–3 lists reference ranges for the enzymes covered in this chapter when measured by commonly used methods.

A second term used to define enzyme activity is **specific activity,** which is the amount of enzyme per milligram of protein. The specific activity is a measure of enzyme purity since this value increases as the enzyme is further purified from its original source. Another term

Table 4–3. Reference Intervals for Serum Enzyme Activities

ENZYME	REFERENCE INTERVAL (U/L, UNLESS OTHERWISE INDICATED)	
ALP (at 37°C)	Adult, 20–50 years	
	Male	53–128
	Female	42–98
	Adult, ≥ 60 years	
	Male	56–119
	Female	53–141
ALT	Newborn/infant	7–40
	Adult	10–35
AMY	Newborn	5–65
	Adult	25–125
	> 70 years	20–160
AST	Newborn	40–120
	Infant	25–95
	Adult	10–40
	With P-5′-P	15–50
CK, total	Newborn	10–200
	Adult	
	Male	15–105
	Female	10–80
	> 70 years	
	Male	22–90
	Female	16–80
GGT	Male	≤ 50
	Female	≤ 30
LD	0–4 days	290–775
	4–10 days	545–2,000
	10 days–24 months	180–430
	24 months–12 years	110–295
	12–60 years	100–190
	> 60 years	110–210
LPS	Adult	30–190
	> 60 years	18–180
SChE		7–19 U/mL[a]
dibucaine inhibition		79–84%
fluoride inhibition		58–64%

[a] *DuPont aca method*

Source: Painter PC, Cope JY, Smith JL. Reference information for the clinical laboratory. In Burtis CA, Ashwood ER (eds.). *Tietz textbook of clinical chemistry,* 3rd ed. Philadelphia: WB Saunders, 1999.

sometimes used to describe the catalytic activity of an enzyme is the **molecular activity** or turnover number. The molecular activity refers to the number of substrate molecules that can be transformed by a single-enzyme molecule, or by a single active site for enzymes having more than one active site, under optimal measuring conditions. Molecular activities vary widely among enzymes; the highest known molecular activity is attributed to catalase at 36 million substrate molecules transformed per minute by a single catalase molecule.

B. ELECTROPHORESIS AND ISOELECTRIC FOCUSING

For many years, electrophoresis was a popular method for the analysis of serum isoenzymes, particularly those of CK, LD, and ALP. Today, however, mass assays have replaced this technique reducing the time and technical expertise that were required for electrophoretic methods. A more current application of electrophoresis is the analysis of enzyme isoforms, particularly those of CK. (Refer to Chapter 12 for a review of the clinical use of isoforms as cardiac markers.)

Isoelectric focusing (IEF) is a specialized technique that may be used to analyze isoenzymes. ALP isoenzymes have been separated by IEF; however, due to the difficulty in interpretation of the numerous bands produced, the practical application of this technique is limited.

Chapter 16 more fully addresses these two separation techniques.

C. ENZYMES AS REAGENTS

Enzymes are important in the clinical laboratory, not only as diagnostic markers, but also as reagents with high specificity. Enzymatic assays for measuring the amount of a substance may be end-point (or equilibrium) or kinetic. In **end-point assays,** sufficient enzyme is ◄ used to allow relatively rapid conversion of substrate (the analyte) to product; readings to measure product are taken after the reaction has stopped (i.e., reached equilibrium). In **kinetic assays,** conditions are ◄ designed to support first-order kinetics so that rate of the reaction is directly proportional to the initial concentration of the substrate (the analyte); thus, the analyte concentration must be small compared to the K_m of the enzyme. Readings are taken at two or more points to measure the rate of change in the product. Both types of assays may be designed to measure the product of the initial enzymatic reaction or a product of a secondary or indicator reaction. Some routine analytes are commonly measured by enzymatic assays, including glucose, urea, and cholesterol. Enzymes are also used as reagents in assays of enzyme activity to catalyze secondary or indicator reactions.

SUGGESTED READINGS

Buehler M, Muench R, Schmid M, et al. Cholestasis in alcoholic acute pancreatitis. Diagnostic value of the transaminase ratio for differentiation between extra- and intra-hepatic cholestasis. *Scand J Gastroenterol* 20:851–57, 1985.

Herrera JL. Abnormal liver enzyme levels. *Postgrad Med* 93:113–16, 1993.

Moss DW, Henderson AR. Enzymes. In Burtis CA, Ashwood ER (eds.). *Tietz textbook of clinical chemistry,* 3rd ed. Philadelphia: WB Saunders, 1999; 735–896.

Niblock AE, Leung FY, Henderson AR. Serum aspartate aminotransferase storage and the effect of pyridoxal phosphate. *J Lab Clin Med* 108:461–69, 1986.

Painter PC, Cope JY, Smith JL. Reference information for the clinical laboratory. In Burtis CA, Ashwood ER (eds.). *Tietz textbook of clinical chemistry,* 3rd ed. Philadelphia: WB Saunders, 1999; 1788–846.

Pappas NJ Jr. Theoretical aspects of enzymes in diagnosis. Why do serum enzymes change in hepatic, myocardial, and other diseases? *Clin Lab Med* 9:595–626, 1989.

Rej R. Aminotransferases in disease. *Clin Lab Med* 9:667–87, 1989.

Items for Further Consideration

1. Your laboratory wishes to switch from an electrophoretic procedure for CK-MB to a mass assay. How do these two techniques correlate? Is one method more sensitive or specific than the other?

2. Should enzymes such as LD and AST be dropped from general biochemical panels due to their lack of tissue specificity? In what cases are they most useful?

3. Your laboratory wishes to convert all testing to plasma sample analysis in order to reduce turnaround time. How would this affect analysis of the enzymes commonly measured?

Metabolic Analytes

Linda C. Gregory

Key Terms

AMMONIA

AZOTEMIA

BILIRUBIN

δ–BILIRUBIN

BILIRUBIN DIGLUCURONIDES

BILIRUBIN MONOGLUCURONIDES

CALCIUM

CONJUGATED BILIRUBIN

CREATINE

CREATININE

CREATININE CLEARANCE

GLOMERULAR FILTRATION RATE

GOUT

JAUNDICE

KERNICTERUS

LACTATE

LACTIC ACIDOSIS

MAGNESIUM

PARATHYROID HORMONE

PHOSPHATE

UDP-GLUCURONYL TRANSFERASE

UNCONJUGATED BILIRUBIN

UREA

UREMIA

URIC ACID

UROBILINOGENS

UROBILINS

VITAMIN D

Chapter Objectives

Upon completion of this chapter, the reader will be able to:

1. Recount the stages in bilirubin metabolism from formation to excretion and predict the effects of defects at each stage on plasma bilirubin fractions.
2. Suggest a consistent diagnosis (i.e., disorder in bilirubin metabolism) given a set of laboratory results, including bilirubin (total and direct), urine bilirubin, urine urobilinogen, and fecal pigments.
3. Describe the forms of bilirubin found in plasma and discuss the measurement of each by various bilirubin analysis methodologies.

4. Compare and contrast the usefulness of serum creatinine and serum urea nitrogen measurements as renal function tests based on: (a) origin in the body, (b) renal handling, and (c) factors that affect concentrations in blood. Explain the usefulness of the blood urea nitrogen (BUN):creatinine ratio.
5. Calculate creatinine clearance, given appropriate data, and interpret the result.
6. Explain the metabolic origin of serum uric acid, situations causing hyperuricemia, and clinical consequences of hyperuricemia.
7. Devise a protocol for the collection and handling of a blood specimen for lactate analysis with minimal preanalytical error.
8. Discuss the processes involved in lactate production under normal circumstances and situations leading to elevated lactate concentrations and lactic acidosis.
9. Characterize the distribution of calcium, phosphate, and magnesium among compartments in the body and among the various forms of each found in the plasma.
10. Discuss the mechanisms involved in calcium, phosphate, and magnesium homeostasis; predict the effects of disturbances in these homeostatic mechanisms.
11. Predict the effect of abnormal serum albumin concentrations on total serum calcium and total serum magnesium results, and abnormal blood pH on free calcium and total calcium results.
12. Describe the commonly used methods of analysis and the expected results for total bilirubin, direct bilirubin, creatinine, creatinine clearance, urea, uric acid, lactate, calcium, phosphate, and magnesium.

Analytes Included

BILIRUBIN:

   TOTAL BILIRUBIN

   DIRECT BILIRUBIN

   INDIRECT BILIRUBIN

   δ-BILIRUBIN (OR Δ-BILIRUBIN)

CREATININE

CREATININE CLEARANCE

UREA (BLOOD UREA NITROGEN)

BUN:CREATININE RATIO

URIC ACID

LACTATE

CALCIUM:

   TOTAL CALCIUM

   FREE (IONIZED) CALCIUM

PHOSPHATE

MAGNESIUM:

   TOTAL MAGNESIUM

   FREE (IONIZED) MAGNESIUM

Pretest

1. Bilirubin is a product of the catabolism of
   ___(1)___ and is transported to the liver
   ___(2)___.

	(1)	(2)
a.	bile acids	complexed to prealbumin
b.	heme	as δ-bilirubin
c.	hemoglobin	as conjugated bilirubin
d.	heme	complexed to albumin

2. Choose the diagnosis most consistent with the following laboratory findings.

   Serum: Total bilirubin   elevated
          Direct bilirubin   0 mg/dL
          Indirect bilirubin   elevated
   Urine: Bilirubin   negative
          Urobilinogen   decreased

   a. uridine diphosphate (UDP)-glucuronyl transferase deficiency
   b. posthepatic bile obstruction
   c. Dubin–Johnson syndrome
   d. intravascular hemolysis

3. A patient with an elevated total serum bilirubin and a positive test for urine bilirubin would also be expected to have
   a. an increased urine urobilinogen
   b. an elevated direct serum bilirubin
   c. a hemolytic disorder
   d. a deficiency in UDP-glucuronyl transferase activity

4. Unconjugated bilirubin
   a. corresponds to direct bilirubin
   b. requires an accelerator to react with diazotized sulfanilic acid
   c. corresponds to bilirubin monoglucuronide
   d. is covalently bound to albumin

5. Comparing the production of creatinine and urea in the body
   a. creatinine production is more affected by high-protein diets
   b. production of urea is a function of muscle mass
   c. production of creatinine is dependent on ammonia formation in the liver
   d. urea production is increased subsequent to gastrointestinal bleeding

6. An advantage to evaluating glomerular filtration rate using the creatinine clearance is that creatinine is
   (1) produced at a relatively constant rate
   (2) filtered but not reabsorbed or secreted by the kidney

   (3) an easily measured endogenous substance
   (4) not affected by urinary tract obstruction
   a. all of the above
   b. (1), (2), and (3)
   c. (1) and (3)
   d. (3) and (4)

7. Given the following data, calculate the creatinine clearance:
   Serum creatinine = 2.4 mg/dL
   Urine creatinine = 120 mg/dL
   Urine volume = 1.1 mL/min
   Patient's estimated body surface area = 1.56 m²
   a. 50 mL/min/1.73 m²
   b. 291 mL/min/1.73 m²
   c. 61 mL/min/1.73 m²
   d. 106 mL/min/1.73 m²

8. The Jaffé reaction is the basis of photometric methods commonly used to measure
   a. uric acid
   b. urea nitrogen
   c. creatinine
   d. lactate

9. Calculate the concentration of urea corresponding to a urea nitrogen result of 27 mg/dL.
   a. 58 mmol urea/L
   b. 13 mg urea/dL
   c. 9.6 mmol urea/L
   d. 76 mg urea/dL

10. In indirect (enzymatic) methods for urea measurement
    a. specimens collected in sodium heparin will inhibit urease in the assay
    b. urea reacts with diacetyl to form a colored product
    c. there should be a mechanism to correct for endogenous ammonia in the sample
    d. all of the above

11. Hyperuricemia may be caused by decreased
    a. renal excretion of urate
    b. production of urate
    c. renal tubular reabsorption of urate
    d. turnover of nucleic acids

12. Recurrent arthritis caused by urate deposition in joint tissue and synovial fluid is called
    a. hypouricemia
    b. gout
    c. rheumatoid arthritis
    d. uremia

13. Lactate
    a. is a waste product of glycolysis and is excreted by the kidneys
    b. can be used as a substrate for gluconeogenesis
    c. accumulation in blood causes a decreased anion gap
    d. production does not occur in healthy individuals

14. When obtaining blood specimens for lactate analysis
    a. use a tourniquet to speed collection
    b. samples are stable at room temperature for 2 hours
    c. the patient should be at complete rest for 2 hours prior to collection
    d. use collection tubes containing ethylenediaminetetraacetic acid (EDTA) to inhibit glycolysis in the sample

15. Measurement of blood lactate may be indicated
    a. in a patient with a metabolic acidosis without ketosis and an elevated anion gap
    b. when impaired renal function is suspected
    c. in cases of unexplained coma
    d. both a and c

16. Hypoparathyroidism may be expected to cause
    a. decreased renal excretion of calcium
    b. decreased blood calcium
    c. increased parathyroid hormone secretion
    d. increased absorption of intestinal calcium

17. An appropriate anticoagulant for collection of a plasma sample for measurement of total calcium is
    a. heparin
    b. EDTA
    c. citrate
    d. oxalate

18. A decreased serum albumin is likely to cause a(n)
    a. increase in serum total calcium
    b. decrease in serum total calcium
    c. increase in blood free (ionized) calcium
    d. decrease in blood free (ionized) calcium

19. A massive transfusion with citrated blood is likely to cause a(n)
    a. increase in serum total calcium
    b. decrease in serum total calcium
    c. increase in blood free (ionized) calcium
    d. decrease in blood free (ionized) calcium

20. The most common cause of hyperphosphatemia is
    a. rhabdomyolysis
    b. vitamin D deficiency
    c. renal failure
    d. acromegaly

21. In the homeostasis of calcium and phosphate
    a. the actions of parathyroid hormone (PTH) tend to increase serum calcium and decrease serum phosphate
    b. the actions of PTH tend to increase serum calcium and phosphate
    c. the active metabolite of vitamin D [$1,25(OH)_2D_3$] decreases intestinal absorption of calcium and phosphate
    d. PTH secretion is stimulated by low serum phosphate

22. Many routine methods for measuring serum phosphate are based on the
    a. formation of diacetyl phosphate complexes
    b. formation of phosphomolybdate complexes
    c. reaction of phosphate with sulfanilic acid
    d. reaction of phosphate with 8-hydroxyquinoline

23. A hemolyzed sample is likely to have a falsely
    a. decreased serum phosphate
    b. increased ionized calcium
    c. increased serum total calcium
    d. increased serum total magnesium

24. Hypomagnesemia is often accompanied by a(n)
    a. decreased serum calcium
    b. increased serum calcium
    c. increased serum potassium
    d. none of the above

25. A reagent that may be added to prevent calcium interference in assays for serum total magnesium is
    a. 8-hydroxyquinoline
    b. calmagite
    c. ethylene glycol-O,O'-bis(2-aminoethyl)-$N,N,$ N',N'-tetraacetic acid (EGTA)
    d. picrate

I. INTRODUCTION

Several of the important routine analytes have been grouped in this chapter under the heading of metabolic analytes: bilirubin is derived from heme catabolism; creatinine, urea, and uric acid are nonprotein nitrogenous metabolites; lactate is the end product of anaerobic glycolysis; and calcium, phosphate, and magnesium, in addition to being involved in bone metabolism, are critical to the functioning of all cells.

II. BILIRUBIN

A. *BIOCHEMISTRY AND PHYSIOLOGY*

Bilirubin is a yellow-orange pigmented waste product from the catabolism of the heme moiety of hemoproteins—hemoglobin, myoglobin, cytochromes, and peroxidases. On degradation of hemoproteins, the amino acids of the apoprotein and the heme iron are reutilized, while the protoporphyrin ring remaining from the heme group is further metabolized. Heme oxygenase catalyzes the opening of the porphyrin ring to produce a green pigment called *biliverdin,* and biliverdin reductase reduces biliverdin to bilirubin.

Most bilirubin (about 85%) is generated in the reticuloendothelial system from hemoglobin reclaimed from senescent erythrocytes. Bilirubin, hydrophobic due to the conformation determined by internal hydrogen bonding, is transported to the liver by forming a complex with albumin in the blood. Upon entering the hepatocytes, bilirubin is released from albumin but complexes with cytosolic binding proteins *ligandin* and *Z protein.* Through this stage, bilirubin is in the *unconjugated* form. Bilirubin becomes *conjugated* with the addition of one or two glucuronic acid groups through the action of the microsomal enzyme **UDP-glucuronyl transferase;** about 90% of conjugated bilirubin is in the diglucuronide form and about 10% in the monoglucuronide form. **Conjugated bilirubin,** now water

---

Key Steps in Bilirubin Metabolism

- Hemoproteins → apoprotein + iron + protoporphyrin
- Protoporphyrin → biliverdin → bilirubin
- (In blood) Bilirubin binds to albumin; transported to liver
- (In hepatocyte) Bilirubin conjugated → bilirubin monoglucuronides + bilirubin diglucuronides
- Conjugated bilirubin enters bile; proceeds to intestine
- (In intestine) Conjugated bilirubin → unconjugated bilirubin → urobilinogen
- (In intestine)
  — (Most) urobilinogen → urobilin → excreted in feces
  — (About 20%) urobilinogen reabsorbed and transported to liver; most excreted in bile; some enters blood and is excreted in urine

soluble due to the added polar glucuronyl groups, is actively transported into the bile canaliculi and proceeds to the intestine via the biliary tree.

▶ In the intestine, bilirubin is enzymatically unconjugated and then reduced to water-soluble, colorless compounds called **urobilinogens** by intestinal bacteria; the three main urobilinogens are *mesobilinogen, stercobilinogen,* and *urobilinogen.* Some (about 20%) of the urobilinogen in the intestine is reabsorbed into the enterohepatic circulation and returned to the liver; most of the reabsorbed urobilinogen is excreted into the bile and the remainder enters the general circulation and is excreted in the urine. Most of the urobilinogen in the intestine is not reabsorbed but oxidizes to form urobilins in ▶ the lower intestine; **urobilins** are the brown pigments eliminated in the feces.

Four bilirubin fractions have been identified in blood:

▶ 1. **Unconjugated bilirubin** (mostly noncovalently complexed to albumin)
▶ 2. **Bilirubin monoglucuronides** (conjugated)
▶ 3. **Bilirubin diglucuronides** (conjugated)
4. A fraction covalently bound to protein (mostly albumin)
▶ called **δ-bilirubin** (or Δ-bilirubin)

B. MEDICAL RELEVANCE

Serum bilirubin is measured to aid in the detection, diagnosis, and monitoring of liver or hepatobiliary disease. *Hyperbilirubinemia* is the accumulation of bilirubin (unconjugated, conjugated, or both) in ▶ the blood. **Jaundice,** the yellow discoloration of skin, mucous membranes, and plasma, occurs when bilirubin concentrations reach about 3 mg/dL (or 50 μmol/L). Abnormalities in bilirubin metabolism, whether inherited or acquired, may be grouped as *prehepatic, hepatic,* or *posthepatic.* Table 5–1 lists some disorders associated with hyperbilirubinemia and the laboratory findings typical of these conditions. Prehepatic hyperbilirubinemia occurs when production of unconjugated bilirubin is increased beyond the conjugating and excretory capacity of the liver; typical laboratory results include elevations in serum total bilirubin, serum unconjugated bilirubin, and urine urobilinogen. Defects or disturbances in the conjugation or excretion steps in the hepatocyte lead to hepatic hyperbilirubinemia; in hepatic jaundice laboratory results will depend on the specific disorder. Posthepatic jaundice is caused by extrahepatic obstruction of bile flow (i.e., extrahepatic cholestasis); cholestatic jaundice may be expected to cause elevations in total, unconjugated, and conjugated bilirubin, as well as serum enzymes that indicate biliary tract disease (e.g., γ-glutamyl transferase and alkaline phosphatase). Bilirubin is not found in urine under normal circumstances, but will be excreted in urine when conjugated bilirubin is elevated since it is a soluble form that can be cleared by the kidney.

The clinical consequences of hyperbilirubinemia are usually dependent on other effects of the particular disease process causing the disturbance in bilirubin excretion. In cases of uncorrected, unconjugated hyperbilirubinemia in neonates, unconjugated bilirubin may accumulate in nerve tissue and the brain, a condition called ▶ **kernicterus,** causing cell destruction and encephalopathy.

Table 5-1. Conditions Associated with Hyperbilirubinemia: Expected Changes in Bilirubin and Bilirubin Metabolites

CONDITION	TOTAL BILIRUBIN (SERUM)	DIRECT BILIRUBIN (CONJUGATED)	INDIRECT BILIRUBIN (UNCONJUGATED)	URINE BILIRUBIN	URINE UROBILINOGEN	FECAL PIGMENTS (UROBILINS)	OTHER TYPICAL RESULTS
Prehepatic							
Increased production of unconjugated bilirubin							
Hemolysis (genetic or acquired)	↑	Normal	↑	—	↑[b]	Normal	↓ Hemoglobin, ↑ Reticulocytes, ↑ LD-1
Ineffective erythropoiesis							
Neonatal (increased turnover of red cells)							
Decreased delivery of unconjugated bilirubin to liver							
Congestive heart failure	↑	Normal	↑	—	Normal	Normal	
Hepatic							
Decreased uptake of unconjugated bilirubin							
Gilbert's syndrome (genetic uptake defect; benign; may also have ↓ UDP-glucuronyl transferase activity)	↑	Normal	↑	—	N - →	Normal	
Decreased conjugation of bilirubin							
Neonatal physiological jaundice (immature liver; ↓ conjugating enzyme)	↑	Normal	↑	—	Normal	Normal	
Hepatocellular dysfunction[a] (hepatitis, cirrhosis, etc.)	↑	↑	↑	+	N - →	N - →	↑ ALT, AST, ↑ LD
Crigler–Najjar syndrome Type I (no UDP-glucuronyl transferase activity)	↑	→	↑↑	—	→	→	
Crigler–Najjar syndrome Type II (↓ UDP-glucuronyl transferase activity)	↑	→	↑	—	→	→	
Decreased secretion of conjugated bilirubin							
Hepatocellular dysfunction[a] (hepatitis, cirrhosis, etc.)	↑	↑	↑	+	N - →	N - →	↑ ALT, AST, ↑ LD
Intrahepatic cholestasis [drug reaction (e.g., anabolic steroids), cholangitis, infiltration of foreign cells (e.g., leukemia), primary biliary cirrhosis]	↑	↑	N - ↑	+	→	→	↑ GGT, ↑ ALP
Dubin–Johnson syndrome and Rotor's syndrome (genetic; excretion defect; benign)	↑	↑	N - ↑	+	N - →	N - →	
Posthepatic							
Extrahepatic obstruction (cholestasis)							
Gall stones, inflammation of gall bladder, stones in common bile duct, tumors in biliary tract or in head of pancreas, narrowing of bile ducts (strictures)	↑	↑	N - ↑	+	→	→	↑ GGT, ↑ ALP

[a] Hepatocellular dysfunction may include decreased conjugation and/or decreased bilirubin excretion.

[b] With increased bilirubin production, more urobilinogen is produced overall causing increased excretion of urobilinogen in urine.

LD, lactate dehydrogenase; ALT, alanine aminotransferase; GGT, gamma-glutamyltransferase; ALP, alkaline phosphatase; AST, aspartate aminotransferase.

C. TEST PROCEDURES

1. Preanalytical Factors. Bilirubin may be measured in serum, plasma, urine, and spinal fluid. Specimens must be protected from light and heat to prevent photo-oxidation of bilirubin; analyze promptly. Hemolysis causes falsely low results when using diazo methods of analysis.

2. Chemical (Diazo) Methods. Bilirubin reacts with diazotized sulfanilic acid (a reagent made from sulfanilic acid and sodium nitrite) to form azobilirubin pigments that are measured spectrophotometrically. Bilirubin that reacts "directly" in this assay is called *direct* bilirubin; direct bilirubin mainly reflects the conjugated (i.e., soluble) forms plus the δ-bilirubin fraction. Unconjugated bilirubin reacts in the presence of an accelerator, a reagent that presumably makes unconjugated bilirubin more soluble; all bilirubin fractions react in the presence of accelerator giving a measure of total bilirubin. The difference between total bilirubin (measured with accelerator) and direct bilirubin (measured without accelerator) is termed *indirect* bilirubin, which corresponds to the unconjugated bilirubin fraction.

The Jendrassik and Grof modification is the most popular of the diazo methods; the accelerator used is caffeine–benzoate reagent, and an alkaline tartrate reagent is added after the reaction to form blue-green azobilirubin that is measured at 600 nm.

3. Spectral Shift Methods. A spectral shift method is available using dry slide technology and reflectance photometry; bilirubin binds to cationic polymers, causing a shift in spectra where the change in reflectance at measured wavelengths is a function of bilirubin concentration; fractionation of conjugated and unconjugated bilirubin is possible because they demonstrate different spectral shifts. δ-Bilirubin can be determined by the difference between total bilirubin and the sum of conjugated plus unconjugated bilirubin.

D. TEST RESULTS AND INTERPRETATION

Bilirubin reference ranges for adults are (Painter, Cope, & Smith, 1999):

Total bilirubin	0.3–1.2 mg/dL	5–21 μmol/L
Direct bilirubin	0–0.2 mg/dL	0–3.4 μmol/L

Bilirubin monitoring may be of critical importance in newborns; while bilirubin concentration is higher in newborns, intervention is necessary to prevent kernicterus in cases of severe elevations. Total bilirubin reference ranges for full-term neonates are (Painter, Cope, & Smith, 1999):

Cord blood	< 2.0 mg/dL	< 34 μmol/L
0–1 day	1.4–8.7 mg/dL	24–149 μmol/L
1–2 days	3.4–11.5 mg/dL	58–197 μmol/L
3–5 days	1.5–12.0 mg/dL	26–205 μmol/L

While not routinely measured, δ-bilirubin is usually found to be elevated in cases of obstructive jaundice (i.e., when conjugated bilirubin is elevated). On resolution of the obstruction, the conju-

---

Bilirubin Fractions Measured by Diazo Methods

Total bilirubin corresponds to conjugated, unconjugated, and δ-bilirubin

Direct bilirubin corresponds to conjugated and δ-bilirubin

Indirect bilirubin corresponds to unconjugated bilirubin

gated form of bilirubin is cleared rapidly by the kidney, while the δ-bilirubin (and thus the total bilirubin) may remain elevated for days since the δ-bilirubin is cleared at the same rate as albumin (half-life = 19 days).

III. CREATININE

A. BIOCHEMISTRY AND PHYSIOLOGY

Creatinine is a nitrogenous waste product derived from **creatine.** ◀ Creatine is synthesized in the liver and kidney from amino acids and then transported via the blood to muscle and brain tissues. The enzyme *creatine kinase* catalyzes the phosphorylation of creatine to creatine phosphate, a high-energy compound used in muscle metabolism.

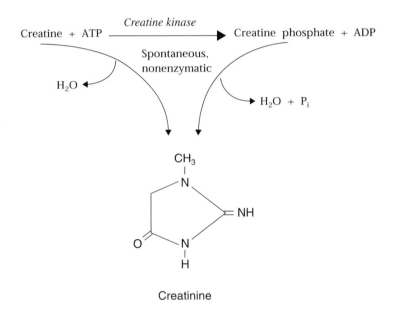

Creatinine

Daily, about 1 to 2% of muscle creatine and creatine phosphate is spontaneously converted to creatinine through a nonenzymatic, irreversible dehydration reaction. Creatinine production within an individual is proportional to muscle mass and relatively constant over time. Once creatinine is produced, it is released into the blood and then excreted by the kidney, where (1) creatinine is freely filtered through the glomerulus; (2) tubular reabsorption of creatinine is negligible, and it is not metabolized in the tubules; and (3) a small amount of creatinine enters the urine through tubular secretion.

B. MEDICAL RELEVANCE

Creatinine, produced endogenously at a constant rate and excreted entirely by the kidney, will accumulate in the blood with any type of renal dysfunction where the **glomerular filtration rate** (GFR) is de- ◀ creased. Serum creatinine and **creatinine clearance** are useful renal ◀ function tests.

1. Serum Creatinine. Elevations in serum creatinine are quite specific for kidney disease, being little affected by diet, hydration, and protein metabolism. However, concentrations of serum creatinine double only after renal function has been reduced by at least 50% of normal and, therefore, may not be sensitive to early, mild, or even moderate impairment unless baseline concentrations for a particular individual are available for comparison.

2. Creatinine Clearance. GFR, a sensitive and specific indicator of renal dysfunction, can be measured by determining (for a timed interval) the volume of plasma cleared of a substance that is freely filtered by the glomerulus but neither metabolized, secreted, nor reabsorbed by the tubules. *Inulin clearance* is a possible "gold standard" for measuring GFR; however, inulin is an exogenous substance that must be administered and is rarely measured in the clinical laboratory. Clearance of certain radiolabeled pharmaceuticals (e.g., ^{125}I-iothalamate) also accurately quantifies GFR.

Creatinine clearance is a far more convenient measure that approximates the GFR. Creatinine clearance, in terms of volume of plasma cleared of creatinine per minute per standard body surface area, can be calculated as:

$$\text{Creatinine clearance (mL/min)} = [(U \times V)/P] \times (1.73/A)$$

where

> U = [creatinine] of timed urine sample
> P = serum or plasma [creatinine] in same units as urine concentration
> V = volume of urine flow in milliliters per minute (from timed urine sample)
> A = body surface area in square meters
> (estimated from nomograms using patient's height and weight)
> 1.73 = defined standard body surface area in square meters

The creatinine clearance is normalized to a defined standard body surface area—a rough approximation to account for the relationship between creatinine excretion and muscle mass.

Creatinine clearance is, however, slightly higher than inulin clearance (and GFR) due to some tubular secretion of creatinine. The discrepancy between GFR and creatinine clearance becomes more significant as renal failure progresses and the relative contribution to creatinine excretion by tubular secretion increases. The accuracy of creatinine clearance values are further compromised by the difficulty in obtaining a properly collected, timed urine sample, and the calculation of the clearance value using three measurements, each with its own associated error.

There is some controversy concerning the most appropriate utilization of these two tests; generally, once reliable measures of creatinine clearance (and associated serum creatinine concentration) have been established for a patient, serum creatinine may be the most useful and convenient test for continued monitoring.

Table 5–2 lists conditions that affect serum creatinine concentrations.

C. TEST PROCEDURES

1. Preanalytical Factors. Creatinine may be analyzed in serum, plasma, or diluted urine; usually a 1:100 dilution of urine will yield a

Table 5–2. Conditions that Affect Serum Creatinine Concentration

Conditions Associated with Elevated Creatinine

Prerenal:

Impaired renal perfusion (i.e., decreased blood flow to the kidneys) caused by congestive heart failure, dehydration, shock, hemorrhage, etc.

High meat intake (mild, transient effect)

Strenuous exercise (mild, transient effect)

Renal (any condition causing a decrease in GFR):

Loss of functional nephrons with acute or chronic kidney diseases such as glomerulonephritis, pyelonephritis, tubular necrosis, nephrotoxicity, etc.

Kidney damage secondary to prerenal and postrenal conditions

Postrenal:

Urinary tract obstruction due to tumors of the genitourinary tract, prostatic hypertrophy, renal stones, strictures, or infections

Nonrenal:

Drugs that compete with tubular secretion (e.g., salicylates, cimetidine)

Analytical positive interference (Jaffé method)

Conditions Associated with Decreased Creatinine

Pregnancy (retention of fluid and increased GFR)

Decreased muscle mass (wasting diseases, starvation, steroid therapy)

---

creatinine concentration within method calibration. Fasting blood samples may be preferred due to possible temporary increases in creatinine following high meat intake. Accuracy of creatinine clearance values depends on the complete and accurate collection of urine for a timed period, often 12 or 24 hours.

2. Jaffé Reaction Methods. Many routine methods are modifications of the classic *Jaffé reaction* in which creatinine reacts with the picrate ion in alkaline solution to form a red-orange adduct with an absorbance measured between 510 and 520 nm. The reaction is non-specific; other picrate-reacting substances (positive interferents) are especially common in blood, including glucose, uric acid, ascorbate, pyruvate, acetone, acetoacetate, protein, guanidine, and cephalosporin antibiotics. Bilirubin and hemoglobin degradation products cause negative interference. Method modifications are aimed at improving specificity. Manual modifications treat a serum protein-free filtrate with a creatinine-adsorbent (e.g., Fuller's earth) and then proceed with the Jaffé reaction by adding alkaline picrate reagent to the pelleted adsorbent.

Various kinetic modifications of the Jaffé reaction are more suited to automation where absorbance can be measured at precisely timed points:

1. The first readings are made after a delay (about 20 seconds) to allow depletion of the fast-reacting interferents.
2. Sequential readings are made in a time window (at about 20 to 80 seconds) in which primarily creatinine reacts.
3. Readings are stopped before slower-reacting substances interfere.

Creatinine concentration is a function of the reaction rate determined for the middle phase.

3. Enzymatic Methods. Enzymatic methods available for measuring creatinine are more specific and suited for automation than the Jaffé methods, but are more expensive. In creatinine iminohydrolase methods, ammonium ions produced in stoichiometric amounts are measurable by ion-selective electrode or the N-methylhydantoin produced is measured by coupled indicator reactions. In creatinine amidohydrolase (also known as creatininase) methods, creatinine is converted to creatine measurable by coupled indicator reactions.

D. Test Results and Interpretation

Since creatinine production is related to body muscle mass, serum creatinine concentration tends to be higher in males than females and higher in adults than children. Creatinine clearance rates remain fairly constant until age 40, when rates begin to decrease with increasing age. See Tables 5–3 and 5–4 for serum creatinine and creatinine clearance reference range values.

IV. UREA (BLOOD UREA NITROGEN)

A. Biochemistry and Physiology

▶ **Ammonia** is generated from the deamination of amino acids during protein catabolism. In the liver, enzymes of the urea cycle convert
▶ ammonia to **urea,** thereby preventing ammonia toxicity to the central nervous system. Urea is the major nitrogen-containing end product of protein catabolism; its production depends on several variables, including dietary protein intake, hepatic function, and catabolic state. After entering the general circulation, urea is largely excreted through the kidneys; urea is freely filtered through the glomerulus, but a significant portion (40 to 50%) passively diffuses back through the renal tubules and reenters the blood; in conditions with decreased urine flow (e.g., hypovolemia or congestive heart failure), a larger proportion (up to 70%) of urea is passively reabsorbed.

B. Medical Relevance

While urea concentration increases in the blood with the loss of renal function, it is neither a specific nor sensitive indicator of renal

Table 5–3. Serum Creatinine Reference Intervals

AGE	CREATININE mg/dL	CREATININE µmol/L
Newborn (1–4 days)	0.3–1.0	27–88
Infant	0.2–0.4	18–35
Child	0.3–0.7	27–62
Adolescent	0.5–1.0	44–88
Adult		
Female	0.6–1.1	53–97
Male	0.7–1.3	62–115

Source: Painter PC, Cope JY, Smith JL. Reference information for the clinical laboratory. In Burtic CA, Ashwood ER (eds.). *Tietz textbook of clinical chemistry,* 3rd ed. Philadelphia: WB Saunders, 1999.

Table 5–4. Creatinine Clearance Values: Reference Intervals and Expected Results with Renal Dysfunction

POPULATION	CREATININE CLEARANCE mL/min/1.73m^2
Adults	
20–29 years of age[a]:	
Female	72–110
Male	94–140
Impairment level:	
Borderline	62.5–80
Slight	52–63
Mild	42–52
Moderate	28–42
Marked	< 28

[a] Creatinine clearance tends to decrease by ~6.5 mL/min/m^2 for each additional 10 years.

Source: Painter PC, Cope JY, Smith JL. Reference information for the clinical laboratory. In Burtic CA, Ashwood ER (eds.). *Tietz textbook of clinical chemistry,* 3rd ed. Philadelphia: WB Saunders, 1999.

impairment. Serum urea is also dependent on nonrenal factors such as protein intake, hydration state, and renal perfusion (Table 5–5) and reflects the balance of urea production and clearance. **Uremia** refers to increased urea in the blood; **azotemia** is the term used to describe elevations in nitrogen-containing end products, including urea, creatinine, and uric acid. Although serum urea is commonly

Table 5–5. Conditions that Affect Serum Urea Concentrations

Conditions Associated with Elevated Urea (i.e., Uremia)
Increased production of urea:
  Prerenal causes[a]:
    Impaired renal perfusion (i.e. decreased blood flow to the kidneys) caused by congestive heart failure, dehydration, shock, hemorrhage, etc.
    High-protein diet
    Gastrointestinal bleeding (absorption of blood proteins)
    Increased protein catabolism as occurs with corticosteroid treatment or increased thyroid hormones
    Excess destruction of cellular proteins as occurs with starvation, muscle wasting diseases, febrile illnesses
Decreased excretion of urea:
  Renal causes:
    Any cause of loss of renal function with a decrease in GFR
    Kidney damage secondary to prerenal and postrenal conditions
  Postrenal causes[a]:
    Mechanical obstruction of urine flow due to renal stones, tumors of genitourinary tract, strictures, or infections

Conditions Associated with Decreased Urea
Advanced liver disease (decreased urea synthesis)
Low-protein diet
Hemodialysis
Normal pregnancy (increased glomerular filtration rate)
Water retention (e.g., syndrome of inappropriate antidiuretic hormone)
Increased protein anabolism (e.g., androgen or growth hormone treatments)

[a] Urea is elevated to a greater degree than creatinine in these conditions.

used as a screening test for renal disease, the ratio of urea to creatinine is more useful.

1. BUN:Creatinine Ratio. In the United States, the concentration of serum urea is expressed as serum urea nitrogen, and the test is (misleadingly) often called *blood urea nitrogen* or BUN, a historical holdover from early urea measurement methods using whole blood. The ratio of BUN to creatinine may provide a rough indication of the possible cause of renal dysfunction (Table 5–6).

2. Urea Clearance. Urea clearance underestimates the GFR and has many limitations as a renal function test compared to creatinine clearance; it is rarely performed.

C. TEST PROCEDURES

1. Preanalytical Factors. Urea may be analyzed in serum, plasma, or urine. When the enzyme *urease* is used in the measurement method, plasma samples must *not* be collected with fluoride (inhibits urease) or ammonium heparin.

2. Enzymatic (Indirect) Methods. Indirect urea determinations first hydrolyze urea using urease.

$$H_2N - \overset{\overset{\displaystyle O}{\displaystyle \|}}{C} - NH_2 \xrightarrow[H_2O]{Urease} 2\,NH_4^+ + CO_3^{2-}$$

Urea

The concentration of produced ammonium ions is subsequently measured by one of several available methods:

1. Ammonium ion is most commonly measured using a coupled enzyme reaction catalyzed by glutamate dehydrogenase (GluDH), where the decrease in absorbance at 340 nm can be monitored in kinetic or end-point mode.

$$NH_4^+ + \alpha\text{-ketoglutarate} + NADH + H^+ \xrightarrow{GluDH} Glutamate + NAD^+ + H_2O$$

Table 5–6. Interpretation of Urea Nitrogen:Creatinine Ratios

UREA NITROGEN:CREATININE RATIO (IN SAME UNITS OF MEASURE)	INTERPRETATION
Above 20:1	*With normal creatinine* indicates: Prerenal causes of uremia (Table 5–5) *With elevated creatinine* indicates: Postrenal causes of uremia (Table 5–5)
10:1 to 20:1	Normal (usually 12:1 to 16:1)
Below 10:1	Acute tubular necrosis Advanced liver disease (decreased urea synthesis) Low protein intake Hemodialysis (urea dialyzed more readily than creatinine)

2. Electrochemical methods measure the rate of change in conductivity with generation of ammonium ion or measure ammonia potentiometrically with an ammonium ion–selective electrode.

3. Methods using spectral changes of pH indicator dyes have also been developed.

For the indirect methods, the contribution by endogenous ammonia may be accounted for by blanking, removal, or use of kinetic analysis.

3. Direct Methods. A chemical method using the Fearon reaction measures urea directly: diacetyl monoxime reagent is hydrolyzed to diacetyl and hydroxylamine; diacetyl and urea condense in acidic solution to form diazine, a yellow product that absorbs at 540 nm.

4. Unit Conversions. It is sometimes necessary to convert results expressed in terms of urea nitrogen to urea (and vice versa) or results expressed in units of milligrams of urea nitrogen per deciliter to millimoles of urea per liter (and vice versa). The conversion factors are:

$$mg \; urea \; N/dL \times 2.14 = mg \; urea/dL$$
$$mg \; urea/dL \times 0.467 = mg \; urea \; N/dL$$
$$mg \; urea \; N/dL \times 0.357 = mmol \; urea/L$$
$$mmol \; urea/L \times 2.80 = mg \; urea \; N/dL$$

Factors are derived from the molecular weight for urea (60 daltons) and the nitrogen content of urea (28 g/mol; i.e., N atomic weight = 14 and there are two N atoms per urea molecule).

D. TEST RESULTS AND INTERPRETATION

The reference range for urea nitrogen in adults is from 6 to 20 mg/dL (2.1 to 7.1 mmol urea/L) (Painter, Cope, & Smith, 1999). Values may range slightly lower for children and slightly higher for older adults (> 60 years). There may be some variation between methods.

V. URIC ACID

A. BIOCHEMISTRY AND PHYSIOLOGY

Uric acid is a nitrogenous end product of the catabolism of purines ◄ from dietary and endogenous sources.

Uric Acid

Free purine bases (i.e., adenine, guanine, hypoxanthine), whether preexisting or generated from degradation of nucleic acids, purine nucleotides, and purine nucleosides, may be used to resynthesize purine nucleotides via salvage pathways or may be further degraded to xanthine and finally to uric acid. At a pH > 5.6, monosodium urate is the predominant form of uric acid. Urate is excreted via the kidneys (about two-thirds of excreted urate) and the gastrointestinal tract (about one-third of excreted urate). The renal excretion process is complex:

1. Plasma urate is completely filtered at the glomerulus.
2. Nearly all of the filtered urate is reabsorbed in the proximal convoluted tubule.
3. About half of the reabsorbed urate is secreted into the lumen in the distal segment of the proximal tubule.
4. A portion of the urate is again reabsorbed, this time in the distal tubule.

The net renal excretion of urate is about 10% of the filtered load.

B. MEDICAL RELEVANCE

Hyperuricemia is the result of overproduction or underexcretion (or both) of uric acid (see Table 5–7). Causes of elevated serum uric acid may be primary (i.e., due to inborn error) or secondary (i.e., due to another condition affecting uric acid production or excretion).

At serum uric acid concentrations only somewhat above the upper limit of the reference range, urate solutions become supersaturated, increasing the likelihood for crystallization of the urate salt. ▶ **Gout** is the clinical manifestation of urate crystal deposits in connective tissue, most notably in the tissue surrounding joints as well as in the synovial fluid, causing an inflammatory response and subsequent recurrent arthritis. Clinical gout seldom occurs due to secondary hyperuricemia, but asymptomatic hyperuricemia should be monitored due to the increased risk for renal damage induced by urate crystallization.

Hypouricemia is not clinically important in and of itself, but may be consistent with or accompany other conditions and pathologies (Table 5–7).

C. TEST PROCEDURES

1. Preanalytical Factors. Uric acid may be measured in serum or heparinized plasma, or in urine that has been collected in sodium hydroxide. Plasma collected in EDTA or sodium fluoride is not acceptable for uricase methods.

2. Phosphotungstic Acid Methods. In chemical methods, the oxidation of urate to allantoin and carbon dioxide in alkaline solution is accompanied by the reduction of phosphotungstic acid to tungsten blue, which can be measured spectrophotometrically (at about 700 nm). Deproteination of the sample is required prior to reaction. This method is somewhat nonspecific; reducing substances such as glucose, ascorbate, glutathione, acetylsalicylic acid, and caffeine will also reduce phosphotungstic acid, causing a positive interference.

Table 5–7. Causes of Abnormal Serum Uric Acid Concentration

Causes of Hyperuricemia
Overproduction of urate:
  PRIMARY (inborn error causing increased purine synthesis)
    Idiopathic (about 10% of cases of primary gout)
    Inherited metabolic error
      Hypoxanthine guanine phosphoribosyl transferase deficiency:
        Lesch–Nyhan syndrome (inoperative salvage pathways)
        Partial deficiency (decreased activity of salvage pathways)
      Phosphoribosylpyrophosphate synthetase variant (increased purine
        nucleotide synthesis)
      Glucose-6-phosphatase (G-6-P) deficiency or von Gierke's disease
        (causes ↑ production and ↓ secretion)
  SECONDARY
    Excessive purine intake (high-protein diet)
    Excess alcohol ingestion
    Increased turnover of nucleic acids: myeloproliferative and lymphopro-
      liferative disorders, cytotoxic drug therapy (i.e., chemotherapy), se-
      vere acute illness, starvation (increased cell destruction), hypercata-
      bolic states, strenuous exercise, psoriasis (increased rate of skin cell
      turnover)
Decreased renal excretion of urate:
  PRIMARY (inborn error causing impaired excretion of urate)
    Idiopathic (90% of cases of primary gout)
  SECONDARY (conditions with ↓ GFR, ↑ renal reabsorption, or ↓ renal
    secretion)
    Chronic renal failure (decreased GFR)
    Drugs
      Diuretic therapy (↓ secretion of urate)
      Salicylate in low doses (↓ secretion of urate)
    Lead poisoning
    Metabolic acidosis or organic aciduria (interferes with tubular secretion)
      Lactic acidosis (tissue anoxia, starvation, G-6-P deficiency, alcohol
        ingestion)
      Ketosis (acetoacetate)

Causes of Hypouricemia
Increased urine urate excretion:
  Overmedication with uricosuric drugs (e.g., probenecid used to ↓ serum
    uric acid)
  Salicylate at high doses (↓ tubular reabsorption)
  Renal tubular disorders (e.g., Fanconi syndrome)
Decreased uric acid production:
  Overmedication with allopurinol (used to ↓ uric acid synthesis)
  Decreased dietary intake of proteins and purines
  Severe liver disease
  Xanthine oxidase deficiency (xanthinuria)

3. Uricase Methods. Specificity is improved in enzymatic methods using *uricase* to catalyze the oxidation of urate. The initial reaction common to all uricase methods,

$$\text{Urate} + O_2 \xrightarrow{\text{Uricase}} \text{Allantoin} + CO_2 + H_2O_2$$

is followed by one of the appropriate measurement methods: (1) measuring the decrease in ultraviolet absorbance in the 282- to 292-nm
region as urate is consumed; (2) measuring the hydrogen peroxide
from the initial reaction with coupled indicator reaction systems us-

ing a peroxidase or catalase reaction coupled with a chromogen producing reaction (most common and most easily automated); or (3) polarographic measurement of the oxygen consumed in the uricase reaction (rarely used).

D. TEST RESULTS AND INTERPRETATION

Uric acid reference ranges are method dependent. The ranges for adults when using a uricase method are (Painter, Cope, & Smith, 1999):

Males	3.5–7.2 mg/dL	0.21–0.42 mmol/L
Females	2.6–6.0 mg/dL	0.15–0.35 mmol/L

There is considerable interindividual variation in serum uric acid:

1. Levels are generally higher in males than females.
2. Levels tend to increase with age.
3. Levels tend to be higher in obese individuals, in those with high-protein diets or with high consumption of alcohol, and in the affluent population (may be diet related).
4. Levels depend somewhat on ethnic origin (i.e., genetic factors).

The risk of gout and urate-induced kidney damage increases with hyperuricemia; however, there are no distinct serum uric acid cutoff values for diagnosis of gout or to predict clinical complications from hyperuricemia. Uric acid concentrations must be used in the context of the clinical presentation. Demonstration of urate crystals in synovial fluid aspirated from an affected joint is definitive for the diagnosis of gout.

Measurement of uric acid excretion in a timed urine may help identify the cause of hyperuricemia; urine uric acid excretion will be elevated with overproduction of uric acid but decreased in cases of underexcretion.

VI. LACTATE (LACTIC ACID)

A. BIOCHEMISTRY AND PHYSIOLOGY

▶ **Lactate** is the end product of anaerobic glycolysis. The glycolytic pathway metabolizes glucose to pyruvate. When oxygen is available for generating NAD^+ in the mitochondria, glycolysis proceeds aerobically and pyruvate is metabolized to carbon dioxide and water through the tricarboxylic acid (TCA or Krebs) cycle. When the oxygen supply is insufficient for NAD^+ production, pyruvate is converted to lactate (i.e., anaerobic glycolysis). During physical activity, blood lactate is generated mostly by skeletal muscle; while at complete rest, most blood lactate is derived from erythrocytes.

Upon sudden demand for energy for muscle contraction, muscle glycogen is catabolized; the amount of lactate produced depends on the extent to which the rate of glycolysis exceeds the rate of the TCA cycle. The lactate, transported in the blood, is cleared mainly by the liver and kidneys, where it is used as a substrate for gluconeogenesis (Cori cycle). Plasma lactate concentrations are low at rest; lactate rises sharply with exercise and normally approaches resting levels soon after exercise.

B. MEDICAL RELEVANCE

Uncorrected accumulation of lactic acid in tissues and/or blood leads to an acidic shift in pH with consumption of bicarbonate causing a metabolic acidosis, specifically a **lactic acidosis.** Lactic acidosis causes ◄ weakness and stupor and, if untreated, may progress to stages that are irreversible resulting in coma and death.

Type A lactic acidosis (the more common form) is caused by tissue hypoxia (an insufficient supply of oxygen); *type B* lactic acidosis is a result of metabolic disease or drugs and toxins (Table 5–8).

Situations in which lactate measurement may be indicated include (1) an unexplained anion gap > 20 when pH is < 7.25 and Pco_2 is not elevated, (2) cases of unexplained coma, and (3) management of hypoxia.

Simultaneous measurement of lactate and pyruvate allows calculation of the *lactate/pyruvate ratio,* sometimes helpful in determining the cause of lactic acidosis. The lactate/pyruvate ratio is typically from about 6:1 to 10:1; the ratio is markedly increased in type A lactic acidosis, increased in type B lactic acidosis, and normal in cases in which lactate concentration is elevated without significant acidosis (e.g., strenuous exercise, hyperventilation, and glycogen storage diseases such as von Gierke's disease).

Indications for Lactate Measurement

- An unexplained anion gap > 20 when pH is < 7.25 and Pco_2 is not elevated
- Unexplained coma
- Hypoxia

C. TEST PROCEDURES

1. Preanalytical Factors. Lactate may be measured in arterial or venous whole blood or plasma. Specimens must be collected and handled using procedures that minimally affect lactate concentration:

1. Patients should be fasting and at complete rest for 2 hours prior to sample collection.
2. Venous samples should be collected without a tourniquet and without fist clenching (i.e., minimal muscle activity).

Table 5–8. Causes of Lactic Acidosis

Causes of Type A Lactic Acidosis
Associated with any condition causing tissue hypoxia
    Shock or prolonged hypotension
    Congestive heart failure (acute)
    Hemorrhage (hypovolemia)
    Severe anemia
    Myocardial infarction
    Severe pulmonary insufficiency

Causes of Type B Lactic Acidosis
Associated with a number of conditions; the causative mechanism is unknown but may be related to insufficient availability of adenosine triphosphate and oxidized nicotinamide-adenine dinucleotide (NAD^+)
    Severe infections (e.g., pyelonephritis)
    Cirrhosis or fulminant hepatic failure
    Diabetes mellitus (diabetic coma without ketoacidosis)
    Thiamine deficiency
    Ethanol
    Methanol
    Salicylates
    Pregnancy (third trimester, toxemia)

3. Venous samples for plasma measurement should be collected in tubes containing sodium fluoride and potassium oxalate (gray top), placed in ice water for transport, and plasma separated from cells within 15 minutes of collection.
4. Arterial or venous samples for whole blood measurement should be collected in heparinized syringes, immediately added to a chilled, premeasured volume of a suitable protein precipitant (e.g., metaphosphoric acid), and centrifuged to obtain a clear supernatant for analysis.

Lactate may also be measured in urine or cerebrospinal fluid.

2. Enzymatic Methods. Lactate dehydrogenase catalyzes the oxidation of lactate to pyruvate with the concomitant reduction of NAD^+ to $NADH + H^+$; the increase in absorbance at 340 nm is a function of lactate concentration. An alternative enzymatic method uses lactate oxidase to convert lactate to pyruvate and hydrogen peroxide; the hydrogen peroxide generated is measured by a coupled peroxidase reaction that produces a colored complex.

D. TEST RESULTS AND INTERPRETATION

The reference ranges for lactate in whole blood from individuals after complete rest are (Painter, Cope, & Smith, 1999):

Venous	0.9–1.7 mmol/L	8.1–15.3 mg/dL
Arterial	< 1.3 mmol/L	< 11.3 mg/dL

High lactate concentrations without acidosis are expected after initiation of exercise. A blood pH < 7.25 with a lactate concentration > 4 mmol/L indicates significant lactic acidosis.

VII. CALCIUM

A. BIOCHEMISTRY AND PHYSIOLOGY

▶ **Calcium** is a divalent cation (atomic mass = 40 g/mole). The body of an average adult human contains about 1 kilogram of calcium. Ninety-nine percent of body calcium is in bone; hydrated calcium phosphate complexes form hydroxyapatite crystals, the mineral component of the skeletal structure. Most (99%) of soluble calcium is in plasma; the calcium inside cells is mostly sequestered in membranous organelles. Cytosolic calcium is kept very low by active transport out of the cell or into organelles; an increase in cytosolic calcium is a signal (i.e., a second messenger) in many cell processes (e.g., changes in cell shape and motility, metabolic changes, secretory functions, and cell division). Calcium also serves as a cofactor for certain enzymes.

Only a small amount (about 0.1 to 0.3%) of total body calcium is in the extracellular fluid compartment, which includes the plasma. There are three forms of calcium in plasma: about 40 to 45% is bound to plasma proteins (about three-fourths of this is bound to albumin and one-fourth to globulins); about 10% is complexed with anions (e.g., phosphate, citrate, bicarbonate, lactate, sulfate); and about 45 to 50% is in the free, ionized form. Plasma *free* (or *ionized*) *calcium* is

---

Actions of Parathyroid Hormone in Calcium Homeostasis

- Stimulates net resorption of bone to mobilize calcium and phosphate
- Decreases renal excretion of calcium
- Increases renal excretion of phosphate
- Stimulates renal production of $1,25(OH)_2D_3$

Net effect = increases plasma calcium and decreases plasma phosphate

---

the physiologically active form; it is crucial in neuromuscular transmission and cardiac function and is a cofactor in blood coagulation.

It is the concentration of free calcium that is maintained within narrow limits by the body's homeostatic mechanisms. Regulation is maintained largely by the actions of two hormones, **parathyroid hormone** (PTH) and 1,25-dihydroxyvitamin D_3 [$1,25(OH)_2D_3$], on three organ systems—bone, kidney, and intestine. PTH, a polypeptide hormone secreted by the parathyroid glands: (1) stimulates net resorption of bone to mobilize calcium and phosphate, (2) decreases renal excretion of calcium, (3) increases renal excretion of phosphate, and (4) stimulates renal production of $1,25(OH)_2D_3$. Overall, PTH tends to increase plasma calcium and decrease plasma phosphate. $1,25(OH)_2D_3$, the active form of **vitamin D:** (1) increases intestinal absorption of calcium and phosphate, (2) indirectly promotes normal bone mineralization by keeping calcium and phosphate available, and (3) stimulates osteoclastic bone resorption when present at high concentration. *Hypocalcemia* (decreased free calcium) stimulates the secretion of PTH which subsequently increases the production of $1,25(OH)_2D_3$; the net effect of the hormones raises the calcium concentration toward normal. PTH secretion is inhibited by *hypercalcemia* (increased free calcium); in the absence of PTH, calcium excretion by the kidney is increased, production of $1,25(OH)_2D_3$ is decreased (decreasing intestinal absorption of calcium), and bone resorption is decreased—the combined actions tending to decrease calcium concentration toward normal.

Actions of $1,25 (OH)_2D_3$ in Calcium Homeostasis

- Increases intestinal absorption of calcium and phosphate
- Indirectly promotes normal bone mineralization
- Stimulates osteoclastic bone resorption

Net effect = increases plasma calcium

B. MEDICAL RELEVANCE

Tight homeostatic regulation of free, ionized calcium is critical. A low free calcium concentration leads to tetany (irregular spasms of the muscles) and cardiac arrhythmias, possibly progressing to seizures and death; chronic hypocalcemia may cause psychiatric symptoms, bone pain and fragility, skin coarseness, dental abnormalities, and cataract formation. A high free calcium leads to gastrointestinal symptoms and psychiatric disturbances, with the potential of progressing to coma and death; chronic hypercalcemia causes renal tubular damage and increases risk of renal stone formation.

Hypercalcemia results from a defect in the calcium homeostatic mechanism itself or when the mechanism cannot handle the load of calcium in the body. A defect in PTH effectiveness (e.g., by de-

Table 5–9. Disorders in Calcium Homeostasis

Causes of Hypercalcemia

Increased mobilization of bone calcium:

Primary hyperparathyroidism: Parathyroid hormone (PTH) is inappropriately elevated for plasma calcium (adenoma, hyperplasia, or carcinoma of the parathyroid gland or multiple endocrine neoplasia syndrome)

Malignancy-associated humoral factors (e.g., PTH-like protein produced by some neoplasms)

Malignancy-associated bone mobilization (osteolytic activity of tumor cells)

Endocrine disorders: cause release of bone calcium by poorly understood mechanisms (hyperthyroidism, adrenal failure, acromegaly, pheochromocytoma)

Prolonged immobilization: associated with decreased bone formation with normal bone resorption (especially in Paget's disease and osteoporosis)

Lithium therapy: stimulates PTH secretion

Increased absorption of intestinal calcium:

Vitamin D toxicity: excessive absorption of calcium

Granulomatous diseases: macrophages in granulomas sometimes able to produce $1,25(OH)_2D_3$ (sarcoidosis, tuberculosis, histoplasmosis, etc.)

Milk–alkali syndrome: chronic ingestion of milk and antacids to control peptic ulcers

Hypophosphatemia

Decreased renal excretion of calcium:

Familial hypocalciuric hypercalcemia: genetic; low calcium excretion rate even with hypercalcemia

Thiazide diuretics

Causes of Hypocalcemia

Decreased PTH secretion:

Hypoparathyroidism (congenital, idiopathic, or autoimmune): inappropriately low PTH secretion

Destruction of parathyroid glands (surgery, radiation, foreign infiltrations)

Hypomagnesemia (Mg required for PTH secretion and action)

Resistance to PTH action:

Vitamin D deficiency (nutritional deficiency, malabsorption, or inadequate exposure to light)

Anticonvulsant therapies (e.g., phenytoin) interferes with 25-hydroxylation of vitamin D_3

Renal disease (acute or chronic) leads to hyperphosphatemia causing decreased production of $1,25(OH)_2D_3$

Vitamin D–resistant rickets: genetic defect in hydroxylation of $25(OH)D_3$

Pseudohypoparathyroidism: hereditary defect in PTH receptor mechanism

Increased sequestration of calcium (free calcium depleted in the blood):

Hyperphosphatemia (from exogenous administration or renal failure) can cause calcium phosphate deposition

Chelation: massive transfusion with citrated blood or use of EDTA as chelating agent

Acute pancreatitis or rhabdomyolysis: cause precipitation of calcium soaps in abdominal cavity

Osteoblastic metastases: may cause bone deposition

Hungry bone syndrome: rapid calcium deposition in bone upon treatment or healing of metabolic bone disease

---

creased secretion or decreased target response) or loss of free calcium from the blood lead to hypocalcemia. Table 5–9 lists causes of abnormalities in blood calcium.

Disturbances in calcium (or phosphate) metabolism may cause defects in bone mineralization leading to metabolic bone diseases: rickets and osteomalacia occur with a deficiency of vitamin D or disturbance in vitamin D metabolism; osteoporosis occurs with a long-term negative calcium balance; demineralization with hyperparathyroidism may occur due to increases in PTH levels; renal osteodystrophy may occur if calcium and phosphate changes that accompany chronic renal failure are untreated.

Other laboratory measurements helpful in differential diagnosis of calcium disorders include serum phosphate, PTH, urinary calcium, and alkaline phosphatase.

C. TEST PROCEDURES

1. Preanalytical Factors. Total calcium may be measured in serum or heparinized plasma promptly separated from red cells; specimens collected with calcium-chelating anticoagulants (i.e., EDTA, oxalate, citrate) are not acceptable. To avoid hemoconcentration in the sample, patients should sit for 5 to 10 minutes prior to collection, and tourniquets should be used sparingly during blood collection. Free calcium is usually measured in heparinized whole blood; samples should be collected anaerobically, transported on ice, and analyzed promptly. Urine samples should be collected in hydrochloric acid to prevent precipitation of calcium salts.

Calcium in blood samples is measured routinely as *total calcium* (free + protein-bound + complexed), but measurement of free (or ionized) calcium, the physiologically active form, is increasingly available and more clinically useful.

2. Spectrophotometric Methods for Total Calcium. Metal-complexing reagents selectively bind calcium, causing a color change that is measured spectrophotometrically. Commonly used calcium-binding reagents include *o*-cresolphthalein complexone and arsenazo III. Addition of 8-hydroxyquinoline helps reduce interference by magnesium. These methods are easily automated and the most popular for measuring total calcium.

3. Fluorescent Titrimetric Methods for Total Calcium. Fluorescent calcium complexes are formed with an appropriate fluorescent indicator (e.g., calcein, a fluorescein derivative); the fluorescent complexes are titrated using a reagent with a higher affinity for binding calcium (e.g., EDTA or EGTA); total calcium is related to the amount of calcium-binding reagent needed to titrate the fluorescent calcium complexes.

4. Atomic Absorption Spectrometry Methods for Total Calcium. Samples are diluted in a solution of lanthanum HCl to dissociate all calcium complexes. The amount of light absorbed at 422.7 nm by the sample after aspiration into an air–acetylene flame is a function of the total calcium concentration. This has been designated as a reference method for measuring total calcium.

5. Potentiometric Methods for Free (Ionized) Calcium. Free calcium is measured by ion-selective electrode; calcium-selective membranes are made using uncharged calcium-selective organic molecules or organophosphate ion exchangers. Instruments are usually designed to measure ionized calcium and pH simultaneously and to calculate ionized calcium corrected to a pH of 7.4.

D. TEST RESULTS AND INTERPRETATION

Plasma albumin concentration and blood pH may affect the interpretation of calcium results.

In general, changes in plasma albumin cause similar changes in total calcium, reflecting the increases and decreases in the protein-bound calcium fraction as albumin increases or decreases; however, free calcium is maintained at normal levels by the action of PTH even with abnormal albumin levels. Protein-bound calcium (and thus total calcium) may also be increased with markedly elevated plasma globulins, as in multiple myeloma. Total calcium results should be interpreted with respect to the concentration of plasma albumin.

Blood pH (i.e., [H$^+$]) affects the equilibrium between free and protein-bound calcium while total calcium remains unchanged. As pH decreases (i.e., [H$^+$] increases), calcium is displaced from proteins, resulting in an increased free calcium; as pH increases (i.e., [H$^+$] decreases), more protein-binding sites are available resulting in a decreased free calcium. pH effects are more likely to be apparent in acute acid–base imbalances; homeostatic mechanisms will tend to correct the concentration of free calcium in cases of chronic acidosis and alkalosis. Free calcium results should be interpreted with respect to blood pH.

Calcium reference ranges are (Painter, Cope, & Smith, 1999):

Total Calcium (Serum)

0–10 days	7.6–10.4 mg/dL	1.90–2.60 mmol/L
10 days–2 years	9.0–11.0 mg/dL	2.25–2.75 mmol/L
2–12 years	8.8–10.8 mg/dL	2.20–2.70 mmol/L
Adult	8.6–10.0 mg/dL	2.15–2.50 mmol/L

Free (Ionized) Calcium (Whole Blood)

Adult (> 18 years)	4.6–5.1 mg/dL	1.15–1.27 mmol/L

Calcium concentrations are occasionally expressed in units of milliequivalents per liter; to calculate calcium in terms of milliequivalents per liter, multiply results in millimoles per liter by 2 (to account for the divalent charge of calcium).

Total calcium *critical values* are ≤ 6 mg/dL and ≥ 13 mg/dL, indicating life-threatening calcium abnormalities that must be immediately reported to the physician.

VIII. PHOSPHATE

A. *BIOCHEMISTRY AND PHYSIOLOGY*

▶ About 85% of total body **phosphate** is in the skeleton as a component of hydroxyapatite, the calcium phosphate crystals forming the bone mineral. Another 14% of body phosphate is found in the intracellular compartment. Phosphate is the major intracellular anion but most intracellular phosphate is in organic form. Intracellular organic phosphates are vital to cell function and structure and include high-energy compounds (e.g., adenosine triphosphate [ATP]), nucleic acids, membrane phospholipids, phosphoproteins, second messengers (e.g., cyclic adenosine monophosphate [cAMP]), and coenzymes. Only about 1% of body phosphate is in the extracellular fluid, a fraction of which is found in plasma. While plasma contains both organic phosphate (in lipids and lipoproteins) and inorganic phosphate, it is the inorganic form that is commonly measured. About 55% of inorganic phosphate in plasma is free (or ionized) in the $H_2PO_4^-$ (monovalent) and the HPO_4^{2-} (divalent) forms, in which the relative proportion of the two forms depends on blood pH. At pH 7.4, the ratio is 1:4 $H_2PO_4^-$ to HPO_4^{2-}; as the pH becomes more acidic, the proportion of $H_2PO_4^-$ rises and as the pH becomes more alkaline the proportion of $H_2PO_4^-$ declines. Of the remaining plasma inorganic phosphate, about 10% is bound to albumin and about 35% is complexed to calcium or magnesium.

Plasma phosphate is not tightly controlled and homeostatic mechanisms are not fully understood. Renal excretion is the primary mechanism regulating phosphate levels. Plasma phosphate is filtered at the glomerulus and about 80% is reabsorbed in the proximal tubule. The capacity for renal phosphate reabsorption is apparently adapted to an individual's recent phosphate intake. Phosphate excretion is also affected by several hormones, especially PTH which inhibits phosphate reabsorption, thereby increasing phosphate excretion. Intestinal absorption of phosphate depends largely on the amount ingested but is stimulated to some extent by $1,25(OH)_2D_3$. Since phosphate and calcium concentrations are maintained in a roughly reciprocal relationship to avoid precipitation of calcium phosphate in soft tissues, calcium homeostatic mechanisms also affect phosphate levels.

In *hypophosphatemia* (low serum phosphate), mechanisms attempt to raise phosphate levels: (1) production of $1,25(OH)_2D_3$ is stimulated, which in turn increases intestinal absorption of phosphate, (2) there is an increase in phosphate reabsorption via intrarenal mechanisms, and (3) there is a reciprocal increase in calcium causing a decline in PTH which enhances renal phosphate reabsorption. These mechanisms are reversed in *hyperphosphatemia* (high serum phosphate) to generate a reduction in phosphate levels.

B. Medical Relevance

Hypophosphatemia may be the result of a depletion in total body phosphate due to insufficient intestinal absorption or inappropriate renal excretion or the result of the redistribution of phosphate from the extracellular to the intracellular compartment with no loss in total body phosphate. Major symptoms of phosphate depletion are caused by the deficit in intracellular ATP (e.g., encephalopathy and muscle weakness) and by diminished oxygen delivery due to the deficit in erythrocyte 2,3-diphosphoglycerate (e.g., cardiac disturbances). Chronic phosphate deficiency may lead to rickets or osteomalacia.

Total body phosphate becomes elevated as the renal excretion of phosphate decreases, most commonly as a result of impaired renal function, but may also be elevated due to excessive phosphate administration. Redistribution of phosphate from the intracellular to the extracellular compartment may cause hyperphosphatemia without an increase in total body phosphate. The consequences of hyperphosphatemia are most likely linked to the secondary hypocalcemia (i.e., the reciprocal change in calcium). Severe hyperphosphatemia may lead to calcium phosphate precipitation in soft tissues.

Causes of hypophosphatemia and hyperphosphatemia are listed in Table 5–10. Serum phosphate concentration may not accurately reflect stores of total body phosphate.

C. Test Procedures

1. Preanalytical Factors. Phosphate may be measured in serum (preferred) or plasma. Collection of a fasting morning specimen is recommended since phosphate exhibits diurnal variation and is affected by meals and exercise. Samples collected in EDTA, citrate, or oxalate are not acceptable due to interference with the formation of

Table 5–10. Disorders in Phosphate Homeostasis

Causes of Hypophosphatemia

Redistribution of phosphate from extracellular fluid to intracellular compartment:

Cellular carbohydrate uptake: as cellular phosphate is utilized during glycolysis more phosphate enters cells (e.g., following administration of insulin with treatment of diabetic ketoacidosis, administration of carbohydrates in a phosphate-deficient patient, transient effect following meals in healthy individuals)

Acute respiratory alkalosis: the decrease in P_{CO_2} activates glycolysis

Decreased intestinal absorption of phosphate:

Vitamin D deficiency

Starvation

Malabsorption, steatorrhea

Prolonged use of phosphate-binding antacids (i.e., antacids containing calcium, magnesium, or aluminum)

Increased renal phosphate excretion:

Hyperparathyroidism (primary or secondary): elevated parathyroid hormone (PTH) increases phosphate excretion

Defective or inhibited phosphate reabsorption (including hypervolemia, Fanconi's syndrome, vitamin D–resistant rickets, and metabolic acidosis)

Causes of Hyperphosphatemia

Redistribution of phosphate from intracellular compartment to extracellular fluid:

Cell lysis syndromes (hemolysis, leukemia, tumor lysis [from chemotherapy], rhabdomyolysis)

Excessive phosphate intake or absorption:

Phosphate-containing laxatives or enemas

IV administration

Decreased renal phosphate excretion:

Renal failure (reduced glomerular filtration rate [GFR])

Hypovolemia (reduced GFR)

Hypoparathyroidism (reduced PTH permits phosphate reabsorption)

Pseudohypoparathyroidism (kidney does not respond to PTH; increased phosphate reabsorption)

Acromegaly (growth hormone increases renal phosphate reabsorption)

Hyperthyroidism (thyroxine increases renal phosphate reabsorption)

---

complexes in phosphomolybdate methods. Blood cells contain a high concentration of organic phosphates; therefore, serum must be separated promptly from cell components and hemolyzed specimens are not acceptable. Phosphate may be measured in urine; specimens should be collected in hydrochloric acid.

2. Phosphomolybdate Methods. Routine methods for measuring serum phosphate begin with the reaction of phosphate ions and ammonium molybdate to form a phosphomolybdate complex. The phosphomolybdate complex is colorless but can be measured by ultraviolet absorption at 340 nm; most automated phosphate assays use this "direct" measurement method. Other methods are available in which the phosphomolybdate complex is subsequently reduced to molybdenum blue, measurable spectrophotometrically at 660 nm. Various reagents are used for reduction of phosphomolybdate, including semidine HCl, ferrous ammonium sulfate, and stannous chloride.

3. Enzymatic Methods. Several enzymatic methods using coupled reaction systems have been described for measuring phosphate. Hydrolysis of organic phosphate esters is prevented by performing these assays at a neutral pH.

D. TEST RESULTS AND INTERPRETATION

Serum phosphate assays measure inorganic phosphate; however, since there are two forms of phosphate found in serum ($H_2PO_4^-$ and HPO_4^{2-}), concentrations in units of mass per volume are commonly expressed as mass of elemental phosphorus per volume (i.e., mg/dL of phosphorus refers to milligrams of phosphorus in phosphate per deciliter). Expression in molar units (i.e., mmol/L) of phosphate avoids confusion.

Serum phosphate levels are affected by food intake and exhibit diurnal variation; reference ranges are generally given for morning fasting specimens. Age-dependent ranges for serum phosphate expressed as phosphorus are (Painter, Cope, & Smith, 1999):

0–10 days	4.5–9.0 mg/dL	1.45–2.91 mmol/L
10 days–2 years	4.5–6.7 mg/dL	1.45–2.16 mmol/L
2 years–12 years	4.5–5.5 mg/dL	1.45–1.78 mmol/L
Thereafter	2.7–4.5 mg/dL	0.87–1.45 mmol/L
Male > 60 years	2.3–3.7 mg/dL	0.74–1.20 mmol/L
Female > 60 years	2.8–4.1 mg/dL	0.90–1.32 mmol/L

Serum phosphate < 1 mg/dL signals critical hypophosphatemia and should be reported to the physician immediately.

IX. MAGNESIUM

A. BIOCHEMISTRY AND PHYSIOLOGY

Magnesium is essential for life, playing roles as activators, cofactors, and prosthetic groups for enzymes involved in lipid, carbohydrate, and protein metabolism, as well as deoxyribonucleic acid (DNA) transcription, oxidative phosphorylation, and transmembrane transport. Magnesium ranks fourth among cations in terms of total body content and second in intracellular cation concentration. Of the approximately 25 g of magnesium in the average adult human body, 50 to 60% is located in bone, 35 to 45% is intracellular, and only 1 to 2% is in the extracellular fluid. There are three forms of magnesium in equilibrium in the plasma: 55 to 60% is in the free (or ionized) form, 25 to 30% is bound to plasma proteins (predominantly albumin), and about 15% is complexed to anions (especially bicarbonate).

The homeostatic mechanisms for magnesium balance are poorly understood. Intracellular magnesium is essential to metabolism; extracellular magnesium provides a ready source for maintaining intracellular magnesium via magnesium-specific membrane transport systems; about one-third of bone magnesium is exchangeable as a source for maintaining extracellular magnesium. Magnesium is also affected by changes in intestinal absorption and renal excretion of magnesium. In the kidney, magnesium is freely filtered at the glomerulus and about 97% is reabsorbed in the tubules; the magnesium reabsorption rate is adjusted (especially in the loop of Henle) to maintain plasma magnesium concentration. There is some evidence that PTH and aldosterone may affect renal handling of magnesium and that $1,25(OH)_2D_3$ may have a role in intestinal absorption.

B. MEDICAL RELEVANCE

Causes of magnesium abnormalities are listed in Table 5–11. *Hypermagnesemia,* increased plasma magnesium, occurs with magnesium loading by intravenous infusion or excessive oral intake or with decreased renal excretion. Moderate increases in magnesium may

> **Functions of Magnesium**
>
> Activator, cofactor, and prosthetic group for enzymes involved in:
> - Lipid metabolism
> - Carbohydrate metabolism
> - Protein metabolism
> - DNA transcription
> - Oxidative phosphorylation
> - Transmembrane transport

Table 5–11. Disorders in Magnesium Homeostasis

Causes of Hypermagnesemia
Administration of magnesium:
  Intravenous infusion for treatment of preeclampsia, eclampsia, or hypertensive emergencies
  Excessive oral intake of magnesium-containing antacids or cathartics (magnesium more likely to become elevated with concomitant chronic renal failure)
Decreased renal excretion:
  Renal failure: acute renal failure, advanced chronic renal failure, chronic renal failure with excessive magnesium intake
  Hypovolemia (reduced glomerular filtration rate)
  Familial hypocalciuric hypercalcemia
  Lithium therapy

Causes of Hypomagnesemia
Decreased intestinal absorption:
  Insufficient intake (severe deficiency of long duration): starvation, alcoholism
  Gastrointestinal disorders: malabsorption syndromes, acute or chronic diarrhea, steatorrhea
Increased renal excretion (renal magnesium wasting):
  Hypervolemia
  Hypercalcemia
  Hypophosphatemia
  Osmotic diuresis
  Loop diuretics (e.g., furosemide)
  Drug toxicity (e.g., cisplatin, cyclosporine, aminoglycosides)
  Metabolic disorders (e.g., diabetes mellitus, hyperthyroidism, primary hyperaldosteronism)
  Congenital disorders of renal magnesium reabsorption (e.g., Bartter's syndrome or Welt's syndrome)

cause lethargy, weakness, nausea, vomiting, decreased reflexes, hypotension, and a predisposition to cardiac arrhythmias; marked elevations lead to coma and cardiac arrest.

Hypomagnesemia occurs with decreased intestinal absorption or with increased renal excretion of magnesium. Urinary magnesium will be decreased in magnesium deficiency due to decreased intestinal absorption and inappropriately increased when deficiency is caused by renal wasting. If the magnesium concentration is insufficient to maintain normal release and action of PTH, hypocalcemia will develop. Low magnesium concentration also hinders release of renin leading to inappropriately high aldosterone and a subsequent hypokalemia (decreased blood potassium). Symptoms of hypomagnesemia may be partially attributed to the accompanying hypocalcemia and hypokalemia (and sometimes hypophosphatemia); patients with hypomagnesemia exhibit muscle weakness, apathy, nausea, vomiting, nerve and muscle irritability, increased reflexes, and a predisposition to ventricular arrhythmias. If there is unsatisfactory response to treatment for hypocalcemia or hypokalemia, the patient should be checked for an underlying hypomagnesemia.

Serum magnesium concentrations are routinely used to assess magnesium status; however, serum magnesium may be normal with depleted intracellular concentrations, especially in chronic magnesium deficiency. Measurement of intracellular magnesium (e.g., in erythrocytes, muscle, or white blood cells), while not yet widely used, may prove to be a better tool to detect magnesium deficiency.

C. TEST PROCEDURES

1. Preanalytical Factors. Total magnesium can be measured in serum or heparinized plasma; samples anticoagulated with magnesium chelators (i.e., citrate, EDTA, oxalate) are not acceptable. Since the concentration of magnesium is significantly higher in erythrocytes than in plasma, serum or plasma must be separated from red cells promptly; hemolyzed specimens are not acceptable. Urine specimens for magnesium analysis should be collected in hydrochloric acid to prevent precipitation of magnesium salts.

2. Spectrophotometric Methods for Total Magnesium. Magnesium complexes with certain metallochromic indicators to form colored products measurable by spectrophotometry. Commonly used indicator reagents include calmagite, methylthymol blue, and a formazan dye. Specific calcium chelators are added in these methods to reduce calcium interference; EGTA is added in calmagite and methylthymol blue methods to bind calcium.

3. Atomic Absorption Spectrometry Methods for Total Magnesium. Samples are diluted 1:50 in a solution of lanthanum HCl to dissociate magnesium-to-anion complexes (especially magnesium phosphate) that interfere with analysis by atomic absorption. The amount of light absorbed at 285.2 nm by the sample after aspiration into an air–acetylene flame is a function of the total magnesium concentration. Atomic absorption is the reference method for measuring total magnesium.

4. Potentiometric Methods for Free (Ionized) Magnesium. Measurements of free, ionized magnesium are becoming more available as magnesium ion–selective electrodes using neutral carrier ionophores are improved; currently these electrodes are significantly affected by free calcium levels. Free magnesium may prove to be more clinically useful than total magnesium measurements.

D. TEST RESULTS AND INTERPRETATION

Abnormalities in serum albumin may affect total magnesium concentrations in a manner similar (but to a lesser extent) to their effect on total calcium concentrations (discussed above). Serum albumin concentrations should be taken into account when interpreting serum magnesium results.

The reference range for serum total magnesium in adults is 1.6 to 2.6 mg/dL (or 0.66 to 1.07 mmol/L) (Painter, Cope, & Smith, 1999). Magnesium concentrations are occasionally expressed in units of milliequivalents per liter; to calculate magnesium in terms of milliequivalents per liter, multiply results in millimoles per liter by 2 (to account for the divalent charge of magnesium).

Serum magnesium results < 1.2 mg/dL (0.5 mmol/L) or > 4.9 mg/dL (2 mmol/L) indicate severe magnesium disorders and should be promptly reported to the physician.

REFERENCE

Painter PC, Cope JY, Smith JL. Reference information for the clinical laboratory. In Burtis CA, Ashwood ER (eds.). *Tietz textbook of clinical chemistry,* 3rd ed. Philadelphia: WB Saunders, 1999; 1788–846.

SUGGESTED READINGS

Adrogué HJ, Tannen RL. Ketoacidosis, hyperosmolar states, and lactic acidosis. In Kokko JP, Tannen RL (eds.). *Fluids and electrolytes,* 3rd ed. Philadelphia: WB Saunders, 1996; 643–74.

Dennis VW. Phosphate disorders. In Kokko JP, Tannen RL (eds.). *Fluids and electrolytes,* 3rd ed. Philadelphia: WB Saunders, 1996; 359–90.

Endres DB, Rude RK. Mineral and bone metabolism. In Burtis CA, Ashwood ER (eds.). *Tietz textbook of clinical chemistry,* 3rd ed. Philadelphia: WB Saunders, 1999; 1395–457.

Kumar R. Calcium disorders. In Kokko JP, Tannen RL (eds.). *Fluids and electrolytes,* 3rd ed. Philadelphia: WB Saunders, 1996; 391–419.

Newman DJ, Price CP. Renal function and nitrogen metabolites. In Burtis CA, Ashwood ER (eds.). *Tietz textbook of clinical chemistry,* 3rd ed. Philadelphia: WB Saunders, 1999; 1204–70.

Rude RK. Magnesium disorders. In Kokko JP, Tannen RL (eds.). *Fluids and electrolytes,* 3rd ed. Philadelphia: WB Saunders, 1996; 421–45.

Sacks DB. Carbohydrates. In Burtis CA, Ashwood ER (eds.). *Tietz textbook of clinical chemistry,* 3rd ed. Philadelphia: WB Saunders, 1999; 787–90.

Tolman KG, Rej R. Liver function. In Burtis CA, Ashwood ER (eds.). *Tietz textbook of clinical chemistry,* 3rd ed. Philadelphia: WB Saunders, 1999; 1125–77.

Items for Further Consideration

1. Predict the effects that a defect in the secretion of conjugated bilirubin from the hepatocytes would cause on laboratory tests for bilirubin and bilirubin metabolites.

2. Discuss the advantages and disadvantages of using each of the following tests to monitor renal function: (a) serum creatinine, (b) creatinine clearance, and (c) serum urea nitrogen.

3. Describe the clinical manifestations of elevations in serum uric acid.

4. Design a protocol for patient preparation and specimen collection for plasma lactate analysis.

5. Explain the rationale for measuring blood pH and free (ionized) calcium simultaneously.

6. Discuss the advantages and disadvantages of measuring total calcium and total magnesium versus the measurement of free (ionized) calcium and magnesium.

Critical Care Analytes

John Toffaletti

Key Terms ◄

ALDOSTERONE

ANGIOTENSIN II

ANION GAP

ANTIDIURETIC HORMONE

BLOOD GASES

CARBOXYHEMOGLOBIN (COHb)

COMPENSATION

DEOXYHEMOGLOBIN (HHb)

DIABETES INSIPIDUS

FRACTIONAL OXYGEN
    SATURATION (FO₂Hb)

HENDERSON–HASSELBALCH
    EQUATION

HYPERCHLOREMIA

HYPERKALEMIA

HYPERNATREMIA

HYPOCHLOREMIA

HYPOKALEMIA

HYPONATREMIA

ION-SELECTIVE ELECTRODE

METABOLIC ACIDOSIS

METABOLIC ALKALOSIS

METHEMOGLOBIN (MetHb)

OSMOLALITY

OXIMETRY

OXYGEN SATURATION (SO₂)

OXYHEMOGLOBIN (O₂Hb)

PCO₂

pH

PO₂

RENIN

RESPIRATORY ACIDOSIS

RESPIRATORY ALKALOSIS

Chapter Objectives

Upon completion of this chapter, the reader will be able to:

1. Describe the importance of serum osmolality as an indicator of the overall state of water balance in blood.
2. Describe the separate mechanisms for regulation of blood volume and blood osmolality.
3. Discuss the causes of hypo- and hypernatremia.

4. Describe the function and regulation of potassium ions and the effect of alkalosis, acidosis, and other factors on plasma potassium concentrations.
5. Discuss the causes and consequences of hypo- and hyperkalemia.
6. List the medical conditions associated with alterations in serum chloride concentration.
7. Describe an ion-selective membrane and how it functions in an ion-selective electrode.
8. Discuss the importance of the bicarbonate ion–carbon dioxide plasma buffer system.
9. Describe the physiological meaning(s) for alterations in pH, partial pressure of carbon dioxide (PCO_2), bicarbonate (HCO_3^-), and partial pressure of oxygen (PO_2).
10. Describe the interrelationships of our metabolic and respiratory acid–base systems.
11. Describe the function of hemoglobin as both an oxygen carrier and a hydrogen ion buffer.
12. Interpret pH, PCO_2, and HCO_3^- results in evaluating ventilatory and metabolic acid–base status.
13. Evaluate PO_2 in determining the adequacy of arterial oxygenation by the lungs.
14. Practice correct techniques for handling samples for blood gas analysis.

Analytes Included ◀

OSMOLALITY	*CHLORIDE (CL⁻)*	*PARTIAL PRESSURE OF CO$_2$ (PCO$_2$)*
SODIUM (NA⁺)	*ANION GAP*	
POTASSIUM (K⁺)	*OXIMETRY (O$_2$HB, SO$_2$, etc.)*	*PARTIAL PRESSURE OF O$_2$ (PO$_2$)*
TOTAL CO$_2$ (bicarbonate)	*PH*	

Pretest

Directions: Choose the single best answer for each of the following items.

1. Which two factors are important in regulation of both blood osmolality and blood volume?
   a. antidiuretic hormone (ADH) and aldosterone
   b. thirst and ADH
   c. renin and aldosterone
   d. angiotensin and ADH

2. Which hormone responds first to a decrease in blood volume?
   a. aldosterone
   b. angiotensin
   c. atrial natriuretic peptide (ANP)
   d. renin

3. Which set of factors is often a mechanism for hyponatremia?
   a. renal failure and acidosis
   b. renin, angiotensin, and aldosterone
   c. fluid loss, ADH, and drinking water
   d. thirst, fluid loss, and aldosterone

4. With which of these functions is potassium strongly associated?
   a. ADH secretion
   b. glomerular filtration rate
   c. myocardial contraction and rhythm
   d. renin secretion

5. Which are the preferred conditions for collection and transport of blood for potassium measurements?
   a. no anticoagulant, transport blood at room temperature
   b. ethylenediaminetetraacetic acid (EDTA) anticoagulant, transport blood at room temperature
   c. heparin anticoagulant, transport blood on ice
   d. heparin anticoagulant, transport blood at room temperature

6. Which of the following is the least likely cause of hyperkalemia?
   a. blood transfusion
   b. running a marathon
   c. acidosis
   d. hyperaldosteronism

7. Why does the bicarbonate/carbonic acid buffer system have such a large capacity to buffer against acidosis?
   a. CO_2 can be lost by pulmonary ventilation
   b. the pK of bicarbonate/carbonic acid is close to the ideal of 6.1
   c. the pK of bicarbonate/carbonic acid is close to the ideal of 7.4
   d. hemoglobin can combine with CO_2

8. Which compound is an indicator of the metabolic acid–base status?
   a. CO_2
   b. H^+ ion
   c. HCO_3^- ion
   d. PO_2

9. Which parameter best corresponds to the rate of breathing?
   a. hemoglobin saturation (SO_2)
   b. arterial PO_2
   c. arterial PCO_2
   d. blood lactate

10. Which parameter could be used to detect excess carbon monoxide in blood?
    a. PO_2
    b. oxygen saturation of functional hemoglobin (SO_2)
    c. oxygen saturation of total hemoglobin (FO_2Hb)
    d. PCO_2

11. If the PO_2 of atmospheric air is 150 mm Hg, how can a person have an arterial PO_2 of 250 mm Hg?
    a. breathing air containing 20% oxygen
    b. hyperventilating
    c. by exposure to carbon monoxide
    d. breathing air containing 50% oxygen

12. Hyperventilation is which type of process?
    a. respiratory alkalotic
    b. metabolic alkalotic
    c. respiratory acidotic
    d. metabolic acidotic

13. Blood gas results of $PCO_2 = 50$ mm Hg and $HCO_3^- = 30$ mmol/L are most consistent with which type of acid–base disorder?
    a. acute respiratory acidosis
    b. acute respiratory alkalosis
    c. metabolic acidosis
    d. chronic respiratory acidosis

14. A patient who has been vomiting for 2 days and hyperventilating for over 12 hours has developed a low blood potassium concentration. What is the most likely acid–base disorder?
    a. metabolic alkalosis
    b. metabolic alkalosis and respiratory alkalosis
    c. respiratory alkalosis and respiratory acidosis
    d. metabolic acidosis and metabolic alkalosis

15. Three patients are breathing 50% oxygen. Based on the following PO_2 results on each patient, which of the three could most likely be taken off oxygen safely?
    a. 230 mm Hg
    b. 180 mm Hg
    c. 275 mm Hg
    d. all three patients

16. If a patient is not getting an adequate supply of oxygen to their tissues, which parameter would best indicate this?
    a. blood lactate
    b. pH
    c. FO_2Hb
    d. SO_2

17. Which condition most likely represents a medical emergency?
    a. a blood lactate of 3 mmol/L following vigorous exercise
    b. an arterial PO_2 of 70 mm Hg in an 82-year-old man at rest
    c. an arterial PO_2 of 72 mm Hg in a 40-year-old man at rest
    d. an arterial PCO_2 of 30 mm Hg in a 40-year-old man at rest

18. What changes are likely to occur when an arterial blood sample in a plastic syringe is stored in ice water for 60 minutes prior to analysis of blood gases?
    a. PO_2 decreases, PCO_2 decreases
    b. K^+ decreases, PO_2 increases
    c. PCO_2 increases, PO_2 increases
    d. PO_2 increases, K^+ increases

I. INTRODUCTION

This chapter covers the analytes sodium (Na^+), potassium (K^+), chloride (Cl^-), bicarbonate (HCO_3^-), pH, partial pressure of CO_2 (PCO_2), and partial pressure of oxygen (PO_2), often collectively referred to as "electrolytes and blood gases." Traditionally, osmolality is included in this group, both because sodium, potassium, chloride, and bicarbonate contribute significantly to the osmolality, and because factors that affect osmolality also affect sodium. The tests listed above, along with ionized calcium and lactate are often considered "critical care analytes" for several reasons:

- They are used for monitoring patients in critical care settings (e.g., intensive care unit, emergency room, operating room, etc.).
- The results are needed quickly to make decisions about treatment; delayed results usually have reduced clinical value.

The number of tests utilized as critical care analytes has been expanding as analysis systems become available for quick and accurate measurements using whole blood. For example, glucose, ionized magnesium, and coagulation tests (prothrombin time [PT]/partial thromboplastin time [PTT]) are being regarded as critical care analytes. Nearly all of these tests have become available on small (point-of-care) analyzers that can be carried to the patient.

II. OSMOLALITY AND VOLUME REGULATION

A. BIOCHEMISTRY AND PHYSIOLOGY

All osmotically active substances in plasma—virtually all ions and neutral solutes—contribute to the plasma **osmolality.** The size or charge of a molecule has little effect on the osmolality because osmolality is dependent on the number of particles in solution rather than size or charge. Thus, each molecule of albumin, alcohol, glucose, or urea contribute equally. Osmolality can be estimated from the concentrations of commonly measured substances as follows:

Calculated osmolality (mOsm/kg) =
$$2 [Na^+] + [BUN(mg/dL)/2.8] + [glucose(mg/dL)/18]$$

where BUN is blood urea nitrogen. Note that BUN and glucose concentrations are converted from milligrams per deciliter to millimoles per liter using appropriate factors. Sodium concentration is multiplied by a factor of 2 because each sodium ion is associated with an anion (i.e., two particles for each sodium ion).

1. Regulation of Osmolality. Osmolality in plasma is the parameter to which the hypothalamus responds. The regulation of osmolality indirectly affects the sodium concentration in plasma because sodium and its associated anions account for ~90% of the osmolality in plasma. The regulation of blood volume also affects the sodium concentration in blood. Osmolality and volume, although regulated by separate mechanisms, are related because osmolality (particularly sodium) is regulated by changes in water balance, whereas volume is regulated by changes in sodium balance.

Normal plasma osmolality (275–290 mOsm/kg of plasma H_2O) is maintained by processes involving thirst and **antidiuretic hormone** ◀ (ADH). ADH, also known as vasopressin, is secreted by the hypothalamus and acts to increase the reabsorption of water in the collecting tubules of the kidneys; its half-life in the circulation is only 15 to 20 minutes. Osmoreceptors in the hypothalamus respond quickly to small changes in osmolality; a 1 to 2% increase in osmolality causes a fourfold increase in the circulating concentration of ADH, and a 1 to 2% decrease in osmolality shuts off ADH production entirely.

If excess water is ingested, plasma osmolality decreases, suppressing both ADH and thirst. In the absence of ADH, as much as 10 to 20 L of dilute urine can be excreted daily, well above any normal intake of water. Therefore, drinking water in excess will cause hypoosmolality and hyponatremia virtually only in patients with impaired renal excretion of water.

If water intake is inadequate, the plasma osmolality increases, activating both ADH secretion and thirst. ADH minimizes renal water loss, while thirst motivates the person to find and ingest water. Although hypernatremia rarely occurs in a person with a normal thirst mechanism and access to water, hypernatremia can occur in infants, unconscious patients, or anyone who is unable to obtain or drink water. In people over the age of 60, osmotic stimulation of thirst is diminished; dehydration becomes increasingly likely in the older patient with illness and diminished mental status. A patient with **dia-** ◀ **betes insipidus** (i.e., no ADH) may excrete 10 L of urine per day; however, since thirst persists, water intake matches output and plasma sodium remains normal, an example of the effectiveness of thirst in preventing dehydration.

2. Regulation of Blood Volume.

Adequate blood volume is essential to maintain blood pressure and ensure perfusion of blood to all tissues and organs. If blood volume decreases, the *renin–angiotensin–aldosterone system* responds to the decreased renal blood flow (i.e., decreased blood volume and/or blood pressure) by secreting renin in the renal glomeruli. **Renin** converts angiotensinogen to angiotensin I, ◀ which then is converted to angiotensin II in the lungs. **Angiotensin II** ◀ causes both vasoconstriction, which quickly increases blood pressure, and secretion of **aldosterone,** which increases renal retention of ◀ sodium and water. The effects of blood volume and osmolality on sodium and water metabolism are shown in Figure 6–1.

Changes in blood volume and pressure are initially detected by stretch receptors in the circulatory system (e.g., in the carotid sinus, the aortic arch, and the glomerular arterioles). These receptors activate mechanisms that restore volume by appropriately varying vascular resistance, cardiac output, and renal sodium and water retention. Some of the responses to a decreased blood volume include:

- Volume receptors and thirst sensation stimulate both thirst and the release of ADH independently of osmolality.
- Secretion of epinephrine and norepinephrine.
- Production of angiotensin II, leading to vasoconstriction, increased renal reabsorption of sodium, and release of aldosterone.
- Secretion of aldosterone, which promotes distal tubular reabsorption of Na^+ and Cl^- in exchange for K^+ and H^+.

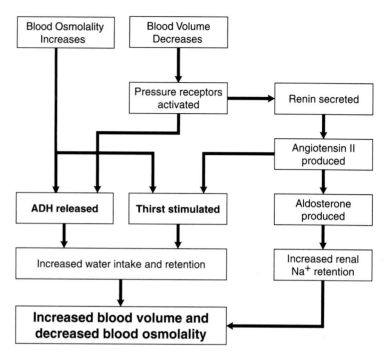

Figure 6–1. Homeostasis of blood osmolality and blood volume.

- Decreased glomerular filtration rate (GFR) due to volume depletion.

Responses to increases in blood volume include:

- Release of atrial natriuretic peptide (ANP) from the myocardial atria, promoting sodium excretion by the kidney.
- Increased GFR due to volume expansion.

Because H_2O follows Na^+ ions, the reabsorption of 98 to 99% of filtered sodium by the tubules is a normal mechanism to conserve nearly all of the 150 L of glomerular filtrate produced daily. Even a 1 to 2% reduction in tubular reabsorption of Na^+ can increase water loss by several liters per day.

3. Urine Osmolality. Urine osmolality varies widely, depending on water intake and the time of collection. Generally, however, urine osmolality is decreased in diabetes insipidus (too little ADH) and polydipsia (chronic thirst) and increased in conditions such as the *syndrome of inappropriate ADH secretion* (SIADH) and hypovolemia. Note that urinary Na^+ is usually decreased in hypovolemia.

III. SODIUM

A. BIOCHEMISTRY AND PHYSIOLOGY

Sodium is the most abundant cation in the extracellular fluid (ECF), representing 90% of all extracellular cations, and largely determines the osmolality of the ECF. Reference ranges for electrolytes and osmolality are listed in Table 6–1.

Table 6–1. Reference Ranges for Sodium, Osmolality, Potassium, Chloride, Total CO_2, and Anion Gap

Sodium, plasma	135–145 mmol/L
Osmolality, plasma	
Children and adults	275–295 mOsm/kg
Adults, > 60 years	280–300 mOsm/kg
Osmolality, urine	300–900 mOsm/kg
(24-hour collection)	
Potassium, plasma	3.5–5.0 mmol/L
Chloride	98–107 mmol/L
Total CO_2	21–28 mmol/L
Anion gap	8–16 mmol/L [$(Na^+) - (Cl^- + HCO_3^-)$]
	11–20 mmol/L [$(Na^+ + K^+) - (Cl^- + HCO_3^-)$]

Sodium concentration is approximately 15-fold higher in the ECF than inside cells. To maintain this gradient, an active transport system involving a Na^+-K^+-ATPase pump moves three sodium ions to the ECF in exchange for moving two potassium ions into cells. Potassium, as discussed later, is the major intracellular cation.

1. Regulation. The plasma sodium concentration depends on the intake and excretion of water and, to a somewhat lesser degree, the renal regulation of sodium. Three processes are important:

1. The intake of water in response to thirst, as stimulated or suppressed by plasma osmolality
2. The excretion of water, largely affected by ADH release in response to changes in either blood volume or osmolality
3. The blood volume status, which affects sodium excretion through aldosterone, angiotensin II, and ANP

The kidneys have the ability to conserve or excrete large amounts of sodium, depending on the sodium content of the ECF and the blood volume. Normally, 60 to 75% of filtered Na^+ is reabsorbed in the proximal tubule. Some sodium is also reabsorbed in the loop and distal tubule and (under the control of aldosterone) is exchanged for K^+ in the connecting segment and cortical collecting tubule.

B. MEDICAL OVERVIEW

1. Hyponatremia. If plasma Na^+ is decreased (< 135 mmol/L), the evaluation of **hyponatremia** should be confirmed by a decreased plasma osmolality. The urine Na^+, and sometimes the urine osmolality, may then help differentiate among the causes of hyponatremia, which are listed in Table 6–2.

The pathogenesis of hyponatremia is perhaps most easily understood by associating the cause of hyponatremia with the blood volume status and ADH response, as listed in Table 6–3. Volume status may be assessed by skin turgor, jugular venous pressure, and urine Na^+ concentration, where a low urine Na^+ indicates aldosterone secretion in response to hypovolemia.

Hypovolemic hyponatremia results from Na^+ loss in excess of water loss. There are several common causes of this condition:

Table 6–2. Differential Diagnosis of Hyponatremia

Plasma Na+ decreased
↓
Plasma osmolality decreased
↓
Measure urine Na+

Urine Na < 15 mmol/L:
    Hypovolemia with hypotonic fluid replacement (diarrhea, vomiting, etc.)
    Polydipsia (chronic thirst)
    Hypervolemia with arterial hypovolemia (congestive heart failure, cirrhosis)

Urine Na > 20 mmol/L:
    Inappropriate excess secretion of ADH (carcinoma of lung, adrenal
       insufficiency, etc.)
    Renal salt loss (thiazides, aldosterone deficiency)
    Reset osmostat
    Renal failure with water overload

- Use of thiazide diuretics (but not loop diuretics) induces Na+ and K+ loss without interfering with ADH-mediated water retention.
- Gastrointestinal (GI) loss of hypotonic fluid by prolonged vomiting or diarrhea, with replacement by a more hypotonic fluid (e.g., water) as thirst and ADH secretion are stimulated by hypovolemia.
- Adrenal insufficiency: deficiency of aldosterone (mineralocorticoid) and cortisol (glucocorticoid) prevents reabsorption of Na+ in the distal tubule.

Normovolemic hyponatremia typically indicates a problem with water balance. Among the causes are:

- SIADH leads to mild hypervolemia and release of ANP, which then leads to excretion of Na+ and water. SIADH may be caused

Causes of SIADH

- Drugs
- Tumors
- CNS disorders
- Endocrine disorders
- Pulmonary conditions

Table 6–3. Hyponatremia Related to Volume Status

Hypovolemic	**Na+ loss in excess of H$_2$O loss**
	Thiazide diuretics
	Loss of hypertonic fluid: GI, burns, sweat
	Potassium depletion
Euvolemic	**Problem with water balance**
	Inappropriate ADH secretion (SIADH)
	Artifactual (due to severe hyperlipidemia when Na+ measured in diluted plasma)
	Hyperosmolar from substance other than sodium
	Adrenal insufficiency
	Reset osmostat
Hypervolemic	**Excess H$_2$O retention (Movement of fluid from intravascular to interstitial space)**
	Advanced renal failure (decreased glomerular filtration rate) with excess water intake
	Congestive heart failure, hepatic cirrhosis
	Nephrotic syndrome (i.e., decreased colloid osmotic pressure); conditions characterized by increased extracellular fluid volume but decreased blood volume

by drugs, tumors, central nervous system (CNS) disorders, endocrine disorders, or pulmonary conditions.

- Artifactual hyponatremia can occur in cases of severe hyperlipidemia or hyperproteinemia if laboratory methods are used that dilute plasma before Na^+ analysis because they measure millimoles of Na^+ per liter of total plasma volume. Methods that do not dilute plasma or whole blood (such as ion-selective electrodes) give accurate Na^+ results because they detect the Na^+ concentration in the plasma water.

- In severe hyperglycemia, the sodium is decreased in order to maintain a normal plasma osmolality.

- Chronic excess intake of water, as with polydipsia, can eventually lead to hyponatremia that is usually mild but occasionally severe.

- Adrenal insufficiency can decrease cortisol and perhaps aldosterone. Cortisol deficiency promotes ADH release and water retention. Although adrenal insufficiency initially causes mild hypovolemia, the ADH-induced water retention typically restores volume status to normal.

- In pregnancy, the hypothalamus regulates plasma Na^+ concentration ~5 mmol/L lower than normal. This process may be initiated by vasodilation, which stimulates ADH secretion and thus retention of water, and lowering of blood pressure.

Hypervolemic hyponatremia is nearly always a problem of water overload, which usually causes edema. The usual therapy is water restriction. Sodium should not be given, because it can increase the severity of edema. For example, congestive heart failure (CHF) or hepatic cirrhosis increases venous back-pressure in the circulation, which promotes movement of fluid from the intravascular space to the interstitium, causing edema. Volume receptors sense hypovolemia, which leads to secretion of ADH and eventual hypervolemia and hyponatremia.

2. Hypernatremia. **Hypernatremia** (serum sodium concentration ◀ > 145 mmol/L) usually results from excessive loss of water relative to sodium (i.e., loss of hypotonic fluid) either by the kidney or through profuse sweating or diarrhea. The measurement of urine osmolality is necessary to evaluate the cause of hypernatremia (Table 6–4). In renal loss, the urine osmolality is low or normal. With extrarenal losses of water, the urine osmolality is increased.

Water loss through the skin and by breathing accounts for about 1 L of water loss per day in adults. Any condition that increases water loss, such as fever, burns, or exposure to heat, will increase the likelihood of developing hypernatremia. Not unexpectedly, hypernatremia occurs more commonly in adults with altered mental status and in infants, both of whom may have an intact thirst mechanism, but who are unable to ask for or obtain water.

Chronic hypernatremia in an alert patient is indicative of hypothalamic disease, usually with a defect in the osmoreceptors rather than from a true resetting of the osmostat. A reset osmostat may occur in primary hyperaldosteronism, in which excess aldosterone induces mild hypervolemia, which retards ADH release, shifting plasma sodium upward by ~3 to 5 mmol/L.

Hypernatremia may be from excess ingestion of salt or administration of hypertonic solutions of sodium, especially in neonates. In

Table 6–4. Evaluation of Hypernatremia

If plasma Na$^+$ is > 145 mmol/L, measure urine osmolality[a]:
< 300 mOsm/kg
   Diabetes insipidus (central or nephrogenic)
300–800 mOsm/kg
   Partial defect in ADH release (partial diabetes insipidus)
   Diuretics
   Osmotic diuresis
> 800 mOsm/kg
   Excess intake of Na$^+$
   Insensible water loss
   GI loss of hypotonic fluid
   Loss of thirst

---

[a] Persons who cannot fully concentrate their urine, such as neonates, young children, the elderly, and some patients with renal insufficiency, may show a relatively lower urine osmolality.

these cases, ADH secretion increases, and urine osmolality is > 800 mOsm/kg (Table 6–4).

Diabetes insipidus is characterized by copious production of dilute urine (3 to 20 L/day). In diabetes insipidus, although ADH secretion is either impaired (central diabetes insipidus) or the kidneys cannot respond to ADH (nephrogenic diabetes insipidus), hypernatremia does not usually occur because increased thirst compensates for renal water loss.

Excess water loss may also occur in renal tubular disease, such as acute tubular necrosis, in which the tubules become unable to fully concentrate the urine.

Treatment of hypernatremia must be gradual because rapid correction of serious hypernatremia (> 160 mmol/L) can induce cerebral edema and death, the maximal decrease in plasma sodium should be 0.5 mmol L^{-1}h^{-1}.

IV. POTASSIUM

A. BIOCHEMISTRY AND PHYSIOLOGY

Potassium is the major intracellular cation, with a 20-fold greater concentration inside cells than in the ECF. Only 2% of the total potassium in the body circulates in the plasma, with the Na$^+$-K$^+$-ATPase pump largely responsible for maintaining the K$^+$-gradient necessary for proper cellular function.

Functions of potassium include:

- Regulation of neuromuscular excitability
- Contraction of the heart and cardiac rhythm
- Affecting acid–base status

Both hypokalemia and hyperkalemia can cause muscle weakness. Severe hypokalemia can disrupt the biochemical and structural integrity of muscle cells.

The potassium ion concentration in plasma has a major effect on the contraction of cardiac muscle. When plasma potassium is high, the heart rate slows because of the decreased resting membrane potential of the cell relative to the threshold potential. A decrease in the plasma potassium concentration increases cardiac muscle excitability and often leads to arrhythmias.

The potassium concentration also affects the hydrogen ion concentration in the blood. For example, in hypokalemia (low plasma potassium), total body potassium is decreased; as a result, sodium and hydrogen ions move into the cell to replace potassium. The hydrogen ion concentration is therefore decreased in the ECF, resulting in alkalemia. The reverse is true for hyperkalemia.

1. Exercise. During exercise, potassium is released from cells, which may increase plasma K^+ by 0.3 to 1.2 mmol/L in mild to moderate exercise and by over 2.0 mmol/L with exhaustive exercise. These changes usually reverse after several minutes of rest. Note that forearm exercise during venipuncture can cause erroneously high plasma K^+ concentrations.

2. Hyperosmolality. Hyperosmolality causes diffusion of water out of cells. In this process, K^+ ions are transported with water, which causes a gradual loss of K^+ from the intracellular fluid.

3. Cellular Breakdown. Cellular breakdown releases K^+ into the ECF. Examples are severe trauma, tumor lysis syndrome, and massive blood transfusions.

4. Regulation. The kidneys are important in regulating potassium balance. Initially, the proximal tubules reabsorb nearly all the potassium. Then, under the influence of aldosterone, potassium is secreted into the urine in exchange for sodium in both the distal tubules and the collecting ducts. Thus, the distal nephron is the principal determinant of urinary potassium excretion.

An acute oral or intravenous (IV) intake of potassium is handled by K^+ uptake from the ECF into the cells. Excess plasma K^+ rapidly enters cells to normalize plasma K^+. As the cellular K^+ gradually leaks back into the plasma, it is removed by urinary excretion. Note that chronic loss of cellular K^+ may result in cellular depletion without appreciable change in the plasma K^+ concentration, because any excess K^+ is normally excreted in the urine.

Among the important factors that influence the distribution of potassium between cells and ECF are:

- Inhibition of the Na^+-K^+-ATPase pump by conditions such as hypoxia, hypomagnesemia, or digoxin overdose.
- Insulin administration, which promotes acute entry of K^+ ions into skeletal muscle and liver by increasing Na^+-K^+-ATPase activity.
- Secretion of catecholamines (e.g., epinephrine), which promotes cellular entry of K^+.
- Alkalemia, which causes cellular release of H^+ and cellular uptake of extracellular K^+ and Na^+ to maintain electroneutrality. Hypokalemia can also induce alkalosis by increasing tubular reabsorption of HCO_3^- and secretion of acid.
- Presence of propranolol, which impairs cellular entry of K^+.

B. MEDICAL OVERVIEW

▶ **1. Hypokalemia. Hypokalemia** (a plasma potassium concentration < 3 mmol/L) can occur with GI or urinary loss of potassium, with increased cellular uptake of potassium, or, rarely, with chronically reduced dietary intake.

GI fluid may be lost through vomiting, diarrhea, gastric suction, or discharge from an intestinal fistula. Renal loss of potassium can result from kidney disorders such as renal tubular acidosis and potassium-losing nephritis. Excess aldosterone can lead to hypokalemia and metabolic alkalosis. Hypomagnesemia can lead to hypokalemia by promoting both urinary and fecal loss of potassium. Magnesium deficiency diminishes the activity of Na^+-K^+-ATPase and enhances the secretion of aldosterone, with effective treatment requiring supplementation with both magnesium and K^+. Reduced dietary intake of K^+ is a rare cause of hypokalemia in healthy persons. However, decreased intake may intensify hypokalemia that is initiated by some other cause, such as use of diuretics.

Increased cellular uptake of potassium occurs with alkalemia and insulin administration. Alkalemia promotes intracellular loss of H^+ to minimize elevation in extracellular pH. To preserve electroneutrality, both K^+ and Na^+ enter cells. Plasma K^+ decreases by ~0.4 mmol/L per 0.1 unit rise in pH. However, other factors such as bicarbonate administration, diuretics, and vomiting can intensify the hypokalemia. Because insulin promotes the entry of K^+ into skeletal muscle and liver cells, insulin therapy often intensifies an underlying hypokalemic state. Plasma K^+ should be monitored carefully in patients on insulin therapy.

Probably by altering the polarization of the cell membrane, hypokalemia can lead to muscle weakness or paralysis, which can interfere with breathing. Hypokalemia may induce a variety of cardiac arrhythmias, including premature atrial and ventricular beats, sinus bradycardia, atrioventricular block, and ventricular tachycardia and fibrillation. These may cause sudden cardiac death in patients who already have compromised cardiac and/or respiratory problems.

Treatment for hypokalemia includes oral or IV replacement of potassium.

▶ **2. Hyperkalemia.** The most common causes of **hyperkalemia** are:

- Decreased renal excretion
- Excessive oral or IV intake of potassium
- Processes that enhance K^+ release from cells to the ECF

Patients with hyperkalemia often have an underlying disorder such as renal insufficiency, diabetes mellitus, or metabolic acidosis that, in combination with another process, causes hyperkalemia. For example, during administration of KCl, a person with renal insufficiency is far more likely to develop hyperkalemia than is a person with normal renal function.

Increased oral or IV intake of K^+ remains a common cause of hyperkalemia. In healthy persons, an acute oral load of potassium will increase plasma K^+ only transiently, because most of the absorbed K^+ will rapidly be transported inside the cell. Normal cellular processes gradually release this excess K$^+$ into the plasma, where it is normally removed by renal excretion. Impairment of urinary K$^+$ excretion is almost always associated with chronic hyperkalemia.

If transcellular shifts of K$^+$ into plasma are too rapid to be removed by renal excretion, acute hyperkalemia develops. In diabetes mellitus, insulin deficiency promotes cellular loss of K$^+$. Hyperglycemia also contributes by producing a hyperosmolar plasma that pulls water and some associated K$^+$ from cells. Since water loss from cells exceeds cellular loss of K$^+$, this action also increases the cellular K$^+$ concentration, which promotes further loss of K$^+$ into the plasma.

In metabolic acidosis, the increased H$^+$ in plasma moves intracellularly, with K$^+$ leaving the cell to maintain electroneutrality. Plasma K$^+$ increases by 0.2 to 1.7 mmol/L for each 0.1 unit reduction in pH. Because intracellular K$^+$ often becomes depleted in cases of acidosis with hyperkalemia (including diabetic ketoacidosis), treatment with agents such as insulin and bicarbonate can cause a rapid intracellular movement of K$^+$, producing severe hypokalemia.

A variety of drugs commonly cause hyperkalemia, especially in patients with either renal insufficiency or diabetes mellitus. These drugs include:

- Captopril (inhibits angiotensin-converting enzyme)
- Nonsteroidal anti-inflammatory agents (inhibit aldosterone)
- Digoxin (inhibits Na$^+$-K$^+$ pump)
- Cyclosporine (inhibits renal response to aldosterone)
- Heparin therapy (inhibits aldosterone secretion)

Potassium may be released into the ECF when tissue breakdown or catabolism is enhanced, resulting in hyperkalemia, especially if renal insufficiency is present. Increased cellular breakdown is associated with:

- Trauma
- Administration of cytotoxic agents
- Massive hemolysis
- Tumor lysis syndrome
- Blood transfusions

During storage of blood, K$^+$ is gradually released from erythrocytes, often resulting in markedly elevated plasma K$^+$ concentrations in blood stored for several weeks. With their small blood volume, infants are especially susceptible to hyperkalemia upon transfusion with stored blood.

Patients on cardiac bypass may develop mild elevations in plasma potassium during warming after surgery. Hypothermia causes movement of potassium into cells, whereas warming causes cellular release of potassium.

Exercise is associated with a rise in plasma K$^+$ concentration in proportion to the intensity of exercise; mild to moderate exercise may increase plasma K$^+$ by 0.3 to 1.0 mmol/L, and physical exhaustion may increase plasma K$^+$ by over 2 mmol/L, possibly causing sudden death.

When plasma K$^+$ reaches ~8 mmol/L, the hyperkalemia can cause muscle weakness by decreasing the ratio of intra- to extracellular potassium, which alters neuromuscular conduction. By altering cardiac conduction, hyperkalemia can lead to cardiac arrhythmias (plasma K$^+$ of 6 to 7 mmol/L) and cardiac arrest (K$^+$ > 10 mmol/L).

Calcium provides immediate but short-lived protection to the myocardium against the effects of hyperkalemia by lowering the resting potential of myocardial cells. Substances that cause potassium movement back into cells, such as sodium bicarbonate, glucose, and/or insulin, may also be administered. Patients treated with these agents must be carefully monitored to prevent hypokalemia if too much K+ moves back into cells. A cation-exchange resin, such as sodium polystyrene sulfonate, lowers plasma K+ by binding K+ ions in the gut. If the preceding measures do not prove successful or if renal failure occurs, peritoneal dialysis and hemodialysis may be necessary.

V. CHLORIDE

A. BIOCHEMISTRY AND PHYSIOLOGY

Although Cl⁻ is the major extracellular anion, its precise function in the body is not well understood. However, Cl⁻ is involved in maintaining osmotic pressure, proper body hydration, and ionic neutrality.

B. MEDICAL OVERVIEW

▶ **Hyperchloremia** may occur with excessive loss of bicarbonate due to:

- GI losses (such as diarrhea)
- Renal tubular acidosis
- Mineralocorticoid deficiency
- Compensated respiratory alkalosis

Hyperchloremia often accompanies hypernatremia. Mild elevations may also be seen in primary hyperparathyroidism.

Though not under the realm of "critical care," it is noteworthy to mention that the analysis of sweat chloride is used in the diagnosis of cystic fibrosis. Here, sweat production is stimulated by pilocarpine iontophoresis and the fluid collected analyzed for chloride and sodium.

▶ **Hypochloremia** may result from:

- Excessive loss of chloride from the body due to GI losses of hydrochloric acid
- Mineralocorticoid excess
- Salt-losing renal disease such as pyelonephritis

A low serum chloride also may be encountered in conditions associated with high serum bicarbonate concentrations, such as compensated respiratory acidosis or metabolic alkalosis. Chloride measurements are useful when interpreting difficult acid–base disorders, because hyperchloremia often indicates an acidotic process, whereas hypochloremia indicates an alkalotic process.

VI. BICARBONATE

A. BIOCHEMISTRY AND PHYSIOLOGY

Bicarbonate is the second most abundant anion in the extracellular fluid. Total CO_2 is composed of the bicarbonate ion (HCO_3^-), carbonic acid (H_2CO_3), and dissolved CO_2, with bicarbonate accounting for more than 90% of the total CO_2 at physiological pH. Because the bicarbonate concentration is an indicator of the buffering capacity of blood, a low bicarbonate indicates that a larger pH change will occur for a given amount of acid or base produced. Bicarbonate is classified as the metabolic component of acid–base balance and may be calculated from the pH and PCO_2 with the Henderson–Hasselbalch equation discussed in Section VIII below on Blood Gases.

Bicarbonate is the major component of the buffering system in the blood. It is produced from CO_2 by carbonic anhydrase in erythrocytes. To buffer against acidosis, bicarbonate combines with excess H^+ to produce CO_2 and H_2O. The loss of excess CO_2 (an acidic gas) by the lungs gives the bicarbonate–carbonic acid system a great capacity to buffer acid production. In the kidneys, most (85%) of the bicarbonate ion is reabsorbed by the proximal tubules, with 15% being reabsorbed by the distal tubules.

B. MEDICAL OVERVIEW

Bicarbonate is decreased in metabolic acidosis as bicarbonate combines with H^+ to produce CO_2, which is exhaled by the lungs. The typical response to metabolic acidosis is hyperventilation, which lowers PCO_2 and returns pH to normal. Elevated total CO_2 concentrations occur in metabolic alkalosis as bicarbonate is retained, often with an increase in PCO_2 as hypoventilation attempts to normalize pH. Other causes of metabolic alkalosis include severe vomiting, hypokalemia, and excess alkali intake.

VII. TEST PROCEDURES

A. SAMPLE COLLECTION AND HANDLING

1. Sodium and Chloride. Specimens for these analytes require no special handling precautions. Either serum or heparinized plasma is acceptable, and heparinized whole blood is suitable for many current analyzers, although the sample collection tube or syringe should contain only dry heparin as an anticoagulant. Relative to plasma, erythrocytes contain 10% of the sodium and 50% of the chloride. Therefore, hemolysis has little effect, although severe hemolysis can lower the measured sodium concentration by a dilutional effect. Posture also has little effect on the concentration of these analytes.

2. Bicarbonate. The same general principles of handling samples for sodium and chloride generally apply to bicarbonate, except that a sample of serum or plasma can lose CO_2 gas if the sample is exposed to the atmosphere. If exposure is prolonged, the bicarbonate concentration will decrease by 3 to 4 mmol/L. However, with typical sample handling and processing conditions, this effect is minimal.

3. Potassium. The potassium concentration is relatively sensitive to improper collection and handling of blood, usually resulting in a falsely elevated potassium.

- Because the coagulation process releases K^+ from platelets, serum K^+ may be 0.1 to 0.5 mmol/L higher than plasma K^+

concentrations. If the patient's platelet count is elevated (thrombocytosis), serum potassium may be further elevated. These situations may be avoided by using a heparinized specimen.

- If a tourniquet is left on the arm too long during blood collection, or if the patient clenches the fist excessively or exercises the forearm before venipuncture, cells may release potassium into the plasma.
- Storing whole blood on ice promotes the release of potassium from cells. Therefore, whole blood samples for potassium determinations should be stored at room temperature and analyzed promptly or centrifuged to remove the cells.
- Any hemolysis that occurs after the blood is drawn can falsely elevate potassium.

B. METHODS OF ANALYSIS

Although flame spectrometric methods (i.e., atomic absorption and flame emission) were once commonly used to quantitate sodium and potassium, modern analyzers are based on electrochemical methods using an **ion-selective electrode** (ISE). The heart of an ISE is a membrane that contains "ionophores" having a specific affinity for the analyte ion. When blood contacts these membranes, analyte ions from the blood (e.g., K^+) bind to one side of the membrane, creating a potential across the membrane. This potential is measured and used to calculate the concentration of potassium ions in the blood.

1. Sodium. Sodium electrodes typically use glass membranes that have a specific affinity for sodium ions. These special glasses are made from various combinations of silicon dioxide, sodium oxide, aluminum oxide, and lithium aluminum silicate.

2. Potassium. For many years, potassium-selective membranes have used the antibiotic valinomycin imbedded in a plastic membrane. Valinomycin has a ring structure containing several oxygen atoms that forms a pocket of the proper size and charge to accept potassium ions.

3. Chloride. Ion-selective electrodes for chloride ion usually have a quaternary ammonium salt as the ionophore, such as tri-*n*-octyl-propyl-ammonium chloride decanol.

4. Bicarbonate or Total CO_2. Methods for bicarbonate or total CO_2 that are used on most modern analyzers utilize an ion-selective electrode that responds to CO_2. Typically, this is a pH electrode covered with a silicone membrane, which is permeable to CO_2. For total CO_2, acid is mixed with the sample to convert bicarbonate to CO_2. The acidic CO_2 gas diffuses across the silicone membrane into a thin film of buffer solution, changing the pH of the buffer. The pH electrode detects the change in pH, which is related to the total CO_2 in the sample.

C. ANION GAP

The **anion gap** is a term for the difference in concentration of the cations (Na^+ and K^+) and anions (Cl^- and HCO_3^-) that are commonly measured in a chemistry test panel. The anion gap is calculated using the measured sodium, chloride, bicarbonate, and sometimes potassium concentrations:

$$(1)\ (Na^+ + K^+) - (Cl^- + HCO_3^-)\ or$$
$$(2)\ Na^+ - (Cl^- + HCO_3^-)$$

The approximate reference ranges for each calculation method are given in Table 6–1.

Note that there is no actual "anion gap" in blood, because the sum of all cationic charges equals the sum of all anionic charges. The anion gap is a means to estimate unusually high concentrations of one or more unmeasured anions; that is, those other than chloride or bicarbonate. An elevated anion gap indicates an elevation of anions such as lactate, ketoacids, or an anionic metabolite of a toxin or drug.

VIII. BLOOD GASES

A. INTRODUCTION

The term **blood gases** refers to the parameters pH, PCO_2, and PO_2, ◄ which are commonly measured in blood. Note that the lowercase "p" in pH stands for negative log, while the uppercase "P" in PCO_2 and PO_2 stands for the partial pressure of each of these gases. Blood gases are commonly measured by electrochemical (potentiometric or amperometric) methods that employ ion-selective or gas-selective electrodes.

Oximetry refers to the measurement of various forms of hemo- ◄ globin, including oxyhemoglobin, from which the term is derived. Co-oximeters are specialized spectrophotometers (usually of several fixed wavelengths) that measure the absorbance of the various hemoglobins present in blood.

B. BIOCHEMISTRY AND PHYSIOLOGY: BUFFER SYSTEMS

1. Bicarbonate–Carbonic Acid. This is the buffer in highest concentration (~24 mmol/L) in the blood plasma, which is also of central importance in acid–base regulation. CO_2 is a volatile acidic gas, is soluble in water, and is produced as a major product of energy metabolism. CO_2 produced by metabolism readily diffuses from cells into blood (with a lower PCO_2), where it combines with H_2O to produce carbonic acid, which immediately dissociates to bicarbonate and hydrogen ions. pH, HCO_3^- (mmol/L), and PCO_2 (mm Hg) are related by the equation:

$$pH = pK + \log \frac{[HCO_3^-]}{(0.03 \times PCO_2)}$$

which is derived from the **Henderson–Hasselbalch equation:** ◄

$$pH = pK + \log \frac{[base]}{[acid]}$$

For this buffer system, the base is bicarbonate and the acid is carbonic acid, whose concentration is calculated from the solubility of carbon dioxide in water: 0.03 mmol/L/mm Hg. The pK for bicarbonate/carbonic acid is 6.1. Note that the pK is defined as the pH at which the HCO_3^- and H_2CO_3 (or $0.03 \times PCO_2$) are in equal concentrations (i.e., if their ratio is 1, the log of 1 is equal to 0).

At the normal concentration ratio in blood of 20:1 ("ideal" would be 1:1), and with a pK of 6.1 ("ideal" would be 7.4), the HCO_3^-–H_2CO_3 buffer system would seem to be a poor buffer. However, this excess of base (HCO_3^-), along with the volatility of CO_2, gives tremendous ability to buffer acid. The lungs effectively make this an open system, with pulmonary regulation of CO_2 providing extensive buffering capacity. Bicarbonate is regulated primarily by the kidneys, with the ratio of HCO_3^- to H_2CO_3 determining the pH. Thus, an HCO_3^-:H_2CO_3 ratio of 10:0.5 has the same pH as a ratio of 20:1.

2. Hemoglobin. The great value of hemoglobin as a buffer lies in its ability to transport acid from the tissues to the lungs. Hemoglobin is remarkable in that it increases its affinity for hydrogen ions (H^+) as it loses oxygen. That is, **deoxyhemoglobin (HHb),** also called *reduced hemoglobin,* has a greater affinity for H^+ than does **oxyhemoglobin (O_2Hb).** In tissue capillaries, O_2Hb enters an environment of low PO_2, high acid, and warmth. These conditions promote release of O_2, which also promotes binding of H^+. In the lungs, HHb encounters an environment of high PO_2, low acid, and relative coolness, which promote gain of O_2 and release of H^+. These relationships are shown in Figure 6–2.

3. Phosphate. The HPO_4^{2-}–$H_2PO_4^-$ buffer pair is of minor importance as a buffer in plasma, with a concentration of ~1 mmol/L (3.1 mg/dL). It is of greater importance, and in higher concentration, as an intracellular buffer.

4. Albumin and Other Proteins. Largely due to the imidazole groups on the amino acid histidine, with a pK of ~7.4, albumin and other proteins also function as pH buffers.

C. ACID–BASE REGULATION

1. Metabolic (Renal) System. When the H^+ concentration deviates from normal, the kidneys respond by reabsorbing or secreting

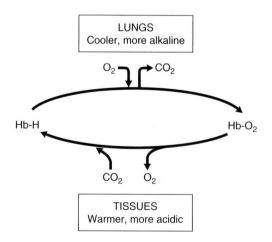

Figure 6–2. Gas exchange in lungs and tissues.

hydrogen, bicarbonate, and other ions to regulate the pH of blood. Because kidneys are the major regulator of bicarbonate, bicarbonate is considered the metabolic component of the bicarbonate/carbonic acid buffer system. (Although hydrogen ions are regulated by the kidney, the liver, through gluconeogenesis, is more important than the kidney in removal of H^+.) Metabolic acidosis may develop either if H^+ accumulates or if bicarbonate ions are lost. Metabolic alkalosis may develop from either loss of H^+ or increase of bicarbonate.

Although a change in ventilatory rate can alter arterial pH in minutes, the kidneys require hours to days to significantly affect pH by altering the excretion of bicarbonate. Thus, metabolic compensation takes several hours to alter the pH significantly.

2. Respiratory (Ventilatory) System. PCO_2 is considered the respiratory component of the bicarbonate–carbonic acid buffer system because arterial CO_2 is influenced greatly by the ventilatory rate. CO_2 is the end product of many aerobic metabolic processes; thus, buffering and removal of CO_2 are continually required for pH regulation. Provided there is a sufficient gradient of PCO_2 between tissues and blood, CO_2 will readily diffuse into the blood. CO_2 quickly combines enzymatically with H_2O to form the unstable H_2CO_3; H_2CO_3 then quickly dissociates into HCO_3^- and H^+ ions, and the H^+ ions are readily accepted by the HHb (see Figure 6–2). Upon entering a region of high PO_2 in the lungs, HHb is oxygenated to O_2Hb, which immediately promotes loss of H^+. H^+ ion quickly combines with bicarbonate to produce dissolved CO_2, which diffuses into the alveolar air for ventilatory removal.

The arterial PCO_2 represents a balance between tissue production of CO_2 and pulmonary removal of CO_2. An elevated PCO_2 usually indicates inadequate ventilation (hypoventilation). Conversely, a decreased PCO_2 usually indicates excessive ventilation (hyperventilation). These conditions can lead to respiratory acidosis (hypoventilation) or respiratory alkalosis (hyperventilation).

D. TISSUE OXYGENATION

Exposing hemoglobin to increasing concentrations of O_2 (i.e., increasing PO_2) will eventually saturate hemoglobin with oxygen. As PO_2 increases from 10 to about 50 mm Hg, the percentage of hemoglobin saturated with oxygen increases somewhat linearly. As PO_2 is increased further, the **oxygen saturation (SO_2)** increases much more gradually as it approaches 100%. This sigmoidal relationship between PO_2 and SO_2 is shown in Figure 6–3.

SO_2 may vary from 0 to 100%. There are two terms for oxygen saturation of hemoglobin, depending on whether the O_2Hb is compared with the functional hemoglobin (all forms of hemoglobin able to bind O_2; see equation 1) or the total hemoglobin (equation 2):

$$SO_2 = O_2Hb / (O_2Hb + HHb) \qquad (1)$$

$$FO_2Hb = O_2Hb / (O_2Hb + HHb + COHb + MetHb) \qquad (2)$$

where COHb is **carboxyhemoglobin,** MetHb is **methemoglobin,** and FO_2Hb is the **fractional oxygen saturation.** COHb is bound with carbon monoxide and is inactive for O_2 transport until the carbon monoxide is released. MetHb contains the ferric ion (Fe^{3+}) in-

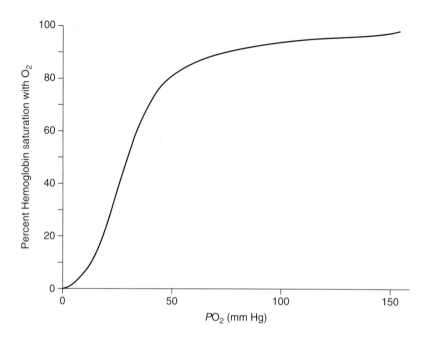

Figure 6–3. Oxygen saturation curve for hemoglobin.

stead of the usual ferrous ion (Fe^{2+}) and is dysfunctional (unable to bind O_2).

Although SO_2 in equation 1 may be used for determining the adequacy of gas exchange in the lungs (as in pulse oximetry), PO_2 is better for this purpose. The FO_2Hb in equation 2 is sometimes not useful for monitoring pulmonary gas exchange, because a heavy smoker could have relatively low FO_2Hb due to a high $COHb$. However, the FO_2Hb (equation 2) must be used for any calculations of O_2 content, O_2 delivery, or O_2 consumption.

E. MEDICAL OVERVIEW

▶ **pH** is an index of the acidity or alkalinity of the blood. If normal arterial pH is 7.35 to 7.45, then a pH < 7.35 indicates an acid state, and a pH > 7.45 indicates an alkaline state. In critical care, the slightly wider range of 7.30 to 7.50 is often clinically acceptable as a guideline. *Acidemia* refers to the condition where blood is too acidic, and *acidosis* refers to the metabolic process within the patient that causes the acidemia. Analagous terms are used for the alkaline state: *alkalemia* and *alkalosis*. Because all enzymes and physiological processes may be affected by pH, pH is normally regulated within extremely close tolerances.

▶ **PCO_2** is a measure of the tension or pressure of carbon dioxide dissolved in the blood and is the respiratory component of acid–base balance. The PCO_2 of blood represents the balance between cellular production of CO_2 and ventilatory removal of CO_2. A relatively stable PCO_2 indicates that the lungs are removing CO_2 at about the same rate as tissues are producing CO_2. A change in PCO_2 indicates an alteration in this balance, usually due to a change in the rate and depth of breathing or ventilation, provided that the rate of metabolic production of CO_2 is constant.

Forms of Hemoglobin

<u>Functional Hb</u>

- Oxyhemoglobin
- Deoxyhemoglobin

<u>Dysfunctional Hb</u>

- Carboxyhemoglobin
- Methemoglobin

PO₂ is a measure of the tension or pressure of oxygen dissolved ◀
in the blood. The PO_2 of arterial blood is primarily related to the
ability of the lungs to oxygenate blood from alveolar air. A decreased arterial PO_2 indicates one or more of the following circumstances:

- Decreased pulmonary ventilation (e.g., airway obstruction,
  pulmonary failure, trauma to the brain, etc.)
- Impaired gas exchange between alveolar air and pulmonary
  capillary blood (e.g., bronchitis, emphysema, pulmonary
  edema, etc.)
- Altered blood flow within the heart or lungs (e.g., congenital
  defects in the heart or shunting of venous blood into the arterial system without oxygenation in the lungs)

1. Causes of Metabolic Acidosis. Metabolic acidosis is a condi- ◀
tion defined by a decreased pH with a decreased HCO_3^- in blood.
There are several noteworthy causes.

- *Ketoacidosis.* Normal oxidative metabolism requires the entry of glucose into cells to ultimately produce adenosine
  triphosphate (ATP). If glucose cannot enter cells, as with insulin deficiency, fatty acids must be utilized for fuel, producing ketoacid by-products.
- *Hypoxic (lactate) acidosis.* If adequate oxygen is not available to tissues, cellular conditions begin to favor conversion
  of pyruvate to lactate, with the overall glycolytic pathway
  producing far less ATP than by oxidative metabolism. Actually, lactate, not lactic acid, is produced. The acid (H^+) is generated from the degradation of large amounts of ATP to
  adenosine diphosphate (ADP).
- *Renal failure.* Depending on whether glomerular or tubular
  failure occurs, accumulation of H^+ and/or loss of HCO_3^- may
  occur. Because renal acidosis usually has a slow onset, hyperventilation usually readily compensates to prevent
  acidemia.

Use of the Anion Gap in Metabolic Acidosis. The anion gap,
especially when elevated, may be useful in diagnosing the type of
metabolic acidosis and in indicating the possibility of a mixed
acid–base disorder, as shown in Table 6–5.

Causes of Metabolic Acidosis

- Ketoacidosis
- Hypoxic (lactate) acidosis
- Renal failure

Table 6–5. Changes of Anion Gap in Various Acid–Base Disorders

DISORDER	DECREASED/LOST	ELEVATED/GAINED	EFFECT ON ANION GAP
Diarrhea	HCO_3^-	Cl^-	Slight
Renal tubular acidosis	HCO_3^-	Cl^-	Slight
Lactate acidosis	HCO_3^-	Lactate	Increased
Ketoacidosis	HCO_3^-	Ketoacids	Increased
Mixed disorder: ketoacidosis with metabolic alkalosis	HCO_3^-	Ketoacids and HCO_3^-	Increased with little change in HCO_3^-

---

Causes of Metabolic Alkalosis

- Hypokalemia
- Administration of excess bicarbonate
- Vomiting or GI suctioning
- Corticosteroid excess

▶ **2. Causes of Metabolic Alkalosis. Metabolic alkalosis** is defined by an increased pH and an increased HCO_3^- in blood. In the alert patient, it is uncommon for metabolic alkalosis to induce marked hypoventilation to compensate for alkalosis. In the unconscious patient, however, metabolic alkalosis may induce hypoventilation, leading to hypoxia. There are several typical causes of metabolic alkalosis.

- *Hypokalemia.* Hypokalemia stimulates distal tubular reabsorption of K^+ ions in exchange for H^+ ions, which are secreted in the urine. If hypokalemia develops in primary mineralocorticoid excess, K^+ ions move out of the cells and are replaced by Na^+ and H^+ ions moving into the cells. These movements lead to intracellular acidosis and extracellular alkalosis.

- *Administration of excess bicarbonate.* Infusion of sodium bicarbonate solution is often part of treatment for cardiopulmonary arrest and other causes of metabolic acidosis, and is a metabolic alkalotic process. If excessive, bicarbonate administration may lead to alkalosis, especially if renal function is compromised, as is often the case in cardiac failure.

- *Vomiting or gastrointestinal suction.* The initial loss of gastric acid causes the metabolic alkalosis, which may be followed by renal loss of potassium in an attempt to conserve H^+.

- *Corticosteroid excess.* High doses of sodium-retaining steroids lead to excretion of both K^+ and H^+ in the distal renal tubules. Excess cortisol, as in Cushing's syndrome, can lead to both metabolic alkalosis and hypokalemia, because cortisol has some mineralocorticoid activity and because mineralocorticoids such as corticosterone may also be in excess.

▶ **3. Causes of Respiratory Acidosis.** A decreased arterial pH and increased PCO_2 are diagnostic for **respiratory acidosis** (ventilatory failure). Some causes are listed below:

- Obstructive lung disease, such as chronic bronchitis or emphysema. Respiratory acidosis is the end result of chronic progressive alveolar destruction.
- Acute airway obstruction by aspiration or blockage of a tube.
- Circulatory failure, which causes insufficient delivery of blood to the lungs.
- Impaired function of respiratory center (head trauma, sedation by a drug or toxin, or anesthesia)
- Mechanical ventilation that does not provide adequate oxygen for the patient. If CO_2 production increases while ventilation rate remains constant, the blood PCO_2 will rise.

▶ **4. Causes of Respiratory Alkalosis. Respiratory alkalosis** (hyperventilation) results from an increased ventilatory rate in which CO_2 is lost faster than it is produced.

- Hypoxemia-induced hyperventilation, such as by breathing oxygen-poor air or exposure to high altitude.
- Pulmonary embolism or pulmonary edema, in which oxygen transport across the alveolar membrane is impaired to a greater extent than is CO_2 transport.

- Anxiety that leads to hyperventilation can rapidly decrease arterial PCO_2.
- Overventilation by mechanical ventilator, usually from aggressive use of the ventilator to increase arterial oxygen tension.
- Several drugs can cause hyperventilation, such as salicylate, nicotine, or progesterone.
- Central nervous system (CNS) disorders that result from conditions such as septicemia or trauma.

F. OXYGENATION

For adequate tissue oxygenation, the following conditions are necessary:

- Sufficient oxygen in alveolar air
- Gas exchange in the lungs between alveolar air and blood
- Binding of oxygen to hemoglobin
- Blood flow to tissues as a result of adequate cardiac output and appropriate tone of blood vessels
- Release of oxygen to the tissues

1. PO_2 of Inspired Air. At a barometric pressure of 760 mm Hg, the PO_2 of atmospheric air (21% oxygen) is ~150 mm Hg, accounting for the contribution of water vapor (47 mm Hg) to the atmospheric pressure:

$$(760 - 47) \times 0.21 = 150 \text{ mm Hg}$$

2. Gas Exchange in the Lungs. A normal adult breathing air with a PO_2 of 150 mm Hg should develop an arterial PO_2 of 80 to 100 mm Hg. The range for arterial PO_2 depends on age; a newborn normally has an arterial PO_2 of 40 to 70 mm Hg, and persons > 60 years will have progressively declining normal PO_2 levels, due to alveolar degeneration. Acceptable arterial PO_2 levels for age groups are shown in Table 6–6. A PO_2 below these minimal tensions is considered evidence of hypoxemia. (Also see the section "Evaluating Arterial Oxygenation.")

3. Blood Flow to Tissues. Although normal blood flow provides sufficient oxygen and nutrients to all organs and tissues, many abnormalities in flow may develop that cause inadequate perfusion of general or localized areas of the body. Such conditions may lead to shock states, tissue necrosis, and ultimately death. Some common causes of poor perfusion are:

- Inadequate cardiac output
- Decreased blood volume (hypovolemia)
- Emboli
- Vasoconstriction
- Shunting of blood

Table 6–6. Acceptable Arterial Oxygen Tensions by Age

AGE (YEARS)	ACCEPTABLE PO_2
1–60	> 80 mm Hg
70	> 70 mm Hg
80	> 60 mm Hg
90	> 50 mm Hg

Both anemia and shunting of blood within the lungs may diminish oxygen content in arterial blood. In anemia, blood with less hemoglobin will carry proportionately less oxygen. In shunting of blood within the lungs, by passing through nonfunctioning alveoli, venous blood is not oxygenated adequately. When blood from these "shunts" mixes with oxygenated blood in the pulmonary vein, blood of lower PO_2 and oxygen content is delivered to the systemic circulation.

4. Release of Oxygen to Tissues. If arterial PO_2 can be raised sufficiently high (to 90 mm Hg) and blood flow is adequate, the conditions in the tissues (i.e., low PO_2 and high PCO_2 and acidity) will usually ensure that oxygen is released from hemoglobin to the tissues. The measurement of blood lactate may be used to determine whether overall oxygen metabolism is adequate.

G. TEST PROCEDURES

1. Collection and Handling of Samples for Blood Gas Analysis.
Blood gas parameters, especially PO_2, are highly susceptible to improper specimen collection and handling. Anaerobic conditions are essential during collection and handling because room air has a PCO_2 of nearly 0 and a PO_2 of ~150 mm Hg. The complete removal of all air bubbles is important to minimize the effect on PO_2, especially when transported by pneumatic systems. The other factors that must be controlled are:

- The type of anticoagulant
- The type of syringe used (glass or plastic)
- The temperature of storage before analysis
- The length of delay between collection and analysis of blood

Although liquid heparin < 10% (vol:vol) of the volume of blood may have little effect on pH, PCO_2, or PO_2, the use of liquid heparin will dilute other constituents in blood, such as electrolytes and glucose, which may be analyzed simultaneously in many current analyzers. For this reason, dry heparin should be used as the anticoagulant whenever possible.

Although plastic syringes are used for nearly all blood gas measurements, they have a potential disadvantage due to their ability to absorb oxygen. When stored in ice water, because of the increased oxygen affinity of hemoglobin at cold temperatures, blood can absorb oxygen dissolved within and transmitted through the plastic wall of the syringe. This effect is most pronounced when samples have a PO_2 of > 100 mm Hg, that is, when hemoglobin is already nearly fully saturated with oxygen and is unable to buffer any added O_2; a PO_2 of 100 mm Hg may increase by 8 mm Hg during 30 minutes of storage on ice. Hemoglobin that is less saturated, for example, at a PO_2 of 60 mm Hg, is better able to buffer the additional oxygen, with little or no measurable change in PO_2. Because glass is not permeable to O_2, PO_2 is not affected in blood stored in glass syringes in ice.

When blood is stored at ambient temperature for > 30 minutes, cellular metabolism can decrease PO_2. In general, storage of blood in plastic syringes at room temperature is acceptable if analysis oc-

---

Parameters of Blood Gas Analysis

Measured Parameters

- pH
- PCO_2
- PO_2
- SO_2

Calculated Parameters

- SO_2
- FO_2Hb
- O_2 content
- HCO_3^-

curs within 15 minutes. In most samples stored at ambient temperature (22 to 24°C) for 15 minutes, PO_2 changes by < 1 mm Hg even at a PO_2 of 100 mm Hg, PCO_2 changes < 1 mm Hg, and pH changes < 0.01 unit. However, pH, PCO_2, and PO_2 (and glucose) in samples from patients with extreme leukocytosis can change dramatically when stored at room temperature. These samples must be analyzed as soon as possible.

2. Methods of Analysis. pH, PCO_2, and PO_2 are all measured by electrodes that are selective for the specific analyte. pH is measured by an ISE that utilizes a pH-sensitive glass. PCO_2 is measured by a modified pH electrode that is covered by a silicone rubber membrane permeable to CO_2. The PO_2 electrode is little more than a platinum wire held at a constant reducing potential. Specificity for oxygen is achieved by covering the electrode with a polypropylene membrane that restricts contaminants but allows oxygen to diffuse into the electrode chamber where the oxygen is reduced.

Calculated parameters:

- Bicarbonate: Rearrangement of Henderson–Hasselbalch equation (see page 155)
- Oxygen saturation: $SO_2 = O_2Hb / (HHb + O_2Hb)$
- Fractional oxygen saturation: $FO_2Hb = O_2Hb/total\ Hb$
- Oxygen content (mL O_2/dL blood):

$$O_2\ ct = (1.34 \times total\ Hb \times FO_2Hb) + 0.003\ PO_2$$

Notes:

a. 1.34 is a factor that converts total oxygen-saturated hemoglobin in g/dL to oxygen in mL/dL. There ate 1.34 mL O_2 per gram of hemoglobin. This factor varies slightly among different studies.

b. 0.003 (mL/dL/mm Hg) is the solubility of oxygen in water and is related to the partial pressure of oxygen in the water.

Oximetry is the measurement of the forms of hemoglobin that circulate in blood. These are commonly oxyhemoglobin, deoxyhemoglobin, carboxyhemoglobin, and methemoglobin. Because each of these compounds has a distinct absorption spectra, multiwavelength spectrophotometers (co-oximeters) are used to determine the relative concentrations of each of these species in a blood sample. Four or more wavelengths are monitored and used to compute the concentration of each form of hemoglobin and the total hemoglobin concentration. Co-oximetry is the most accurate means to assess oxygenation status, particularly in cases where dysfunctional hemoglobins are present.

H. TEST RESULTS

1. Reference Ranges. The reference ranges in arterial blood are shown in Table 6–7 for the common blood gas and acid–base parameters. The clinically acceptable range for arterial blood is also shown, which may be used as a guideline in critical care. Note that the PO_2 of arterial blood varies with age (Table 6–6).

2. Clinical Evaluation of Blood Gas and Acid–Base Results. Ideally, the evaluation of acid–base disorders, including mixed acid–base disorders, would include simultaneous information on the patient's clinical history, the electrolyte results, and the blood gas results. However, in practice, the blood gas results are usually available before the electrolyte results.

The acid–base and blood gas results (pH, PCO_2, PO_2, HCO_3^-) may be evaluated in several steps, as follows:

1. Evaluate the pH.
2. Evaluate the ventilatory (PCO_2) and metabolic (HCO_3^-) status.
3. Determine if a mixed disorder is present.
   a. For a primary metabolic (or respiratory) disturbance, is respiratory (or metabolic) compensation appropriate?
   b. Do any other parameters, such as electrolytes, lactate, or anion gap indicate that an acid–base disturbance is present?
   c. Does the patient's history indicate other conditions associated with an acid–base disorder?
4. Evaluate arterial PO_2 and the cause of any abnormality, such as trauma, drugs, poor perfusion, airway obstruction, and so on.

▶   **Compensation** is a homeostatic response to an acid–base disorder in which the body attempts to restore the $(HCO_3^-):(PCO_2)$ ratio to normal. Compensation involves either a ventilatory response (change in PCO_2) to a metabolic abnormality, or a metabolic response (change in HCO_3^-) to a ventilatory abnormality. Some facts to remember about compensation are:

- It is driven by and is very sensitive to changes in pH.
- As compensation returns pH to normal, the pH-driven compensation process slows, then stops.
- Metabolic compensation by renal loss of HCO_3^- is much slower than is respiratory compensation by loss of CO_2.
- The compensatory response is in the same direction as the change from the primary abnormality.
- A failure to compensate as expected suggests that a mixed disorder is present.

Step 1: Evaluating the pH. An abnormal pH confirms that an acidosis or alkalosis has occurred. Furthermore, the pH indicates the extent of the acid–base disorder and may suggest that compensation has occurred. However, the pH by itself does not indicate whether a mixed disorder is present. Consider the following examples:

Table 6–7. Reference Ranges for Arterial Blood Gases

MEASUREMENT	REFERENCE RANGE (ARTERIAL)	CLINICALLY ACCEPTABLE RANGE (ARTERIAL)
pH	7.35–7.45	7.30–7.50
PCO_2	35–45 mm Hg (4.7–6.0 kPa[a])	30–50 mm Hg (4.0–6.7 kPa)
HCO_3^-	21–28 mmol/L	Variable
PO_2[b]	83–103 mm Hg (11.1–14.4 kPa)	> 80 mm Hg (> 10.7 kPa)
SO_2 (%)	> 95	> 90

[a] kPa = kilopascals; International unit for pressure; 1 mm Hg = 0.133 kPa.

[b] PO_2 of arterial blood varies with age.

- pH 7.20 confirms that a severe acidosis is present and that compensation is either not present or has not controlled the acidosis. Further investigation is required to determine the metabolic or respiratory origin of the acidosis and whether more than one acidotic process is present.
- pH 7.48 indicates that a mild alkalosis is present. Further information is required to determine the cause of the alkalosis and whether the alkalosis is in its early stages (and may get worse), has been nearly compensated, or is part of a mixed disorder.

Although a normal pH may indicate that the patient has no acid–base disorder, a normal pH must be evaluated carefully. This is because the patient may have a mix of acidotic and alkalotic events (primary or compensatory) that have offset each other. The patient's history, HCO_3^-, and PCO_2 must be examined along with the pH. Consider these examples:

- pH 7.45 suggests that a primary alkalosis has been compensated by an appropriate acidotic response. Inspection of the HCO_3^- and PCO_2 would determine whether the primary alkalosis was metabolic or respiratory.
- pH 7.40 with both HCO_3^- and PCO_2 abnormal indicates that a mixed acidosis and alkalosis are present that are fortuitously offsetting each other to give normal pH. This is because the compensatory response shuts down as pH approaches the outer limit of normal, either 7.35 or 7.45.

Step 2: Evaluating the Ventilatory and Metabolic Status. The simplest approach to evaluating the PCO_2 and HCO_3^- is to consider each parameter separately as indicating either acidosis, alkalosis, or normal, then evaluate them together to determine whether the primary disorder is respiratory or metabolic and whether a mixed disorder or compensation is present.

A decreased PCO_2 indicates a respiratory alkalotic process (hyperventilation). An increased PCO_2 indicates a respiratory acidotic process (hypoventilation). These may be either a primary abnormality or an appropriate compensation process.

A decreased HCO_3^- indicates a metabolic acidosis. An increased HCO_3^- indicates a metabolic alkalosis. These may be either primary, compensatory, or mixed.

A normal PCO_2 or normal HCO_3^- could indicate normal ventilatory or metabolic status, respectively. However, if one of these parameters is normal when the other is abnormal, then a mixed disorder may be present. That is, a lack of an appropriate compensatory response is considered a mixed disorder.

Consider two examples, each with a PCO_2 result of 25 mm Hg, indicating that a respiratory alkalotic process is present. Whether it is a primary abnormality or an appropriate compensatory change can be determined from the HCO_3^- and pH associated with this PCO_2:

- If the HCO_3^- is 14 mmol/L and the pH is about 7.32, the PCO_2 of 25 mm Hg probably indicates compensating hyperventilation (a respiratory alkalotic process) for a primary metabolic acidosis.

- If the HCO_3^- is 23 mmol/L and the pH is about 7.55, the PCO_2 of 25 mm Hg indicates a primary respiratory alkalosis, which could be either acute or chronic hyperventilation, depending on the duration of the hyperventilation. If acute, the metabolic compensatory response to eliminate HCO_3^- may not yet have had enough time to respond. If the hyperventilation is chronic, the inability of the kidney to compensate by eliminating HCO_3^- indicates that a metabolic alkalosis is also present.

Step 3: Determining Whether a Mixed Disorder Is Present

Is the Compensation Adequate for the Primary Disorder?

METABOLIC ACIDOSIS. Metabolic acidosis is diagnosed as a decreased pH with a decreased HCO_3^-. The expected compensatory response is hyperventilation, which effectively removes acid by eliminating CO_2 (an acidic gas). After several hours, for a given HCO_3^- (mmol/L), the expected PCO_2 can be approximated by the following rule:

$$\text{Expected } PCO_2 \ (\pm 2 \text{ mm Hg}) = (1.5 \times HCO_3^-) + 8$$

Example: An HCO_3^- of 15 mmol/L indicates a metabolic acidosis. From the above equation, the PCO_2 should be 28 to 33, if the expected ventilatory response is occurring. Patients with a $PCO_2 < 28$ mm Hg may have an underlying respiratory alkalosis, or patients with a $PCO_2 > 33$ mm Hg may have an underlying respiratory acidosis.

METABOLIC ALKALOSIS. Metabolic alkalosis is diagnosed as an increased pH with an increased HCO_3^-. The expected compensatory response is hypoventilation, which increases PCO_2 to normalize pH. By several hours after onset of metabolic alkalosis, for a given HCO_3^-, the expected PCO_2 can be approximated by:

$$\text{Expected } PCO_2 \ (\pm 2 \text{ mm Hg}) = (0.9 \times HCO_3^-) + 9$$

Example: With a HCO_3^- of 35 mmol/L, PCO_2 should be 38 to 43 mm Hg, if the expected ventilatory response is occurring. Patients with a $PCO_2 < 38$ mm Hg may have an underlying respiratory alkalosis, or patients with a $PCO_2 > 43$ mm Hg may have an underlying respiratory acidosis.

RESPIRATORY ACIDOSIS. Respiratory acidosis (ventilatory failure) results from decreased ventilatory removal of CO_2, which causes blood PCO_2 to increase. In the acute phase (< 24 hours), metabolic compensation is quite limited; plasma bicarbonate increases by no more than ~2 mmol/L. Patients with acute ventilatory failure and an elevated plasma HCO_3^- may also have an underlying metabolic alkalosis, and patients with acute ventilatory failure and a decreased plasma HCO_3^- may also have an underlying metabolic acidosis.

After 1 to 4 days, chronic ventilatory failure should increase retention of HCO_3^- and return pH to normal. For a given PCO_2 in mm Hg, the plasma bicarbonate should be approximately:

$$\text{Expected } HCO_3^- \ (\pm 3 \text{ mmol/L}) = (0.43 \times PCO_2) + 7$$

Example: If PCO_2 increases from 40 to 70 mm Hg, then HCO_3^- should be ~37 mmol/L. Therefore, patients with a bicarbonate of 34 to 40 mmol/L are having the expected renal response to increased PCO_2 in chronic ventilatory failure. Patients whose bicarbonate is < 34 mmol/L may also have a primary metabolic acidosis, and patients whose bicarbonate is > 40 mmol/L may have an underlying metabolic alkalosis.

RESPIRATORY ALKALOSIS (HYPERVENTILATION). Hyperventilation enhances ventilatory removal of CO_2 from the blood, resulting in a respiratory alkalosis in which PCO_2 is decreased and pH is increased. In the acute phase of hyperventilation, plasma bicarbonate usually remains normal or only slightly decreased. If the plasma bicarbonate is abnormal, the patient may have an underlying primary metabolic acidosis (bicarbonate decreased) or primary metabolic alkalosis (bicarbonate increased).

In chronic hyperventilation, the pH may be normalized, or compensated, by renal excretion of bicarbonate with a corresponding increase in Cl^-. For a given PCO_2 in mm Hg, the HCO_3^- should be about:

$$\text{Expected } HCO_3^- \text{ (mmol/L)} = (0.5 \times PCO_2) + 5$$

Example: If PCO_2 decreases from 40 to 20 mm Hg, then HCO_3^- should be ~15 mmol/L.

Do Other Laboratory Results Suggest That an Acid–Base Disorder Is Present?

In addition to the guidelines described in the previous section for determining whether HCO_3^-, PCO_2, and pH results indicate appropriate compensation or an additional underlying primary disorder, other common tests may help determine if two or three primary acid–base disorders are present.

POTASSIUM. Hypokalemia suggests the possibility of metabolic alkalosis.

CHLORIDE. Hyperchloremia may indicate a primary metabolic acidosis or a respiratory alkalosis with metabolic acidotic compensation. Hypochloremia may indicate metabolic alkalosis either as a primary disorder (as HCO_3^- increases) or as compensation for chronic respiratory acidosis (as Cl^- is excreted with NH_4^-).

ANION GAP (AG). An elevated AG [$(Na^+ + K^+) - (Cl^- + CO_2)$] of > 20 mmol/L indicates metabolic acidosis. AG is also useful in differentiating the causes of metabolic acidoses:

- An elevated AG occurs in uremic acidosis, ketoacidosis, lactate acidosis, and ingestion of toxin (e.g., glycol or salicylate).
- A normal AG occurs in renal tubular acidosis and diarrhea, among other conditions.

As a calculated value, AG is subject to variation from the measurements of three (or four) electrolytes and is a relatively insensitive indicator of metabolic acidosis. Some reports claim that, in ~33% of cases, anion gaps even well above 20 mmol/L did not indicate that a metabolic acidosis was present and that about half of patients with a clearly elevated blood lactate did not have an elevated AG.

Causes of Elevated Anion Gap

- Uremic acidosis
- Ketoacidosis
- Lactate acidosis
- Ingestion of toxin

Table 6–8. Examples of the Expected Acid–Base Disorder Associated with Various Clinical Conditions

CLINICAL CONDITION	EXPECTED ACID–BASE DISORDER
Cardiac arrest	Metabolic acidosis
Pulmonary arrest	Respiratory acidosis
Hyperventilation	Respiratory alkalosis
Blockage of endotracheal tube (after removal of blockage from tube)	Respiratory acidosis (respiratory alkalosis)
Congestive heart failure leading to hyperventilation	Respiratory alkalosis
Vomiting	Metabolic alkalosis
Diarrhea	Metabolic acidosis
Shock state (inadequate perfusion)	Metabolic acidosis
Pulmonary edema	Respiratory alkalosis (hypoxia leads to hyperventilation)
Severe pulmonary edema	Respiratory acidosis
Diuretic therapy	Metabolic alkalosis
Drug intoxication	Respiratory acidosis (respiratory arrest)
Bicarbonate therapy	Metabolic alkalosis
Poor perfusion	Metabolic acidosis

LACTATE. A rising blood lactate is a sensitive indicator of metabolic acidosis resulting from poor oxygenation of tissues.

CREATININE. An elevated creatinine indicates renal insufficiency and possible uremic metabolic acidosis.

Does the Patient Have Other Conditions Associated with an Acid–Base Disorder? An evaluation of blood gas results must always include the patient's historical and physical findings. Many conditions are frequently associated with an acid–base disorder (some common ones are listed in Table 6–8), especially if the blood gas results cannot readily be explained by the patient's "primary" disorder. The presence of more than one such condition should alert one to the possibility of a mixed acid–base disorder. Mixed acid–base conditions are quite common.

3. Clinical Evaluation of Oxygenation Status

Step 1: Evaluating Arterial Oxygenation. Hypoxemia is defined as an arterial PO_2 less than the minimally acceptable limit for persons of a given age group breathing room air (Table 6–9).

Table 6–9. Evaluation of Hypoxemia in Patients < 60 Years Old (Breathing Room Air)

ARTERIAL PO_2[a]	CONDITION
> 80 mm Hg	Adequate oxygenation
60–80 mm Hg	Mild hypoxemia
40–60 mm Hg	Moderate hypoxemia
< 40 mm Hg	Severe hypoxemia

[a] For each year over 60, subtract 1 mm Hg from the limits for mild and moderate hypoxemia. A PO_2 < 40 mm Hg at any age indicates severe hypoxemia.

Source: Bishop ML, Duben-Engelkirk JL, Fody EP. Clinical chemistry: Principles, procedures, correlations, 3rd ed. Philadelphia: JB Lippincott, 1996.

Table 6–10. Effect of Breathing Oxygen-Enriched Air

F_{IO_2} (%)	PREDICTED MINIMAL PO_2
30	150 mm Hg
40	200 mm Hg
50	250 mm Hg
80	400 mm Hg
100	500 mm Hg

Source: Shapiro BA, Peruzzi WT, Templin R. *Clinical applications of blood gases,* 5th ed. St. Louis: CV Mosby, 1994.

It is important to note that the actual diagnosis of hypoxemia can be made only when the patient is breathing room air; however, oxygen therapy must not be interrupted simply to assess hypoxemia. Instead, the likelihood of hypoxemia in a patient breathing oxygen-enriched air may be evaluated with use of the predicted minimal PO_2 for a given percent oxygen being breathed [F_{IO_2} (%)]. If this minimally acceptable PO_2 is not achieved, the patient may be assumed to have been hypoxemic on room air (see Table 6–10). A simple way to remember the minimally acceptable PO_2 in millimeters of mercury (mm Hg) is to multiply F_{IO_2} (%) by 5.

Step 2: Evaluating Tissue Oxygenation. Assessment of tissue oxygenation requires clinical assessment of cardiac output, peripheral perfusion, and blood oxygen transport.

Although cardiac output may be crudely assessed by measurements such as blood pressure, heart rate, and electrocardiogram, the definitive method is by a pulmonary artery catheter with a thermodilution tip that measures blood flow; a known quantity of cold solution is injected, with the change in blood temperature measured and related to blood flow.

Peripheral perfusion may also be crudely assessed by pulse pressure, skin color and temperature, and urine output. More sophisticated assessment is provided by determination of the blood lactate.

Blood oxygen transport is assessed by measurements of blood oxygen content. Anemia will dramatically affect this, as will hypoxemia (low PO_2), hypercapnia (high PCO_2), acidemia (low pH), and hyperthermia, all of which cause hemoglobin to be less saturated for a given PO_2.

SUGGESTED READINGS

Astles JR, Lubarsky D, Loun B, Sedor FA, Toffaletti JG. Pneumatic transport exacerbates interference of room air contamination of blood gas samples. *Arch Pathol Lab Med* 120:642–47, 1996.

Astrup A, Severinghaus JW. *The history of blood gases, acids and bases.* Copenhagen: Munksgaard, 1986.

Bishop ML, Duben-Engelkirk JL, Fody EP (eds.). *Clinical chemistry: Principles, procedures, correlations,* 4th ed. Philadelphia: JB Lippincott, 2000.

Burtis CA, Ashwood ER. *Tietz textbook of clinical chemistry,* 3rd ed. Philadelphia: WB Saunders, 1999.

Ehrmeyer SS, Shrout JB. Blood gases, pH, and buffer systems. In Bishop ML, Duben-Engelkirk JL, Fody EP (eds.). *Clinical chemistry: Principles, procedures, correlations,* 4th ed. Philadelphia: JB Lippincott, 2000.

Fleisher M, Gladstone M, Crystal D, Schwartz MK. Two whole-blood multi-analyte analyzers evaluated. *Clin Chem* 35:1532–35, 1989.

Gabow PA, Kaehny WD, Fennessey PV, et al. Diagnostic importance of an increased anion gap. *N Engl J Med* 303:854–58, 1980.

Henry JB. *Clinical diagnosis and management by laboratory methods*, 19th ed. Philadelphia: WB Saunders, 1996.

Iberti TJ, Leibowitz AB, Papadakos PJ, Fischer EP. Low sensitivity of the anion gap as a screen to detect hyperlactatemia in critically ill patients. *Crit Care Med* 18:275–77, 1990.

Lambertsen CJ. Transport of oxygen, carbon dioxide, and inert gases by the blood. In Mountcastle VB (ed.). *Medical physiology*, 14th ed. St. Louis: CV Mosby, 1980; 1725.

Mahoney TJ, Harvey JA, Wong RJ, Van Kessel AL. Changes in oxygen measurements when whole blood is stored in iced plastic or glass syringes. *Clin Chem* 37:1244–48, 1991.

Narins RG (ed.). *Maxwell and Kleeman's clinical disorders of fluid and electrolyte metabolism*, 5th ed. New York: McGraw-Hill, 1994.

Rimmer JM, Horn JF, Gennari FJ. Hyperkalemia as a complication of drug therapy. *Arch Intern Med* 147:867–69, 1987.

Rose BD. *Clinical physiology of acid–base and electrolyte disorders*, 3rd ed. New York: McGraw-Hill, 1989.

Shapiro BA, Peruzzi WT, Templin R. *Clinical applications of blood gases*, 5th ed. St. Louis: CV Mosby, 1994.

Tuchschmidt J, Oblitas D, Fried JC. Oxygen consumption in sepsis and septic shock. *Crit Care Med* 19:664–71, 1991.

Walmsley RN, White GH. *A guide to diagnostic clinical chemistry*, 3rd ed. Oxford: Blackwell Scientific, 1994; 79–127.

Items for Further Consideration

1. Should serum osmolality replace the measurement of serum sodium?

2. How should the laboratory handle a hemolyzed blood sample sent for measurement of potassium?

3. Why would a blood gas result that is reported 60 minutes after collection be useless for medical treatment?

4. Are blood gas and electrolyte results used mostly for diagnosis or monitoring?

5. How has and will point-of-care testing affect all laboratory testing for blood gases and electrolytes?

Endocrine Function

Janine Denis Cook

Key Terms

ADDISON'S DISEASE

ALDOSTERONE

ANDROGENS

CATECHOLAMINES

CONN'S SYNDROME

CORTISOL

CUSHING'S SYNDROME

DIURNAL VARIATION

ESTROGENS

EUTHYROID SICK SYNDROME

FEEDBACK CONTROL
POSITIVE FEEDBACK
NEGATIVE FEEDBACK

GLUCOCORTICOIDS

GRAVES' DISEASE

GROWTH HORMONE

HASHIMOTO'S THYROIDITIS

HORMONES

HYPERGONADISM

HYPERTHYROIDISM

HYPOGONADISM

HYPOTHALAMIC–PITUITARY–
TARGET ORGAN AXIS

HYPOTHYROIDISM

MINERALOCORTICOIDS

PHEOCHROMOCYTOMAS

PITUITARY

PROLACTIN

RELEASING FACTORS

TROPIC HORMONE

Chapter Objectives

Upon completion of this chapter, the reader will be able to:

1. List the major hormones from the hypothalamus, pituitary, and target gland/organ that are clinically important in regard to thyroid, adrenal, and reproductive function.
2. Discuss the functioning of the hypothalamic–pituitary axis.
3. Explain how positive and negative feedback mechanisms control hormone concentration in the peripheral circulation.
4. Differentiate between free and bound hormone and indicate which form is more physiologically important.
5. Distinguish between primary, secondary, and tertiary disease states.
6. State the biological functions of the major thyroid, adrenal, and reproductive hormones.
7. Describe the etiology of the major disease states associated with thyroid, adrenal, and reproductive function.

8. Summarize expected laboratory findings in regard to thyroid, adrenal, and reproductive disorders.
9. Summarize expected hormonal changes in pregnancy, menstruation, and menopause.
10. Discuss the metabolism and clinical function of growth hormone and prolactin.
11. Discuss laboratory testing strategies including techniques such as immunoassay and dynamic testing, and screening and confirmatory protocols.

Analytes Included

ADRENOCORTICOTROPIC HORMONE (ACTH)

ALDOSTERONE

ANTIDIURETIC HORMONE/VASOPRESSIN (ADH)

CATECHOLAMINES (EPINEPHRINE, NOREPINEPHRINE, DOPAMINE)

CORTICOTROPIN-RELEASING HORMONE (CRH)

CORTISOL

DEHYDROEPIANDROSTERONE (DHEA)

ESTROGEN/ESTRADIOL

FOLLICLE-STIMULATING HORMONE (FSH)

GONADOTROPIN-RELEASING HORMONE (GnRH)

GROWTH HORMONE (GH)

HUMAN CHORIONIC GONADOTROPIN (hCG)

LUTEINIZING HORMONE (LH)

METANEPHRINES

OXYTOCIN

PROGESTERONE

PROLACTIN (PRL)

PROLACTIN-INHIBITING FACTOR (PIF)

RENIN

TESTOSTERONE

THYROID ANTIBODIES (TPO, TGAB, TSI, TBII)

THYROID-STIMULATING HORMONE (TSH)

THYROTROPIN-RELEASING HORMONE (TRH)

TETRAIODOTHYRONINE/THYROXINE (T_4)

TRIIODOTHYRONINE (T_3)

VANILLYLMANDELIC ACID (VMA)

Pretest

Directions: Choose the single best answer for each of the following items.

1. Which of the following hormone pairs exhibits positive feedback control?
   a. TSH–T_3
   b. ACTH–cortisol
   c. LH–estradiol
   d. CRH–testosterone

2. In the negative feedback loop between the adrenal glands and the pituitary, an increase in cortisol results in a(n)
   a. decrease in aldosterone
   b. increase in renin
   c. decrease in ACTH
   d. increase in catecholamines

3. Releasing hormones are produced by the
   a. anterior pituitary
   b. hypothalamus
   c. target organ
   d. cerebellum

4. Which of the following thyroid hormones is more potent biologically?
   a. free T_3 (FT_3)
   b. reverse T_3 (rT_3)
   c. total T_4 (TT_4)
   d. thyroglobulin

5. Which of the following hormones exhibits diurnal variation in regard to its release into the peripheral circulation?
   a. TSH
   b. progesterone
   c. hCG
   d. cortisol

6. The measurement of which of the following hormones is affected by posture and salt intake?
   a. ACTH
   b. aldosterone
   c. testosterone
   d. catecholamines

7. The precursor of the adrenal steroid hormones is
   a. tyrosine
   b. bilirubin
   c. cholesterol
   d. vitamin A

8. Which of the following laboratory results is consistent with a diagnosis of pheochromocytoma?
   a. increased urine cortisol, decreased plasma ACTH
   b. increased plasma renin, increased urine aldosterone
   c. increased serum testosterone, decreased dehydroepiandrosterone sulfate (DHEAS)
   d. increased plasma catecholamines, increased urine vanillylmandelic acid (VMA)

9. The most abundant adrenal androgens are
   a. testosterone and dihydrotestosterone
   b. DHEA and DHEAS
   c. androstenedione and pregnenolone
   d. estradiol and 17-OH-progesterone

10. The most potent natural estrogen produced by non-pregnant women is
    a. estriol
    b. progesterone
    c. estradiol
    d. prolactin

11. Changes in protein concentration affect which of the following thyroid hormones the most?
    a. TSH
    b. rT_3
    c. FT_4
    d. TT_4

12. Which of the following laboratory results is consistent with the presence of Graves' disease?
    a. increased TSH
    b. increased FT_4
    c. decreased FT_3
    d. decreased thyroxine-binding globulin (TBG)

13. Which of the following autoimmune disorders results in hypothyroidism?
    a. Hashimoto's thyroiditis
    b. Graves' disease
    c. euthyroid sick syndrome
    d. goiter

14. The recommended screening test for thyroid disease in ambulatory nonhospitalized patients is
    a. FT_4
    b. TT_4
    c. FT_3
    d. TSH

15. In secondary Cushing's syndrome, or Cushing's disease, which of the following laboratory results may be expected?
    a. increased ACTH, increased cortisol
    b. increased cortisol, decreased ACTH
    c. decreased aldosterone, increased ACTH
    d. increased catecholamines, decreased ACTH

16. The laboratory findings listed below are consistent with which of the following disease states?

    Increased—ACTH and potassium
    Decreased—cortisol, aldosterone, and sodium

    a. primary Cushing's syndrome
    b. Conn's syndrome
    c. Addison's disease
    d. pheochromocytoma

17. In regard to a woman in menopause, which of the following laboratory results would be expected?
    a. increased progesterone
    b. decreased estradiol
    c. decreased LH
    d. decreased FSH

18. During pregnancy, which of the following hormones is produced and secreted in increased concentration?
    a. LH
    b. progesterone
    c. FSH
    d. DHEAS

19. For TSH immunoassays, functional sensitivity is the
    a. limit of detection defined as the lowest concentration measurable that is distinguishable from zero
    b. reference interval
    c. limit of quantification at a specified level of imprecision
    d. dynamic range of the assay

20. An example of a hormone that is lipid soluble and thus highly bound to carrier protein is
    a. hCG
    b. T_4
    c. LH
    d. GnRH

I. INTRODUCTION

▶ **Hormones,** the chemical substances that are produced and secreted into the bloodstream by an organ or tissue, have specific effects on their target organs that are typically located some distance from the site of hormone production. Hormones are usually organized into a chain of chemical messengers. The primary configuration of this chain is the hypothalamus–pituitary–target organ axis. An overview of the hormones in this axis is given in Table 7–1. Through this chain of hormonal communication, the concentrations of individual hormones are tightly controlled to elicit the normal desired effects.

II. HORMONE METABOLISM

A. *BIOCHEMISTRY*

Hormones can be grouped into three major classes based on structure—protein or polypeptide, steroid, and aromatic amino acid derivatives. The peptide or glycoprotein hormones are gene products produced by the hypothalamus, pituitary, and certain target organs. In response to stimulation, these hormones are released into the bloodstream and circulate freely, without being bound to a carrier protein. Examples of this hormone class include the hypothalamic releasing hormones and the hormones from the anterior pituitary.

The steroid hormones are polycyclic cholesterol–derived molecules that are lipid soluble and generally hydrophobic. They are synthesized by certain target organs and released into circulation, where they are predominantly transported bound to carrier plasma proteins. It is the small percentage of hormone remaining unbound or free that is biologically active and interacts with its specific cellular receptor to exert its biological effects. Examples of this hormone class include cortisol, aldosterone, testosterone, estradiol, and progesterone.

The aromatic amino acid–derived hormones are synthesized from the common precursor tyrosine. Both thyroid hormones, triiodothyronine (T_3) and tetraiodothyronine (T_4), and the adrenal medullary catecholamines are derived from tyrosine. Whether these hormones are found free or protein bound in circulation is dependent on the hormone's solubility.

To communicate with the body's cells, hormones rely on specialized proteins, known as cellular receptors, which are capable of binding the hormone molecules with great affinity and avidity, a vital characteristic since hormones are present in the bloodstream in extremely low concentration. Moreover, the biological half-lives of circulating hormones are typically short, thus limiting the window for hormones to produce their desired effects.

B. *HORMONAL REGULATION THROUGH FEEDBACK CONTROL MECHANISMS*

Maintaining the balance between hormone levels and their effects is
▶ vital to supporting homeostasis within the body. **Feedback control** systems tightly regulate hormone concentrations in the body to sustain this balance. Two types of feedback exist: negative and positive. Figure 7–1 provides a generalized schematic for these two feedback

---

Classes of Hormones

- Protein or polypeptide
- Steroid
- Aromatic amino acid derivatives

Table 7–1. Hypothalamic–Anterior Pituitary–Target Organ Axis

HYPOTHALAMIC HORMONES	ANTERIOR PITUITARY HORMONES	TARGET GLAND	TARGET HORMONES
TRH	TSH	Thyroid	T_4/T_3
CRH	ACTH	Adrenals	Glucocorticoids (cortisol)
			Mineralocorticoids (aldosterone)
			Medullary (catecholamines)
GnRH	LH/FSH	Ovaries/Testes	Sex steroids (estradiol, testosterone, progesterone)
GHRH/GHIH	GH	Bone/Cartilage	Insulin-like growth factors
PIF/PRL	PRL	Mammary glands	

mechanisms. **Negative feedback,** either directly or indirectly, is the
principal mechanism of control of the **hypothalamic–pituitary–target organ axis.** In negative feedback, hormone A stimulates the
production of hormone B that subsequently acts to diminish the production of hormone A. Typically, the target organ hormone, when
present in excessive amounts, inhibits the production of its hypothalamic releasing factor and its pituitary hormone. Production of the
target hormone is subsequently decreased, thus keeping all hormone
concentrations within normal levels. **Positive feedback** is a specialized method of regulation whereby hormone A stimulates the production of hormone B that in turn stimulates further the production
of hormone A. Estradiol is an example of positive feedback when its
rapidly rising levels in the menstruating female at midcycle stimulate
the release of luteinizing hormone (LH). These rising LH concentrations then further stimulate the release of estradiol.

The hypothalamus is a structure located in the brain that synthesizes and secretes hormones referred to as **releasing factors** and in-

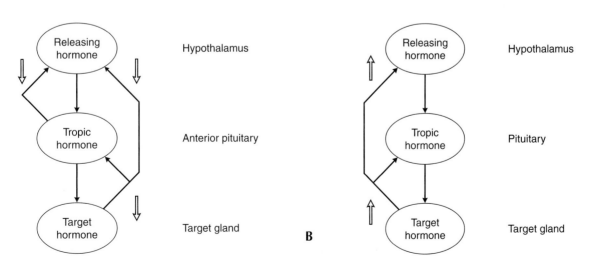

Figure 7–1. **A.** Endocrine negative feedback regulation (⇓ represents negative feedback). **B.** Endocrine positive feedback regulation (⇑ represents positive feedback).

hibiting factors. Eight hypothalamic hormones have been identified. Central nervous system (CNS) stimuli, as well as target organ hormones through negative and positive feedback loops, regulate the release of hypothalamic hormones. Most of the hormones produced by the hypothalamus are transported to the anterior pituitary through a portal circulatory system. Antidiuretic hormone (ADH) and oxytocin are transported from the hypothalamus to the posterior pituitary, which serves as a reservoir for these hormones until they are required.

▶ The **pituitary** gland, or hypophysis, is a small gland located in a bone cavity, known as the sella turcica, found at the base of the skull. In the adult human, the gland is composed of two different lobes—the anterior lobe (adenohypophysis) and the posterior lobe (neurohypophysis). The anterior pituitary synthesizes and releases both tropic and nontropic hormones in response to signals from the

▶ hypothalamus. A **tropic hormone,** such as TSH, stimulates its target organ, the thyroid gland, to release target hormones, T_3 and T_4, to exert the desired effects. In contrast, the nontropic hormone prolactin acts directly without the production of a target hormone. The hormones produced and synthesized by the anterior pituitary are shown in Table 7–1.

The target gland hormones are produced and released in response to their specific tropic hormones from the anterior pituitary. The target hormones, their site of production, and their ultimate site of action are also listed in Table 7–1.

C. HORMONE METABOLISM IN THYROID FUNCTION

Thyrotropin-releasing hormone (TRH) is a tripeptide synthesized and released from the hypothalamus that is responsible for stimulating the synthesis and secretion of the glycoprotein thyroid-stimulating hormone (TSH) from the anterior pituitary gland. TSH, in turn, is responsible for stimulating the thyroid gland to synthesize and secrete T_4 (thyroxine) and T_3. T_4 is synthesized exclusively in the thyroid and may be a reservoir or prohormone for T_3. Thyroid hormone metabolism is outlined in Figure 7–2. T_4 is predominantly metabolized in the peripheral circulation to T_3. T_4 can also be metabolized to reverse T_3 (rT_3), an inactive form of the hormone. The remainder of the unmetabolized T_4 is secreted via the biliary tract. Some T_3 is synthesized directly in the thyroid gland. T_3 is the most active thyroid hormone, being three to four times more potent than T_4; T_3 is the predominant thyroid negative feedback regulator.

Both T_4 and T_3 circulate predominantly (> 99%) bound to the serum proteins, thyroxine-binding globulin (TBG), thyroxine-binding prealbumin (TBPA), and albumin. The protein-bound hormone, whether T_4 or T_3, is not biologically active; it is the free hormone that exerts all physiological effects and is tightly regulated through the feedback mechanism of the hypothalamic–pituitary–target gland axis. These free hormones, free T_3 (FT_3) and free T_4 (FT_4), represent a very small percentage of the total thyroid hormones (FT_4 0.03%, FT_3 0.3%). FT_3 is the more biologically active hormone, mediating about 80% of the thyroid negative feedback system. Both TRH and TSH release is controlled through negative feedback from the target organ hormones, FT_4 and FT_3.

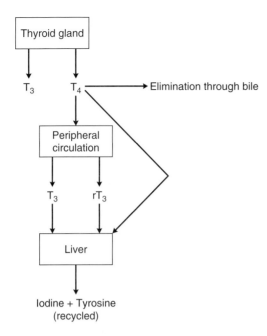

Figure 7–2. Metabolism of the thyroid hormones.

D. HORMONE METABOLISM IN ADRENAL FUNCTION

The polypeptide corticotropin-releasing hormone (CRH), synthesized and released from the hypothalamus, stimulates the synthesis and release of the polypeptide adrenocorticotropic hormone (ACTH) from the anterior pituitary. ACTH subsequently stimulates the secretion of the steroid cortisol from the adrenal glands. Adrenal hormone homeostasis is maintained through a cortisol negative feedback loop at the level of both the hypothalamus and the pituitary. Additionally, CRH release is subjected to CNS controls and negative feedback from ACTH. ACTH exhibits a **diurnal variation,** with the highest levels ◀ occurring in the early morning, typically at about 8 A.M.

Anatomically, the adrenal glands are two small, paired, triangular-shaped glands located atop the kidneys and are composed of an outer cortex and inner medulla. The hormones secreted by the adrenal cortex are classified as **glucocorticoids** and **mineralocorti-** ◀ **coids.** In addition, a small portion of the steroid sex hormones—the androgens, estrogens, and progestins—are produced in the adrenal cortex. The adrenal medulla produces the **catecholamines.** Figures ◀ 7–3 and 7–4 depict the general metabolic pathways of the adrenocortical and medullary hormones.

Cortisol is the major glucocortical steroid hormone. Its secre- ◀ tion into the bloodstream exhibits diurnal variation, with levels in highest concentration at approximately 8 A.M. and in lowest concentration in late evening. Disruption of the normal sleep–wake cycle affects this diurnal variation pattern. Stress is also a powerful regulator of cortisol and overrides cortisol's diurnal variation pattern and negative feedback regulation.

Because of their lipid solubility, glucocorticoids in circulation are typically bound to specific carrier proteins. The small portion of cortisol that is free in circulation represents the biologically active

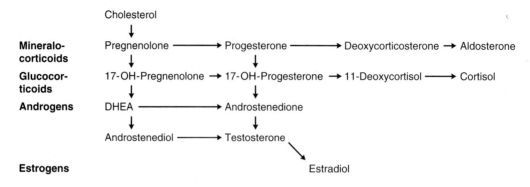

Figure 7–3. Biosynthetic pathway of the cortical adrenal steroids.

fraction. The glucocorticoid hormones have no specific target organ but instead exert their influence throughout the body.

▶   **Aldosterone** is the major adrenal mineralocorticoid hormone. In contrast to cortisol, aldosterone is poorly bound to steroid-binding proteins and thus predominantly exists in the biologically active free form. Aldosterone acts to maintain electrolyte balance and extracellular fluid volume. Regulation of aldosterone secretion is threefold, with primary control through the renin–angiotensin system. The remaining two regulators are extracellular potassium and sodium and, to a much lesser extent, ACTH. In the renin–angiotensin system, the juxtaglomerular apparatus, a volume/osmoreceptor located in afferent arterioles of the renal glomeruli, regulates aldosterone secretion. A decrease in blood volume, an increase in plasma potassium, or a negative sodium balance causes the release of the proteolytic enzyme renin from the kidney. Figure 7–5 illustrates the general mechanism of how renin acts to initiate the secretion of aldosterone.

▶   The **androgens** and **estrogens** are the sex steroid hormones that affect male and female primary and secondary sex characteristics. The most abundant adrenal androgens are dehydroepiandrosterone (DHEA) and dehydroepiandrosterone sulfate (DHEAS). The regulation of sex hormones produced in the adrenal cortex is under the control of ACTH. A more detailed discussion of the sex hormones is presented in a later section of this chapter.

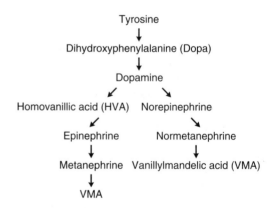

Figure 7–4. Metabolic pathway of the adrenal medullary catecholamines.

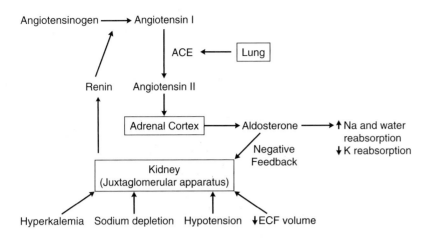

Steps:
1. Renin, an enzyme, is released from the kidney in response to sodium depletion, hyperkalemia, hypotension, or a decreased extracellular fluid (ECF) volume.
2. Renin, a proteolytic enzyme, catalyzes the conversion of angiotensinogen to angiotensin I in the peripheral circulation.
3. Angiotensin I is converted to angiotensin II by angiotensin-converting enzyme (ACE) in the lungs.
4. Angiotensin II, also a potent vasoconstrictor, acts on the adrenal cortex to stimulate the production of aldosterone.
5. Aldosterone acts on the renal tubules to increase the reabsorption of sodium and water and increase the excretion of potassium, thus working to restore proper electrolyte balance and ECF volume. Aldosterone also acts in a negative feedback mechanism on the juxtaglomerular cells in the kidney.

Figure 7–5. Renin–angiotensin system for the release of aldosterone.

The catecholamine hormones—epinephrine, norepinephrine, and dopamine—are produced by the adrenal medulla. Epinephrine, which comprises about 80 to 90% of the total adrenal catecholamines produced, functions as a neurotransmitter and an effector of metabolism to provide energy in times of stress. Both norepinephrine and dopamine function solely as neurotransmitters, with norepinephrine preparing muscles for "fight or flight" and dopamine being involved with signal transmission. Catecholamines are produced elsewhere in the body besides the adrenals; for instance, the CNS nerves synthesize dopamine. Stressors, such as fear and pain, stimulate the release of catecholamines from the adrenals. As shown in Figure 7–4, the catecholamines are catabolized to metanephrine, normetanephrine, and vanillylmandelic acid (VMA), which are excreted into the urine.

E. HORMONE METABOLISM IN REPRODUCTIVE FUNCTION

Gonadotropin-releasing hormone (GnRH), a decapeptide synthesized and released from the hypothalamus, stimulates the production and release of LH and follicle-stimulating hormone (FSH) by the anterior pituitary. LH, a glycoprotein hormone, is released as a steady basal secretion with superimposed spikes. In the adult female, LH is responsible for promoting ovulation through its menstrual midcycle surge, for transforming the ruptured follicle into the corpus luteum, and for stimulating production of estradiol and progesterone from the corpus luteum. In the adult male, LH stimulates the Leydig cells of the testes to synthesize and secrete testosterone. Like LH, FSH is also released in a pulsatile fashion in response to GnRH. In the adult female, FSH promotes the growth and development of the ovarian follicles and stimulates estradiol production and secretion by these maturing follicles. In the adult male, FSH promotes spermatogenesis.

The CNS and sex steroid hormones control GnRH release; testosterone in males and the estrogens in females provide negative feedback control while estrogen regulates via positive feedback just prior to ovulation in menstruating females. The pulsatile secretion of GnRH by the hypothalamus is necessary to prevent the pituitary from becoming sensitized to the releasing hormone's signal. The frequency of the GnRH signal codes for the selective release of LH and FSH, thus allowing one releasing hormone to stimulate two different pituitary hormones.

Regulation of FSH and LH release in the adult female is provided through negative feedback from estradiol produced by the developing follicles and the corpus luteum. Also, the secretion of LH is controlled by positive feedback from the rapidly rising levels of estrogen just prior to ovulation. In males, testosterone from the testicular Leydig cells controls FSH and LH release through negative feedback. The peptide inhibin produced by the seminiferous tubules and the ovaries regulate FSH secretion also.

The male androgenic sex steroids include testosterone, dihydrotestosterone (DHT), androstenediol, androstenedione, DHEA, and DHEAS. The most important androgen is testosterone. This steroid is synthesized in the male by the testes under the hormonal control of LH. In males, testosterone is responsible for the masculine differentiation of the fetal genital tract and the development and maintenance of the male secondary sex characteristics and spermatogenesis. In the female, testosterone is produced by the adrenals and ovaries and through the peripheral metabolism of weak adrenal androgens, such as androstenedione, in adipose tissue. Testosterone serves as an estrogen precursor in the female. Testosterone exhibits a diurnal variation with concentrations being highest between 6 and 9 A.M. Because this hormone is lipid soluble, testosterone is predominantly bound in circulation to carrier proteins; about 60% of testosterone is bound to sex hormone–binding globulin (SHBG), about 40% to albumin, and less than 1% of the hormone is found in the free state. This unbound or free testosterone is biologically active.

Estradiol is the most potent natural estrogen that is produced mainly by the testes in males and the ovaries in nonpregnant females. Estradiol is responsible for the development of the female secondary sex characteristics. In the adult female, it also causes the proliferation of the endometrium, increases the thickness of the vaginal epithelium, increases the vascularity of the cervix, increases cervical mucous elasticity to allow sperm penetration, and dilates the cervical os. Like testosterone, it circulates predominantly bound to SHBG.

The ovarian corpus luteum, the placenta, and the adrenals synthesize progesterone. Its function is to prepare for and maintain pregnancy by thickening and vascularizing the endometrium, to in-

Androgenic Sex Steroids

- Testosterone
- Dihydrotestosterone
- Androstenediol
- Androstenedione
- DHEA
- DHEAS

duce the differentiation of estrogen-primed ductal breast tissue, to support the breast's secretory function during lactation, and to increase basal body temperature.

F. MISCELLANEOUS HORMONES: METABOLISM OF GROWTH HORMONE AND PROLACTIN

Growth hormone (GH) is a protein hormone synthesized and released from the anterior pituitary in response to stimulation from the hypothalamic growth hormone–releasing hormone (GHRH). Stimuli such as exercise, physical and emotional stress, hypoglycemia, increased amino acid levels (particularly arginine), and hormones such as testosterone, estrogens, and thyroxine stimulate GH secretion. Abnormally high levels of glucocorticoids suppress GH secretion. Serum GH influences the release of its hypothalamic hormone through a short negative feedback loop. Growth hormone mediates growth in many soft tissues, muscle, cartilage, and bone. It affects metabolic processes by stimulating protein synthesis and fat and glucose metabolism. The effects of GH are exerted directly and indirectly through insulin-like growth factors, previously known as somatomedins, that are under GH control. GH is secreted episodically with a circulation half-life of 20 minutes and fluctuates, with levels being stable and relatively low during most of the day. One to two secretion spikes occur 3 to 4 hours after a meal. There is a marked rise to peak GH values within the first 2 hours of sleep due to stimulation by serotonin.

Prolactin is a polypeptide that is under inhibitory control by hypothalamic dopamine. Prolactin exhibits a diurnal variation with highest levels occurring during sleep. Though it has no known function in males, in adult females it is responsible for the initiation and maintenance of lactation. Prolactin acts directly without the production of a target gland hormone. Thus, it is a nontropic and self-regulating hormone with no feedback mechanism. Prolactin also inhibits the pulsatile secretion of GnRH and the positive feedback of estradiol on LH release.

III. CLINICAL OVERVIEW

Endocrine disorders occur due to either hyperfunction or hypofunction of an endocrine gland. Classification of the hyper- or hypofunction is primary, secondary, or tertiary, depending on the site of the defect: *Primary* disease refers to the target organ, *secondary* disease to the pituitary, and the rare *tertiary* disease to the hypothalamus. When considering any hormonal disturbance, the patient presentation and test results help delineate the disorder as either hypo- or hyperfunction. Figure 7–6 depicts the classification of defects based on target and tropic hormone levels.

Endocrine Disorder Classification
• Primary disease—target organ
• Secondary disease—pituitary
• Tertiary disease—hypothalamus
• Ectopic

A. THYROID FUNCTION

The thyroid gland is composed of two lobes connected by a narrow isthmus located on either side of the trachea at the base of the throat. This gland is responsible for the synthesis, storage, and release of the thyroid hormones, T_4 and T_3. The physiological func-

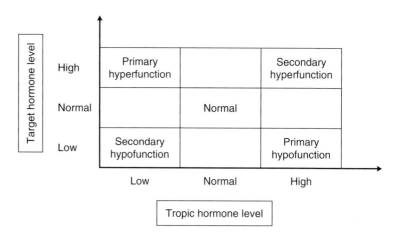

Figure 7–6. Differentiation of hormonal disease.

tions of the thyroid hormones, which include maintaining overall body metabolism, are listed in Table 7–2.

The most common cause of thyroid disease is autoimmune mediated. It is estimated that about 6% of the population shows evidence of autoimmune thyroid disease, which is threefold more likely to occur in females than males. Thyroid disease increases in incidence with advancing age and exhibits a female preference.

Overt thyroid disease, evidenced by a characteristic clinical presentation and abnormal thyroid test results, is referred to as clinical thyroid disease. Early thyroid disease with either no or minimal clinical symptoms and normal to slightly abnormal thyroid test results is referred to as subclinical thyroid disease. Thyroid disease is most often classified into the categories of euthyroidism, hyperthyroidism, hypothyroidism, or nonthyroidal illness that is referred to as the **euthyroid sick syndrome.** Table 7–3 provides an overview of the laboratory results in the various thyroid disorders.

Since thyroid hormones circulate in the blood extensively (> 99%) bound to protein carriers, changes in the concentration of these serum-binding proteins will alter the total hormone concentrations, but the free hormone concentrations remain unaffected by changes in carrier protein status. Abnormal levels of total T_4 (TT_4) and total T_3 (TT_3) may be due to changes in the hormone production rate or to alterations in the binding protein levels. Protein-binding abnormalities that affect levels of total thyroid hormones are caused by clinical conditions that influence the hepatic production or renal clearance of binding proteins. For example, pregnancy, estrogen administration, acute hepatitis, chronic active hepatitis, and the

Table 7–2. Physiological Functions of the Thyroid Hormones

- Necessary for fetal brain and skeletal muscle development
- Causes an increase in oxygen consumption and heat production
- Causes an increased blood flow, cardiac output, and heart rate
- Necessary for skeletal growth and development
- Affects carbohydrate and lipid metabolism
- Affects the development of the CNS
- Affects other endocrine systems (e.g., reproductive)

Table 7–3. Correlation of Laboratory Results with Thyroid Disease

	TSH	TT_4	FT_4	TT_3	ANTIBODIES
Overt primary hyperthyroidism[a]	↓↓	↑	↑	→↑	Often
Overt primary hypothyroidism	↑↑	↓	↓	NA[b]	Often
Subclinical hypothyroidism	↑	→	→	NA[b]	+/–
Euthyroid sick syndrome	↑↓	→↓	→	↓	–
Subclinical hyperthyroidism	↓	→	→	→	+/–

[a] In cases of T_3 toxicosis, T_4 is normal while T_3 is increased.

[b] Not useful in hypothyroidism; results are typically normal.

administration of certain drugs increase binding proteins. Conversely, chronic liver disease, malnutrition, androgen administration, glucocorticoids, nephrosis, decreased levels of thyroid-binding globulin (TBG), acromegaly, and hereditary TBG deficiency decrease thyroid-binding proteins.

1. Hyperthyroidism. Hyperthyroidism, or thyrotoxicosis, is defined as an increased concentration of the thyroid hormones. Symptoms of this disorder include nervousness, sweating, heat intolerance, moist and warm skin, tremor, angina, tachycardia, and menstrual irregularities.

Causes of hyperthyroidism include autoimmune disease **(Graves' disease)** and toxic adenoma(s). Of these, Graves' disease is the major cause of hyperthyroidism, constituting about 80% of all hyperthyroid cases. This autoimmune disorder typically has thyroid antibodies present, with 90% of the diagnosed patients having TSH receptor antibodies (stimulatory or inhibitory) and 75% having thyroperoxidase (TPO) antibodies. Goiter, the enlargement of the thyroid gland, is frequently noted, and infiltrative ophthalmopathy (exophthalmos) may be present. The demographics of the disease show a female-to-male ratio of 5:1, with the age of onset usually between 20 and 40 years. Subclinical hyperthyroidism is not as important to detect because these patients rarely progress to overt disease.

2. Hypothyroidism. Diminished levels of the thyroid hormones characterize **hypothyroidism.** Symptoms of this disorder include lethargy, slow speech and thought, weakness, fatigue, dry and cold skin, cold intolerance, hypercholesterolemia, constipation, weight gain, hoarseness, menstrual irregularities, and bradycardia. The severity of hypothyroidism ranges from mild, subclinical disease to an overt severe presentation referred to as cretinism in children and myxedema in adults. Subclinical hypothyroidism is often difficult to detect clinically because early symptoms are nonspecific and mimic the effects of aging, stress, and so on. However, more emphasis is being placed on the detection of subclinical disease since approximately 30 to 50% of those with this disorder go on to become symptomatic. Early detection in high-risk groups of subclinical hypothyroidism through screening programs interrupts the progression to overt disease.

The incidence of hypothyroidism increases with age, with up to 5% of the population over 65 years having this condition, and exhibits a female preference. Because of the nonspecific symptoms, this condition often is misdiagnosed. Causes of hypothyroidism are

Symptoms of Hyperthyroidism

- Nervousness
- Sweating
- Heat intolerance
- Moist, warm skin
- Tremor
- Angina
- Tachycardia
- Menstrual irregularities

Symptoms of Hypothyroidism

- Lethargy
- Slow speech and thought
- Weakness
- Fatigue
- Dry, cold skin
- Cold intolerance
- Hypercholesterolemia
- Constipation
- Weight gain
- Hoarseness
- Menstrual irregularities
- Bradycardia

varied and include autoimmune disease, congenital disorders, treatment of Graves' disease, and iodine deficiency, among others.

▶ The major cause of hypothyroidism is **Hashimoto's thyroiditis,** an autoimmune disorder known also as *chronic lymphocytic thyroiditis.* Demographics show that 95% of those affected are female and the incidence of the disease increases with age. Ninety percent of patients with this disorder have thyroid antibodies, notably TPO. Goiter, a thyroid gland typically two to four times its normal size, is frequently noted in those with Hashimoto's thyroiditis.

3. Nonthyroidal Illness (Euthyroid Sick Syndrome). Abnormal thyroid tests are frequently noted in severely ill euthyroid patients, as illustrated in Figure 7–7. In such cases, illness results in a decline in FT_3 and TT_3, caused by the decreased peripheral conversion of T_4 to T_3, an increased peripheral conversion of T_4 to rT_3, and a decreased clearance of rT_3. If the illness is very severe, TT_3 is very low and TT_4 declines while FT_4 remains normal. TSH becomes abnormal, either elevated or suppressed, in about 15% of these patients. During recovery from an illness, the thyroid hormones and TSH begin to normalize. Because of these hormonal changes in the severely ill, the most reliable indicator of thyroid status is FT_4, especially in hospitalized patients.

B. ADRENAL FUNCTION

As mentioned previously, the adrenal cortical hormones include the mineralocorticoids, glucocorticoids, and a small amount of androgens and estrogens. The adrenal medulla predominantly secretes the catecholamine epinephrine.

The primary mineralocorticoid, aldosterone, is responsible for the regulation of electrolyte and water balance. Cortisol, the major glucocorticoid, functions in the regulation of carbohydrate, protein, and lipid metabolism; maintenance of blood pressure; and suppression of the immune response. Though the ovaries in the female and the testes in the male synthesize the majority of the sex hormones,

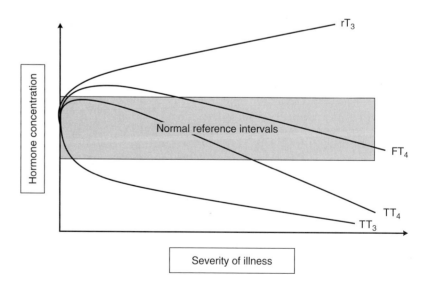

Figure 7–7. The effect of euthyroid sick syndrome on thyroid test results.

the adrenals produce a small amount of these hormones. The medullary catecholamines, which act mainly as neurotransmitters, are responsible for the "fight or flight" response. Table 7–4 provides a summary of the major disease states associated with each of these hormone classes and the typical laboratory findings.

1. Cortical Hyperfunction. The primary disorders associated with adrenal cortical hyperfunction include the excessive production of the glucocorticoid and mineralocorticoid hormones, cortisol and aldosterone.

In regard to the glucocorticoids, **Cushing's syndrome** refers to the physiological effects of excessive cortisol. The clinical presentation of Cushing's syndrome is caused by excessive amounts of circulating adrenal hormones. Notable features include truncal obesity, "moon face," bruising, muscle weakness, poor wound healing, glucose intolerance, osteoporosis, acne, and hypertension. Excessive glucocorticoids can originate endogenously or exogenously. Causes of endogenous Cushing's syndrome include an ACTH-producing pituitary adenoma, a glucocorticoid-producing adrenal neoplasm, or an ectopic ACTH-producing neoplasm. Exogenous sources of glucocorticoids originate most commonly from chronic long-term glucocorticoid therapy.

Because cortisol is secreted in a pulsatile fashion, a single serum cortisol determination lacks the requisite sensitivity and specificity for diagnosis. In most patients with Cushing's syndrome, the cortisol diurnal variation pattern is lost, resulting in no difference between morning and afternoon cortisol levels.

Primary Cushing's syndrome results from excessive glucocorticoid secretion from an adrenal adenoma or carcinoma. Secondary Cushing's syndrome, also known as Cushing's disease, results from an ACTH-secreting pituitary tumor that stimulates the adrenal cortex to release cortisol. In Cushing's disease, the normal negative feedback mechanism fails and bilateral adrenal hyperplasia ensues, causing constant production of cortisol. In some patients with Cushing's disease, the excessive secretion of cortisol is intermittent. Repeat testing will ultimately confirm the diagnosis in this group of patients. Another cause of Cushing's syndrome, though not classified as primary or secondary, is an ectopic, nonpituitary, ACTH-secreting tumor. The most common source of this ectopic ACTH production is the lung.

> **Clinical Features of Cushing's Syndrome**
>
> - Truncal obesity
> - Moon face
> - Bruising
> - Muscle weakness
> - Poor wound healing
> - Glucose intolerance
> - Osteoporosis
> - Acne
> - Hypertension

Table 7–4. Adrenal Disorders and Associated Laboratory Findings

Cortical hyperfunction	
Cushing's syndrome	↑ Cortisol, ACTH variable[a]
Cushing's disease	↑ Cortisol, ↑ ACTH
Primary aldosteronism	↑ Aldosterone, ↑ serum sodium and potassium, ↓ renin, ↓ pH
Cortical hypofunction	
Addison's disease	↓ Aldosterone, ↓ cortisol, ↑ ACTH, ↓ serum sodium, ↑ serum potassium
Medullary hyperfunction	
Pheochromocytoma	↑ Catecholamines, metanephrine, VMA

[a] ACTH production depends on the disease origin. Causes of Cushing's syndrome include adrenal tumors (↓ ACTH), ectopic nonpituitary production of ACTH (i.e., lung tumor) (↑ ACTH), and exogenous long-term administration of glucocorticoids (↓ ACTH).

The major disease associated with excess mineralocorticoid hyperfunction is primary hyperaldosteronism. In primary hyperaldosteronism, the clinical presentation includes hypertension, hypokalemia, and kaliuresis. Preferential reabsorption of sodium and water at the expense of potassium and hydrogen ions is related to the excessive aldosterone levels. The causes of primary hyperaldosteronism include aldosterone-producing adrenal adenomas **(Conn's syndrome),** bilateral idiopathic hyperplasia of the zona glomerulosa cells, and adrenal carcinoma. Laboratory findings for adrenocortical hyperfunctions are listed in Table 7–4.

2. Cortical Hypofunction. Adrenal insufficiency is classified as primary, secondary, or tertiary, depending on the site of the defect. Primary adrenal insufficiency is the most common and results from the destruction of the entire adrenal cortex, thus causing a deficiency of all the adrenal steroids. This destruction of the adrenal cortex may be either chronic or acute. Causes of acute adrenal destruction include adrenal hemorrhage, infection, uncontrolled anticoagulant therapy, and adrenalectomy. Chronic primary adrenal insufficiency, a rare disorder known as **Addison's disease,** is most often caused by the autoimmune-mediated destruction of the adrenal cortex. Other causes include tuberculosis, metastatic carcinoma, and metabolic disorders such as amyloidosis. Addison's disease typically has a slow presentation of symptoms, the severity of which is related to the amount of functioning gland remaining. The progressive loss of the adrenocortical hormones, most notably cortisol, aldosterone, and the adrenal sex hormones, causes the clinical features of Addison's disease, which include weakness, weight loss, dehydration, postural hypotension and salt cravings. During times of stress, the body normally produces cortisol and aldosterone. In acute adrenal insufficiency, the adrenals are not able to counteract the stress, and the patient may precipitate into a life-threatening condition known as *addisonian crisis*. Features such as shock, hypotension, fever, abdominal pain, nausea, vomiting, anorexia, weakness, and apathy characterize this crisis. Laboratory findings for adrenocortical hypofunction are listed in Table 7–4.

3. Medullary Hyperfunction. Tumors of mature adrenal chromaffin cells are known as **pheochromocytomas.** Pheochromocytoma tumors produce excess levels of catecholamines, which cause the most characteristic clinical symptom, increased blood pressure. Other symptoms associated with a pheochromocytoma include diaphoresis, headache, palpitations, pallor, nausea, and tremor. Symptoms may be chronic or episodic; if episodic, the patient has the sensation of an exaggerated "fight or flight" response.

C. REPRODUCTIVE FUNCTION

The gonads are the reproductive target organs. In the male, the testes are responsible for the synthesis of the sex steroids, while in the female the ovaries fulfill this function. Table 7–5 provides a summary of the laboratory findings typically observed in normal and abnormal reproductive function. Reproductive hormones are employed to differentially diagnose a myriad of reproductive endocrinology disorders. For example, testosterone is used to evaluate hirsutism

---

Clinical Features of Addison's Disease

- Weakness
- Weight loss
- Dehydration
- Postural hypotension
- Salt cravings

Clinical Features of Addisonian Crisis

- Shock
- Hypotension
- Fever
- Abdominal pain
- Nausea
- Vomiting
- Anorexia
- Weakness
- Apathy

Table 7–5. Laboratory Assessment of Reproductive Function

Female normal function	
Pregnancy	↑ hCG, ↑ progesterone, ↑ estrogens, ↓ LH, ↓ FSH
Menopause	↓ Estradiol, ↑ LH, ↑ FSH, FSH:LH > 1.0, ↓ progesterone
Female hypogonadism	
Primary	↓ Estradiol, ↑ LH, ↑ FSH
Secondary	↓ Estradiol, ↓ LH, ↓ FSH, ↓ progesterone
Female hypergonadism	
Primary	↑ Estradiol, ↓ LH, ↓ FSH
Secondary	↑ Estradiol, ↑ LH, ↑ FSH
Male hypogonadism	
Primary	↓ Testosterone, ↑ LH, ↑ FSH
Secondary	↓ Testosterone, ↓ LH, ↓ FSH
Miscellaneous disorders	
Ectopic pregnancy	↓ hCG
Gynecomastia	↑ Estrogen:androgen ratio
Hirsutism	↑ DHEAS and/or ↑ testosterone
Luteal-phase defect	↓ Progesterone
Nonviable pregnancy	↓ hCG, ↓ progesterone
Polycystic ovary disease	LH:FSH > 2.0
Precocious puberty	↑ Estradiol or ↑ testosterone, ↑ LH, ↑ FSH
Prolactinoma	↑ Prolactin, ↓ LH, ↓ FSH, ↓ estradiol or ↓ testosterone
Trophoblastic tumor, hydatidiform mole, choriocarcinoma, or testicular teratoma	↑ hCG
Virilization	↑ DHEAS and/or ↑ testosterone

and virilization in females and primary hypogonadism in males. Estradiol measurements are used clinically for the differential diagnosis of amenorrhea, for the evaluation of precocious puberty in girls, for the assessment of follicular maturation in ovulation, as a prognostic indicator of assisted reproductive technology treatment protocols, and the evaluation of gynecomastia in males. Progesterone is used clinically for ovulation detection and confirmation, the detection of luteal-phase defects, the verification of ovulation induction, monitoring progesterone replacement therapy, and the evaluation of patients at risk for early-gestation abortion. hCG determinations are used to detect pregnancy, to assess pregnancy viability, and to evaluate a suspected ectopic pregnancy. hCG is also detectable in some cases of trophoblastic tumor, hydatidiform mole, choriocarcinoma, and testicular teratoma.

1. Normal Adult Female Reproductive Endocrinology. The adult female menstrual cycle is divided into three phases—follicular, ovulatory, and luteal. The cycle, beginning with the first day of uterine bleeding, generally lasts 28 days.

The follicular phase begins with the first day of menstrual bleeding and ends just prior to the ovulatory surge. During the follicular phase, FSH stimulates the maturation of an ovarian follicle. The follicle secretes increasing amounts of estrogen as it grows in size. The estrogen, in turn, stimulates the proliferation of the uterine endometrium. LH levels begin to rise slowly in the early follicular phase, usually 1 to 2 days after the rise in FSH. LH is responsible for stimulating the synthesis of androstenedione and testosterone, which are subsequently converted to estrogens. Overall, sex steroid production is low and constant during the first half of the follicular phase. Seven to eight days before the LH surge, the developing follicle is producing ever-increasing amounts of estradiol. Maximum estradiol levels are achieved 1 day prior to the LH surge. In the later part of the follicular phase, FSH secretion from the pituitary is selectively inhibited because of the high estradiol levels and the release of inhibin from the developing follicle.

The ovulatory phase encompasses 1 day on either side of the LH surge. Estradiol positive feedback causes the preovulatory pulsatile surge of LH, which causes the mature follicle to release its ovum. The ovum-deficient follicle, now known as the corpus luteum, decreases its estradiol production early in the luteal phase, causing a simultaneous small increase in FSH. At this point, the corpus luteum begins to release progesterone and prostaglandins, signaling the beginning of the luteal phase. This final phase lasts for the life span of the corpus luteum. The progesterone produced by the corpus luteum to maintain the uterine lining reaches a maximum concentration 6 to 8 days after the LH surge. In the middle of the luteal phase, estradiol levels increase while LH and FSH levels decrease until just prior to menstruation. In the absence of fertilization, the corpus luteum degenerates and estradiol and progesterone levels decline and eventually result in endometrial bleeding.

Pregnancy. During pregnancy, cells in the placenta produce hCG. This hormone has several functions, including maintenance of the corpus luteum in early pregnancy, stimulation of fetal gonadal development, promotion of steroidogenesis in the fetoplacental unit, and stimulation of fetal testicular secretion of testosterone.

Menopause. Menopause is the cessation of the cyclic ovarian function that usually occurs in the fifth decade of life. The signs and symptoms of menopause result from the waning of ovarian follicular activity and decreased estradiol production. During the perimenopausal transition, menstrual cycles become irregular, usually shortened, and may be ovulatory or anovulatory. FSH levels are increased, while LH is normal and estradiol and progesterone are decreased when compared to normal ovulatory cycles. With true menopause, both FSH and LH are greatly increased while estrogen levels are markedly reduced.

2. Female Reproductive Endocrinology Disorders. Female reproductive disorders consist of either hypofunction or hyperfunction conditions. Increased levels of FSH and LH suggest gonadal dysfunction that results in an absence of negative feedback to the pituitary. Decreased levels of LH and FSH are due to either hypothalamic or pituitary disease. Normal menstrual periods in an adult female imply normal levels of FSH, LH, and estradiol and thus no reproductive endocrinology disorder.

▶ *Female Hypogonadism.* The clinical presentation of female **hypogonadism** depends on the age of onset and the degree of estrogen deficiency. In primary female hypogonadism (hypergonadotropic hypogonadism), the defect is due to ovarian hypofunction. Causes include Turner's syndrome (45 XO), menopause, testicular feminization, and polycystic ovary disease.

The level of the defect for secondary hypogonadism (hypogonadotropic hypogonadism) is at the level of either the pituitary or the hypothalamus. Causes of female secondary hypogonadism include

---

Causes of Female Hypogonadism

Primary:
- Turner's syndrome
- Menopause
- Testicular feminization
- Polycystic ovary disease

Secondary:
- Hypopituitarism
- Hypothalamic disorders
- Hypothyroidism
- Pregnancy
- Central disorders

hypopituitarism, hypothalamic disorders, hypothyroidism, pregnancy, and central disorders (anorexia nervosa, stress, intense physical training, weight loss, and malnutrition).

Female Hypergonadism. Estrogen-secreting tumors are the primary cause of female primary **hypergonadism** (hypogonadotropic hypergonadism).

The major causes of female secondary hypergonadism (hypergonadotropic hypergonadism) are precocious puberty and the premature maturation of the CNS.

3. Male Reproductive Endocrinology Disorders. In males, most clinical dysfunction is due to hypofunction.

Male Hypogonadism. The presentation of male hypogonadism depends on the age of onset and the degree of testosterone deficiency. Variations in presentation include ambiguous genitalia, delayed puberty, postpubertal gonadal failure, gynecomastia, and infertility.

In male primary hypogonadism (hypergonadotropic hypogonadism), the disorder is caused by gonadal failure. Causes include Klinefelter's syndrome (XXY), male climacteric (Leydig cell failure in older men), mumps, castration, complete androgen insensitivity (testicular feminization), and incomplete androgen sensitivity.

In secondary hypogonadism (hypogonadotropic hypogonadism), the abnormality is at the level of the pituitary or the hypothalamus.

D. Dysfunction Involving Miscellaneous Hormones

1. Growth Hormone. Causes of increased GH include GH-secreting pituitary tumors, hypothalamic or ectopic GHRH-secreting tumors, or the very rare ectopic GH-secreting tumor. Excess GH results in *gigantism* in children and *acromegaly* in adults. Conversely, causes of depressed GH levels include a congenital or acquired pituitary lesion, a pituitary tumor and its metastases, or cranial irradiation. GH deficiency causes the rare condition of pituitary dwarfism in children and negligible symptoms in adults.

2. Prolactin. Increased levels of prolactin result in hypogonadism due to the hormone's ability to inhibit GnRH pulsatile secretion. A pituitary adenoma is the predominant cause of hyperprolactinemia. Females with a prolactinoma present with amenorrhea, galactorrhea, and infertility while adult males exhibit impotency, decreased libido, and infertility. Clinically, serum prolactin is monitored to assess infertility and to diagnose and monitor prolactinomas.

IV. TEST PROCEDURES

A. Immunoassays

Hormone testing requires high sensitivity because hormones are typically present in very low concentrations in serum. Specificity is also required because of the structural similarities between the various chemical classes of hormones. For example, LH and FSH are glycoproteins comprised of both alpha and beta subunits. The alpha subunits of LH

Causes of Female Hypergonadism

- Estrogen-secreting tumors
- Precocious puberty
- Premature maturation of the CNS

Causes of Male Hypogonadism

Primary:
- Klinefelter's syndrome
- Male climacteric
- Mumps
- Castration
- Complete androgen insensitivity
- Incomplete androgen sensitivity

Secondary:
- Pituitary disorders
- Hypothalamic disorders

and FSH have considerable homology to the alpha subunits of TSH and hCG. It is the beta subunit that is immunologically distinct in each of these hormones and confers the biological activity. The use of intact two-site immunoassays allows for the most reliable measurement of LH and FSH. Typically, immunoassays have sensitive labels and specific antibodies to quantify the various hormones that are present in very low concentrations in the bloodstream. Refer to Chapter 12 for a more detailed discussion of immunoassay techniques.

B. DYNAMIC TESTING

Dynamic testing, such as stimulation or suppression testing, may be used to evaluate the endocrine gland's response to its usual control mechanisms. With hormonal excess, suppression testing is employed. In cases of hormonal deficiency, stimulation testing is performed. Because of the enhanced sensitivity of today's immunoassays in assessing low hormone levels, stimulation testing is less frequently used. Also, advances in imaging techniques, such as computed tomography (CT) scan and magnetic resonance imaging (MRI), have supplanted the need for stimulation testing. Furthermore, because patients undergoing dynamic testing are typically hospitalized during and after the procedures, recent medical cost containment policies have discouraged the use of these types of tests.

Examples of dynamic testing include the dexamethasone suppression test and the CRH stimulation test that may be used in the evaluation of Cushing's syndrome.

C. LABORATORY ASSESSMENT OF THYROID FUNCTION

1. Recommended Screening Protocols. The American Thyroid Council (ATC) has published its recommendations for thyroid disease screening protocols. According to the ATC, high-risk patients, including the elderly, neonates, postpartum women, and those with a history of autoimmune disease as well as those with a family history of thyroid disease, should be screened for thyroid disorders. TSH is the screening test of choice unless central hypothyroidism is suspected. For the diagnosis of hypothyroidism, TSH and FT_4 are suggested. For the diagnosis of hyperthyroidism, TSH and FT_4 are again recommended with the caveat that the sensitivity of the TSH method be < 0.1 mIU/L. Thyroid testing of the sick should be delayed, if possible, until the resolution of the illness. If thyroid assessment is required, FT_4 is the most reliable indicator of thyroid status in the sick patient.

TSH. It is recommended that TSH assay sensitivity be determined using *functional sensitivity* as opposed to *analytical sensitivity* because of lower-limit imprecision issues. Functional sensitivity, as known as the *limit of quantification,* is defined as the analyte concentration at 20% coefficient variation (CV) and clearly delineates the imprecision of the assay. In contrast, the limit of detection, defined by analytical sensitivity, yields an unacceptably high CV for TSH methods.

Different *"generations"* of TSH immunoassays exist and differ in their sensitivities. First-generation TSH methods are usually radioimmunoassay (RIA) methods with sensitivity limits of 1 to 2 mIU/L; these methods easily detect the elevated TSH levels of hypothyroidism but are unable to detect the suppressed TSH levels associated with hyperthyroidism. Second-generation tests are usually immunoradiometric assay (IRMA) methods with 20% CV at 0.1 mIU/L. Third-generation TSH assays are typically chemiluminescent methods with sensitivity limits of 0.01 mIU/L at 20% CV; these methods easily detect hyperthyroidism.

The ideal method for analyzing TSH is a two-site immunometric assay that forms a "sandwich" to measure the biologically active intact TSH molecule. In this sandwich, one antibody is directed against the alpha subunit while the other is against the beta subunit.

FT_4. The concentration of FT_4 can be measured directly or calculated indirectly as the free thyroxine index (FTI). The use of the FTI is now an anachronism and is not recommended, however, because more accurate immunoassays are available. FTI is calculated by multiplying the TT_4 by the T-uptake. The most common method for the quantification of FT_4 is the direct measurement assay, while the gold standard method for measuring FT_4 directly is equilibrium dialysis. In equilibrium dialysis, T_4 is partitioned across a dialysis membrane whose pores allow passage of the free hormone but not that bound to proteins. Though regarded as the reference method, it is technically difficult, labor intensive, and not readily available. The most common method for the quantification of FT_4 is the direct measurement assay. This assay is available in two formats—the two-step and one-step techniques. The two-step format involves the incubation of the patient's serum with the anti-T_4 antibody immobilized on a solid support, removal of unbound FT_4, and incubation of the immobilized FT_4–antibody complex with labeled T_4 in order to estimate the unoccupied antibody-binding sites. The one-step format uses labeled T_4 analogs that bind to anti-T_4 antibody but not to endogenous serum-binding proteins.

2. Other Thyroid Markers

TT_4. Though not recommended, many laboratories still perform the TT_4 assay because the immunoassay is readily automated and rather inexpensive to perform. TT_4 levels are affected by both changes in thyroid function and thyroid hormone–binding proteins, thus making the diagnosis of thyroid disease difficult. The TT_4 assay measures the total hormone, including protein bound plus the free hormone.

T-Uptake. The T-uptake, now considered to be an obsolete test, provides an indirect estimation of TBG, or the number of thyroid hormone–binding sites available. Because of the ease of automation and low cost, the T-uptake immunoassay method is performed in conjunction with the TT_4 assay to yield a calculated FTI.

Thyroid Antibodies. Thyroid disease can result from antibodies that produce autoimmune disorders. Antithyroid antibodies are detected in 10 to 15% of the population, especially in elderly women. Antibodies often present before TSH and FT_4 are abnormal and thus may serve as an important indicator of future thyroid disease. The antibodies associated with autoimmune thyroid disorders include antibody to thyroperoxidase (TPO), antibody to thyroid microsomal antigen (Mic), antibody to thyroglobulin (Tg), antibody to thyrotropin receptor, thyroid-stimulating immunoglobulins (TSI), and thyrotropin-binding inhibitory immunoglobulins (TBII). In general, the TPO antibody test has replaced the MicAb and TgAb tests for the routine assessment of autoimmune thyroid disease.

Thyroid Tumor Markers. Two compounds found in the thyroid gland are established as tumor markers. Thyroglobulin, produced solely by the thyroid, serves as a marker of differentiated thyroid cancer (papillary or follicular). The blood levels of thyroglobulin are related to the mass of the cancer and provide an index of inflammation. Calcitonin, a peptide hypocalcemic hormone produced by the thyroidal parafollicular cells, may be used as a marker for medullary thyroid cancer.

D. LABORATORY ASSESSMENT OF ADRENAL FUNCTION

1. Cortical Function. The primary screening test for cortical function is plasma cortisol. The most sensitive direct test is a 24-hour urinary cortisol, which represents the free or unbound fraction. The 24-hour urine collection normalizes the cortisol diurnal variation pattern. Plasma ACTH levels can be used to differentiate between the various types of Cushing disorders. Cortisol, whether plasma or urine, and ACTH are measured by immunoassays. The normal reference range for plasma cortisol is dependent on the time of collection. Values are highest upon waking and lowest in early sleep.

A plasma aldosterone measurement is of limited diagnostic utility in the diagnosis of hyperaldosteronism. A 24-hour urinary aldosterone specimen, which estimates the aldosterone production rate, is effective at distinguishing primary hyperaldosteronism from essential hypertension, especially when used in conjunction with a urinary sodium level. Interpretation of aldosterone levels depends on the patient's salt intake and posture at the time of collection. The measurement of aldosterone in blood and urine is routinely performed by immunoassays. Plasma renin activity (PRA) is measured via a kinetic enzyme assay whereby its product, angiotensin I, is quantified typically by immunoassay. PRA reference ranges are dependent on salt intake, posture, and certain medications, such as diuretics.

2. Medullary Function. The diagnosis of a pheochromocytoma is done through laboratory tests and imaging techniques. Fractionated urine and plasma catecholamines and their major metabolites, urinary VMA and the metanephrines, comprise the battery of laboratory diagnostic analyses available. Most patients with pheochromocytoma hypersecrete catecholamines constantly, so it is not necessary to collect specimens only when the patient is symptomatic. Stress adversely affects the secretion of catecholamines, causing elevations. Special requirements, including patient preparation regarding diet and medications, may be necessary to stabilize plasma catecholamines in the specimen since VMA is found in common foods such as chocolate and bananas.

Methods for the measurement of catecholamines and their metabolites include spectrophotometric and fluorometric techniques that are relatively nonspecific and high-pressure liquid chromatography (HPLC), gas chromatography/mass spectrophotometry (GC/MS), and radioenzymatic immunoassay, which are highly specific and sensitive.

E. LABORATORY ASSESSMENT OF REPRODUCTIVE FUNCTION

Testing for all the reproductive hormones is done by immunoassay. Some of the reproductive hormones require serial testing for a definitive diagnosis. For example, in a case of suspected stable ectopic pregnancy, hCG is monitored serially to assess whether hCG production is doubling as expected every 2 days.

SUGGESTED READINGS

Cook JD. Thyroid disorders. In Dufour DR (ed.). *Professional practice in clinical chemistry: A review.* Washington, DC: American Association for Clinical Chemistry, 1999; 29-1–29-9.

Cook JD. Adrenal disorders. In Dufour DR (ed.). *Professional practice in clinical chemistry: A review.* Washington, DC: American Association for Clinical Chemistry, 1999; 30-1–30-16.

Cook JD. Pituitary, hypothalamic and gonadal disorders. In Dufour DR (ed.). *Professional practice in clinical chemistry: A review.* Washington, DC: American Association for Clinical Chemistry, 1999; 28-1–28-15.

Demers LM. Pituitary function. In Burtis CA, Ashwood ER (eds.). *Tietz textbook of clinical chemistry,* 3rd ed. Philadelphia: WB Saunders, 1999; 1470–75.

Demers LM, Whitley RJ. Function of the adrenal cortex. In Burtis CA, Ashwood ER (eds.). *Tietz textbook of clinical chemistry,* 3rd ed. Philadelphia: WB Saunders, 1999; 1530–69.

Dufour DR. Laboratory evaluation of adrenal function. In Boyd JC, Christenson RH, Dufour DR (eds.). *Professional practice in clinical chemistry: A review.* Washington, DC: American Association for Clinical Chemistry, 1999; FF1–FF8.

Dufour DR. Laboratory evaluation of pituitary and gonadal function. In Boyd JC, Christenson RH, Dufour DR (eds.). *Professional practice in clinical chemistry: A review.* Washington, DC: American Association for Clinical Chemistry, 1999; Y1–Y9.

Dufour DR. Laboratory evaluation of thyroid function. In Boyd JC, Christenson RH, Dufour DR (eds.). *Professional practice in clinical chemistry: A review.* Washington, DC: American Association for Clinical Chemistry, 1999; Z1–Z9.

Gronowski AM, Landau-Levine M. Reproductive endocrine function. In Burtis CA, Ashwood ER (eds.). *Tietz textbook of clinical chemistry,* 3rd ed. Philadelphia: WB Saunders, 1999; 1601–41.

Whitley RJ. General endocrine function. In Burtis CA, Ashwood ER (eds.). *Tietz textbook of clinical chemistry,* 3rd ed. Philadelphia: WB Saunders, 1999; 1458–69.

Whitley RJ. Thyroid function. In Burtis CA, Ashwood ER (eds.). *Tietz textbook of clinical chemistry,* 3rd ed. Philadelphia: WB Saunders, 1999; 1496–529.

Items for Further Consideration

1. You have been asked to prepare educational materials for your physician clients outlining recommended protocols for thyroid testing. Adhering to the American Thyroid Association guidelines, prepare thyroid-testing algorithms for:

   a. High-risk individuals (list high-risk criteria)
   b. Euthyroid sick patients

2. Your laboratory is considering offering assisted reproductive technology testing services. List which analytes should be included in your test menu and the service requirements needed (turnaround time, setups, sensitivity, etc.).

3. For each of the following adrenal disorders, prepare a table indicating the respective diagnostic hormones; whether their values are increased, decreased, or normal; and the major clinical symptoms.

   a. Cushing's syndrome
   b. Cushing's disease
   c. Addison's disease
   d. Conn's syndrome
   e. Pheochromocytoma

Therapeutic Drug Monitoring

Larry A. Broussard

Key Terms

ANTIBIOTICS

ANTICONVULSANTS

ANTINEOPLASTIC

BIOAVAILABILITY

BRONCHODILATOR

CARDIOACTIVE

FIRST-ORDER KINETICS

FIRST-PASS ELIMINATION

HALF-LIFE

IMMUNOSUPPRESSANT

PEAK CONCENTRATION

PHARMACOKINETICS

PSYCHOTROPICS

STEADY STATE

THERAPEUTIC INDEX

THERAPEUTIC RANGE

TROUGH VALUE

Chapter Objectives

Upon completion of this chapter, the reader will be able to:

1. State the reasons for performing therapeutic drug monitoring (TDM).
2. Discuss the key principles for interpretation of plasma drug concentrations.
3. Explain the factors affecting drug absorption, metabolism, and excretion.
4. Discuss the concept of half-life, steady-state concentration, and therapeutic concentrations of drugs.
5. Compare peak and trough measurements.
6. State proper timing of sample collection for various TDM measurements.
7. Analyze and interpret drug concentrations of case studies, including factors that may affect them.
8. List drugs in the following classes which are therapeutically monitored: anticonvulsant, cardioactive, bronchodilator, antibiotic, psychotropic, antineoplastic, and immunosuppressant.
9. Compare the specific requirements (including time of collection after dosing, time to steady state, therapeutic concentration, and elimination rate) for monitoring each of the following drugs: phenobarbital, phenytoin, valproic acid, primidone, carbamazepine, ethosuximide, digoxin, lidocaine, quinidine, procainamide, disopyramide, propranolol, theophylline, caffeine, gentamicin, tobramycin, amikacin, lithium, tricyclic antidepressants, methotrexate, cyclosporine, and tacrolimus and sirolimus.
10. Name five drugs for which active metabolites should be monitored and discuss the appropriate strategy for monitoring each.

Analytes Included ◀

AMIKACIN	ETHOSUXIMIDE	PRIMIDONE
AMITRIPTYLINE	GENTAMICIN	PROCAINAMIDE
CAFFEINE	IMIPRAMINE	QUINIDINE
CARBAMAZEPINE	KANAMYCIN	SIROLIMUS
CHLORAMPHENICOL	LIDOCAINE	TACROLIMUS
CYCLOSPORINE	LITHIUM	THEOPHYLLINE
DESIPRAMINE	METHOTREXATE	TOBRAMYCIN
DISOPYRAMIDE	NORTRIPTYLINE	VALPROIC ACID
DIGOXIN	PHENOBARBITAL	VANCOMYCIN
DOXEPIN	PHENYTOIN	

Pretest

Directions: Choose the single best answer for each of the following items.

1. Which of the following statements is true with respect to blood sampling?
   a. the best time to draw routine blood specimens for trough values for therapeutic drug monitoring (TDM) is immediately prior to the next dose
   b. serum drug concentrations must be interpreted with respect to the total clinical status of the patient
   c. at the time of peak serum concentration, drug absorption and distribution phases are essentially complete
   d. all of the above

2. Routine monitoring following procainamide therapy should
   a. measure both procainamide and its metabolite N-acetyl procainamide (NAPA)
   b. measure NAPA only
   c. measure procainamide only
   d. measure peak and trough levels of procainamide

3. The measured concentration of a drug in plasma reflects
   a. the saturation kinetics of the drug
   b. free drug circulating in the extracellular fluid
   c. unless specified, the sum of free-drug plus bound-drug concentrations in the plasma
   d. the activity of the drug at the receptor site

4. As a rule of thumb, the steady-state concentration of a drug is achieved after how many half-lives?
   a. 3
   b. 4
   c. 5½
   d. 8

5. Concentrations of antibiotics greater than the upper limit of the therapeutic range may cause
   a. convulsions and seizures
   b. ototoxicity and nephrotoxicity
   c. metastic tumors
   d. liver failure

6. Digoxin concentration may become elevated with no change in patient dosage due to initiation of
   a. quinidine
   b. valproic acid
   c. phenytoin
   d. digitoxin

7. Patients using lithium long term are also routinely monitored for the development of
   a. hyperthyroidism and liver dysfunction
   b. hypothyroidism and renal dysfunction
   c. arrhythmias
   d. convulsions

8. In neonates, the theophylline metabolite that is also monitored is
   a. caffeine
   b. nortriptyline
   c. phenobarbital
   d. imipramine

9. Therapeutic monitoring of the anticonvulsant primidone should include analysis of
   a. phenobarbital and ethosuximide
   b. primidone and phenobarbital
   c. primidone and carbamazepine
   d. primidone and phenytoin

10. Physiologically, both tacrolimus and cyclosporine are
    a. cardioactive
    b. immunosuppressive
    c. antibiotic
    d. anticonvulsant

11. Nortriptyline is a metabolite of
    a. procainamide
    b. amitriptyline
    c. imipramine
    d. desipramine

12. The best time to collect a trough level of a drug for therapeutic monitoring is
    a. one-half hour after the oral dose
    b. one-half hour after an IV is discontinued
    c. immediately before the next dose is to be given
    d. 2 hours after the oral dose

13. Phenobarbital has a half-life of approximately 96 hours. If the medication is discontinued, 97% of the phenobarbital will be eliminated after
    a. 48 hours
    b. 72 hours
    c. 22 days
    d. 5½ days

14. A patient who overdosed on digoxin was given the antidote Digibind. The physician ordered a digoxin level to monitor the level of digoxin. Which of the following is the most appropriate laboratory response?

a. recommend ordering a digitoxin level also
b. perform the digoxin and report the result
c. inform the physician that a digoxin concentration cannot be accurately determined in the presence of Digibind
d. recommend ordering a quinidine level also

15. A patient can show signs and symptoms of a digoxin overdose when the measured concentration is within the therapeutic range with which of the following conditions?
    a. hypothyroidism and hypopituitarism
    b. hypocalcemia and hyperphosphatemia
    c. hyperuricemia and uremia
    d. hypokalemia and hypomagnesiumia

16. Most drugs must be in what state in order to elicit a pharmacologic response?
    a. free (unbound to protein)
    b. protein bound
    c. conjugated to a glucuronide
    d. demethylated

17. A TDM sample drawn immediately before the patient takes the next dose should be compared to a reference range based on
    a. peak values
    b. trough values
    c. peak and trough values
    d. urine levels

18. The site in the body where most of the drug metabolism occurs is the
    a. mouth
    b. stomach
    c. liver
    d. kidneys

19. TDM of methotrexate consists of collecting samples at
    a. 1 hour after the completion of IV infusion
    b. 8 hours after the dose for the trough level and 1 hour after the dose for the peak
    c. 24, 48, and 72 hours after high-dose therapy
    d. one-half hour after oral ingestion of the drug

20. The specimen of choice for TDM of cyclosporine, tacrolimus, and sirolimus is
    a. urine
    b. serum
    c. saliva
    d. whole blood

I. INTRODUCTION

The use of plasma or blood concentrations of drugs to adjust dosage regimens in order to avoid toxic effects and achieve the desired clinical effect is known as *therapeutic drug monitoring* (TDM). TDM is important when there is a clear association between plasma drug concentrations and the clinical effects of the specific drug. The therapeutic effect of drugs is achieved only after plasma concentrations have reached a specific concentration. There is an optimal plasma concentration range for these drugs in which most patients will achieve the desired therapeutic response. Side effects, including toxicity, may become clinically evident when concentrations are above this therapeutic range. The evolution of TDM was accelerated with the introduction of reliable and timely analytical techniques for quantitating drugs and metabolites. These techniques initially were colorimetric or chromatographic (high performance liquid chromatography [HPLC] and gas chromatography [GC]) but are currently almost exclusively immunoassays on automated analyzers.

There are several indications for performing TDM. Drugs with narrow, well-defined therapeutic ranges should be monitored. Also, TDM is important when plasma drug concentrations are not predictable from dosage alone because the desired therapeutic effect may not be achieved or symptoms of toxicity may be observed. In these cases, plasma drug concentrations provide crucial information to the caregiver. Further, when noncompliance (including patient noncompliance, dosage error, and wrong medication) is suspected, TDM helps to identify the problem. Other indications for TDM include alteration of drug utilization due to disease, physiologic states such as pregnancy, interindividual variability, or drug interactions. Finally, TDM may be used in situations requiring medicolegal verification of treatment.

II. BASIC PHARMACOKINETICS

Although the ultimate goal is to optimize the dosage for each individual patient, several general principles of **pharmacokinetics** apply when discussing the basis of TDM. An understanding of the principles of drug absorption, distribution, metabolism, and excretion is necessary for the interpretation of TDM results. Routes of drug administration include intramuscular, intravenous, and inhalation, as well as the most commonly used, oral ingestion. Absorption of orally ingested medication is affected by its **bioavailability,** or the fraction of drug absorbed into systemic circulation. When generic formulations that may have different bioavailablity are substituted, TDM may be used to determine the dosage of the generic formulation necessary to achieve the desired plasma concentration. Also, other factors (e.g., stomach or intestinal conditions, drug–drug interaction, composition of food) that may affect the absorption of the drug must be considered when interpreting TDM results.

In general, ingested drugs enter the vascular system and are bound to proteins to a variable extent. Albumin is the principal protein that binds acidic drugs, and α_1-acid glycoprotein, an acute-phase reactant, is the principal protein that binds basic drugs. Any condition that affects proteins and/or the protein–drug interaction must be

---

Indications for Performing TDM

- Drugs with narrow, well-defined therapeutic ranges
- Plasma drug concentrations not predictable from dosage alone
- Noncompliance
- Alteration of drug utilization due to disease
- Physiologic states such as pregnancy
- Interindividual variability
- Drug interactions
- Situations requiring medicolegal verification of treatment

considered when interpreting TDM results. For all drugs, there is equilibrium between the bound and free forms; many factors can affect this equilibrium. Because free drug is available for crossing membranes and interacting at sites of action, the free-drug concentration is considered the most clinically relevant measurement. Even so, most drug assays measure the total concentration (i.e., both free and bound). However, there are assays available to measure the free-drug fraction only. Though the total drug level may remain constant, when there is a change in the protein–drug equilibrium, there may be a pronounced clinical effect if there is an alteration in the ratio of bound-to-free concentrations. Compounds such as other medications, fatty acids, urea, bilirubin, vitamins, ethanol, and hormones, some of which may compete with the drug of interest for protein-binding sites, may cause an increase in the free-drug level with no change in total concentration. In this situation, the patient may exhibit toxic symptoms even though the measured total drug concentration is within the therapeutic range. Any condition affecting protein status or protein–drug interaction (such as acid–base imbalance) may cause similar effects.

The primary site of drug metabolism is the liver. Drugs absorbed through the intestinal mucosa enter the hepatic portal system and go to the liver before entering systemic circulation. The term **first-pass** ◀ **elimination** refers to the rapid metabolism of certain drugs as they circulate through the liver. Metabolism of drugs usually involves reactions that convert the drug or metabolite to a more water-soluble compound that can be excreted in the urine. Individuals may metabolize drugs at different rates, and the same individual will metabolize drugs differently depending on age, disease condition, and other variables. An infant with immature hepatic and renal function will metabolize and eliminate drugs slowly, particularly in the first month of life. An example of this is the metabolism of theophylline to caffeine in neonates. As drug metabolizing systems mature, the rate of metabolism increases, reaching a peak rate during prepubescence. At puberty, there is a decrease in metabolism, and drug levels in these children should be closely monitored. For example, a child on anticonvulsant medication whose dosage has remained fairly constant may show behavioral changes such as listlessness and poor performance in school at the onset of puberty. These changes may be due to decreased metabolism of the anticonvulsant, resulting in toxic blood concentrations with the same dose. Another group of patients who should be closely monitored is the elderly, in whom, as in pubescent children, the metabolic rate of drug clearance decreases. Some drugs and other factors, including smoking and alcohol consumption, may increase the rate of metabolism by induction of the enzymes responsible for their metabolism. For example, phenobarbital induces several of these enzymes, including cytochrome P_{450} reductase and cytochrome P_{450}. For some drugs, the metabolite(s) may be a physiologically active compound, thus requiring TDM of both parent drug and the metabolite.

Although drugs and/or their metabolites are primarily excreted in urine, excretion may occur in the bile, feces, sweat, and expired air. Because the pH of the urine affects drug solubility, any ingested compounds such as foods or coadministered substances that alter the urinary pH will affect the rate of elimination. Further, both hepatic and renal disease may alter drug metabolism and elimination, which must be considered when reviewing TDM results.

Drug absorption, distribution, metabolism, and excretion factors affect interpretation of TDM results and are components of the overall pharmacokinetics of drugs. As a general overview of pharmacokinetics, consider the concepts of TDM illustrated in Figure 8–1. This figure shows that there is a change in the plasma concentration of a drug as each succeeding dose is taken. The dosing interval shown is the half-life of the drug. **Half-life** is defined as the time necessary for the drug concentration to reach one-half of its peak concentration. For example, the half-life of the anticonvulsant drug phenobarbitol is 96 hours in adults. Thus, if a single standard phenobarbitol dose achieves a maximum value of 20 µg/mL, the plasma phenobarbitol concentration will be 10 µg/mL 96 hours later.

The clearance of many drugs follows **first-order kinetics,** a term meaning that the rate of clearance follows the rules of a unimolecular reaction, and therefore clearance is directly proportional to drug concentration. The half-life expression for first-order reactions is: $t_{\frac{1}{2}} = 0.693/K$, where K is a constant for the specific drug. (Note that the half-life for first-order reactions is independent of drug concentration.) Phenytoin, salicylates, ethanol, and theophylline are exceptions because they do not follow simple first-order elimination at all concentrations. Instead, their elimination is *capacity limited* or *nonlinear,* which means that the apparent half-life changes with changes in concentration. This variability may be due to saturation of an enzyme involved in metabolism or renal capacity for excretion.

The usual oral dosing for drugs is every half-life. After each dose, the drug concentration rises to a **peak concentration** and then falls to a **trough value** one half-life later, before the next dose is ingested. With each succeeding dose, the peak level rises until it reaches a plateau or steady-state level after 5½ half-lives. The **steady state** reflects equilibrium between the rate of absorption and distribution and the rate of elimination.

The **therapeutic range** of a drug is between the minimum effective concentration at the lower limit, and the concentration at which there are minimum toxic effects at the upper limit. In this

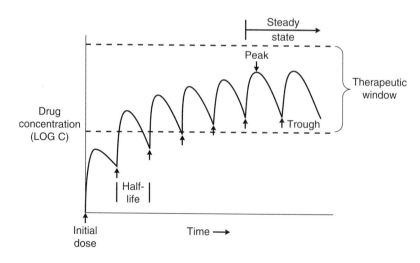

Figure 8–1. Drug concentration as a function of time.

way, drug concentrations below the therapeutic range indicate ineffective clinical activity whereas drug concentrations above the therapeutic range indicate risk of possible toxic effects. At steady state, both the peak and trough levels should fall within the therapeutic range. Therapeutic ranges for most drugs have been established using samples drawn at the trough levels. It is difficult to accurately determine when the peak concentration of a drug will occur; however, collecting samples at the trough concentration is relatively simple due to the small change in concentration with time during the trough phase of metabolism. Unless informed otherwise, therapeutic range should be used to evaluate trough values. Theophylline is a notable exception to this rule in that its therapeutic ranges were established using peak concentrations. For most antibiotics, both peak and trough therapeutic ranges are routinely monitored, although only trough values are advocated when the larger doses are given once a day (see discussion in antibiotics section).

The **therapeutic index** is a term that describes the concentration difference between the maximum plasma concentration that is clinically effective and that which is toxic. For many drugs, this concentration span is narrow, and thus the therapeutic index is small.

The most commonly encountered problem with TDM is ensuring that the specimen is collected at the correct time. Because most therapeutic ranges represent trough levels, the rule of thumb for the proper time of collection is immediately before the next dose. If peak concentrations are desired, the proper time of collection varies with the drug and the route of administration. Generally, collection of samples for peak values are: 1 hour after oral ingestion, one-half hour after completion of an intravenous infusion, and one-half to 1 hour following intramuscular injection. Before using these guidelines, however, an authoritative source should be consulted, particularly for orally ingested drugs. The 1 hour after oral ingestion guideline assumes a half-life of 2 hours or longer. Nonetheless, there are several drugs that are slowly absorbed such as sustained-release formulations of theophylline, and the proper time of collection for peak levels could be as long as 6 to 9 hours after ingestion. The principles illustrated in Figure 8–1 provide the answers to the following commonly asked questions regarding sample collection for TDM.

1. When a patient begins taking a drug, when should therapeutic monitoring begin? *Answer:* Steady state must be reached (5½ half-lives) before beginning TDM.
2. When a patient discontinues medication, how long before the drug is eliminated? *Answer:* After 5½ half-lives, more than 97% of the drug has been eliminated.
3. When should a specimen be collected for TDM? *Answer:* Assuming steady state, for most orally ingested drugs TDM specimens should be collected immediately before the next dose is given to obtain trough concentrations.

General Guidelines for Sample Collection

Trough levels
• Immediately prior to next dose

Peak levels
• 1 hour after oral injection
• ½ hour after completion of an IV
• ½–1 hour after an intramuscular injection

III. TEST PROCEDURES

Therapeutic ranges for drugs are method dependent and can vary between institutions. In general, chromatographic procedures such as GC and HPLC methods measure the parent drugs and metabolites individually, whereas immunoassays have variable responses to the metabolites depending on antibody cross-reactivity. For accurate interpretation of drug results, particularly those with active metabolites, it is essential to know the specificity of the method used to obtain the results. Most drugs are monitored using immunoassay techniques (see Chapter 12).

IV. DRUG CLASSES, INTERFERENCES, TOXICITY

Drugs that are commonly monitored will be discussed briefly. The major classes, grouped by pharmacological use, are: anticonvulsants, cardioactive drugs, bronchodilators, antibiotics, psychotropics, antineoplastics, and immunosuppressants. Table 8–1 lists the characteristics of specific drugs in each class. For this discussion, drugs are identified by chemical or generic name, with the common brand name listed in parentheses. Interpretation of TDM results for individual patients requires evaluation of laboratory data as well as clinical presentation.

A. ANTICONVULSANTS

▶ **Anticonvulsants,** also known as antiepileptics, are used to treat epilepsy as well as seizure disorders secondary to other diseases. Often, these drugs are used in combination, in which case drug interactions must be considered when interpreting results.

Phenobarbital is effective in the treatment of almost all seizure disorders, except absence seizures, and is sometimes administered in combination with other drugs. It has a long half-life of 24 to 120 hours and has a therapeutic range of 15 to 40 mg/L. Absorption is relatively slow and the time of occurrence of peak concentrations is variable at 4 to 18 hours after dosing. The elimination half-life is about 30% shorter for children than for adults. Phenobarbital induces liver enzymes and thus enhances the metabolism of other drugs. For this reason, there may be a decrease in the plasma concentrations of the other drugs due to an increased rate of metabolism when phenobarbital is added to a patient's regimen. Conversely, some drugs, such as valproate, salicylate, and phenytoin, inhibit the metabolism or elimination of phenobarbital, causing an increase in plasma concentrations.

Phenytoin (Dilantin) is used for the treatment of generalized tonic–clonic and other seizures. Phenytoin use is widespread because it does not have the hypnotic properties of phenobarbital. The therapeutic range for phenytoin is 10 to 20 mg/L. Because phenytoin elimination does not follow first-order pharmacokinetics, apparent half-life values vary from 6 to 60 hours in adults and 7 to 29 hours in children. Phenytoin is extensively protein bound, at approximately 90%, and consequently TDM of this drug is critical. The free (unbound) phenytoin plasma level may be affected by drugs such as valproate and salicylate, which compete for protein binding; also, physiological conditions that cause changes in protein status (e.g., uremia, cirrhosis, pregnancy, etc.) may affect the free phenytoin concentration. In some situations, the total concentration of phenytoin may remain constant, but the therapeutic effect is altered because of changes in the proportion (amount) of free drug present. Phenytoin is not readily soluble in aqueous solutions; it is therefore

---

Classes of Commonly Monitored Drugs

- Anticonvulsants
- Cardioactive drugs
- Bronchodilators
- Antibiotics
- Psychotropics
- Antineoplastics
- Immunosuppresants

Table 8–1. Characteristics of Commonly Monitored Drugs

DRUG	HALF-LIFE (HOURS)	THERAPEUTIC RANGE (mg/L)	COMMENTS
Anticonvulsants (AEDs)			
Carbamazepine	24–48	8–12	Active epoxide metabolite not monitored
Ethosuximide	27–39	40–100	Often coadministered with other anticonvulsants
Phenobarbital	24–120	15–40	Enhances metabolism of other drugs
Phenytoin	6–60 adult; 7–29 children	10–20	Highly protein bound
Primidone	3–12	5–12	Also monitor active metabolite, phenobarbital
Valproic acid	9–18	50–120	Causes increase in phenobarb levels
Cardioactive Drugs			
Disopyramide	5.0–7.0	2.0–5.0	Concentration-dependent binding
Digoxin	23–61	0.5–2.0 ng/mL	Digoxin-like immunoreactive substances interfere; increased sensitivity when low K, Mg, or high Ca
Lidocaine	1.4–2.2	1.5–5.0	Also monitored with ECG; rapidly metabolized
Procainamide	2.4–3.6	4.0–8.0	Monitor active metabolite, N-acetylprocainamide (NAPA)
NAPA	5.5–6.2	10–20	Toxicity at procainamide + NAPA ≥ 30 mg/L
Quinidine	4.4–9.0	2.0–5.0	Causes increased digoxin level
Bronchodilators			
Caffeine	40–230	6–11	Treatment of neonatal apnea
Theophylline	6–8 adults; 20–30 infants	10–20 adults	Monitor metabolite caffeine in neonates; reduced half-life in smokers
Antibiotics			
Amikacin	1.4–2.6	20–25 peak 5–10 trough	Toxic ranges: 35 peak; 10 trough
Chloramphenicol	1.9–3.5	20–30 peak 10–20 trough	Toxic range: 25 trough
Gentamicin	2–3	5.0–8.0 peak 1.0–2.0 trough	Toxic ranges: 12 peak; 2 trough
Kanamycin	1.9–2.3	20–25 peak 5–10 trough	
Tobramycin	2.1–2.3	5.0–8.0 peak 1.0–2.0 trough	Toxic ranges: 12 peak; 2 trough
Vancomycin	5–6	20–30 peak	Toxic range: 40 trough
Psychotropics			
Lithium	14–30; 48–72	0.8–1.2 mmol/L	Biphasic elimination; toxicity at ≥ 1.5 mmol/L
Amitriptyline	9–46	120–250 µg/L	Range includes amitriptyline and its metabolite nortriptyline
Nortriptyline	18–56	50–150 µg/L	Metabolite of amitriptyline and may be given separately
Imipramine	6–28	180–350 µg/L	Range includes imipramine and its metabolite desipramine
Desipramine	12–28	115–250 µg/L	Metabolite of imipramine and may be given separately
Doxepin	8–36	150–250 µg/L	Range includes doxepin and its metabolite nordoxepine
Immunosuppressants			
Cyclosporine	3–7; 18–25	100–300 ng/mL	Range method dependent; whole blood specimen of choice
Tacrolimus	4–41	5–20 ng/mL	Whole blood is specimen of choice
Sirolimus	46–78	5–15 ng/mL	Whole blood; when concomitant use with cyclosporine
		5–25 ng/mL	Whole blood; when used alone

crucial to shake suspensions of this medication to ensure proper dosage. If proper mixing is not achieved, under- or overmedication becomes a risk because initial doses contain decreased concentrations of phenytoin, and subsequent doses contain inappropriately high concentrations. In fact, phenytoin overdose has occurred in children due to failure of parents to adequately shake suspensions of the medication.

Valproic acid (Depakene, Depakote) is a drug that is used for treatment of absence seizures as well as other seizure disorders. Valproic acid is approximately 90% protein bound and, as previously mentioned, competes with other drugs for protein-binding sites. The half-life is 9 to 18 hours; the therapeutic range is 50 to 120 mg/L. Coadministration of valproate can cause increased phenobarbital blood concentration due to decreased nonrenal clearance. The interaction of valproate and phenytoin is complex, and there is usually an initial decrease of total phenytoin levels with either no change or an increase in free phenytoin concentrations. The dosage of coadministered drugs may need to be adjusted, as guided by the results of TDM.

Primidone (Mysoline) is a drug with anticonvulsant activity due to its own activity and that of both its major metabolites, phenobarbital and phenylethylmalonamide (PEMA). Primidone has a half-life of 3 to 12 hours and a therapeutic range of 5 to 12 mg/L. Patients taking primidone should have both primidone and phenobarbital concentrations monitored.

Carbamazepine (Tegretol) is an anticonvulsant that is sometimes coadministered with phenytoin. It has a half-life of 24 to 48 hours and a therapeutic range usually listed as either 4 to 12 or 8 to 12 mg/L. One of the metabolites of carbamazepine, the 10,11-epoxide, has anticonvulsant activity but is not routinely monitored, possibly because such monitoring requires the use of HPLC methodology. The epoxide concentration is approximately 20 to 25% of the parent carbamazepine concentration, but this may vary considerably. Carbamazepine is highly protein bound at approximately 80%.

Ethosuximide (Zarontin) has a half-life of 27 to 39 hours, with an average of 30 hours for children and approximately 50 hours for adults. The therapeutic range for ethosuximide is 40 to 100 mg/L. Ethosuximide is effective against absence (petit mal) seizures only, and is frequently coadministered with other anticonvulsant drugs such as phenytoin or carbamazepine that are used to treat grand mal seizures, which may accompany the absence seizures.

B. CARDIOACTIVE DRUGS

▶ The **cardioactive** glycoside digoxin (Lanoxin) is an isolate from the digitalis plant, foxglove (*Digitalis purpurea*), and in most laboratories is among the highest-volume monitored therapeutic drugs. Another cardioactive glycoside from digitalis is digitoxin, which is infrequently used clinically and will not be discussed. Digoxin is administered for treatment of supraventricular tachycardia and to enhance cardiac contraction in congestive heart failure (CHF). Digoxin's half-life is 23 to 61 hours; the therapeutic range is 0.5 to 2.0 µg/L. It is important to note that after administration 8 to 10 hours are needed for distribution between plasma and tissue before digoxin concentrations are interpretable.

Immunoassays are commonly used to measure digoxin concentrations. Several clinical conditions, including renal insufficiency, CHF, and complicated third-trimester pregnancies, are characterized by the production of substances termed digoxin-like immunoreactive substances (DLIS), which cross-react with the antibodies in these assays. Background interference due to DLIS has produced digoxin concentrations up to 0.5 μg/L. Each laboratory should be aware of the degree of interference for their assay and patient population.

Symptoms of digoxin toxicity include nausea, vomiting, anorexia, and visual disturbances including yellow–green distortion and the corona effect. In fact, it was postulated that Vincent van Gogh suffered from digoxin overdose due to the foxglove used to treat his seizures. His predominant use of yellow and the corona effect in paintings such as *Starry Night* are offered in support of this theory. Patients taking digoxin who also have hypokalemia, hypercalcemia, or hypomagnesemia must be monitored closely because they are more likely to develop toxic symptoms even though digoxin concentrations remain in the therapeutic range. When quinidine is coadministered, digoxin concentrations become increased apparently due to decreased clearance of digoxin.

Digoxin overdose may be treated using Digibind, the Fab fragment of antidigoxin antibodies. When a patient is treated with Digibind, plasma digoxin levels are misleading because the assays measure total digoxin, which also includes that fraction rendered inactive by binding to Digibind. If reported, these elevated levels falsely indicate that the Digibind was ineffective. In this instance, it is suggested that free digoxin levels be monitored using procedures that involve ultrafiltration followed by immunoassay analysis of the filtrate. Also, serum potassium levels should be closely monitored when Digibind is administered because digoxin intoxication causes hyperkalemia due to release of potassium from cells. When Digibind reverses the overdose effects, potassium can rapidly shift back into cells, possibly leading to hypokalemia.

Lidocaine (Xylocaine) is used to treat ventricular arrhythmias and to prevent ventricular fibrillation. It is administered intravenously or intramuscularly and is rapidly metabolized, with a half-life of 1.4 to 2.2 hours. The therapeutic range is 1.5 to 5.0 mg/L, and concentrations above 8.0 mg/L are considered toxic. Samples for TDM may be collected 30 minutes after a loading dose or 5 to 7 hours after initiation of therapy, when steady state is reached.

Quinidine, available as quinidine sulfate or the slow-release preparation quinidine gluconate, is used to treat various arrhythmias. Quinidine sulfate is rapidly absorbed and has a half-life of 4.4 to 9.0 hours. The therapeutic range is dependent on the method of analysis with an approximate range of 2.0 to 5.0 mg/L for immunoassay methods. Coadministration of quinidine to a patient taking digoxin can cause a doubling in the digoxin concentration, apparently due to competition for active renal tubular secretion and release of digoxin from tissue stores.

Disopyramide (Norpace) is used in many of the same situations as quinidine and has a similar mechanism of action. The half-life is 5 to 7 hours and the therapeutic range is 2.0 to 5.0 mg/L. The binding of disopyramide to protein is concentration dependent, with a decreased proportion of binding as the plasma disopyramide concentration increases. Toxic effects include apnea, loss of consciousness, cardiac arrhythmias, hypotension, and respiratory depression.

Procainamide (Pronestyl) is used to treat ventricular and atrial arrhythmias. The major metabolite of procainamide, N-acetyl procainamide (NAPA), has antiarrhythmic activity and accumulates in patients with impaired renal function and in fast acetylators. Both

---

Drugs with Active Metabolites

Primidone (phenobarbital and PEMA)
Procainamide (NAPA)
Theophylline (caffeine)
Amitriptyline (nortriptyline)
Imipramine (desipramine)
Doxepin (nordoxepin)

procainamide and NAPA must be monitored when procainamide is administered. The half-life of procainamide is 2.4 to 3.6 hours and that of NAPA is approximately 8 hours. In fast acetylators, NAPA accumulates to levels often exceeding those of procainamide, and in slow acetylators, the reverse is true. Therapeutic ranges are 4 to 8 mg/L for procainamide and 10 to 20 mg/L for NAPA. There are some advocates of a higher therapeutic range for procainamide because of cases in which concentrations as high as 14 mg/L were required to achieve therapeutic effect. The sum of procainamide plus NAPA should not exceed 30 mg/L because of the severity of adverse cumulative effects. Toxic effects include increased ventricular extrasystoles, ventricular tachycardia or fibrillation, hypotension, tremors, and central nervous system or respiratory depression.

Other, less frequently monitored cardioactive drugs include amiodarone, mexiletine, propranolol, tocainide, and verapamil.

C. BRONCHODILATORS

▶ Theophylline is a **bronchodilator** that acts by relaxing bronchial smooth muscle and is used both prophylactically and to treat acute attacks of asthma. Various theophylline formulations, including sustained-release preparations, are available. The half-life is 6 to 8 hours for adults and 20 to 30 hours for preterm infants. The therapeutic range is 10 to 20 mg/L for adults and 5 to 11 mg/L for neonatal apnea. In neonates, theophylline is rapidly metabolized to caffeine and both theophylline and caffeine are monitored. Toxic effects include nausea, vomiting, headaches, tachycardia, hypotension, and convulsions.

Caffeine is preferred over theophylline for the treatment of apnea in preterm infants due to its longer 40- to 230-hour half-life, more reliable enteral absorption, and lower toxicity. Adverse effects include tachycardia, vomiting, feeding intolerance, and seizures. The therapeutic range for caffeine in neonates is 8 to 20 mg/L.

D. ANTIBIOTICS

▶ The aminoglycoside **antibiotics** include gentamicin, tobramycin, amikacin, netilmicin, streptomycin, neomycin, and kanamycin. These drugs are used to treat or prevent systemic infections of gram-negative aerobic bacteria. Anaerobic bacteria are resistant to these drugs because the transport mechanism of the aminoglycoside antibiotics across the bacteria cell wall is oxygen dependent. The aminoglycosides inhibit protein synthesis by binding to the 30S ribosomal subunit of the bacteria. Hospital infection control committees closely monitor the use of these antibiotics to minimize evolution of resistant bacterial strains and to minimize expense. These drugs are poorly absorbed and are therefore administered intravenously or intramuscularly. The major toxic effects of the aminoglycosides are nephrotoxicity and ototoxicity (permanent hearing loss). Baseline serum creatinine concentration and, if possible, creatinine clearance determinations should be performed prior to initiation of therapy to monitor renal function and ensure adequate clearance of the drug. To avoid toxic effects, the drug levels must fall below a specified trough concentration before the next dose is administered. Peak levels are also monitored in order to ensure rapid achievement of therapeutic

---

Toxic Effects of Procainamide and NAPA

- Increased ventricular extrasystoles
- Ventricular tachycardia or fibrillation
- Hypotension
- Tremors
- Central nervous system or respiratory depression

Toxic Effects of Theophylline

- Nausea
- Vomiting
- Headaches
- Tachycardia
- Hypotension
- Convulsions

(bacteriostatic) concentrations. Accurate interpretation of both trough and peak therapeutic ranges requires careful timing of specimen collection. Hence, for the antibiotics, it is particularly important that phlebotomy staff be instructed to collect the sample in a timely fashion and record the collection time accurately. The most commonly encountered problem in TDM is failure to collect samples at the proper time, and this problem is magnified with TDM of antibiotics.

Samples for trough levels should be collected prior to administration of the next intravenous or intramuscular dose. For peak levels, samples should be collected 30 to 60 minutes after completion of an intravenous infusion and 60 to 90 minutes following intramuscular injection. Recently, pulse dosing of aminoglycoside antibiotics, also known as once-a-day dosing (even though the interval may exceed 24 hours), has been advocated as an alternative to the routine practice of administration two or three times daily. Advocates argue that since toxicity depends on the accumulation of the drug at trough concentration, the higher peak concentration attained with pulse dosing enhances the bacteriocidal activity without harming the patient. When pulse dosing is used, baseline and periodic (1 to 3 days) serum creatinine levels are monitored. In this case, a timed sample is collected 8 to 12 hours after the infusion and the concentrations obtained are compared to those on nomograms developed for dosing regimens.

The therapeutic ranges of various antibiotics are listed in Table 8–1. In addition to the aminoglycosides previously discussed, vancomycin and chloramphenicol are antibiotics that are therapeutically monitored. Principal uses of vancomycin include treatment of staphylococcal and streptococcal infections in patients with penicillin allergies, management of methicillin-resistant staphylococcal infections, and treatment of dialysis shunt infections. Chloramphenicol is a drug that is used for treatment of *Haemophilus influenzae* type b, the most common cause of meningitis in infants and children.

E. PSYCHOTROPICS

Psychotropics that are commonly monitored therapeutically include lithium and various tricyclic antidepressants and their metabolites. Lithium is administered as either the carbonate or citrate formulation and is used for the treatment of manic, also termed *bipolar,* depression. Elimination of lithium is biphasic, with an initial half-life of 14 to 30 hours and a longer second phase of 48 to 72 hours. The therapeutic range is 0.8 to 1.2 mmol/L with levels greater than 1.5 mmol/L associated with toxicity. Early signs of lithium intoxication include diarrhea, vomiting, drowsiness, muscular weakness, and lack of coordination. Effects of toxic lithium levels include ataxia, giddiness, tinnitus, and blurred vision. The renal function of patients receiving lithium long term is typically evaluated because of the medication's reported association with dimunition in renal concentrating ability. Patients on chronic lithium therapy have shown glomerular and interstitial fibrosis and nephron atrophy. However, it is of interest that similar morphologic changes have been seen in manic–depressive patients who have never been prescribed lithium. The electrolyte and hydration status of patients receiving lithium must be monitored because of the increased risk of toxicity in conjunction with dehydration and sodium loss. Chronic lithium administration has also been

Early Signs of Lithium Toxicity

- Ataxia
- Giddiness
- Tinnitus
- Blurred vision

associated with hypothyroidism. Consequently, the thyroid status of these patients must be monitored periodically. Methods available for the analysis of lithium include ion-selective electrodes, atomic absorption photometry and, less frequently, flame emission photometry.

The tricyclic antidepressants include amitriptyline (Elavil), nortriptyline (Pamelor), imipramine (Tofranil), desipramine (Norpramin), and doxepin (Adapin). Amitriptyline, imipramine, and doxepin are tertiary amines that are metabolized to their respective secondary amines nortriptyline, desipramine, and nordoxepin. In addition, nortriptyline and desipramine are also administered as antidepressants. When the tertiary amines are administered, the corresponding metabolite is monitored also. For example, when amitriptyline is ingested, both the parent compound and nortriptyline are monitored. The therapeutic range for nortriptyline is 50 to 150 µg/L and for amitriptyline is 120 to 250 µg/L when these medications are coadministered. Likewise, the therapeutic range for desipramine is 115 to 250 µg/L and for imipramine is 180 to 350 µg/L with coadministration. The therapeutic range for doxepin, which includes nordoxepin, is 150 to 250 µg/L. Toxic effects include hypotension, seizures, coma, respiratory depression, and cardiotoxic symptoms such as tachycardia and cardiac arrhythmias.

The tricyclic antidepressants inhibit the reuptake of biogenic neurotransmitters, resulting in mood elevation. Urinary levels of the norepinephrine metabolite, 3-methoxy-4-hydroxyphenylglycol (MHPG), may be used to predict responses to tricyclic antidepressants. Patients with low urinary MHPG levels respond better to imipramine, desipramine, and nortriptyline, and patients with higher levels respond more favorably to amitriptyline.

F. ANTINEOPLASTICS

▶ Methotrexate, a folic acid antagonist, is an **antineoplastic** used to treat various neoplasms including leukemia, lymphoma, osteosarcoma, breast carcinoma, and choriocarcinoma. Methotrexate interferes with DNA synthesis, repair, and cellular replication by inhibiting dihydrofolate reductase. Because it is cytotoxic to normal cells, the methotrexate antagonist leucovorin must be given to prevent destruction of, or to "rescue," the normal cells when methotrexate clearance is delayed. For this reason, TDM for methotrexate actually monitors drug elimination following high-dose therapy. Specified concentrations must be achieved at various times. For patients with normal elimination, methotrexate levels should be below 10 µmol/L 24 hours after therapy; at 48 hours, levels should be below 1 µmol/L; and by 72 hours after treatment, the methotrexate concentration should be less than 0.1 µmol/L. If values exceed these concentrations, leucovorin rescue is administered. Leucovorin acts by serving as a substrate for dihydrofolate reductase, thus allowing resumption of the DNA-related processes.

G. IMMUNOSUPPRESSANTS

Cyclosporine is used to prevent host-graft rejection for all types of transplantation. It can be administered with olive or corn oil vehicles (e.g., Sandimmune), or in a microemulsion (e.g., Neoral) which has increased bioavailability and decreased interpatient variability. The

---

Toxic Effects of Doxepin

- Hypotension
- Seizures
- Coma
- Respiratory depression
- Tachycardia and cardiac arrhythmias

elimination is biphasic, with half-lives of 3 to 7 hours and 18 to 25 hours. The specimen of choice is anticoagulated whole blood because approximately two-thirds of cyclosporine is found in red cells. In serum or plasma, there is variable partitioning between the cells and extracelluar fluid depending on temperature, time, and drug concentration. Methods available include HPLC and various immunoassays (radioimmunoassay, enzyme, fluorescence polarization), and the therapeutic ranges are method dependent. Immunoassays have different antibody specificity and measure parent drug along with some of the metabolites of cyclosporine. HPLC measurement of cyclosporine is considered the reference method, and the therapeutic range with this technique is 100 to 300 µg/L. Trough whole blood concentrations greater than 400 to 600 µg/L are associated with hepatic, renal, and neurologic complications and immunocompromise. Because nephrotoxicity is associated with cyclosporine administration, renal function monitoring (e.g., creatinine and blood urea nitrogen [BUN]) must be performed in addition to TDM in these patients. Rising concentrations of creatinine and BUN in renal transplant patients present a complex problem in distinguishing between cyclosporine toxicity and rejection. Monitoring liver function is also clinically necessary because of potential hepatic complications associated with cyclosporine therapy. In most clinical situations in the context of organ transplant, the cyclosporine level provides essential information to caregivers for patient evaluation.

FK-506 or tacrolimus (ProGraf), another **immunosuppressant** ◄ administered to organ transplant patients, has a half-life ranging from 4 to 41 hours. Tacrolimus use is gaining favor among transplant physicians because the toxicity of this drug is less than that of cyclosporine. The specimen of choice for tacrolimus measurements is whole blood; a therapeutic range of 5 to 20 µg/L has been established for this specimen. Plasma measurement is also used (therapeutic range 0.5 to 2.0 µg/L); the red cell to plasma ratio of tacrolimus appears to vary with the specific organ transplanted, so this specimen is less desirable. In addition, less active metabolites typically accumulate in plasma that may cross-react in common assays. The adverse effect of principal concern is renal damage, with other effects including tremors, headache, diarrhea, and hypertension. As with cyclosporine, renal and hepatic function must also be monitored in transplant patients receiving tacrolimus.

Sirolimus (Rapamune) is a new immunosuppressant agent recently approved for use in combination with cyclosporine and corticosteroids in renal allograft transplantation. Whole blood is the biological matrix of choice for TDM because sirolimus is primarily partitioned into red blood cells. Sirolimus acts to suppress T-lymphocyte proliferation by inhibiting the progression of the cell cycle from G_1 to the S phase. Quantitative analysis of sirolimus in clinical trials used HPLC with UV (HPLC/UV) detection. This measurement strategy is plagued with many challenges including low blood concentrations (ng/mL), low recovery during the isolation procedure, sirolimus' low UV absorbance, and interference in the UV region from substances endogenous to whole blood that are difficult to remove without further reducing the recovery of sirolimus. Because of these problems, clinical laboratories have turned to liquid chromatography with mass spectrometry (LC/MS) or tandem MS (LC/MS-MS) detection to provide the reproducibility, accuracy, and sensitivity required for sirolimus monitoring. While LC/MS and LC/MS-MS methods are not subject to interferences and sensitivity problems posed by HPLC/UV, the instrumentation is expensive, requires substantial technical expertise, and is of limited availability. It is anticipated that a more generally applicable immunoassay will be available in the near future. The therapeutic range for sirolimus is 5-15 ng/mL when co-administered with full-dose cyclosporine. If the cyclosporine dose is reduced, it is recommended that the sirolimus dosage be adjusted to maintain the level closer to the upper end of the therapeutic range. If only sirolimus is used, the upper limit of the therapeutic range is often extended to 25 ng/mL. As with other immunosuppressants such as tacrolimus and cyclosporine, sirolimus dosing must be monitored carefully due to the risk of toxicity.

SUGGESTED READINGS

AACC TDM/TOX Laboratory Improvement Committee. *Drug monitoring data pocket guide II.* Washington, DC: AACC Press, 1994.

Kaplan LA, et al. NACB Symposium Standards of Laboratory Practice: Therapeutic Drug Monitoring. *Clin Chem* 44(5):1072–140, 1998.

Moyer TP, Pippenger CE. Therapeutic drug monitoring. In Burtis CA, Ashwood ER (eds.). *Tietz textbook of clinical chemistry,* 3rd ed. Philadelphia: WB Saunders, 1999.

Professional practice in toxicology: A review, 4th ed. Washington, DC: AACC Press, 1996.

Thorne DP. In Bishop ML, Duben-Engelkirk JL, Fody EP (eds.). Therapeutic drug monitoring. In *Clinical chemistry: Principles, procedures, correlations,* 4th ed. Philadelphia: JB Lippincott, 2000.

Warner AM. Therapeutic drug monitoring. In Dufor DR (ed.). *Professional Practice in Clinical Chemistry: A Companion Text.* Washington, DC: AACC Press, 1999.

Items for Further Consideration

1. A patient who has been receiving phenobarbital for seizure control for years begins taking valproic acid in addition to the phenobarbital. What are some of the considerations to be considered for TDM of this patient?

2. A patient comes to the emergency department showing symptoms of digoxin toxicity. When the digoxin concentration is determined, it is within the therapeutic range. What other factors should be considered and what laboratory tests can be performed to evaluate these possibilities?

Toxicology

Larry Broussard

Key Terms ◀

BLOOD ALCOHOL CONCENTRATION

CHAIN OF CUSTODY

CLINICAL TOXICOLOGY

CONFIRMATION

CUTOFF CONCENTRATIONS

DONE NOMOGRAM

MEDICAL REVIEW OFFICER

NATIONAL INSTITUTE ON DRUG
ABUSE (NIDA)

OSMOLAL GAP

RUMACK–MATTHEW NOMOGRAM

SCREENING

SPOT TESTS

SUBSTANCE ABUSE AND MENTAL
HEALTH SERVICES
ADMINISTRATION (SAMHSA)

TOXICOLOGY

WORKPLACE DRUG TESTING

Chapter Objectives

Upon completion of this chapter, the reader will be able to:

1. Compare and contrast clinical and forensic drug testing purpose and protocols.
2. State the sample(s) of choice for clinical and forensic drug testing protocols.
3. Discuss the components of federally mandated (SAMHSA) workplace drug testing.
4. List the five drugs tested under the federal (SAMHSA) guidelines.
5. Distinguish between screening and confirmatory tests for detecting and identifying drugs of abuse.
6. List drugs for which antidotes are available and identify the appropriate antidote for each drug.
7. Discuss commonly identified drugs and metabolites for the following drug classes: alcohols, analgesics, antidepressants, hallucinogens, hypnotics, stimulants, tranquilizers, metals, and pesticides.
8. Describe basic symptomatology associated with toxic concentrations of the drug classes listed above.
9. Discuss the applications of the following laboratory methods and techniques used in toxicology: spot tests, osmolal gap, immunoassays, thin-layer chromatography, gas chromatography, high-performance liquid chromatography, mass spectrometry, and atomic absorption.
10. Discuss the use of the Rumack–Matthew and the Done nomograms in assessing acetaminophen and salicylate toxicity.
11. Describe procotols suggested to prevent adulteration of samples during urine collection for workplace drug testing.
12. List conditions under which a federally certified laboratory may report a sample as dilute, substituted, or adulterated.

Analytes Included ◄

ALCOHOLS:

ETHANOL

ETHYLENE GLYCOL

ISOPROPANOL

METHANOL

ANALGESICS:

ACETAMINOPHEN

OPIATES (MORPHINE, CODEINE, HYDROMORPHONE, HEROIN, ETC.)

SALICYLATE

ANTIDEPRESSANTS:

TRICYCLIC ANTIDEPRESSANTS (AMITRIPTYLINE, NORTRIPTYLINE, DESIPRAMINE, IMIPRAMINE, DOXEPIN)

HALLUCINOGENS:

CANNABINOIDS (MARIJUANA)

PHENCYCLIDINE (PCP)

HYPNOTICS:

BARBITURATES

STIMULANTS:

AMPHETAMINES

COCAINE

TRANQUILIZERS:

BENZODIAZEPINES

METALS:

ARSENIC

LEAD

MERCURY

PESTICIDES:

CARBAMATES

ORGANOPHOSPHATES

Pretest

Directions: Choose the single best answer for each of the following items.

1. Drug testing designed to detect drug overdoses and formulate a treatment plan is referred to as
   a. forensic toxicology
   b. clinical toxicology
   c. medicolegal toxicology
   d. workplace toxicology

2. Federal regulations require that screening for the presence of illicit drugs be performed by which of the following methods?
   a. colorimetric spot tests
   b. thin-layer chromatography
   c. mass spectrometry
   d. immunoassay

3. For workplace drug programs, which of the following body fluids is used for drugs of abuse testing?
   a. urine
   b. saliva
   c. plasma
   d. capillary blood

4. Which of the following methods is used to detect possible adulteration or substitution of urine samples collected for workplace drug testing?

   a. quantitation of total protein
   b. measurement of temperature
   c. measurement of optical density
   d. cell count determination

5. The antidote for methanol toxicity is
   a. formatran
   b. ethanol
   c. physostigmine
   d. atropine

6. The antidote for acetaminophen toxicity is
   a. naloxone
   b. N-acetylcysteine
   c. physostigmine
   d. nortran

7. The most commonly encountered substance in emergency situations is
   a. marijuana
   b. cocaine
   c. phencyclidine (PCP)
   d. ethanol

8. The ingestion of poppy seeds may give a positive urine screen for
   a. cannabinoids
   b. cocaine metabolite
   c. opiates
   d. amphetamines

9. A patient is brought into the emergency room with the following symptoms: coma, respiratory depression, and pinpoint pupils. He is given naloxone (Narcan) and responds. The expected result on the urine toxicology screen is positive for
   a. PCP
   b. opiates
   c. amphetamines
   d. cannabinoids

10. Acetaminophen toxicity may lead to
    a. renal failure
    b. blindness
    c. liver failure
    d. paralysis

11. An individual inhaled what was represented to be marijuana; however, the subsequent effects were much more exaggerated than expected. In addition to feeling euphoric, symptoms included anxiety, agitation, muscle rigidity, hypertension, and an overwhelming sense of being able to physically dominate anyone or anything. Which of the following drugs of abuse was most probably inhaled?
    a. cocaine
    b. psilocybin
    c. cocaethylene
    d. PCP

12. Delta-9-tetrahydrocannabinol (THC) is the primary psychoactive component of
    a. benzodiazepines
    b. tricyclic antidepressants
    c. marijuana
    d. opiates

13. The following results were obtained from a plasma sample of a patient in the emergency room:
    Sodium—140 mmol/L   Glucose—180 mg/dL
    BUN—28 mg/dL   Osmolality—
                          340 mOsm/kg

    The most likely explanation for these results is
    a. patient is in renal failure
    b. patient has ingested ethanol or some other alcohol
    c. one of these results given is in error
    d. patient has salicylate toxicity

14. The metabolite of cocaine that may be detected for up to 72 hours postingestion is
    a. benzoylecgonine
    b. ethchlorvynol
    c. phenocolazine
    d. codiazepoxide

15. The method of choice for heavy metal analysis is
    a. high-pressure liquid chromatography (HPLC)
    b. fluorometry
    c. immunoassay
    d. atomic absorption spectrophotometry

16. Which of the following metals, when in toxic concentration, leads to decreased intelligence quotient (IQ) and growth in early childhood?
    a. mercury
    b. lead
    c. arsenic
    d. copper

17. Which of the following pesticides inhibits the enzyme cholinesterase?
    a. lysergine
    b. carbamate
    c. atropine
    d. pralidoxime

18. For workplace or forensic drug testing, the individual who verifies the validity of the testing and specimen handling process and evaluates all results reported is the
    a. Clinical Director
    b. Laboratory Regulations Officer
    c. Medical Review Officer
    d. Technical Specialist

19. Which of the following general test methods may be useful in clinical toxicology to screen for the presence of drugs such as acetaminophen, salicylates, and phenothiazines?
    a. immunoassay
    b. urine osmolality
    c. spot tests
    d. thin-layer chromatography

20. Ethanol abuse may be evaluated without the specific testing for ethanol in the blood by use of the
    a. cation gap
    b. osmolal gap
    c. Done nomogram
    d. Rubrick calculation

I. INTRODUCTION

▶ **Toxicology** is the study of the adverse effects of chemical agents on biologic systems. The field of toxicology encompasses many overlapping disciplines, including clinical, industrial, environmental, and forensic toxicology.

The focus of this chapter is clinical toxicology and workplace drug testing. The basic principles of pharmacokinetics and descriptions of many types of prescription medications were discussed in Chapter 8, "Therapeutic Drug Monitoring." Discussion is presented on the effects of toxic substances on the human body and the analytical methods used for their detection and/or quantification.

II. MEDICAL OVERVIEW

A. Clinical versus Workplace Drug Testing

In the clinical setting, for instance, in the emergency department, toxicology is employed for the detection and treatment of overdose situations. Workplace drug testing is designed for the identification of illicit drug use. Though the actual methodologies for workplace drugs of abuse testing is similar to that used for patients in the clinical setting, the forensic aspect of workplace drugs of abuse testing adds an additional layer of confirmatory testing, review, and extensive documentation. Moreover, workplace drug testing is federally
▶ regulated by the **Substance Abuse and Mental Health Services Administration (SAMHSA);** in addition, certain state agencies may have further stipulations that apply. Many state and private-sector employers use drug testing protocols that mimic the federal regulations. Table 9–1 provides a comparison of clinical and workplace toxicology.

B. Clinical (Emergency) Drug Testing

▶ **Clinical toxicology** includes the analysis of body fluids for the detection of chemicals (drugs, toxins, poisons, industrial products, etc.) having adverse effects on the patient. The reference range for the presence of toxic substances in the body is *zero!*

The scope of clinical toxicology services offered by an institution will vary, depending on factors such as available equipment and

Table 9–1. A Comparison of Clinical (Emergency) and Workplace Drug Testing

CATEGORY	CLINICAL	WORKPLACE
Purpose	Overdose detection and treatment	Detection of drug use
Result reporting	Rapid, TAT in few hours	MRO involved; TAT in few days
Tests performed	Limited only by methodology	Federally mandated (NIDA 5)
Methodology	Unlimited	Federally mandated: immunoassay screen, GC/MS confirmation
Cutoff levels	Methodology dependent	Federally mandated
Sample handling	Same as other clinical specimens	Chain of custody
Sample type(s)	Urine, gastric, blood	Urine (breath/saliva for ethanol)

TAT, turnaround time; MRO, Medical Review Officer; NIDA 5, amphetamines, cannabinoids, cocaine metabolite, opiates, phencyclidine (PCP); GC, gas chromatography; MS, mass spectrometry.

trained personnel or access to a reference or specialty toxicology laboratory. Emphasis is on the rapid assessment and identification of ingested drugs in order to assist the physician in treating patients in the emergency department. Hence, the turnaround time for reporting results is typically within hours after receipt of the specimen. Initial treatment of suspected drug overdose or poisoning in the emergency department centers on supportive care based on symptoms and situational information (i.e., empty prescription bottles, witness statements) when available. Supportive therapy includes maintenance of fluid and electrolyte balance and respiratory function. Antidotes are available for a few drugs, but frequently detoxification is through the metabolism and excretion of the drug(s) by the patient. Some of the antidotes available include N-acetylcysteine for acetaminophen toxicity, atropine for organophosphate poisoning, desferoxamine to chelate iron, ethanol for methanol and ethylene glycol overdose, naloxone for opiate toxicity, and physostigmine to reverse the effects of anticholinergic agents.

Urine, gastric, and blood samples should be collected for drug testing in suspected overdosed patients. Initial screening for drugs other than ethanol is typically performed using urine and/or gastric samples. When drugs are detected in urine and/or gastric specimens, blood (serum or plasma) is used for quantitation. Blood drug levels help the physician assess the severity of the overdose and determine appropriate treatment. Blood is also the specimen of choice when screening for ethanol abuse.

In the clinical setting, most substances detected fall into one of the following major drug groups: alcohols, analgesics (both narcotic and nonnarcotic), antidepressants, hallucinogens, hypnotics, stimulants, tranquilizers, metals, and pesticides.

1. Alcohols. Alcohols detected include ethanol, methanol, isopropanol, and ethylene glycol. Ethanol is the most common toxic substance encountered in emergency situations. The correlation between **blood alcohol concentration** (BAC) and physiological effects has been demonstrated and is listed in Table 9–2. In this chart, BAC is expressed in units of grams per deciliter or percent. The other unit of measure commonly encountered is milligrams per deciliter. States have established BACs from 0.08 to 0.10% (80 to 100 mg/dL) as cutoff concentrations for driving under the influence (DUI) laws.

Antidotes Available for Certain Drugs or Toxic Substances

- Acetaminophen → N-acetylcysteine
- Organophosphates → atropine
- Iron → desferoxamine
- Methanol, ethylene glycol → ethanol
- Opiates → naloxone
- Anticholinergics → physostignine

Major Drug Groups

- Alcohols
- Analgesics
- Antidepressants
- Hallucinogens
- Hypnotics
- Stimulants
- Tranquilizers
- Metals
- Pesticides

Table 9–2. Correlation of Blood Alcohol Concentration (BAC) and Physiological Effects

BAC	EFFECTS
0.05%	Changes in mood and behavior; judgment and restraint are somewhat impaired, thinking dulled
0.10%	Walking, speech, hand movements clumsy; blurred, split, or tunnel vision may occur
0.20%	Behavior greatly affected; person may become loud, easily angered, fearful
0.30%	Brain responses very slow and dulled; disorientation and confusion, in vision and hearing
≥ 0.40%	Unconsciousness, depression, breathing difficulty, and slowed heartbeat; sometimes death

The preferred specimen for ethanol analysis is whole blood preserved with sodium fluoride. The ratio of whole blood ethanol concentration to serum or plasma ethanol is dependent on the hematocrit, with an average ratio of 1:1.18. Thus, serum or plasma ethanol values are higher than the whole blood ethanol, and a conversion may be necessary. Laws generally refer to blood ethanol concentrations, not serum or plasma values. Ethanol may be measured on site in breath samples using breath alcohol analyzers or by disposable kits. Kits are also available for the analysis of ethanol in saliva. Metabolism of ethanol is dependent on several variables including frequency of use, ingestion of other drugs, and liver function. As a rule of thumb, ethanol is cleared at an hourly rate of 18 mg/dL in women and 15 mg/dL in men.

Methanol, isopropanol, and ethylene glycol are metabolized by liver alcohol dehydrogenase to toxic or longer-acting metabolites. Methanol (found in products such as canned fuel and as a contaminant in illegally distilled spirits) is oxidized first to formaldehyde and subsequently to formic acid (formate). This generation of formic acid can cause severe acidosis and optic nerve damage leading to blindness. Formate levels correlate with toxic effects more closely than methanol levels. Isopropanol (rubbing alcohol) is rapidly metabolized to acetone. Often, the acetone concentration exceeds that of isopropanol originally ingested. Ethylene glycol (antifreeze) is metabolized to several products, notably oxalic, glycolic, and hippuric acids. Toxic effects include acidosis and acute renal failure characterized by the formation of oxalate crystals in the kidneys. Treatment of both methanol and ethylene glycol (and less frequently isopropanol) intoxication includes the administration of ethanol to saturate the alcohol dehydrogenase in the liver and thus prevent formation of the toxic metabolites. This treatment is augmented with hemodialysis to remove the toxins and with sodium bicarbonate therapy to neutralize the acidosis in methanol and ethylene glycol intoxication.

2. Analgesics. Analgesics are another classification of drugs that may be detected in overdose situations. Narcotic analgesics include opiates and other miscellaneous drugs such as propoxyphene (Darvon), meperidine (Demerol), and methadone. Nonnarcotic analgesics include acetaminophen and salicylates.

Opiates or opioids are narcotic analgesics derived from opium, the juice of the poppy seed plant. This class of naturally occurring and synthetic drugs includes morphine, codeine, hydromorphone (Dilaudid), hydrocodone (Vicodin), oxymorphone (Numorphon), and oxycodone (Percodan). Heroin (diacetylmorphine) is rapidly metabolized to 6-(mono)acetylmorphine (6-AM or 6-MAM), which is either conjugated or metabolized to morphine. Codeine is metabolized to morphine and there is a small amount of reverse metabolism from morphine to codeine when morphine is taken. In addition to analgesia, opiates also cause euphoria and sedation. Chronic use may lead to tolerance and addiction. Opiate overdose is characterized by the classical clinical findings of coma, respiratory depression and pinpoint pupils. Naloxone (Narcan) is a narcotic antagonist that reverses the pharmacologic effects of the opiates. It does not affect central nervous system (CNS) depression of nonnarcotic origin, so it may be safely administered without laboratory confirmation of the presence of opiates.

Acetaminophen is a common nonnarcotic analgesic and antipyretic that is a component of more than 200 over-the-counter medications. The role of the laboratory in acute acetaminophen overdose is crucial because of the triphasic clinical course that occurs. In the first 24 hours, the patient suffers from nausea, vomiting, malaise, pallor, diaphoresis, and anorexia. Though the patient appears to improve during the second phase (24 to 48 hours after ingestion), hepatic necrosis begins. At this time, there is a rise in bilirubin and hepatic enzymes (aspartate transaminase [AST], alanine transaminase [ALT]) and a prolonged prothrombin time. The final phase occurs 3 to 5 days postingestion and includes hepatic failure, which may result in death. When overdose levels of acetaminophen are ingested, the liver enzymes that metabolize acetaminophen to the nontoxic glucuronide and sulfate products become saturated and more drug is converted to a toxic metabolite. This unconjugated metabolite, which is unable to be excreted in the urine, accumulates and binds to hepatocytes, causing cell death and ultimately hepatic necrosis. Hepatic necrosis can be prevented by administration of the antidote N-acetylcysteine (Mucomyst) within the first 8 hours postingestion. N-acetylcysteine acts as a glutathione substitute, thus allowing conjugation and excretion of the toxic metabolite. Physicians rely on plasma acetaminophen concentrations to determine if the antidote should be given.

The **Rumack–Matthew nomogram** (Rumack & Peterson, 1978) ◄ was developed to predict acetaminophen overdose severity and is divided into zones of no hepatic toxicity, possible hepatic toxicity, and probable hepatic toxicity. Samples should not be collected before 4 hours after ingestion to ensure complete absorption and the attainment of peak concentration. When the hours of ingestion are unknown or cannot be conservatively estimated, two or more samples may be collected at 2- to 3-hour intervals for acetaminophen analysis and an elimination half-life calculated. A half-life > 4 hours is indicative of probable hepatotoxicity and a half-life > 12 hours is predictive of impending hepatic coma.

Serum salicylate is measured when aspirin (acetylsalicylic acid) intoxication is suspected. Aspirin is a nonnarcotic analgesic, antipyretic, and anti-inflammatory drug available in many over-the-counter preparations. Moderately high doses are used for the treatment of rheumatoid arthritis. Aspirin is rapidly absorbed (within 2 hours) but sustained release or enteric-coated formulations may have delayed absorption. After absorption, acetylsalicylic acid is rapidly hydrolyzed to salicylate (salicylic acid), which is the metabolite measured by the laboratory. Because of delayed absorption with sustained-release formulations and metabolism of aspirin to salicylate, it is possible for the salicylate concentration in a second sample to be higher than that in a prior sample even though no aspirin is ingested between sampling. Symptoms of overdose include tinnitus (ringing in the ears) and acid–base abnormalities (respiratory alkalosis and metabolic acidosis).

The **Done nomogram** (Done, 1960) is used to assess the se- ◄ verity of acute salicylate intoxication. This chart is a plot of serum salicylate concentrations versus hours after ingestion. To ensure completion of absorption, samples should be collected 6 hours postingestion, with a second sample collected 2 to 3 hours after the initial sample. Toxic manifestations usually occur at concentrations >

Early Signs of Acetaminophen Overdose

- Nausea
- Vomiting
- Malaise
- Pallor
- Diaphoresis
- Anorexia

30 mg/dL, although symptoms may occur at concentrations as low as 20 mg/dL.

3. Antidepressants. Antidepressant drugs may also be detected in clinical overdose situations. Prescribed tricyclic antidepressants (TCAs) include amitriptyline, nortriptyline, desipramine, imipramine, and doxepin. TCA intoxication includes CNS, anticholinergic, adrenergic, and/or cardiac manifestations. The cardiotoxicity of these drugs receives the most attention because of the often lethal consequences. Management of the TCA-poisoned patient includes identification and quantitation of the drugs and continuous cardiac monitoring.

4. Hypnotics. Clinically detected hypnotic drugs include the barbiturates. Barbiturates are often divided into classes based on duration of effect (short-acting such as secobarbital to long-acting such as phenobarbital). Barbiturates have been replaced to some extent by the safer benzodiazepines as sedative–hypnotics; however, they continue to be combined with other drugs (such as antihypertensives, antispasmodics, antidiuretics, etc.) in formulations. Toxic effects of barbiturates are due to progressive CNS depression, which includes the respiratory and cardiovascular systems. Management of the overdosed patient includes supportive therapy and may include forced alkaline diuresis if the drug is a long-acting barbiturate.

> ## Symptoms of Benzodiazepine Overdose
>
> - Excitement
> - CNS depression
> - Respiratory depression
> - Hypotension
> - Coma

5. Tranquilizers. Benzodiazepines are tranquilizers that are widely prescribed in the treatment of anxiety. Examples include chlordiazepoxide (Librium), diazepam (Valium), clonazepam (Clonopin), flurazepam (Dalmane), lorazepam (Ativan), and oxazepam (Serax). Benzodiazepines undergo extensive metabolism, often yielding common metabolites, which also have pharmacological properties. Detection of urinary metabolites may not allow identification of the parent benzodiazepine ingested because of the complex metabolic pathways. The laboratory report may simply indicate the presence of benzodiazepines as a class of drugs. Symptoms of benzodiazepine overdose include excitement followed by CNS depression, respiratory depression, hypotension, and coma. Chronic ingestion may lead to dependence and withdrawal symptoms upon cessation. Treatment of overdose is primarily supportive.

6. Hallucinogens. The two main hallucinogens detected in clinical toxicology are cannabinoids (marijuana) and phencyclidine. Phencyclidine (PCP) is administered by inhalation, intravenous injection, or oral ingestion; PCP may be misrepresented to the user as some other substance such as marijuana, mescaline, or psilocybin. In addition to euphoria and hallucinations, other effects include anxiety, agitation, panic, rigidity, hypertension, and an altered perception of danger and physical capabilities. There have been reports of flashback episodes possibly due to release of drug stored in tissues.

Cannabinoids are composed of several chemicals derived from the marijuana plant, *Cannabis sativa,* including delta-9-tetrahydrocannabinol (THC), the primary psychoactive component. Inhalation or ingestion of these components causes euphoria, sedation, loss of short-term memory, and impairment of psychomotor skills. THC by prescription is used as an antiemetic for patients undergoing chemotherapy or other immunosuppressive therapy and for relief of symptoms in patients with glaucoma. Ingestion of hemp oil has been shown to cause positive results and at present a satisfactory resolution to this specificity problem is not in existence. Absorption of THC is very rapid after inhalation, with peak psychological effects occurring within 2 hours. THC is metabolized in the liver, with approximately 85% of a single dose excreted in the feces and urine within 5 days. Chronic use may result in storage of THC in fatty tissues due to its lipophilic properties. Stored THC may be gradually released into the blood for eventual excretion, thus resulting in a longer period of detection for chronic smokers than for infrequent users.

7. Stimulants. Stimulants found in emergency toxicology include amphetamines and cocaine. The drug class amphetamines includes both amphetamine and methamphetamine. These drugs are CNS stimulants used to treat obesity, attention deficit–hyperactivity disorders, and narcolepsy. They often produce euphoria, alertness, decreased sense of fatigue, and the perception of heightened mental and physical capacity. Street names include "speed" and "ice." Amphetamine and methamphetamine have D- (dextro) and L- (levo) isomers with the D-isomers having the most pharmacological actions. L-Methamphetamine is the active ingredient in Vicks inhalers, and its use could cause a positive result for amphetamines. Prescription medications that contain drugs metabolized to amphetamine or methamphetamine include Eldepryl (selegiline), used to treat Parkinson's disease, and the obesity drug Didrex (benzphetamine).

The major metabolite of cocaine that is detected is benzoylecgonine, which may be detected for up to 72 hours. Cocaine produces CNS stimulation that includes euphoria and a sense of increased alertness. The free base form of cocaine is more volatile than the hydrochloride salt and smoking of the free base form ("crack") results in a rapid onset of action. Sudden death associated with cocaine use is due to cardiotoxicity, and overdoses may produce arrhythmias, hypertension, respiratory depression, and seizures. Another metabolite of cocaine is ecgonine methyl ester, and in the presence of ethanol, the metabolite cocaethylene (benzoylecgonine ethyl ester) is produced. Cocaethylene may cause enhanced stimulation and may contribute to the co-abuse of ethanol and cocaine. Cocaine may be detected in body fluids for a longer period of time in chronic users due to storage of the drug in tissue. Cocaine has been used as a local anesthetic and vasoconstrictor for dental and nasal surgical procedures; a drug screen could be positive following such procedures.

8. Metals. The clinical laboratory may also be asked to include testing for certain metals as part of a toxicology screen. Metals may be encountered as environmental toxins or intentionally administered poisons. The metals most frequently included in a screen (sometimes called a heavy metal screen) are arsenic, mercury, and lead. The primary method for metal detection and quantification is atomic absorption spectrophotometry (see Chapter 11).

The most familiar of the heavy metals is arsenic, which is the most frequently mentioned poison in literature. In the past, it was used in the treatment of syphilis and is now found in products such as pesticides and insecticides. Arsenic is found as inorganic (As^{+3}) and organic arsenic (As^{+5}) and as arsine gas (AsH_3), which is the most toxic form. Within 2 hours of ingestion, symptoms of acute ar-

---

Symptoms of Cocaine Overdose

- Arrhythmias
- Hypertension
- Respiratory depression
- Seizures

Symptoms of Acute Arsenic Toxicity and Chronic Exposure

Acute Toxicity
- Transverse white lines on nails
- Garlic breath odor
- Vomiting
- Bloody diarrhea
- Abdominal pain
- Cardiac abnormalities

Chronic Exposure
- Hair loss
- Nausea
- Vomiting
- Diarrhea
- Neuropathy
- Jaundice
- Anemia

senic toxicity develop, including the characteristic transverse white lines on nails (Aldrich–Mees lines) and garlic breath odor. Other symptoms include vomiting, bloody diarrhea, abdominal pain, and cardiac abnormalities. Symptoms of chronic exposure include hair loss, nausea, vomiting, diarrhea, neuropathy, jaundice, and anemia. The specimen of choice for acute toxicity is urine. For chronic toxicity, hair or nail clippings should be analyzed.

Exposure to mercury can occur from occupational (mining, manufacturing, etc.), household (thermometers), and environmental (seafood, dental amalgams) sources. Methyl-mercury is the most common organic form and was identified as the causative agent of the poisoning of the fish-eating residents in Minimata, Japan, in the 1950s. Symptoms of acute mercury toxicity include nausea, vomiting, bloody diarrhea, metallic taste, tremors, and convulsions. Chronic exposure affects the CNS, causing peripheral neuropathy, numbness, hearing loss, headache, and tremors. The term *mad as a hatter* originated because of the neurologic syndrome suffered by workers exposed to mercuric salts used in the manufacture of felt hats. Specimens for mercury analysis include blood and urine.

Sources of lead exposure include older paint (pre-1972), leaded gasoline, lead water pipes, pottery glaze, and working in occupations such as shipbuilding, welding, and battery manufacturing. Accumulation of lead is particularly dangerous in early childhood when it can cause developmental problems, including decreased IQ and growth. Because lead binds to erythrocytes, whole blood is the specimen of choice as an indicator of lead exposure. Acceptable whole blood levels for children have been lowered to less than 10 µg/dL as more information concerning lead toxicity has been obtained. For those industries utilizing lead, employee blood sample and environmental concentrations in the workplace are monitored and corrective actions mandated based on specified levels. Lead inhibits enzymes involved with heme synthesis resulting in elevated aminolevulinic acid (ALA) and free erythrocyte protoporphyrin (FEP, zinc protoporphyrin). A hematofluorometer is a portable instrument that measures FEP and has been used as a screen for lead intoxication. In addition to lead intoxication, elevated FEP levels may indicate iron deficiency anemia or protoporphyria. Though these supportive tests are useful, the laboratory diagnosis of lead intoxication is primarily based on whole blood lead concentration.

9. Pesticides. Finally, the clinical toxicologist may be asked to assess pesticide exposure. Two classes of insecticides are inhibitors of cholinesterase: organophosphates (diazinon, malathion, parathion) and carbamates (aminocarb, carbaryl-Sevin, and landrin). The organophosphate insecticides are potent inhibitors, and the carbamate insecticides are reversible inhibitors. Thus, with carbamate intoxication, the symptoms are generally less severe and of shorter duration. Effects of toxicity include headache, weakness, dizziness, blurred vision, respiratory difficulty, involuntary urination and defecation, convulsions, and coma. Exposure to both types of pesticides may be detected by determination of red cell (true) cholinesterase or serum (pseudo) cholinesterase activity by spectrophotometric methods. Decreased activity is observed as a result of pesticide toxicity. Most commonly, the serum enzyme activity is measured to assess possible inhibition. Workers exposed to organophospates should be regularly monitored to de-

---

Symptoms of Mercury Toxicity

- Nausea
- Vomiting
- Bloody diarrhea
- Metallic taste
- Tremors
- Convulsions
- Peripheral neuropathy
- Numbness
- Hearing loss
- Headaches

Effects of Pesticide Toxicity

- Headache
- Weakness
- Dizziness
- Blurred vision
- Respiratory difficulty
- Involuntary urination and defecation
- Convulsions
- Coma

tect pesticide toxicity. Antidotes available include atropine and pralidoxime (2-PAM), a cholinesterase-regenerating agent.

C. WORKPLACE (FORENSIC) DRUG TESTING

Regulated **workplace drug testing** refers to the testing of urine ◀
samples for the presence of specific drugs when conducted under
the auspices of governmental agencies. Federally mandated testing
includes the testing of federal employees in certain jobs and private-
sector employees in jobs regulated by federal agencies such as the
Department of Transportation (DOT) and Nuclear Regulatory Commission (NRC). The requirements for such testing were established
as part of the original mandate (published in the *Federal Register*) or
in rules promulgated initially by the **National Institute on Drug** ◀
Abuse (NIDA) and now by SAMHSA. Although the number of employees in these categories is not exceptionally large, many private-
sector employers have adopted parts of the requirements for their
drug screening policies.

> ### Drugs Tested under SAMHSA (NIDA 5)
>
> - Amphetamine
> - Cannabinoids
> - Cocaine (benzoylecgonine)
> - Opiates
> - Phencyclidine (PCP)

Components of federally mandated forensic urine drug testing
include:

- *Controlled collection.* Specimens must be collected under
  guidelines, which include collection of split (duplicate) samples for DOT-regulated testing. Protocols must be in place to
  prevent substitution or adulteration of urine samples during
  the collection process. Direct witnessed collections may be
  required.
- ***Chain of custody*** *(COC).* COC protocol stipulates that each ◀
  specimen container be sealed with a tamper-proof seal that is
  initialed by the donor. An audit trail of the specimen and its
  aliquots from the time of collection to final reporting and
  storage is maintained using chain-of-custody documentation.
  This documentation allows for tracking of the specimen
  throughout the entire testing and handling process.
- *Testing and storage guidelines.* All testing must be performed
  at certified laboratories. Specimens may be tested only for the
  following five specified drugs or classes of drugs: amphetamines, cannabinoids, cocaine metabolite, opiates, and PCP.
  These drugs, commonly referred to as the *NIDA 5,* along with
  the established mandated cutoffs are listed in Table 9–3. The
  approved testing methodologies are immunoassay **screening** ◀

Table 9–3. Cutoff Concentrations for Federally Regulated Drug Testing

DRUG/CLASS	SCREENING CUTOFF	GC/MS CUTOFF
Amphetamines	1,000 ng/mL	500 ng/mL (amphetamine/ methamphetamine)
Cannabinoids	50 ng/mL	15 ng/mL
Cocaine metabolite	300 ng/mL	150 ng/mL
Opiates	2,000 ng/mL	2,000 ng/mL (codeine/ morphine) 10 ng/mL (6-AM)
PCP	25 ng/mL	25 ng/mL

DOT also requires testing for ethyl alcohol in breath or saliva samples.

▶ followed by gas chromatography/mass spectrometry (GC/MS) **confirmation** of all presumptive positives. In addition to testing for drug presence, testing for sample dilution, substitution, or adulteration may also be performed. All confirmed positive samples must be stored frozen for one year.

- *Certification of laboratories.* Laboratories wishing to be certified for federally mandated testing must undergo a process administered by SAMSHA. This process includes requirements to score satisfactorily on proficiency surveys and inspections (every 6 months). The labs must pay a fee that is dependent on the daily volume of regulated samples tested but is a minimum of $27,000 per year. Personnel, security, and technical requirements are also specified.

▶ - ***Medical Review Officer*** *(MRO).* Laboratories must report results to a physician designated by the employer or agency as their MRO. The duties of the MRO include verification of the validity of the testing and handling processes. The MRO may request further testing or repeat testing and should contact the employee to determine if there is an explanation (e.g., prescription, ingestion of products) before reporting a result to the employer or agency. The MRO may choose to invalidate positive results and report them as negative.

For nonfederally regulated workplace drug testing, employers often have the freedom to test for other drugs and at other cutoff concentrations. Some states regulate drug testing and have established their own guidelines. Another type of certification for laboratories is provided by the College of American Pathologists and is known as the College of American Pathologists Forensic Urine Drug Testing (CAPFUDT) certification. In order to obtain this certification, labs must meet proficiency survey, inspection, personnel, and security requirements as established in published guidelines.

Interpretation of positive drug screening results is not always simple, particularly if the laboratorian is asked to state how long drugs may be detected in body fluids analyzed. There is considerable individual variation in the metabolism of drugs. Some of the factors that influence metabolism are the amount and potency of the drug, the frequency of drug use, the clinical condition, and the metabolism of the individual. When considering results regarding a urine sample, one of the most critical factors is the amount of fluid intake prior to specimen collection. Hence, the detection times listed in Table 9–4 for the federally regulated drugs are only approximate and can vary between individuals.

III. TEST PROCEDURES

Many different analytical methods are utilized in the detection, identification, and quantitation of drugs and other poisons. Techniques used for the screening of samples for multiple drugs or classes of drugs include immunoassay and chromatographic methods (see Chapters 11 and 12 for principles of analysis). In the clinical setting, identification of drugs or classes of drugs using these screening techniques may provide sufficient information to the physician so that confirmation by a second technique is not required. For workplace

Table 9–4. Length of Time Drugs May Be Detected Following Use

DRUG	APPROXIMATE DETECTION TIME (DAYS)
Amphetamines	2–3
Cannabinoids	5–7 (single use); up to 4 weeks (chronic)
Cocaine metabolite	2–3
Opiates	3–4
PCP	3–8

or forensic drug testing, screening by immunoassay and confirmation by GC/MS is mandated. Please refer back to Table 9–1 for a comparison of clinical and workplace drug testing practices and requirements.

The complexity of the emergency drug screen offered by a clinical laboratory depends on many factors, including equipment availability and analyst expertise. For example, a simple test panel requiring no instrumentation could screen for ethanol and several drugs or classes of drugs using small disposable detection kits. A complex workup might include urine/gastric screening using combinations of automated immunoassays, spot tests, thin-layer chromatography (TLC), gas chromatography (GC), high-performance liquid chromatography (HPLC), and/or GC/MS and quantitation of drugs including ethanol in blood/serum. Most clinical laboratories perform a panel developed in collaboration with caregivers at their institution.

Some of the simplest and least expensive screening tests in use clinically are the colorimetric **spot tests.** Spot tests are available for the detection of acetaminophen, salicylates, ethchlorvynol (Placidyl), phenothiazines, and some tricyclic antidepressants (imipramine, desipramine, and trimipramine). These tests are rapid and generally involve the addition of reagents followed by visualization of a colored product.

Another simple and inexpensive technique used for drug detection is the use of the **osmolal gap,** which is the difference between the calculated and measured serum osmolality, performed by freezing point depression methodology. Osmometers based on vapor pressure depression are available but are not applicable for this use because they do not detect volatile substances. A normal osmolal gap is 10 mOsm/kg; however, the presence of certain toxins in significant quantity results in an increase in the actual or measured serum osmolality, and thus an increase in the osmolal gap. The major contributors to serum osmolality are: sodium, chloride, glucose, bicarbonate, and blood urea nitrogen (BUN). Several formulas for the calculation or estimation of serum osmolality using some of these analytes have been developed. Listed below are two of these formulas:

$$1.86\ Na + \frac{Glucose}{18} + \frac{BUN}{2.8} + 9 = \text{Estimated serum osmolality}$$

Or

$$2\ Na + \frac{Glucose}{20} + \frac{BUN}{3} = \text{Estimated serum osmolality}$$

Thus, the laboratory may measure the sodium, glucose, BUN, and serum osmolality and then calculate the osmolality and the os-

---

Spot Tests Currently Available

- Amphetamine
- Salicylates
- Ethchlorvynol
- Phenothiazines
- Some tricyclic antidepressants (imipramine, desipramine, trimipramine)

mol gap. As stated previously, a gap of > 10 mOsm/kg indicates the presence of other compounds, possibly toxins. Such compounds include ethanol, methanol, acetone (and other serum ketones present in ketoacidosis), isopropanol, ethylene glycol, and diuretics such as glycerol, sorbitol, and mannitol.

The development of immunoassays (see Chapter 12) for the detection of drugs provided methodologies easily adaptable to automation or point-of-care–type manual "visual-read" kits. These assays were initially developed for high-volume drug screening of military personnel and used later for workplace drug testing. The capability of using these reagents on automated chemistry analyzers allows laboratories of any size to perform drug screening on an emergency and routine basis. The principle of these assays is the specific binding of an antibody to a drug or its metabolite. In most types of immunoassays, drug in the urine sample competes with labeled drug for a limited number of antibody-binding sites. Labels include enzymes, isotopes, and fluorescent and chemiluminescent compounds. Particle immunoaggregation is another detection technique used in several types of immunoassays. For most of these assays, arbitrary ▶ **cutoff concentrations** are set at levels required for workplace drug screening and the assay components are designed for optimum precision at the cutoff. Other cutoff concentrations may be used for some of these assays, but in general the results are, at best, semi-quantitative. For workplace drug testing, the cutoffs are absolute. Values at or above the cutoff are reported as *presumptive positives* and values below the cutoff are reported as negative. Workplace drug testing regulations require confirmation of presumptive positives by GC/MS procedures. For clinical toxicology situations, the laboratory may choose lower cutoff concentrations and may adopt a policy allowing reporting of presumptive positives without confirmation or require confirmation by a second method, not necessarily GC/MS.

Immunoassays are available for amphetamines, barbiturates, benzodiazepines, cannabinoids, cocaine metabolite, methadone, methaqualone, opiates, PCP, propoxyphene, and tricyclic antidepressants. The specificity of the assay is determined by the specificity of the antibody; different manufacturers produce assays with differing specificities. Manufacturers' assays for more than one drug or a class of drugs, such as amphetamines, barbiturates, benzodiazepines, and opiates, often have different specificity for compounds within each class. Laboratorians should consider the purpose of the testing when reviewing the specificity and sensitivity of the available tests.

Issues of immunoassay cross-reactivity and specificity are exemplified in the testing of opiates and benzodiazepines. Immunoassays for opiates are primarily designed to detect morphine and codeine with variable cross-reactivity to the other opiates. The ingestion of poppy seeds can cause a positive result for opiates due to the presence of morphine in the seeds. To avoid such positives for workplace testing, the screening cutoff for opiates is set at 2,000 ng/mL. The criteria for reporting a positive morphine result includes GC/MS confirmation of total morphine at a cutoff of 2,000 ng/mL and analysis for the detection of the metabolite, 6-acetylmorphine, with a cutoff of 10 ng/mL.

In the case of benzodiazepines, immunoassay screens may have different specificities and sensitivities depending on the hapten used to obtain the antibody. Antibodies have been constructed using nordiazepam, oxazepam, and glucuronides of other benzodiazepines. Laboratorians should be aware of the specificity and sensitivity for benzodiazepines and their metabolites of the immunoassay and confirmation procedures (GC/MS, HPLC, GC) used in their institution. Unless the screening and confirmation assays have matching characteristics, the laboratory may not be able to confirm drugs in samples which screen positive.

Whenever absolute cutoff concentrations are used, as in workplace drug testing, the hydration state of the individual may influence the results, particularly if the drug is present in concentrations near the cutoff. The slogan "dilution is the solution" is commonly used by those attempting to "cheat" on their drug test. Consequently, laboratories often include urine creatinine and/or specific gravity determinations as part of a drug screen. For federally regulated testing, when the creatinine is below 20 mg/dL and the specific gravity is less than 1.003, laboratories are allowed to report the specimen as being dilute. These results can be caused by excessive hydration or by dilution due to the addition of water to the sample. If a specimen has a creatinine of ≤ 5.0 mg/dL and a specific gravity of ≤ 1.001 or ≥ 1.020, the laboratory does not perfom drug testing but instead reports that the specimen is substituted and is not consistent with human urine. Laboratories may report that a specimen is adulterated if the substance or effect can be identified. Examples of adulteration include the presence of nitrites (≥ 500 µg/mL) or urine pH ≤ 3.0 or ≥ 11.0.

Recommended collection procedures designed to prevent addition of water or substitution of a "clean" urine include requiring that the donor remove coats and leave purses, briefcases, packages, and the like outside the collection room. Toilets should contain bluing agent, and there should not be hot water in the bathroom. To prevent substitution or dilution of the specimen, the temperature is checked using a temperature-sensitive strip attached to the collection cup. In addition to creatinine and specific gravity, other tests that may be used to detect potential adulteration of the specimen include appearance, odor, pH, and gluturaldehyde and nitrite determinations (ingredients in UrinAid and Klear, products designed to interfere with drug screens).

Chromatographic methods (TLC, GC, HPLC, GC/MS) offer the advantage of the ability to detect hundreds of drugs. Because TLC requires no instrumentation, it is less expensive than other chromatographic methods. A commercial TLC system (Toxilab) is available and includes a compendium of expected drug and metabolite patterns. Automated HPLC systems such as the Biolab Remedi are also available. Manufacturers of GC, HPLC, and GC/MS equipment and supplies frequently provide procedures for drug detection and quantitation.

For legal purposes, alcohol analysis should be performed using headspace GC (see Chapter 11), which separates ethanol from other volatile compounds, including acetone, methanol, and isopropanol. These methods separate many volatile compounds and may also be used for the analysis of inhalants whose abuse is a concern in adolescents.

Ethanol analysis may also be performed using an enzymatic method that is easily automated and available on most automated

Sample Collection Procedures Designed to Detect Dilution, Adulteration, or Substitution

- Removal of coats, purses, briefcases, packages, etc.
- Bluing agent in toilet water
- Absence of hot water in specimen collection room
- Temperature check of specimen
- Measurement of urine creatinine and specific gravity
- Notation of specimen appearance and odor
- Measurement of urine pH, nitrite, and glutaraldehyde

chemistry analyzers. Depending on the specificity of the enzyme, other alcohols may be inclusively detected in the assay.

As previously mentioned, federally mandated drug testing programs require that positive screening results be confirmed by GC/MS. These confirmation procedures often include hydrolysis if the drugs/metabolites (cannabinoids and opiates) are excreted as conjugates, extraction using solid phase columns or liquid solvents, and chemical derivitization in order to enhance separation and spectral identification.

REFERENCES

Done AK. Salicylate intoxication: Significance of measurements of salicylate in blood in cases of acute ingestion. *Pediatrics* 26:800, 1960.

Rumack BH, Peterson RG. Acetaminophen overdose: Incidence, diagnosis and management in 416 patients. *Pediatrics* (Suppl. 62): 898–903, 1978.

SUGGESTED READINGS

Casarett LJ, Amdur M, Doull J, Klassen C (eds.). *Casarett and Doull's toxicology: The basic science of poisons,* 5th ed. New York: McGraw-Hill, 1996.

Cook JD, Caplan YH, LaDico CP, Bush DM. The characteristics of human urine for specimen validity determination in workplace drug testing: a review. *J Annal Tox* 24:579-588, 2000.

Levine B (ed.). *Principles of forensic toxicology.* Washington DC: AACC Press, 1998.

Porter WH, Moyer TP. Clinical toxicology. In Burtis C, Ashwood E (eds.). *Tietz textbook of clinical chemistry,* 3rd ed. Philadelphia: WB Saunders, 1998.

Thorne DP. Toxicology. In Bishop ML, Duben-Engelkirk JL, Fody EP (eds.). *Clinical chemistry: Principles, procedures, correlations,* 4th ed. Philadelphia: JB Lippincott, 2000.

Warner AM. Toxicology testing. In Dufour DR (ed.). *Professional Practice in Clinical Chemistry: A Companion Text.* Washington, DC: AACC Press, 1999.

Items for Further Consideration

1. Discuss situations other than illegal use of drugs that might explain positive results on samples tested under federally mandated drug testing regulations. How can these situations be investigated and resolved by a Medical Review Officer?

2. Compare and contrast the methodologies available to the clinical toxicology laboratory for drug testing.

3. A 50-year-old man presents to his physician with symptoms including recurring abdominal cramps, hair loss, and nausea. Upon examination, the physician notes neuropathy and faint white lines on the patient's fingernails. What do these symptoms indicate, and what test should be ordered to confirm this diagnosis?

Disease-Specific Analytes

Markers of Viral Disease

10A

Patrick J. Cummings

Key Terms

ACQUIRED IMMUNE DEFICIENCY
SYNDROME **(AIDS)**

β₂ MICROGLOBULIN

ENZYME-LINKED
IMMUNOSORBENT ASSAY
(ELISA)

FLOW CYTOMETRY

HEPATITIS A VIRUS **(HAV)**

HEPATITIS B CORE ANTIGEN
(HBcAg)

HEPATITIS B "E" ANTIGEN
(HBeAg)

HEPATITIS B SURFACE ANTIGEN
(HBsAg)

HEPATITIS B VIRUS **(HBV)**

HEPATITIS C VIRUS **(HCV)**

HEPATITIS D VIRUS **(HDV)**

HEPATITIS E VIRUS **(HEV)**

HEPATITIS G VIRUS **(HGV)**

HUMAN IMMUNODEFICIENCY
VIRUS **(HIV)**

INDIRECT FLUORESCENT ANTIBODY
(IFA)

LINE IMMUNOASSAY **(LIA)**

RADIOIMMUNOPRECIPITATION
(RIPA)

SEROCONVERSION

T-HELPER LYMPHOCYTE

VIRAL LOAD

WESTERN BLOT

Chapter Objectives

Upon completion of this chapter, the reader will be able to:

1. Describe the temporal appearance of HIV serological markers from initial infection to the development of acquired immune deficiency syndrome (AIDS).
2. Name two confirmatory tests used in HIV serology.
3. Describe the serological testing algorithm used to diagnose HIV-infected individuals.
4. Illustrate the role and order of addition of patient serum, conjugate, and substrate in HIV enzyme-linked immunosorbent assay (ELISA) and Western blot tests.
5. Discuss the experimental and physiological conditions that can result in HIV false-positive and false-negative test results.

6. Name the viruses that cause viral hepatitis.
7. Describe the use of liver function tests in the diagnosis of hepatitis.
8. Discuss the serological markers that are used to identify and monitor patients infected with hepatitis viruses.
9. Illustrate the temporal appearance of the serological markers used to diagnose viral hepatitis.
10. Describe the epidemiological and clinical characteristics of viral hepatitis.

Analytes Included ◀

ANTI-GP41

ANTI-GP120/160

β_2 MICROGLOBULIN

CD4/CD8 RATIO

HAV (IgM ANTI-HAV, TOTAL ANTI-HAV [IgM, IgG])

HBV (IgM ANTI-HBc, TOTAL ANTI-HBc, HBsAg, HBeAg, ANTI-HBs)

HCV (ANTI-HCV, HCV VIRAL LOAD [RT-PCR])

HDV (TOTAL ANTI-HD, HDAg)

HEV (IgM ANTI-HEV, IgG ANTI-HEV)

p24

VIRAL LOAD

Pretest

1. The ELISA screening test for HIV infection detects serum antibody to
   a. CD4+ and CD8+
   b. DNA polymerase
   c. gp120 and gp41
   d. tat and rev

2. A confirmatory test used in HIV serology is
   a. flow cytometry
   b. Western blot
   c. ELISA
   d. CD4/CD8 ratio

3. The first serological marker that can be detected following HIV infection is
   a. anti-gp160
   b. gp41 antigen
   c. anti-gp120
   d. p24 antigen

4. Which of the following laboratory tests is used to monitor the effectiveness of antiviral therapy?
   a. CD4/CD8 ratio and viral load
   b. Western blot and ELISA

   c. p24 and complement levels
   d. CD4+ count and immunoglobulin levels

5. The HIV serological marker that is detected during acute disease and late disease and whose presence correlates with active viral replication is
   a. gp120
   b. p24
   c. viral DNA
   d. anti-gp41

6. Individuals who are seronegative but still infected with HIV-1 can be identified by
   a. polymerase chain reaction (PCR)
   b. ELISA
   c. Western blot
   d. CD4 counts

7. Which of the following is the most common cause of parenterally transmitted viral hepatitis?
   a. HAV
   b. HBV
   c. HCV
   d. HDV

8. Liver function tests, alanine aminotransferase and aspartate aminotransferase, are used in the diagnosis of viral hepatitis to determine the
   a. viral load
   b. status of liver function
   c. viral agent involved
   d. appropriate treatment

9. The first serological marker to appear following HBV infection is
   a. HBeAg
   b. Anti-HBc
   c. HBsAg
   d. Anti-HBs

10. The HAV serological marker that is an indicator of recent acute infection with HAV is
    a. HBeAg
    b. IgM anti-HAV
    c. HBsAg
    d. anti-HAV total

11. The hepatitis virus that causes mild or no symptoms in infected individuals but has a 70 to 80% chance of causing chronic liver disease is
    a. HAV
    b. HBV
    c. HCV
    d. HDV

I. HIV

A. MEDICAL OVERVIEW

Acquired immune deficiency syndrome (AIDS) is caused by infection with a retrovirus called **human immunodeficiency virus (HIV).** HIV-1 is the cause of the worldwide AIDS epidemic. Based on the genetic diversity of HIV-1 isolates, HIV-1 is classified into the dominant M group and the less common O group. Within the M group there are 10 subtypes or *clades,* A through J. Subtype B predominates in America and Europe; E in Asia; and A, C, and D in Africa. HIV-1 group O infections are found in Cameroon and other west-central African nations. HIV-2, which causes a less severe disease, is confined primarily to sub-Saharan Africa. Currently, there are approximately 30 to 40 million people infected with HIV worldwide, and the disease has caused the deaths of 5 million adults and 1.4 million children since the emergence of the epidemic in 1981. Early in the epidemic, individuals most at risk of becoming infected with HIV included homosexual and bisexual men, hemophiliacs, blood transfusion recipients, intravenous drug users, and infants born to infected mothers. More recently, minorities and heterosexual women represent a greater proportion of the infected population.

HIV is a member of the retrovirus family, lentivirus subfamily, of human retroviruses, which are nontransforming and cytopathic for host cells. The virus is characterized by a genome consisting of two identical single-stranded ribonucleic acid (RNA) molecules associated with the enzyme reverse transcriptase, a capsid (core), and an outer envelope derived from the host cell membrane (Figure 10A–1). HIV structural genes are classified into three groups: the group antigen (*gag*), envelope (*env*), and polymerase (*pol*) genes (Table 10A–1). The GAG gene encodes proteins (p7, p9, p15, p18, p24, p55) that are components of the viral capsid, which surrounds the RNA genome. The *env* gene encodes glycoprotein gp160, which is found in infected cells, and is cleaved to gp120 and gp41 for assembly into the viral envelope. The viral envelope encircles the capsid and also mediates entry of the virus into host cells. The *pol* gene products

---

HIV Structural Genes

- Group antigen (*gag*)
- Envelope (*env*)
- Polymerase (*pol*)

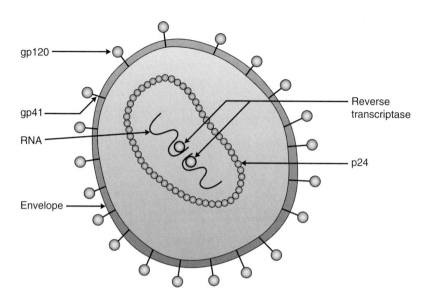

Figure 10A–1. Structure of HIV.

(p66, p31, p10) include enzymes involved in the replication, processing, and integration of the viral genome. There are also several regulatory gene products (vif, vpr, tat, rev, tev, nef, vpu) that control expression and replication of the virus.

▶ The CD4⁺ **T-helper lymphocyte** (CD4⁺ lymphocyte, helper cells) is the primary target for HIV infection because of the affinity of the gp120 viral envelope glycoprotein for the CD4 cell receptor and chemokine coreceptor (CCR5, CXCR4) on helper cells. Other cell types that can also become infected with HIV include macrophage, dendritic, and neuronal cells. The disease is characterized by progressive loss of CD4⁺ lymphocytes, resulting in impairment of immune functions. The impaired immune system allows for development of various disease manifestations, including infections with *Pneumocystis carinii,* cytomegalovirus, *Mycobacterium avium* complex (MAC), *Cryptococcus neoformans, Toxoplasma gondii, Cryptosporidium,* and *Microsporidia.* Also, the defective immune system places individuals at increased risk for Kaposi's sarcoma, B cell lymphoma, and cervical cancer.

The immunological profile following HIV infection is characterized by the appearance of p24 antigen in serum shortly following infection. Presence of this viral antigen is an indication of rapid viral replication. During this "*window period*," HIV antibodies are absent,

Table 10A–1. HIV-1 Genes and Gene Products

HIV GENES	GENE PRODUCTS	FUNCTION
env	gp120, gp41, precursor gp160	Envelope
gag	p24, p18, p7	Core (capsid)
pol	p66 (reverse transcriptase)	Replication, insertion, protein
	p31 (intergrase)	processing
	p10 (protease)	
Regulatory	vif, vpr, vpu, tat, rev, tev, nef	Regulation of gene expression

although the individual is infected with the virus. The level of p24 antigen decreases over the next few weeks to undetectable levels, which corresponds with the appearance of serum antibodies to HIV structural proteins (Figure 10A–2). **Seroconversion,** which is indicated by the appearance of HIV antibodies, occurs within a few weeks to a few months after exposure to the virus. Antibodies to the envelope gene products (gp160, gp120, and gp41) are detected in all HIV-infected individuals; however, the time of appearance and level of antibody responses may vary from person to person. In addition, antibodies to the *pol* gene products (p31 and p66) are also produced during this stage of infection. The level of HIV antibodies remain stable during the clinical latent stage of the disease; however, the level of anti-p24 antibody decreases and the level of p24 antigen increases at later stages of the disease as the patient progresses to full-blown AIDS (Figure 10A–2).

B. HIV ANTIBODY SCREENING TESTS

HIV screening tests are based on an indirect **enzyme-linked immunosorbent assay (ELISA)** procedure, in which HIV antigens (HIV-1/HIV-2) from disrupted virus particles (first-generation assays) or recombinant proteins or peptides (second-generation assays) are immobilized onto a solid matrix, such as plastic beads or microtiter wells. Diluted serum is added to the solid matrix and incubated, followed by washing to remove unbound serum proteins. An enzyme labeled anti–human immunoglobulin conjugate is added to the reaction, incubated, and then washed to remove unbound conjugate. Addition of an appropriate chromogenic substrate permits detection of bound conjugate by formation of a colored end product that is detected spectrophotometrically.

Interpretation. Test results are interpreted by comparing the absorbance values obtained from positive and negative control specimens to the test specimen. An absorbance cutoff value is used to determine reactivity. If the serum specimen tests positive, the test is

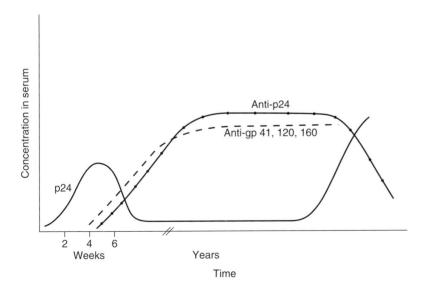

Figure 10A–2. Serological profile following infection with HIV-1.

Causes of False-Positive ELISA Results for HIV Antibody

- Autoimmune conditions
- Malignancies
- Hepatic disease
- History of multiple pregnancies
- Epstein–Barr virus infection
- Recent history of influenza vaccination
- Heat inactivation of serum prior to testing

HIV Antibody Confirmatory Tests

- Western blot
- Indirect fluorescent antibody
- Radioimmunoprecipitation
- Line immunoassay

repeated in duplicate with a freshly collected serum specimen. If the test is reactive on repeat, it is referred to as *repeatedly reactive,* and a confirmatory Western blot test is performed. False-positive ELISA results can occur in individuals with autoimmune conditions, malignancies, hepatic disease, history of multiple pregnancies (due to antibodies to human lymphocyte antigens), Epstein–Barr virus infection, and recent history of influenza vaccination. In addition, serum samples must not be heat inactivated prior to testing, because false-positive test results can occur. A negative ELISA result will be obtained if the serum specimen is collected during the "window period" (1 to 3 weeks following exposure) before seroconversion.

C. HIV ANTIBODY CONFIRMATORY TESTS

Serum samples that are repeatedly reactive by ELISA are tested by a *confirmatory test.* The confirmatory tests currently available are **Western blot, indirect fluorescent antibody (IFA), radioimmunoprecipitation (RIPA),** and **line immunoassay (LIA).** The most commonly used confirmatory test is the Western blot. For Western blot antigen preparation, a mixture of HIV proteins from disrupted virus particles is separated into specific bands by sodium dodecyl sulfate polyacrylamide gel electrophoresis (SDS-PAGE). The proteins will migrate based on their individual molecular weights, with smaller protein migrating the fastest through the gel matrix. The separated viral proteins are then transferred to a solid matrix, such as nitrocellulose or nylon blotting paper, with the protein bands in the same position as they appeared on the gel. The Western blot immunodetection phase involves adding diluted patient serum to an individual antigen strip, followed by incubation and washing to remove unbound serum proteins. If antibodies are present in the patient serum, they will form antibody–antigen complexes at specific bands on the antigen strip. Antibody–antigen complexes that have formed can be visualized by the addition of an enzyme labeled anti–human immunoglobulin conjugate and appropriate substrate. If serum antibodies have complexed with viral proteins on the strip, a color reaction will develop at the band sites.

Interpretation. Each patient serum is tested with an individual antigen strip. In addition, negative and positive control sera (strongly reactive and weakly reactive) are assayed in parallel. Following development, antigen strips are visually inspected and compared to control strips. If the control strips produce the expected results, there are three possible interpretations: negative, positive, or indeterminant. Negative test results indicate no reactive bands visible on the antigen strip, or no bands corresponding to the molecular weights of the known viral proteins. A positive Western blot test has reactive bands for at least two of the following viral proteins: p24, gp41, and gp120/gp160 (Centers for Disease Control and Prevention [CDC] criteria). An indeterminant result has a single band or combination that does not fit the criteria. The reason for an indeterminant Western blot includes incomplete seroconversion (early in infection), declining humoral response (late infection), presence of autoantibodies, or possible infection with HIV-2 (due to cross-reactive antibody responses to *gag* and *pol* gene products, but not the *env* gene product). Patients with an indeterminant Western blot result should be retested in 3 to 6 months. An algorithm for the sequence of serological testing used in diagnosis of HIV infection is shown in Figure 10A–3.

D. HIV ANTIGEN DETECTION

HIV p24 antigen can be detected 1 to 2 weeks following infection, and later in infection after anti-p24 antibody declines. p24 antigenemia is detected by a sandwich ELISA format, where anti-p24 monoclonal antibody is immobilized on polystyrene beads or plastic microtiter wells. The serum specimen is incubated with the sensitized solid matrix, washed, and incubated with enzyme-conjugated polyclonal anti-p24 to form a sandwich. After washing, substrate is added for color development. Pretreatment of serum specimens with acid dissociates immune complexes (anti-p24/p24) and increases the sensitivity of the assay. The p24 sandwich ELISA can be used to screen sera to detect patients in the window period, and to assess the therapeutic efficacy of antiviral agents.

E. OTHER MARKERS

1. CD4/CD8 Ratio. CD4+ T-helper lymphocytes (CD4+ cells) have a major role in the ability of the body to respond to a variety of infectious agents. HIV infection results in the gradual depletion of CD4+ cells during the course of disease progression. **Flow cytome-** ◀

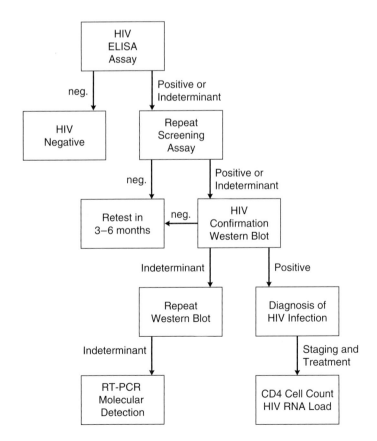

Figure 10A–3. Algorithm used for serological testing to diagnose HIV infection.

try is used to determine the number of T lymphocytes in circulation, with the number of cells representing a good indicator of the immune status of the individual. The T-lymphocyte number is expressed as a ratio of the number of CD4+ cells/μL to CD8+ cells/μL of blood (CD4/CD8 or T4/T8 ratio). The CD4/CD8 ratio is used as a prognostic indicator of AIDS progression, and is also used to monitor the effectiveness of antiviral therapy. The normal CD4/CD8 ratio is ~2.0. HIV-infected patients with severe immunodeficiency or who are progressing to full-blown AIDS have a CD4/CD8 ratio < 1.0.

▶ **2. Viral Load. Viral load** refers to the quantitation of viremia in HIV infection by measuring the amount of viral RNA in plasma using molecular biology techniques, such as nucleic acid hybridization and/or amplification. Most assays include nucleic acid release, isolation, amplification, and detection. Internal quantitation standards are added to the test samples to monitor the efficiency of extraction and amplification.

One of the most common methods for viral load determination is reverse transcriptase–polymerase chain reaction (RT-PCR). Using RT-PCR, viral RNA in plasma is converted into cDNA by reverse transcriptase, followed by amplification of the DNA template by the polymerase chain reaction (PCR) using virus specific DNA primers. The amount of PCR product generated can be directly related to the concentration of HIV RNA in the plasma. Viral load measurement by RT-PCR is the most sensitive method for detection and quantitation of virus in plasma. Along with the CD4/CD8 ratio, viral load provides the most useful test for staging the disease, assessing the immune status of the individual, and monitoring the effectiveness of antiviral therapy (see Chapter 14).

▶ **3. β_2 Microglobulin (β_2M). β_2 Microglobulin** is a low-molecular-weight plasma membrane protein associated with major histocompatability complex (MHC) Class I molecules on the surface of nucleated cells. Presence of free β_2M is an indirect indicator of cell lysis due to HIV infection. β_2M levels < 2.6 mg/L are normal. Patients with rapid cell death usually have β_2M levels of > 5 mg/L. Elevation of blood β_2M levels occurs late in HIV disease as the patient's immune system collapses.

II. VIRAL HEPATITIS

A. MEDICAL OVERVIEW

▶ There are six distinct viruses known to cause hepatitis: **hepatitis A**
▶ **virus (HAV), hepatitis B virus (HBV), hepatitis C virus (HCV;** formerly designated non-A, non-B hepatitis virus–blood borne
▶ [NANB–blood borne]), **hepatitis D virus (HDV), hepatitis E virus**
▶ **(HEV;** formerly designated NANB–enteric), and **hepatitis G virus (HGV;** GBV-C). Other viruses that can cause acute liver inflammation include human cytomegalovirus, Epstein–Barr virus, rubella virus, herpes simplex virus, and enteroviruses. The symptoms caused by viral hepatitis vary depending on the virus and the individual's response to infection; however, some generalized symptoms include fatigue, myalgia, anorexia, nausea, vomiting, diarrhea or constipation, fever, weight loss, jaundice, and hepatomegaly. As the viral dis-

---

Viral Infections of the Liver

Hepatitis Viruses
- HAV
- HBV
- HCV
- HDV
- HEV
- HGV

Other viruses
- Cytomegalovirus
- Epstein–Barr
- Rubella
- Herpes simplex
- Enteroviruses

ease progresses, liver function is compromised as indicated by development of jaundice, due to accumulation of bilirubin in the skin, and excretion of amber-colored urine and pale feces, due to defective bilirubin metabolism. In addition, elevated levels of blood bilirubin and liver-specific enzymes, alanine aminotransferase (ALT) and aspartate aminotransferase (AST), are the first indicators of liver inflammation, with the blood levels somewhat in proportion to the degree of liver damage. Acute disease is often very severe with HBV and HBV–HDV coinfections; however, HAV, HEV, HCV, and HGV infections are usually less severe in the acute stage. Persons infected with HBV, HBV–HDV, HCV, or HGV can become chronic viral carriers, whereas HAV and HEV do not establish a chronic carrier state in the host (Table 10A–2).

HAV (infectious hepatitis) and HEV (NANB–enteric) are very similar in their mechanism of transmission and clinical course. Both viruses are transmitted person-to-person by the fecal–oral route of transmission through contaminated food or water, or by other conditions of poor hygiene and sanitation. HAV has worldwide incidence and is endemic in many developing countries. Children in day care centers, residents of custodial institutions, travelers, and male homosexuals are at high risk for HAV infection. HEV occurs rarely in the United States; however, it causes occasional waterborne outbreaks in India and central Asia. The incubation period for HAV is approximately 28 days (range: 10 to 50 days), followed by an abrupt onset of symptoms in adults, whereas most children remain asymptomatic. No chronic disease or carrier state develops. Serum antibody appears 1 to 3 weeks following infection with HAV, with IgM anti-HAV detectable during acute infection. Within 1 to 2 months, IgG anti-HAV can be detected in serum, and it will persist and provide lifelong immunity (Table 10A–2). The clinical features of HEV infection can be more severe than HAV infection, with the potential for cholestatic liver disease and fulminant hepatitis during pregnancy.

HBV, which causes serum hepatitis, is a double-stranded DNA virus surrounded by a central core, which contains the **hepatitis B** ◀ **core antigen (HBcAg)** and the **hepatitis B "e" antigen (HBeAg).** ◀ The hepatitis B virus particle was originally called the *Dane particle.* The viral core is surrounded by an envelope containing the **hepatitis** ◀

Table 10A–2. Summary of the Clinical and Epidemiological Characteristics of Viral Hepatitis

VIRUS	HAV	HBV	HCV	HDV	HEV	HGV
Nucleic acid	RNA	DNA	RNA	RNA	RNA	RNA
Disease/former name	Infectious hepatitis	Serum hepatitis	NANB–blood borne	Delta agent	NANB–enteric	Non–A-E
Route of transmission						
Fecal–oral	Yes	No	No	No	Yes	No
Parenteral	Rare	Yes	Yes	Yes	No	Yes
Sexual	No	Yes	Yes	Yes	No	?
Perinatal	No	Yes	Possible	Possible	No	?
Water/food	Yes	No	No	No	Yes	No
Incubation period	2–6 weeks	8–26 weeks	2–15 weeks	3–13 weeks	3–6 weeks	?
Chronicity	None	Yes	Yes	Yes	?	Yes
Vaccine	Yes	Yes	No	No	No	No
Immune globulin prophylaxis	Yes	Yes (HBIG)	No	No	No	No

Persons at High Risk of HBV Infection

- Parenteral drug abusers
- Persons with multiple sex partners
- Infants born to infected mothers
- Hemodialysis patients
- Health care workers
- Persons from HBV endemic areas

B surface antigen (HBsAg; previously known as the *Australian antigen*). There are four antigenic determinants (subtypes) of HBsAg, *adw, adr, ayw,* and *ayr,* defining nine serotypes of HBV. HBV antigens and their corresponding antibodies are used for the serological assessment of HBV infection (Figure 10A–4). HBV is transmitted by three major routes of transmission: parenteral, perinatal, or sexual. Persons at high risk of HBV infection include parenteral drug abusers, persons with multiple sex partners, infants born to infected mothers, hemodialysis patients, health care workers, and persons from areas where HBV is endemic such as Africa, Eastern Europe, and South America. The incubation period for HBV is 120 days (range: 45 to 160 days). Chronic HBV infection occurs in 5 to 10% of infected adults and 50 to 90% of infected children and infants. Approximately 25% of carriers develop chronic active hepatitis, which can progress to cirrhosis or liver cancer. A recombinant HBsAg vaccine is available to prevent infection with HBV, and it is recommended that all health care workers be vaccinated. In addition, a postexposure passive immunotherapy, using hepatitis B immunoglobulin (HBIG), is recommended following acute exposure to body fluids, such as a needle-stick injury.

HDV (Delta agent) is a viral particle consisting of an RNA molecule and inner protein core (HDAg) surrounded by HBsAg. Thus, HDV can infect only individuals who are already infected with HBV. Simultaneous infection with HBV and HDV (coinfection) results in a more severe acute phase of infection, often characterized by a biphasic ALT serum profile that is indicative of two episodes of liver damage. HDV infection of HBV carriers (superinfection) often results in progressive liver cirrhosis or fulminant hepatitis.

HCV consists of an RNA genome, six nonstructural proteins (NS2, NS3, NS4a, NS4b, NS5, NS5b), a core protein (C), and two envelope proteins (E1 and E2/NS1). Portions of these viral antigens form the basis for the serum antibody test. HCV is the most common

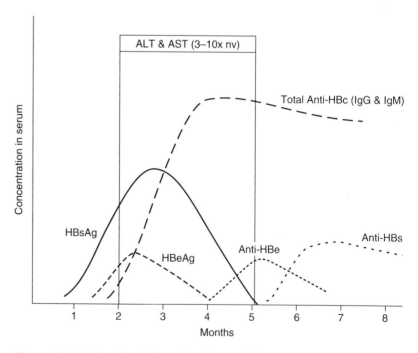

Figure 10A–4. Serological profile observed during acute HBV infection.

cause of parenterally transmitted viral hepatitis. Individuals most at risk for HCV infection include IV drug users, recipients of blood products, organ transplant recipients, and health care workers. The incubation period ranges from 5 to 10 weeks (range: 2 weeks to 4 months). Clinically, HCV infection is similar to HBV infection except 70 to 80% of patients develop chronic liver disease, and 20% of these individuals develop cirrhosis. Diagnostic tests for HCV include an enzyme immunoassay (EIA) for screening sera for anti-HCV, a supplemental recombinant immunoblot assay (RIBA) for resolving indeterminant EIA results, and a confirmatory RT-PCR test for detection of viral nucleic acid. Currently, there is no vaccine available to prevent infection with HCV.

HGV is a recently discovered hepatitis virus associated with parenteral transmission. Individuals most at risk include IV drug abusers, hemodialysis patients, hemophiliacs, and blood transfusion recipients. Infection with HGV is associated with mild disease. However, dual infections with HGV and HBV or HCV appear to be common in exposed groups, and often progresses to chronic liver disease.

Persons at High Risk of HCV Infection
• IV drug users
• Recipients of blood products
• Organ transplant recipients
• Health care workers

B. CLINICAL DIAGNOSIS

Standard RIA, EIA, or ELISA tests are used to detect the antigens and antibodies associated with viral hepatitis (see Chapter 12). Following HAV infection, serum antibody can be detected in 3 to 4 weeks, with IgM antibody (IgM anti-HAV) the first antibody to appear, which is indicative of an acute infection. IgM anti-HAV can be detected for 8 to 10 weeks, at which time the concentration drops to an undetectable level. IgG anti-HAV appears in serum shortly after IgM, and will persist in serum and provide life long immunity.

HBV infection is characterized by the appearance of HBsAg 2 to 4 weeks before the development of liver dysfunction or clinical symptoms, and it is the first serological marker detected in serum. The highest level of HBsAg is found with the onset of clinical symptoms; however, the level gradually declines to undetectable levels within 4 to 6 months. HBeAg, which can also be detected during this stage of infection, is a reliable marker for the presence of high levels of virus and a high degree of infectivity. Following the disappearance of HBsAg and HBeAg from the serum, antibodies to HBeAg (anti-HBe) and HBsAg (anti-HBs) can be detected. Anti-HBs persists in serum and provides protection against reinfection with HBV. Within two weeks of the appearance of HBsAg, IgM and IgG antibodies to HBcAg (IgM anti-HBc and IgG anti-HBc) can be detected. IgG anti-HBc persists following infection, and can be used as an indication of past infection. In patients who become HBV chronic carriers, HBsAg and HBeAg will persist in the serum, and the corresponding antibody response will not develop.

Diagnostic tests for HCV include an EIA for screening sera for anti-HCV, a supplemental recombinant immunoblot assay (RIBA) for resolving indeterminate EIA results, and a confirmatory RT-PCR test for detection and quantitation of viral nucleic acid. Third-generation EIA and RIBA assays for HCV antibodies provide increased sensitivity by incorporating c22, c200, and NS 5 viral antigens. Anti-HCV antibodies appear 5 to 7 weeks following infection; however, seroconversion can be delayed in some patients. Because of the low titer of circulating viral antigen, direct assays for viral antigens are not available. However, molecular techniques are available to detect and quantitate viral nucleic acid in plasma. The RT-PCR test involves amplification of reverse transcribed HCV RNA. Using RT-PCR, viremia can be detected 1 to 3 weeks following infection (before seroconversion), chronically infected individuals can be identified, and patient response to interferon therapy monitored. Viral nucleic acid can be detected 1 to 3 weeks following infection with HCV. Following determination of seropositivity, HCV infection can be staged as an acute, chronic, or past infection using the RT-PCR test (Table 10A–3).

Following acute coinfection with HDV and HBV, HDAg and HDV nucleic acid are detected in serum, usually after the appearance of HBsAg. HDAg is detected for a few weeks following infection, but decreases to undetectable levels in 3 to 4 weeks. Total anti-HD and IgM anti-HD are detectable during acute illness, but wane 6 months postinfection. Identifying an HDV infection as a coinfection (with HBV) or superinfection can be determined by careful analysis of viral antigens, antibodies, and nucleic acid (Table 10A–3).

Following HEV infection, serum antibody can be detected in 3 to 4 weeks, with IgM (IgM Anti-HEV) and IgG (IgG anti-HEV) appearing in serum about the same time. IgM anti-HEV is indicative of an acute infection, and IgG anti-HEV without IgM is indicative of past infection with HEV. Finally, HGV infections are detected by molecular methods, such as RT-PCR.

Table 10A–3. Summary of Antigens, Antibodies, and Stages of Disease Associated with Viral Hepatitis Infection

STATUS	HAV	HBV	HCV	HDV/HBV	HEV
Acute infection	IgM anti-HAV	HBsAg IgM anti-HBc	Anti-HCV HCV RNA Positive	—	IgM anti-HEV
Chronic infection	None	HBsAg HBeAg Persistence	Anti-HCV HCV RNA Persistence	—	None
Past infection	IgG anti-HAV	Anti-HBs Anti-HBc	Anti-HCV HCV RNA Neg. Anti-HBs	Anti-HD Anti-HBc	IgG anti-HEV
Coinfection* (acute)	—	—	—	HBsAg IgM anti-HD IgM anti-HBc	—
Superinfection** (acute HDV/ chronic HBV)	—	—	—	HBsAg IgM anti-HD	—
Vaccinated	IgG anti-HAV	Anti-HBs	None	None	None

HGV-1 coinfections with HBV or HCV are common in high-risk groups. HGV infections detected by molecular methods.

SUGGESTED READINGS

Centers for Disease Control. Interpretation and use of the Western blot assay for serodiagnosis of human immunodeficiency virus type 1 infection. *Morbid Mortal Weekly Rep* 38:1–7, 1989.

Constantine NT, Callahan JD, Watts DM. *Retroviral testing: Essential for quality control and laboratory diagnosis.* Boca Raton, FL: CRC Press, 1992.

Davey RT, Vasudevachari MB. Serological evaluation of patients with human immunodeficiency virus infection. In Rose N, Conway de Macario E, Folds JD, Lane HC, Nakamura RM (eds.). *Manual of clinical laboratory immunology.* Washington, DC: ASM Press, 1997; 364–70.

Gretch DR. Diagnostic tests for hepatitis C. *Hepatology* 26:43S–46S, 1997.

Hollinger FB, Dreesman GR. Hepatitis viruses. In Rose N, Conway de Macario E, Folds JD, Lane HC, Nakamura RM (eds.). *Manual of clinical laboratory immunology.* Washington, DC: ASM Press, 1997; 702–17.

Stine GJ. *AIDS Update 1998.* Upper Saddle River, NJ: Prentice Hall, 1998.

Items for Further Consideration

1. Discuss the temporal appearance of the viral and immunological markers of HIV infection during disease progression.

2. Compare and contrast the sensitivity and specificity of the ELISA and Western blot procedures used in HIV diagnosis.

3. Discuss the role of CD4:CD8 ratio and viral load in managing patients infected with HIV.

4. Speculate as to the earliest indicators of each type of viral hepatitis infection and predict technologies that will be used for identification and monitoring.

Tumor Markers

10B

Patrick J. Cummings

Key Terms ◀

CANCER

DIFFERENTIATION

ONCOGENES

PROLIFERATION

PROTO-ONCOGENES

TUMOR MARKERS

TUMOR SUPPRESSOR GENES

Chapter Objectives

Upon completion of this chapter, the reader will be able to:

1. Describe the multistep progression a normal cell undergoes as it converts into a cancer cell.
2. Explain the role of tumor markers in diagnosis and treatment of cancer.
3. Describe the differences between proto-oncogenes and oncogenes.
4. Name major cancers and their associated tumor markers.
5. Define the following terms: oncofetal antigen, carcinoma (tumor)-associated antigen, and tumor-specific antigen.
6. Classify the various tumor markers as hormones, oncofetal antigens, mucin-like glycoproteins, or genetic markers.

Analytes Included ◄

Adrenocorticotropic hormone (ACTH)	**CA 549**	**MCA**
	CA 72-4	**Monoclonal paraprotein**
Alpha-fetoprotein (AFP)	**Calcitonin**	**myc**
bcl-2	**Carcinoembryonic antigen (CEA)**	**p53**
Bence Jones protein		**Prostate-specific antigen (PSA)**
BRCA-1 and -2	**DU-PAN-2**	
CA 125	**Estrogen/progesterone receptors**	**ras**
CA 15-3		**Rb**
CA 19-9	**HER-2/neu**	**WT**
CA 242	**Human chorionic gonadotropin (hCG)**	
CA 27.29		

Pretest

Directions: Choose the single best answer for each of the following items.

1. Which of the following genetic markers is associated with breast cancer?
   a. Rb
   b. BRCA-1 and -2
   c. WT1
   d. CA 125

2. The oncofetal antigen associated with liver cancer is
   a. CEA
   b. hCG
   c. AFP
   d. PSA

3. Elevation in serum calcitonin is associated with cancer of the
   a. colon
   b. stomach
   c. thyroid
   d. ovary

4. Tumor associated antigens composed of high-molecular-weight mucin-like glycoproteins include
   a. BRCA-1 and -2
   b. CA 125 and CA 19-9
   c. PSA and AFP
   d. p53 and Rb

5. The most specific tumor marker for prostatic cancer is
   a. prostate specific antigen
   b. prostatic acid phosphatase
   c. α fetoprotein
   d. human chorionic gonadotropin

6. AFP and CEA are examples of which type of tumor marker?
   a. hormone
   b. enzyme
   c. oncofetal antigen
   d. carcinoma-associated antigen

7. The carcinoma associated tumor antigen most useful for following patients with ovarian cancer is
   a. CA 72-4
   b. CA 15-3
   c. CA 125
   d. CEA

8. CEA is most commonly associated with
   a. breast cancer
   b. ovarian cancer
   c. colon cancer
   d. myelocytic leukemia

I. BIOCHEMISTRY, PHYSIOLOGY, AND MEDICAL OVERVIEW

The basic unit of living organisms is the cell. Groups of identical and similar cells organized into a defined structure with a specific function are known as tissues. The millions of cells that constitute different tissues are derived from different lineages and have different functions, but together they contribute to the overall structure and function of organs and systems. The two major processes involved in normal cell growth are differentiation and proliferation. **Differentiation** involves the orderly progression of undifferentiated cells down a developmental pathway to produce terminally differentiated cells with a specific function. **Proliferation** is the process whereby cells increase in number by cell division. Cell division occurs during progression down the differentiation pathway, and following terminal cell differentiation. Both cellular processes require expression of various genes in a finely controlled and coordinated fashion to maintain normal tissue structure and function. When normal cells lose the ability to control the differentiation and proliferation pathways, a transformed cell develops. If this abnormal cell is not recognized and destroyed by the immune system, it has the potential to develop into cancer.

Cancer is the uncontrolled proliferation of cells that leads to tissue and organ dysfunction, and eventually to system failure and death. Development of cancer is a multistep process that involves the gradual loss of biological control over cell division. The genetic changes that contribute to cell transformation involve mutations in genes normally involved in controlling cell division known as **proto-oncogenes** and **tumor suppressor genes.** Proto-oncogenes are involved in promoting normal cellular division, and tumor suppressor genes suppress normal cell division. Proto-oncogene products include proteins that normally function as growth hormones, growth hormone receptors, signal transduction molecules, cell cycle progression proteins, and deoxyribonucleic acid (DNA) transcription factors. Changes in the quantity or quality of proto-oncogene products can occur due to gene amplifications, deletions, rearrangements, or point mutations. Defective proto-oncogenes found in cancer cells are known as **oncogenes.** Various types of cancer will have different oncogenes that arise at different stages of the oncogenic pathway, which are associated with an abnormal cell genotype. In addition, loss of tumor suppressor genes, which are involved in constraining cell division, due to deletion or mutational inactivation also contributes to the conversion of a normal cell to a transformed phenotype. Thus, cancer results from a multistep progression of a normal cell to a cancer cell that involves the gradual accumulation of mutations in proto-oncogenes and tumor suppressor genes. The accumulation of these genetic changes results in the phenotypic conversion of a normal cell from a benign growth to a malignant, invasive, and metastatic tumor.

II. TUMOR MARKERS

Tumor markers are defined as cellular characteristics or substances produced by cancer cells that can be used to differentiate tumor cells from normal tissue. The characteristics of an ideal tumor marker include specificity for a single type of cancer, high sensitivity for cancerous growths, correlation of marker level with tumor size, and short half-life of marker in circulation. Tumor markers include enzymes, hormones, fetal antigens, carcinoma-associated antigens, proteins, and genetic defects. Various tumor markers are measured qualitatively or quantitatively by chemical, immunological, or molecular biological techniques. Most tumor markers that are secreted into the systemic circulation are not specific or sensitive enough to be used for detection or screening of cancer, because they are also produced in normal and benign tissues. Tumor markers are most useful for assisting in differential diagnosis, staging of cancer, estimating tumor volume, monitoring disease progression, detecting recurrence of cancer, and evaluating response to drug treatment regimens.

A. HORMONES

Elevation of hormones can occur in various cancers due to excess production by endocrine (eutopic) or nonendocrine tissues (ectopic). Eutopic hormones produced in excess include human chorionic gonadotropin (hCG) by trophoblastic tumors, and calcitonin by thyroid–medullary tumors. An ectopic hormone produced by nonendocrine tissue is adrenocorticotropic hormone (ACTH) by small-cell carcinoma of the lung. hCG is a glycoprotein secreted by the trophoblastic cells of the normal placenta. Increased serum and urine levels of hCG are found in pregnancy, trophoblastic carcinoma, and germ cell tumors. Calcitonin is produced by the C cells of the thyroid and is involved in controlling the serum calcium level. An elevated level of calcitonin is associated with carcinoma of the thyroid. ACTH is normally produced by the corticotropic cells of the anterior pituitary gland. Ectopic ACTH production occurs primarily in small-cell carcinoma of the lung; however, elevations can also occur in pancreatic, breast, gastric, and colon cancer. Excess synthesis and release of hormones by tumors have severe systemic effects on the patient and are referred to as the *paraneoplastic endocrine syndromes.*

B. TISSUE-SPECIFIC ANTIGEN

Prostate-specific antigen (PSA) is a sensitive and accurate indicator of prostate cancer. PSA is a serine protease secreted by normal prostatic epithelial cells. In serum, PSA can exist free or complexed with antichymotrypsin to form PSA-ACT. High levels of PSA are found in prostate cancer and benign prostatic hyperplasia (BPH); however, the level of PSA-ACT is usually higher in prostate cancer and thus the free PSA fraction is smaller. PSA is especially useful for monitoring for recurrence of prostate cancer following prostatectomy.

C. ONCOFETAL ANTIGENS

Oncofetal (oncodevelopmental) antigens are proteins produced normally during fetal development. However, following malignant transformation, oncofetal proteins are again expressed by certain types of tumors. These antigens represent the most specific tumor markers available for clinical diagnosis and disease management. Alpha-fetoprotein (AFP) is the most specific marker for primary carcinoma of the liver. AFP levels are also useful for monitoring drug therapy and assessing prognosis. Maternal AFP levels are also elevated during normal pregnancy, with peak levels occurring during the third trimester. Carcinoembryonic antigen (CEA) is the most widely used marker for colon (gastrointestinal) cancer, with CEA levels useful for clinical staging following diagnosis. Because CEA has low specificity for colon cancer, however, its primary application involves monitoring therapy and detecting recurrent disease following surgery.

D. Carcinoma-Associated (CA) Antigens

The carcinoma-associated (CA) antigens are high-molecular-weight, mucin-like glycoproteins that are overexpressed by various types of cancer cells. Monitoring CA 125 levels is useful for assessing treatment, determining tumor volume, and predicting prognosis in ovarian cancer. The CA antigens elevated in breast cancer include CA 549, CA 27.29, and mucin-like carcinoma-associated antigen (MCA). In addition, CA 15-3 is used to monitor the clinical course of patients with breast cancer and to determine the presence of metastatic disease. CA 19-9 is a mucin-like tumor marker related to the Lewis blood group antigens, which is elevated in pancreatic, lung, colorectal, and gastric adenocarcinomas. Two markers elevated in pancreatic cancer are CA 242 and DU-PAN-2. The primary use of the CA antigens is to monitor disease progression and manage therapeutic treatments following cancer diagnosis.

E. Serum Proteins

Elevation of certain serum proteins above their normal levels have been associated with certain types of cancer. Presence of a monoclonal paraprotein in serum and Bence Jones protein in urine is characteristic of multiple myeloma (see Chapter 3). Elevations in serum immunoglobulin M (IgM) is associated with the B cell cancer Waldenström's macroglobulinemia. Additional serum proteins that are elevated in various types of cancer include ferritin in leukemia, β_2 microglobulin in B cell cancers, and C peptide in insulinomas.

Table 10B–1 provides a summary of the individual tumor markers and the associated diseases.

F. Combined Markers

Different types of tumors at different stages of development produce a variety of different tumor markers at various concentrations. Because of this heterogeneity in marker expression, most individual tumor markers lack the sensitivity and specificity required to identify a specific type of cancer. However, quantitative assays for multiple tumor markers are becoming an increasingly valuable tool in the clinical and differential diagnosis of cancer. Various studies have demonstrated that certain patterns of tumor marker expression are associated with certain types of cancer. These patterns have emerged when using multiple markers as well as by determining the ratio of two markers. Table 10B–2 summarizes the use of multiple tumor markers in the identification of specific types of cancer.

Most tumor markers released into serum are quantitated by standard automated enzyme immunoassay systems such as enzyme-linked

Table 10B–1. Individual Tumor Markers and Associated Cancers

MARKER	PRINCIPAL ASSOCIATED CANCER	OTHER ASSOCIATED CANCERS
Hormones		
Calcitonin	Medullary thyroid	Liver, renal, lung, breast
ACTH	Small-cell carcinoma of the lung	Pancreatic, colon, breast
hCG	Trophoblastic tumors	Choriocarcinoma, testicular, ovarian, breast, lung
Tissue-Specific Antigen		
PSA	Prostate	BPH, prostatitis
Oncofetal Antigens		
AFP	Liver	Ovary, testes
CEA	Colon	Breast, lung
SCC	Squamous cell	Various
Carcinoma-Associated Antigens		
CA-125	Ovarian and endometrial	Various
CA 15-3	Metastatic Breast	
CA 549	Breast	Various
CA 27.29	Breast	
MCA	Breast	
CA 72-4	Stomach, ovary	Various
DU-PAN-2	Pancreatic	
CA 242	Pancreatic and colorectal	
CA 19-9	Pancreatic and colorectal	Various
Serum Proteins		
Immunoglobulins	Multiple myeloma	Lymphoma
Ferritin	Myelocytic leukemia	Spleen tumors
IgM	Waldenström's macroglobulinemia	
β_2 microglobulin	B cell cancers	Leukemia
C peptide	Insulinoma	

ACTH, adrenocorticotropic hormone; hCG, human chorionic gonadotropin; PSA, prostate-specific antigen; BPH, benign prostatic hyperplasia; AFP, alpha-fetoprotein; CEA, carcinoembryonic antigen; SCC, squamous cell carcinoma antigen; MCA, mucin-like carcinoma-associated antigen.

immunosorbent assay (ELISA), microparticle enzyme immunoassay (MEIA), and immunoradiometric assay (IRMA) (see Chapter 12).

G. ESTROGEN AND PROGESTERONE RECEPTORS

Detection of estrogen and progesterone receptors on breast cancer tissues can be used as a prognostic indicator of disease progression, and help determine appropriate treatment. Patients with receptor-

Table 10B–2. Combined Tumor Markers and Disease

MARKERS	MAJOR CLINICAL APPLICATION
hCG and AFP	Staging and monitoring testicular cancer
CEA and CA 19-9	Pancreatic cancer
CEA and CA 15-3	Breast cancer
CA 125 and CA 15-3	Ovarian cancer
CA 72-4/CA 19-9 ratio	Differentiate colon from pancreatic cancer
CA 125/CA 50 ratio	Differentiate between ovarian and colorectal cancer
CA 125/CEA ratio	Differentiate ovarian from colon cancer

positive breast cancer tend to respond to hormonal therapy, while patients with receptor-negative tumors respond better to chemotherapy.

H. GENETIC MARKERS

In the multistep progression of normal cells to cancer, various changes occur to genes involved in cell division. The targets for these genetic changes are proto-oncogenes and tumor suppressor genes. Changes in the proto-oncogenes include point mutation and amplification, which results in defective or overexpressed onco-proteins. Mutated versions of proto-oncogenes found in transformed cells are known as oncogenes. Deletion or mutational inactivation of tumor suppressor genes also contributes to the transformed phenotype of cancer cells. The average number of genetic defects in proto-oncogenes and tumor suppressor genes that converts a cell to a fully malignant phenotype is four to five. Proto-oncogene products function as growth hormones, growth hormone receptors, signal transduction molecules, cell cycle progression proteins, and DNA transcription factors. Clinically, genetic markers are used to determine the genetic predisposition of patients with a family history of cancer, early detection of cancer, staging individuals diagnosed with cancer, and determining prognostic outcomes. Table 10B–3 contains a summary of oncogenes and tumor suppressor genes that have been associated with various types of cancer.

Because the genetic changes occur in DNA, whole cells must be obtained to isolate genomic DNA for genetic testing. The specific genetic defects can be detected using molecular techniques such as polymerase chain reaction (PCR), single-stranded conformational polymorphism (SSCP) analysis, DNA sequencing, and DNA probes (see Chapter 14).

> ## Functions of Proto-oncogene and Tumor Suppressor Gene Products
>
> - Growth hormones
> - Growth hormone receptors
> - Signal transduction molecules
> - Cell cycle progression proteins
> - DNA transcription factors

Table 10B–3. Genetic Tumor Markers

TYPES	FUNCTION	MAJOR CANCER
Oncogenes		
N-ras	Signal tranduction	Leukemia, neuroblastoma
K-ras	Signal transduction	Pancreatic, leukemia, lymphoma
c-myc	Transcription regulation	Lymphoma, small-cell carcinoma
c-abl/bcr	Signal transduction	Chronic myelogenous leukemia
bcl-2	Controls apoptosis	Lymphoma, leukemia
HER-2/neu (c-erbB-2)	Growth factor receptor	Breast, ovarian
Tumor Suppressor Genes		
p53	Control cell division	Breast, colon, lung, brain, leukemia
RB1	Suppress DNA synthesis	Retinoblastoma
WT1	Suppress transcription	Nephroblastoma
WT2	?	Wilms' tumor, rhabdomyosarcoma
BRCA-1	?	Breast and ovarian
BRCA-2	?	Breast
APC	?	Colorectal cancer
MTS1	CDK4 inhibitor	Melanoma

SUGGESTED READINGS

Butch AW, Pappas AA. *Tumor markers.* In Bishop ML, Duben-Engelkirk JL, Fody EP (eds.). *Clinical chemistry: Principles, procedures, correlations,* 3rd ed. Philadelphia, PA: JB Lippincott, 1996; 483–99.

Chan DW, Sell S. *Tumor markers.* In Burtis CA, Ashwood ER (eds.). *Tietz textbook of clinical chemistry,* 3rd ed. Orlando, FL: WB Saunders, 1999; 722–49.

Christensen SE. *Tumor markers.* In Anderson SC, Cockayne S (eds.). *Clinical chemistry: Concepts and applications.* Orlando, FL: WB Saunders, 1993; 323–36.

Coleman WB, Tsongalis GJ. *Molecular mutations in human neoplastic disease.* In *Molecular diagnostics for the clinical laboratorian.* Totowa, NJ: Humana Press, 1997.

Wu J, Nakamura R. *Human circulating tumor markers: Current concepts and clinical applications.* Chicago, IL: American Society of Clinical Pathologists, 1997.

Items for Further Consideration

1. Discuss the multistep progession to cancer.

2. Differentiate markers used for screening, prognosis, and monitoring disease progression.

3. Describe the role of oncogenes and tumor suppressor genes in the development of cancer.

4. Discuss the role of tumor markers in the clinical management of patients with cancer.

Biochemical Markers of Cardiac Injury

Hassan M.E. Azzazy

Key Terms ◀

ACUTE MYOCARDIAL INFARCTION

ANGINA

TROPONIN I

TROPONIN T

CK-MB

CK-MB RELATIVE INDEX

CREATINE KINASE MB

ENZYMATIC MARKERS

ISCHEMIA

MYOGLOBIN

NECROSIS

NONENZYMATIC MARKERS

RISK STRATIFICATION

TOTAL CK

Chapter Objectives

Upon completion of this chapter, the reader will be able to:

1. List the criteria for diagnosis of acute myocardial infarction.
2. Identify characteristics of the ideal marker of cardiac injury.
3. List four different markers for diagnosis of cardiac injury and sketch their release patterns.
4. Discuss why creatine kinase isoenzyme MB (CK-MB) activity assays were replaced with CK-MB mass assays.
5. Discuss the different applications of cardiac markers to assess their utility to guide intervention and to assess reperfusion, risk stratification, and reinfarction in patients with acute coronary syndromes.

Analytes Included ◀

TROPONIN I (TnI)

TROPONIN T (TnT)

CREATINE KINASE ISOENZYME MB (CK-MB)

MYOGLOBIN

259

Pretest

Directions: Choose the single best answer for each of the following items.

1. Which of the following markers has the highest reported specificity for the heart?
   a. myoglobin
   b. CK-MB
   c. cardiac troponin I
   d. lactate dehydrogenase (LDH)

2. Which of the following proteins is released first following necrosis of heart tissue?
   a. myoglobin
   b. CK-MB

   c. cardiac troponin T
   d. CK-MM

3. Levels of which of the following proteins remain elevated for up to 14 days after myocardial infarction?
   a. myoglobin
   b. cTnT
   c. cTnI
   d. CK-MB

I. ACUTE MYOCARDIAL INFARCTION

Heart disease is the number one cause of death in the United States. ▶ About 1.25 million Americans annually suffer from **acute myocardial infarction** (AMI). Mortality following AMI can be as high as 25% before patients reach the hospital, with 5 to 10% of the survivors being expected to die within the next year. AMI is caused by vasospasm or occlusion of one or more of the major coronary arteries that feed blood to the heart.

Rapid assessment of AMI is necessary to (1) facilitate early intervention with thrombolytic therapy or angioplasty in AMI patients, (2) identify patients who require admission to the intensive care unit, and (3) identify low-risk patients who can be safely followed as outpatients. It is of note that in 4 to 8% of MI patients the diagnosis is missed, and there is high mortality within this group. In addition, misdiagnosis of MI represents the highest outlay of malpractice dollars among emergency medicine physicians.

AMI has been defined by the World Health Organization (WHO) based on patients fulfilling at least two of the following three criteria: history of chest pain, abnormal electrocardiogram, and rise and fall of cardiac markers in serum. Unfortunately, none of these criteria alone produce a definitive diagnosis of AMI. Biochemical markers play a critical role in diagnosing AMI because the patient's history may be misleading. About one-fourth of all patients with AMI experience atypical chest pain; for example, gastrointestinal symptoms. The electrocardiogram is not diagnostic in 50% of AMI patients.

There are essentially two kinds of biochemical cardiac markers: ▶ (1) **enzymatic markers** such as creatine kinase (CK) and lactate de- ▶ hydrogenase (LDH); and (2) **nonenzymatic markers** such as myo- ▶ globin, **troponin I** (TnI), and **troponin T** (TnT). Figure 10C–1 shows the temporal release of various cardiac markers. The use of LDH will not continue in the future as TnT or TnI will be the marker of choice for assessing patients presenting late after infarction. The ideal biochemical marker of myocardial injury should have the following characteristics: abundance in cardiac cells, exclusivity to myocytes, rapid

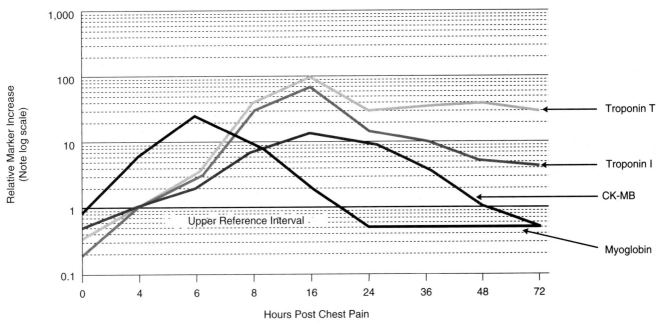

Figure 10C–1. Temporal release patterns of biochemical markers after myocardial infarction. Markers are expressed as multiples of the upper limit of the reference interval. Therefore, the relative increase will vary depending on the normal reference interval utilized.

release after myocardial injury, persistence in the plasma for several hours but not for so long as to mask recurrent infarction, ease of measurement, and availability of a rapid test methodology. There are several criteria that determine a marker's clinical specificity and sensitivity. Some of these include:

- *Size*—Low-molecular-weight protein is released faster into the bloodstream and its level peaks earlier than high-molecular-weight protein.
- *Cellular location*—In general, cytoplasmic proteins are released earlier than are structural proteins.
- *Release ratio*—Some macromolecules may undergo local degradation after release.
- *Clearance*—Smaller proteins pass out of the bloodstream more quickly.
- *Specificity for the heart*—The marker must exist exclusively in the myocytes.

The characteristics of biochemical markers for diagnosis of AMI are shown in Table 10C–1.

Table 10C–1. Characteristics of Biochemical Markers for the Diagnosis of AMI

MARKER	MW (kDa)	INCREASE (h)	PEAK (h)	NORMALIZATION (d)
Myoglobin	17.8	2–6	6–12	1
CK-MB mass	86	2–6	12–24	3
cTnI	24	2–8	12–24	5–10
cTnT	37	2–8	12–96	5–14

(KDa) = kilodaltons; (h) = hours; (d) = days

II. MARKERS OF CARDIAC INJURY

A. MYOGLOBIN

▶ **Myoglobin,** a low-molecular-weight protein (17.8 kilodaltons) abundant in cardiac and skeletal muscle but not smooth muscle, is detectable as early as 1 hour following AMI and peaks at 3 to 9 hours after onset of symptoms. It is a sensitive, but not a specific marker for cardiac injury because elevated myoglobin levels are also detected in patients with skeletal muscle injury and renal failure. Myoglobin, which is rapidly released from the necrotic heart, undergoes rapid renal clearance; therefore, renal impairment can result in elevated myoglobin lev-

▶ els. In patients with unstable **angina,** elevated levels of myoglobin were reported, which may indicate small areas of undetected myocardial cell death. Although myoglobin is a sensitive and early indicator marker of cardiac injury, it is also present in high concentrations in skeletal muscle. Thus, most experts believe that myoglobin is most useful for its negative predictive value rather than for its clinical sensitivity and specificity. Currently, several rapid myoglobin assays are commercially available both for laboratory and point-of-care measurement. The temporal release of myoglobin is displayed in Figure 10C–1.

B. CREATINE KINASE MB (CK-MB)

▶ **Creatine Kinase MB** (CK-MB) has been shown to be an important tool in the evaluation of AMI. CK-MB is one of three dimeric isoenzymes comprising total creatine kinase (CK) activity. All cytoplasmic CK is composed of M and/or B subunits that associate to form CK-MM, CK-MB, and CK-BB isoenzymes. CK-MM predominates in striated muscle, both skeletal and myocardial. In patients having significant myocardial disease (i.e., aortic stenosis, coronary artery disease, or both), the CK-MB isoenzyme comprises approximately 20% of the total CK in this tissue, whereas CK-MB comprises only 0 to 3% of CK

▶ in skeletal muscle. **Total CK** refers to the cumulative activity of the MM, MB, and BB isoenzymes in patient samples.

Currently, CK-MB must be considered the benchmark of biochemical markers of myocardial injury and, as such, has been the basis for comparison of other markers. Although CK-MB is diagnostically specific for myocardial injury, skeletal muscle has higher total CK activity per gram of tissue and may have up to 3% CK-MB. This potentiates nonspecificity, particularly in patients with concomitant myocardial and skeletal muscle injury. To confer greater cardiac

▶ specificity to CK-MB measurements, a **CK-MB relative index** is frequently calculated according to the following equation:

$$\text{CK-MB index} = 100\% \ (\text{CK-MB/Total CK})$$

Although CK-MB index values exceeding 2.5% are often associated with myocardial injury, values as low as 2% and as high as 5% may be seen depending on the variability of both the numerator and denominator terms of the relative index.

CK-MB's characteristic rise and fall in serial measurements is nearly pathognomonic for diagnosing MI. The first rise in CK-MB after MI requires 4 to 6 hours after onset of symptoms. For diagnosis with high clinical sensitivity and specificity, serial sampling over a period of 8 to 12 hours is required.

CK-MB activity assays have been replaced by *CK-MB mass assays,* which measure the protein concentration rather than catalytic activity. Using CK-MB mass assays, analytical interferences that can lead to false-positive test results are less frequent than with CK-MB activity assays. Also, the enhanced analytical performance of CK-MB mass assays compared with CK-MB activity assays results in a higher clinical sensitivity of CK-MB. Following AMI, elevated CK-MB mass concentrations are often detected 1 hour earlier than increased CK-MB activities.

C. TROPONINS

TnT and TnI are considered the best markers for MI in patients presenting with symptoms of **ischemia.** Along with troponin C, troponin T and troponin I are essential components of the contractile complex of both skeletal and myocardial striated muscle. Troponin T functions to bind the troponin complex to the tropomyosin strand; troponin I functions to inhibit the activity of actinomyosin adenosine triphosphatase (ATPase); and troponin C serves to bind four calcium ions, thus regulating contraction (Figure 10C–2). Clinical interest in the proteins of the troponin complex is driven by cardiac-specific isoforms of troponin T and troponin I that have been purified, allowing antibody production and development of immunoassays that are cardiac specific. The amino acid sequence for troponin C is identical in cardiac and skeletal muscle tissue, precluding use of this protein as a specific cardiac marker.

Currently, there is one quantitative and one qualitative assay for cTnT; both of these assays use the same antibody pairs and results correlate well. cTnT increases have been documented in patients with end-stage renal disease; however, the clinical meaning of these elevations is unclear pending completion of properly designed outcome studies. There are numerous cTnI qualitative or quantitative immunoassays that have demonstrated excellent potential for clinical use in the diagnosis of MI and that have been cleared by the Food and Drug Administration (FDA). cTnI may have an important role in strategies for evaluating AMI patients, an area that has been of in-

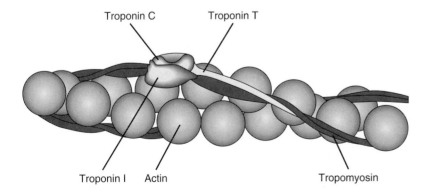

Figure 10C–2. Schematic representation of the thin filament of muscle. The troponin complex consists of three subunits: troponin T, which binds to tropomyosin; troponin I, which binds to actin; and troponin C, which is the calcium-binding subunit.

tense interest, discussion, and study over recent years. Data indicate that cTnI is a specific marker in cases involving skeletal muscle injury and renal failure. Troponins T and I may represent a new standard for diagnosis of AMI and assessment of the acute coronary syndromes.

III. OTHER APPLICATIONS

A. *Risk Stratification of Acute Coronary Syndrome Patients*

In addition to confirming or ruling out AMI, biochemical markers are also used to discriminate chest pain patients with no or stable coronary artery disease from those with acute coronary syndromes. It is important to identify high-risk patients with unstable angina because 5 to 20% of these patients have a poor prognosis with progression to AMI or cardiac death within the first year. Much interest has been generated by outcome-based studies indicating that biochemical markers are useful for stratifying risk in acute coronary syndrome patients. Recent studies indicated that elevated levels of troponins T and I represent an independent risk factor associated with increasing rates of cardiac events. The admission troponin concentration provided essential **risk stratification.**

B. *Noninvassive Assessment of Infarct Zone Reperfusion*

Early detection of failure of recanalization may allow for mechanical intervention to reperfuse ischemic tissue. Relief of chest pain, normalization of ST-segment changes, and reperfusion arrythmias have been found to indicate coronary recanalization; however, all three criteria occur in only 15% of patients. Thrombolytic therapy effectiveness can be better assessed by monitoring the relative increase or early rate of increase of all cytosolic proteins including troponins during thrombolytic therapy, which has higher diagnostic accuracy than other noninvasive methods (e.g., ST monitoring, reperfusion arrythmias). If acute coronary angiography is not available, it is possible to identify patients with failed reperfusion early by utilizing biochemical markers. For best results, combined evaluation of the early rate of increase or relative increase of a biochemical marker between blood samples drawn before and after thrombolytic therapy with electrocardiographic ST-segment monitoring is recommended. CK-MB, TnI, and TnT can be used to assess reperfusion, although they reach peak levels at a slow rate. Myoglobin reaches peak levels within a few hours after therapy and may be considered an early marker of successful reperfusion.

C. *Assessment of Reinfarction*

The rapid clearance of myoglobin from the circulation following AMI has allowed the utilization of this marker in the detection of reinfarction and myocardial extensions. CK-MB is also an essential component in assessing reinfarction or infarct extensions in patients.

D. REDEFINITION OF MI

A joint committee of the European Society of Cardiology (ESC) and the American College of Cardiology (ACC) recently released a consensus document suggesting a new definition of myocardial **necrosis.** ◀ According to this document, biochemical indicators for detecting myocardial necrosis are: (1) "maximal concentration for troponin T or I exceeding the decision limit (99th percentile of the values for a reference control group) on at least one occasion during the first 24 h after the index clinical event," (2) "maximal value of CK-MB (preferably CK-MB mass) exceeding the 99th percentile of the values for a reference control group on two successive samples, or maximal value exceeding twice the upper limit of normal for the specific institution on one occasion during the first hours after the index clinical event."

Either one of the following criteria satisfies the diagnosis of acute, evolving, or recent MI:

1. Typical rise and gradual fall (troponin) or more rapid rise and fall (CK-MB) of biochemical markers of myocardial necrosis with a least one of the following:
   a. ischemic symptoms;
   b. development of pathologic Q waves on the ECG;
   c. ECG changes indicative of ischemia (ST segment elevation or depression); or
   d. coronary artery intervention (e.g., coronary angioplasty).
2. Pathologic findings of an acute MI.

These criteria differ significantly from those established by the WHO (discussed earlier). These criteria are significant because a patient who was previously diagnosed as having severe unstable angina might now be diagnosed as having had an MI.

SUGGESTED READINGS

Adams J III, Abendschein DS, Jaffe AS. Biochemical markers of myocardial injury: Is MB the choice for the 1990's? *Circulation* 88:750–63, 1993.

Alpert JS, Thygesen K. Myocardial infarction redefined—A consensus document of the European Society for Cardiology/American College of Cardiology commitee for the redefinition of myocardial infarction. *J Am Coll Cardiol* 36:959–69, 2000.

Christenson RH, Azzazy HME. Biochemical markers of the acute coronary syndromes. *Clin Chem* 44:1855–64, 1998.

Christenson RH, Duh SH, Newby LK, et al. Cardiac troponin T and cardiac troponin I: Relative values in short-term risk stratification of patients with acute coronary syndromes. *Clin Chem* 44:494–501, 1998.

Christenson RH, Ohman EM, Topol EJ, et al. Combining myoglobin, creatine kinase-MB and clinical variables for assessing coronary reperfusion after thrombolytic therapy. *Circulation* 96:1776–82, 1997.

Keffer JH. Myocardial markers of injury: Evolution and insights. *Am J Clin Pathol* 105:305–20, 1996.

Newby KN, Gibler WB, Ohman EM, Christenson RH. Biochemical markers in suspected acute myocardial infarction: The need for early assessment. *Clin Chem* 41:1263–65, 1995.

Ohman EM, Armstrong PW, Christenson RH, et al. Risk stratification with admission cardiac troponin T levels in acute myocardial ischemia. *N Engl J Med* 335:1333–41, 1996.

Puleo PR, Meyer D, Wathen C, et al. Use of a rapid assay of subforms of creatine kinase MB to diagnose or rule out acute myocardial infarction. *N Engl J Med* 331:561–66, 1994.

Ryan TJ, Anderson JL, Antman EM, et al. ACC/AHA guidelines for the management of patients with acute myocardial infarction: Executive summary. *Circulation* 94:2341–50, 1996.

Wu AHB, Clive JM. Impact of CK-MB testing policies on hospital length of stay and laboratory costs for patients with myocardial infarction or chest pain. *Clin Chem* 43:326–32, 1995.

Wu AHB, Feng YJ, Contois JH, Azar R, Waters D. Prognostic value of cardiac troponin I in patients with chest pain. *Clin Chem* 42:651–52, 1996.

Zabel M, Hohnloser SH, Koster W, et al. Analysis of creatine kinase, CK-MB, myoglobin and troponin T time-activity curves for early assessment of coronary artery reperfusion after intravenous thrombolysis. *Circulation* 87:1542–50, 1993.

Items for Further Consideration

1. Compare the WHO criteria and ESC/ACC criteria for diagnosing MI.

2. Sketch the release patterns of the following cardiac markers following myocardial necrosis: myoglobin, TnT, TnI, and CK-MB.

3. Discuss the utility of other cardiac markers to enhance the specificity of myoglobin as an early marker for diagnosing MI.

4. Explain why CK-MB activity assays were replaced by CK-MB mass assays.

Biochemical Markers of Bone Metabolism

10D

Hassan M.E. Azzazy

Key Terms

ACID PHOSPHATASE

ALKALINE PHOSPHATASE (ALP)

CANCELLOUS (TRABECULAR) BONE

CORTICAL BONE

C-TELOPEPTIDES

DEOXYPRIDINOLINE

LACUNA

N-TELOPEPTIDES

OSTEOBLASTS

OSTEOCALCIN (Ocn)

OSTEOCLASTS

OSTEOCYTES

OSTEOID

OSTEOPOROSIS

PARATHYROID HORMONE (PTH)

PAGET'S DISEASE

PYRIDINOLINE

Chapter Objectives

Upon completion of this chapter, the reader will be able to:

1. Describe the bone formation cycle.
2. List two biochemical markers of bone formation.
3. List four different biochemical markers of bone resorption.
4. Discuss the effect of each of the following hormones on bone turnover: parathyroid hormone, estrogen, insulin, triiodothyronine (T_3), testosterone, cortisol, calcitonin, vitamin D.
5. Discuss the utilization of bone biochemical markers in the assessment of osteoporosis, Paget's disease, and cancer.

Analytes Included ◀

CALCITONIN	*N-TELOPEPTIDE*
C-TELOPEPTIDE	*VITAMIN D*

PreTest

Directions: Choose the single best answer for each of the following items.

1. Which of the following is utilized as a marker of bone formation?
   a. acid phosphatase
   b. alkaline phosphatase
   c. pyridinoline
   d. N-telopeptide

2. A deficiency in which of the following hormones increases bone turnover?
   a. parathyroid hormone
   b. estrogen
   c. testosterone
   d. insulin

3. Type I collagen, the major component of bone organic matrix, has a high proportion of the amino acids
   a. glycine and alanine
   b. cystine and cysteine
   c. tyrosine and phenylalanine
   d. proline and hydroxyproline

4. Which of the following methods is considered a mass assay for measuring bone alkaline phosphatase?
   a. wheat germ lecithin
   b. immunoassay
   c. electrophoresis
   d. heat stability

5. Which of the following diseases is characterized by low bone mass?
   a. osteoporosis
   b. Paget's disease
   c. Cushing's syndrome
   d. cancer

I. BIOCHEMICAL THEORY, PHYSIOLOGY, AND MEDICAL OVERVIEW

Diseases of the bone, particularly osteoporosis, represent a major health problem, affecting approximately one-half of all U.S. women over age 50 and one-fourth of all U.S. men over age 50, at an annual cost of over $14 billion.

Biochemical markers of bone metabolism are one aspect of cost-effective monitoring of the aging population. A relatively recent development in the clinical arena, bone markers are more responsive markers of resorption, formation, and turnover than bone mineral density measured by dual-energy x-ray absorptiometry (DEXA). While DEXA provides an accurate ($\geq$ 95%) and precise means for diagnosing decreased bone mass and predicting fracture, the efficacy of any intervention can be measured reliably only after 2 to 3 years of treatment. In contrast, biochemical bone markers can help physicians make clinical decisions regarding patient therapy and monitoring within several months of initiating treatment.

A. CELLS INVOLVED IN BONE METABOLISM

Osteoclasts and **osteoblasts** carry out bone metabolism at the fundamental *bone metabolic unit* (BMU). Osteoclasts function early in the bone remodeling cycle (Figure 10D–1) to resorb existing bone and make way for new bone matrix. They are usually multinucleated with apical and basolateral poles that differ both morphologically and functionally. The osteoclast's apical pole has a fenestrated membrane that attaches to the bone's surface, isolating a microenvironment or "sealing zone" where pH is lowered and potent enzymes such as acid phosphatase are released to erode the bone underneath. The osteoclast's basolateral pole has receptors for hormones and other substances active in physiological, regulatory, and other cellular functions.

Osteoblasts are involved in bone formation. After tunneling a depression termed a **lacuna** (see Figure 10D–1) by the osteoclast, these cells lay down replacement bone matrix, termed **osteoid,** at BMU sites. Osteoblasts have one nucleus and an extensive network of rough endoplasmic reticulum, the organelle that is responsible for synthesis of the bone matrix protein. After osteoblasts have filled the BMU's lacunae with osteoid, the process of bone formation is halted. Some of the osteoblasts become entrapped in the bone matrix where they become **osteocytes.** In the past, osteocytes were considered metabolically inactive, but new studies indicate they may play an important role in activating osteoclasts by detecting microfractures or other flaws in bone structure.

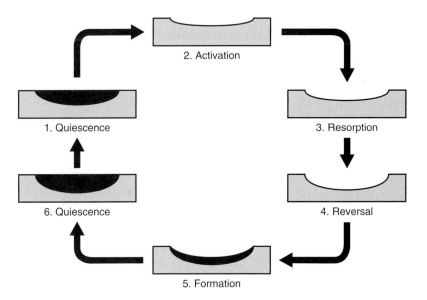

Figure 10D–1. Bone remodeling cycle for a single lacuna: (1) the quiescence phase; (2) activation of the bone cycle; (3) the resorption phase, carried out by osteoclasts; (4) reversal when osteoblasts are recruited; (5) bone formation, in which the lacuna is filled with osteoid by osteoblasts; (6) early quiescence phase, in which there is mineralization of the osteoid with hydroxyapatite; return to the quiescent phase.

B. BASIC BONE STRUCTURE AND COMPOSITION

The skeleton consists of two basic types of bone. The first type, comprising 80% of the skeleton, is **cortical bone.** This bone type is well suited for mechanical, structural, and a protective function because it is 80 to 90% calcified and is quite dense. Cortical bone comprises the outside protective surfaces of all bones and is the major component of long bones. The second type of bone, comprising 20% of the skeleton, is called **cancellous** or **trabecular bone.** Because only 5 to 20% of cancellous bone mass is calcified, it is much less dense than cortical bone. Cancellous bone has a microscopic honeycombed appearance because it is laced with trabeculae. Since bone metabolism occurs only at surface sites, this honeycomb structure confers high metabolic activity to cancellous bone. Even though cortical bone comprises fourfold more of the skeleton than cancellous bone, their metabolic activity is approximately equal.

The organic matrix of bone is mostly composed of type I collagen. Although connective and some other tissues also contain type I collagen, bone has a far higher proportion of this protein and a much higher rate of collagen turnover.

Type I collagen has a high proportion of the amino acids proline and hydroxyproline, and its precursor protein has relatively large extension peptides on both the carboxyterminal and aminoterminal ends (Figure 10D–2). These extension peptides are cleaved during the secretion and fibril-formation process.

In the case of type I collagen, the ends are each linked to a helical portion of a nearby molecule by a pyridinium cross-link. The amino- and carboxy- nonhelical ends are termed the **N-telopeptides** and **C-telopeptides,** respectively.

Bone degradation by osteoclasts during the resorption process releases fragments of various sizes—some still attached to helical portions of a nearby molecule by a pyridinium cross-link—for metabolism or excretion in urine. With further degradation in the liver and kidneys, the fragments are finally broken down to their constituent modified and unmodified amino acids as well as the modified pyridiniums: **pyridinoline** (Pyr) and **deoxypyridinoline** (D-Pyr). Assays for peptides of the N-terminal and C-terminal region, as well as for the Pyr and D-Pyr molecules, have been developed for use in monitoring bone resorption (Table 10D–1).

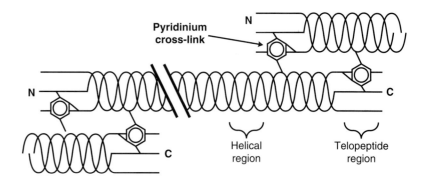

Figure 10D–2. Fibrils of collagen showing the nonhelical carboxyterminal and aminoterminal ends bonding to the helical areas of adjacent fibrils by pyridinium cross-links.

Table 10D–1. Biochemical Markers of Bone Formation and Degradation

MARKER	SPECIFICITY FOR BONE	METHOD	SAMPLE
Bone Formation			
Bone AP	+++	ENZ/IRMA	Serum
Osteocalcin	+++	RIA/IRMA	Serum
Bone Resorption			
Pyr	++	HPLC/ELISA	Urine
D-Pyr	+++	HPLC/ELISA	Urine
N-telopeptide	+++	ELISA	Urine
C-telopeptide	+++	ELISA	Urine

AP, acid phosphatase; ENZ, enzymatic; IRMA, immunoradiometric assay; RIA, radioimmunoassay; ELISA, enzyme-linked immunosorbent assay; HPLC, high-pressure liquid chromatography.

C. REGULATION OF BONE METABOLISM

Bone metabolism is regulated by a complex interaction of many hormones and factors that affect progenitor cells, osteoblasts, and/or osteoclasts. Regulators of bone metabolism include **parathyroid hormone (PTH),** vitamin D, estrogen, and calcitonin, among others (Table 10D–2).

Increased PTH stimulates bone metabolism and resorption. Glucocorticoid therapy may magnify osteoclastic sensitivity to the resorbing effects of circulating PTH.

The biologically active form of *vitamin D*, 1α, 25-dihydroxy vit-

Regulators of Bone Metabolism

- Parathyroid hormone
- Vitamin D
- Estrogen
- Calcitonin
- Thyroid hormone
- Prolactin
- Cortisol and related steroids

Table 10D–2. Hormones and Factors Involved in Bone Metabolism

HORMONE OR FACTOR	EFFECT ON BONE TURNOVER	CELLS AFFECTED	MECHANISM OF EFFECT
Parathyroid hormone	Increase	Progenitor, osteoblasts	Osteoblasts stimulated causing increased osteoclast activity and increased activation frequency and accelerated bone loss
Thyroxine (T₄)	Increase	Osteoclasts	Differential effects on cortical and cancellous bone; cortical bone lost preferentially
Estrogen	Decrease	Osteoblasts	Osteoblasts stimulated causing increased osteoclast activity and increased activation frequency and accelerated bone loss
Testosterone	Decrease	Osteoblasts	Osteoblasts stimulated causing increased osteoclast activity and increased activation frequency and accelerated bone loss
Vitamin D (calcitriol)	Decrease	Osteoblasts	Increased activation frequency but also inhibits mineralization of newly synthesized osteoid matrix
Cortisol	Increase	Progenitor, osteoblasts, osteoclasts	Profound effect by both increasing bone resorption and inhibiting bone formation, leading to extremely accelerated bone loss
Calcitonin	Decrease	?	Inhibits bone resorption; used therapeutically to treat increased bone loss (e.g., Paget's disease, high-turnover osteoporosis)
Insulin	Decrease	Osteoblasts	Causes increased insulin-like growth factor-I synthesis in liver, resulting in increased collagen synthesis by osteoblasts

amin D3 (calcitriol), is synthesized from the major circulating metabolite 25-hydroxy vitamin D3. Calcitriol is a pleiotropic hormone that is involved in the regulation of calcium homeostasis and control of bone cell differentiation. Decreased concentrations of vitamin D are associated with increased BMU activation and, as a result, a greater rate of bone turnover.

Estrogen is a critical regulator of bone metabolism because it decreases production of osteoid matrix and may promote formation of trabecular bone. The estrogen deficit that naturally occurs with the onset of menopause promotes bone resorption and increases bone turnover.

In contrast to estrogen, calcitonin is a calcitropic hormone that is an effective inhibitor of bone resorption. Although the mechanism of calcitonin's action is unknown, the hormone has been used to treat patients with high bone turnover, osteoporosis, Paget's disease, and hypercalcemia of malignancy.

Other hormones that play a role in bone metabolism include thyroid hormone and prolactin. Hyperthyroidism or therapeutic administration of thyroid hormone may cause increased bone turnover. Also, cortisol and related steroids increase bone turnover by directly stimulating both bone resorption and formation. Osteoporosis can be a common and serious secondary side effect that may go undetected until substantial bone loss has occurred in patients receiving long-term glucocorticoid treatment.

II. MARKERS OF BONE FORMATION

A. ALKALINE PHOSPHATASE (ALP)

▶ **Alkaline phosphatase (ALP)** is associated with the cell plasma membrane and appears to play a role in the transport of substances from the intracellular compartment across the cell membrane to the extracellular region. Four ALP isoenzymes may be found in blood, each of which is relatively specific for the respective liver, bone, placental, and intestinal tissues with which it is associated. The liver-, renal-, and bone-specific isoenzymes are coded by the same gene; therefore, differences between these isoenzymes are the result of post-translational modification.

Bone alkaline phosphatase (B-ALP) is produced by the osteoblast. B-ALP is produced in extremely high amounts during the bone cycle formation phase and is therefore a good indicator of overall bone formation activity.

For therapeutic monitoring of patients, B-ALP measurements are a good indicator of the metabolic activity of bone. Rising B-ALP concentrations may indicate estrogen deficiency.

All assays for B-ALP are serum-based (Table 10D–1). Although numerous methods including electrophoresis, wheat germ lecithin precipitation, heat stability, and immunoassay (also termed *mass assay*) are available, most experts consider immunoassay the method of choice because these assays have better analytical sensitivity and lower imprecision. Because of the similar epitopes, however, tissue specificity of B-ALP immunoassays may be confounded by potential cross-reactivity with ALP from liver. However, clinical performance of B-ALP assays appears to be acceptable.

B. OSTEOCALCIN (OCN)

Osteocalcin (Ocn) is a relatively small protein (5,800 daltons) produced by osteoblasts during the matrix mineralization phase. It is released during bone formation into blood and also incorporated into the bone matrix, where it is the most abundant noncollagenous protein. Ocn synthesis is dependent on vitamin K, which post-translationally modifies the gene product with γ-carboxyglutamate (Gla) residues. Ocn is secreted into the blood and, in individuals having normal renal function, it is also excreted in urine.

The body degrades Ocn during bone resorption, with up to 70% entering the circulatory system. Because circulating Ocn may be either newly synthesized during bone formation or released during resorption, there is some question whether Ocn should be considered a marker of osteoblast activity or an indicator of bone matrix metabolism and turnover.

III. MARKERS OF BONE RESORPTION

A. ACID PHOSPHATASE

Five isoenzymes of the lysosomal enzyme **acid phosphatase** (AP) circulate in blood, the major sources of which are bone, prostate, platelets, erythrocytes, and spleen. Bone AP is presumably released into circulation by "leakage" during resorption as well as after detachment of the osteoclast's sealing zone. Assays for AP are serum or plasma based. Data for the use of AP as a marker of bone metabolism are incomplete. This may be the result of numerous analytical problems and the fact that the enzyme is released into the sealed osteoclast microenvironment rather than directly into the bloodstream.

B. N-TELOPEPTIDE

Fragments from the N-terminus are released into circulation as a result of osteoclast degradation of type I collagen during the resorption process. Because the vast majority of these fragments are relatively small, they readily pass through the glomerulus into urine. N-telopeptides are very specific for bone tissue breakdown because other tissues composed of type I collagen (e.g., skin) are not actively metabolized by osteoclasts.

A urine assay for N-telopeptides that monitors bone resorption has been commercially developed. This assay is based on a monoclonal antibody that specifically recognizes the α-2 chain N-telopeptide fragment. The recommended specimen is a second morning spot urine or a 24-hour collection. A serum test that quantitatively measures cross-linked N-telopeptides is also available. Compared to their urine-based counterparts, serum assays are considered to be more convenient, have less biologic variability (especially variability in creatinine excretion), and have little diurnal variation.

C. C-TELOPEPTIDE

Fragments derived from the C-terminus are also released into circulation as a result of the osteoclast-mediated degradation of type I collagen during the bone resorption process. As with the N-telopeptide, these fragments are highly bone specific because osteoclasts are not normally active in degrading other type I collagen-containing tissues. Diagnostic companies have developed both serum and urine assays for these peptides.

D. PYRIDINOLINE (PYR) AND DEOXYPYRIDINOLINE (D-PYR) CROSS-LINKS

Pyr and D-Pyr cross-links, resulting from post-translational processing of lysine and hydroxylysine residues, are essential for stabilizing the mature forms of collagen fibers and elastin. Although both Pyr and D-Pyr are found in bone, D-Pyr is more bone specific. Fluorimetric detection of both the free and protein-bound forms of Pyr and D-Pyr by high-performance liquid chromatography (HPLC) provides a highly specific marker of bone resorption, but can be technically demanding and are not routinely offered in all clinical laboratories. Immunoassays for free Pyr and free D-Pyr in urine are available. These assays provide a quantitative measure of free cross-links that reflect bone resorption.

IV. UTILIZATION OF BONE MARKERS IN VARIOUS DISEASES

A. OSTEOPOROSIS

▶ According to the World Health Organization, **osteoporosis** is "a disease characterized by low bone mass and micro-architectural deterioration of bone tissue, leading to enhanced bone fragility and a consequential increase in fracture risk." Bone biochemical markers hold the promise for determining whether the patient is responding to therapy for the disease.

B. PAGET'S DISEASE

▶ **Paget's disease** is characterized by gross distortion of normal bone remodeling that is reflected in elevated levels of bone markers. Skeletal ALP is reported to have the highest diagnostic accuracy in Paget's disease patients. Collagen type I cross-linked C-telopeptide and collagen type I cross-linked N-telopeptide are also elevated in Paget's disease and respond rapidly to therapy.

C. CANCER

Clinical studies are incomplete, but biochemical bone markers show promise for the evaluation of cancer patients to determine whether skeletal metastases are present and respond to therapy.

SUGGESTED READINGS

Christenson RH. Biochemical markers of bone metabolism: An overview. *Clin Biochem* 30:573–93, 1997.

Consensus Development Conference. Diagnosis, prophylaxis, and treatment of osteoporosis. *Am J Med* 94:646–50, 1993.

Delmas PD. Biochemical markers for the assessment of bone turnover. In Riggs BL, Melton LH III (eds.). *Osteoporosis: Etiology, diagnosis, and management.* Philadelphia: Lippincott-Raven, 1995; 319–33.

Demers LM. Clinical usefulness of markers of bone degradation and formation. *Scand J Clin Lab Invest* 57(Suppl 237):12–20, 1997.

Gamero P, Hausheff E, Chapuy M-C, et al. Markers of bone resorption predict hip fracture in elderly women: The EPIDOS prospective study. *J Bone Miner Res* 11:1531–37, 1996.

Hanson DA, Weis-Bollen A-M, Masian SL, et al. A specific immunoassay for monitoring human bone resorption: Quantitation of type I collagen cross-linked N-telopeptides in urine. *J Bone Miner Res* 7:1251–58, 1992.

Kleerekoper M, Edelson GW. Biochemical studies in the evaluation and management of osteoporosis: Current status and future prospects. *Endocrin Pract* 2:13–19, 1996.

Miller PD, Baran DT, Bilezikian JP, et al. Practical clinical applications of biochemical markers of bone turnover. *J Clin Densit* 2:323–42, 1999.

Panigrahi K, Delmas PD, Singer F, et al. Characteristics of a two-site immunoradiometric assay for human skeletal alkaline phosphatase in serum. *Clin Chem* 40:822–28, 1994.

Risteli L, Risteli J. Biochemical markers of bone metabolism. *Ann Med* 25:385–93, 1993.

Items for Further Consideration

1. Sketch the bone cycle. List the major cell types that are involved in bone metabolism. Explain their specific roles.

2. Compare the following:

   a) Cortical bone and cancellous bone.

   b) Acid phosphatase and alkaline phosphatase.

   c) PTH and OCN.

   d) Osteoblasts and osteoclasts.

3. Compare and contrast markers of bone formation and bone resorption.

4. How are the C- and N-telopeptides formed? Why do they have high bone specificity?

5. List five issues that need to be investigated to further the role of bone biochemical markers in patient care.

part

II

Instrumentation and Analytical Techniques

Photometric and Electrochemical Measurements

John S. Davis

Key Terms

ABSORBANCE	**ION-SELECTIVE ELECTRODES**	**REFERENCE ELECTRODE**
AMPEROMETRY	**LUMINOMETER**	**REFLECTANCE SPECTROPHOTOMETRY**
ATOMIC ABSORPTION SPECTROPHOTOMETRY	**MONOCHROMATIC LIGHT**	
	MONOCHROMATOR	**SCINTILLATION COUNTERS**
BEER–LAMBERT LAW	**NEPHELOMETERS**	**SPECTRAL BANDWIDTH (BANDPASS)**
BEER'S LAW	**PERCENT TRANSMITTANCE (%T)**	
CHEMILUMINESCENCE	**PHOTOMULTIPLIER TUBE**	**SPECTROPHOTOMETRIC MEASUREMENTS**
DENSITOMETER	**POLYCHROMATIC LIGHT**	**STRAY LIGHT**
DETECTORS	**POTENTIOMETRY**	**TURBIDIMETER**
FLUORESCENCE	**QUENCHING**	**WAVELENGTH**
FREQUENCY		

Chapter Objectives

Upon completion of this chapter, the reader will be able to:

1. Describe the quantitative nature and clinical applications of spectrophotometric and electrochemical measurements.
2. Describe the properties of radiant energy and corresponding relationships to quantitative measurements.
3. State Beer's law in terms of absorbance, transmittance, analyte concentration, and absorptivity.
4. Given appropriate information, calculate (a) analyte concentration using Beer's law and (b) a factor for determining analyte concentration from a standard.

5. Identify the basic functional components of a spectrophotometer and state the purpose of each.
6. Describe the procedures necessary for the maintenance and calibration of a spectrophotometer.
7. Define the following terms as they relate to photometric instruments: *dark current, stray light,* and *spectral band-pass.*
8. Discuss the principle, instrument schematics, and clinical applications for spectrophotometry, atomic absorption spectrophotometry, turbidimetry, nephelometry, fluorometry, chemiluminescence/luminometry, and scintillation counting.
9. Explain the principles of potentiometric and amperometric measurements.
10. Discuss the utilization of electrochemical techniques such as potentiometry and amperometry in the clinical laboratory.
11. Describe functional components and principles of electrochemical instrumentation.
12. Suggest basic troubleshooting steps regarding the instrumentation and analytical techniques discussed in this chapter.

Instruments/Techniques Included ◀

SPECTROPHOTOMETRY (SINGLE-BEAM, DOUBLE-BEAM, AND REFLECTANCE)

DENSITOMETRY

TURBIDIMETRY

NEPHELOMETRY

SCINTILLATION COUNTING (β- AND γ-COUNTERS)

CHEMILUMINESCENCE/LUMINOME-TRY

FLUOROMETRY

ATOMIC ABSORPTION SPECTROPHOTOMETRY

POTENTIOMETRY

AMPEROMETRY

Pretest

Directions: Choose the single best answer for each of the following items.

1. Which of the following correctly states the relationship between the wavelength and energy of light?
   a. the longer the wavelength the lower the energy
   b. the longer the wavelength the higher the energy
   c. changes in energy and wavelength are independent of each other
   d. no relationship exists between wavelength and energy

2. Absorbance
   a. is directly proportional to %T
   b. varies linearly with %T
   c. is an inverse function of %T
   d. varies linearly with wavelength

3. To measure absorbance at 405 nm
   a. a quartz cuvet must be used
   b. a tungsten lamp should be used
   c. a neutral density filter must be used
   d. a light path of 1.0 inches must be used

4. Calculate the concentration of an unknown sample given the following information (assume the data follow Beer's Law):

Unknown absorbance: 0.133	Wavelength: 550 nm
Standard absorbance: 0.207	Light path: 1.0 cm
Standard concentration: 85 mg/dL	Absorptivity: 2.7×10^{-3} mol/cm²

   a. 36 mg/dL
   b. 55 mg/dL
   c. 132 mg/dL
   d. 196 mg/dL

5. In spectrophotometry, the device most often used for wavelength selection is the
   a. didymium filter
   b. photomultiplier tube
   c. chopper
   d. diffraction grating

6. In an assay measuring the transmittance (%T) of the colored end product spectrophotometrically, readings of 70%T for the highest standard and 98%T for the sample would indicate that the analyte concentration in the sample was
   a. low
   b. mid-range
   c. high
   d. above the linear range of the assay

7. Spectral bandpass reflects the quality of which component in a spectrophotometer?
   a. energy source
   b. monochromator
   c. cuvet
   d. detector

8. Which of the following devices is used most often to verify wavelength selection in a spectrophotometer?
   a. cutoff filter
   b. deuterium lamp
   c. holmium oxide filter
   d. solution of $NaNO_2$

9. Which of the following techniques is used to measure light scattered by particles in solution?
   a. turbidimetry
   b. nephelometry
   c. densitometry
   d. reflectance spectrophotometry

10. Immune complexes that form large particles in solution are often measured by
    a. turbidimetry, in which the detector is in a straight line with the incident light beam
    b. nephelometry, in which the detector is at a 30° angle to the incident light beam
    c. fluorescence, in which the detector is at a 90° angle to the incident light beam
    d. densitometry, in which the detector is at a 45° angle to the incident light beam

11. Which of the following instruments would be used to detect radiation emitted by an analyte labeled with the isotope [125]I?
    a. crystal scintillation counter
    b. liquid scintillation counter

    c. chemiluminescence luminometer
    d. NaI luminometer

12. Which component of a scintillation counter converts the γ-radiation to flashes of light?
    a. liquid cocktail solution
    b. photomultiplier tube
    c. lead detector
    d. NaI/thallium crystal

13. Which of the following compounds would release chemiluminescent energy after being oxidized in a chemical reaction?
    a. fluorescein
    b. tritium
    c. acridinium ester
    d. thallium iodide

14. Chemiluminescent energy is detected by which of the following instruments?
    a. fluorometer
    b. luminometer
    c. scintillation counter
    d. nephelometer

15. In fluorometry, the excitation wavelength is _____ the emission wavelength.
    a. longer than
    b. shorter than
    c. one-half
    d. the same as

16. Which instrument requires two wavelength selection devices?
    a. densitometer
    b. atomic absorption spectrophotometer
    c. luminometer
    d. fluorometer

17. In fluorescence measurements, quenching is the loss of energy due to
    a. fluorophor interaction with solvent or solute molecules
    b. increased reaction temperature
    c. a solvent pH that is too acidic
    d. the deterioration of the energy source

18. The flame in an atomic absorption spectrophotometer
    a. converts the analyte to atoms in the ground state
    b. serves as the high-energy source of radiation
    c. excites analytes to be measured to a higher-energy state
    d. produces monochromatic light for analyte absorption

19. Which component in an atomic absorption spectrophotometer is analogous to the cuvet in a spectrophotometer?
    a. hollow cathode lamp
    b. chopper
    c. nebulizer
    d. flame

20. In calcium analyses by atomic absorption spectroscopy, which compound is used to reduce chemical interference by phosphate?
    a. holmium oxide
    b. thallate
    c. lanthanum
    d. acetone

21. Which of the following electrodes is often used in clinical applications as the reference in a potentiometric system?
    a. valinomycin
    b. hydrogen
    c. high-actinic glass
    d. silver–silver chloride

22. Which of the following parameters is measured in a potentiometric determination?

    a. current
    b. electrical potential
    c. capacitance
    d. resistance

23. In potentiometry, analyte concentration can be determined from the electrochemical potential using the
    a. Henderson–Hasselbalch equation
    b. Farraday conversion
    c. Impedance quotient
    d. Nernst equation

24. Which of the following parameters is measured by an amperometric electrode?
    a. current
    b. voltage
    c. electrical potential
    d. resistance

25. In an amperometric oxygen electrode, increased oxygen in the blood sample would correlate with
    a. increased capacitance of the cathode
    b. decreased voltage
    c. increased voltage
    d. increased current

I. INTRODUCTION

Even after decades of development in analytical technology, photometric and electrochemical techniques still account for the majority of measurements made in the clinical laboratory. Photometric techniques have evolved from the old, colorimetric methods to newer, more sensitive methods including those based on the measurement of chemiluminescent end products. While automation and computerization have greatly changed the way clinical laboratory measurements are performed, the principles of radiant and electrical energy are still the analytical core of these newer, automated systems.

II. PHOTOMETRIC MEASUREMENTS

A. RADIANT ENERGY

Light is *electromagnetic* (EM) *radiation,* a form of radiant energy that travels in waves at a constant speed in a given medium. The characteristics used to describe types of EM radiation include wavelength and frequency. **Wavelength** (λ), often measured in nanometers, is the distance between adjacent peaks or troughs in a continuous wave. **Frequency** (υ) is the number of waves that pass a given point per unit time (e.g., in units of cycles per second). Frequency can be expressed as:

$$\upsilon = c/\lambda$$

where c is the speed of light in a vacuum (3×10^{10} cm/s). Light energy (E), in units called *photons,* can be expressed as:

$$E = h(c/\lambda)$$

where h is Planck's constant (6.62×10^{-27} erg seconds) and the energy of the photon is in ergs. Thus, as wavelength decreases, energy increases and vice versa. The *electromagnetic spectrum* ranges from short, high-energy cosmic waves measuring 3×10^{-12} meters to the long, low-energy radio waves measuring 3×10^{4} meters. The visible and near visible light used in clinical measurements is only a small portion (i.e., from 280 to 750 nm) of the entire electromagnetic spectrum.

If the energy of a photon corresponds to the natural frequencies of vibration and rotation of electrons and/or atoms within a molecule, radiation may be absorbed. Unique molecular structures have very specific energy requirements for excitation, and this energy can be provided by specific wavelengths of EM radiation that are absorbed by a given molecule. This selective absorption of photons of certain frequencies results in an absorption spectrum that is characteristic for each molecular species. The amount of energy absorbed depends on the amount of substance present. As a specific fraction of light or quantum of energy is absorbed, the remaining energy is either reflected or transmitted. Energy not absorbed by a substance in solution is transmitted, and the wavelengths of the transmitted light combine to make up the color that the eye observes; wavelengths that are absorbed do not contribute to the observed color. The solution is colorless ("white") when all wavelengths in the visible region are transmitted; conversely, a substance appears black when all wavelengths are absorbed. **Polychromatic light,** composed of many different wavelengths, can be dispersed into separate wavelengths by refraction. For example, rainbows occur when sunlight passes through rain, just as the color spectrum is produced when polychromatic light from the sun or a light bulb is directed through a prism (Figure 11–1).

Spectrophotometric measurements are based on detection and quantification of energy that is transmitted after passing an incident beam of light through the solution being analyzed. Transmitted energy is expressed in terms of **percent transmittance (%T)** by the following equation:

$$\%T = I_s/I_o \times 100$$

where I_s = the intensity of transmitted light, and I_o = the intensity of

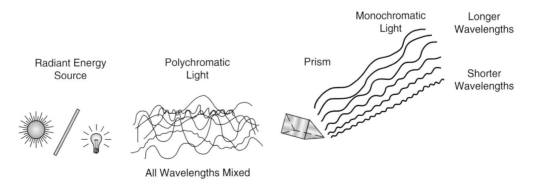

Figure 11–1. Separation of polychromatic light into individual wavelengths.

incident light (original). Consequently, if all light is absorbed (i.e., none of the incident light is transmitted), then %T = 0; conversely, if none of the source light is absorbed (i.e., all of the incident light is transmitted), then %T = 100. Figure 11–2A shows graphically the nonlinear relationship between %T and the concentration of an absorbing substance in solution; %T varies inversely and logarithmically with concentration.

▶ Since nonlinear relationships are inconvenient to use in practice, %T can be expressed as **absorbance** (A), which is directly proportional to the concentration of the absorbing substance (Figure 11–2B). The relationship is expressed by the following equation:

$$A = 2 - \log \%T$$

Thus, at 100 %T, A = 0 (i.e., no light was absorbed); conversely, at 0 %T, A = 2 (i.e., 100% of the light was absorbed). Absorbance can be used for analyte quantitation since transmitted energy relates to the energy absorbed by the molecules in solution. As the number of absorbing particles present increases, the amount of light transmitted through the solution decreases.

▶ This phenomenon of specific and quantitative absorption forms the basis of Beer's law. **Beer's law** states that absorbance is directly proportional to the concentration of the absorbing matter. Bouguer's law or Lambert's law states that absorbance is directly proportional to the length of the light path through the sample; the light path is equal to 1.0 cm in most spectrophotometers. These two laws are most com-
▶ monly combined into the **Beer–Lambert law,** which states:

$$A = abc$$

where A = absorbance, a = the *absorptivity constant* in units of moles/cm², b = path length in centimeters, and c = concentration. Since the molar absorptivity constant (a) does not change for a given analyte at a specified wavelength and the light path (b) is constant (e.g., usually 1 cm), the equation describes the direct relationship between absorbance (A) and concentration (c). Consequently, if a plot of absorbance versus concentration produces a straight line for the substance of interest, Beer's law is obeyed and the following relationship between a sample with an unknown analyte concentration and a *standard* with a known analyte concentration exists:

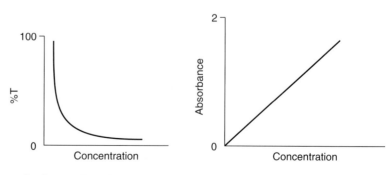

A. Percent T vs. Concentration   B. Absorbance vs. Concentration

Figure 11–2. **A.** Nonlinear relationship between %T and concentration. **B.** Linear relationship between absorbance and concentration.

$$\frac{\text{Conc. of Std. }(C_s)}{\text{Abs. of Std. }(A_s)} = \frac{\text{Conc. of Unk. }(C_u)}{\text{Abs. of Unk. }(A_u)}$$

or, more simply:

$$C_u = C_s \times A_u/A_s$$

If the absorbance curve is linear for an analyte over the range of expected concentrations, it is possible to use one standard and apply the above equation to determine unknown concentrations.

The relationship between absorbance and concentration may deviate from Beer's law when:

1. The concentration of the analyte is very high.
2. The solvent of the analyzed solution absorbs a significant amount of light.
3. There is a chemical reaction between the analyte and another substance in solution.
4. The incident light beam is not monochromatic.
5. There is significant stray light in the incident beam of the spectrophotometer.

Therefore, in most cases, two or more standards or calibrators must be run and the absorbance of the unknown is related to the corresponding concentration on the standard curve; standards should cover the entire absorbance range in which readings of unknowns are expected to fall. Unknowns having an absorbance beyond that of the highest standard must be diluted and reanalyzed. If the relationship between absorbance and concentration of a substance is nonlinear, enough standards must be used to define the curve over the range of expected concentrations.

B. SPECTROPHOTOMETRY

All spectrophotometric instruments include the components found in the single-beam spectrophotometer shown in Figure 11–3. Common enhancements used to achieve better optic and readout quality include lenses, mirrors, and fiber optics. In addition, advances in technology have allowed design options such as miniaturization and computerization.

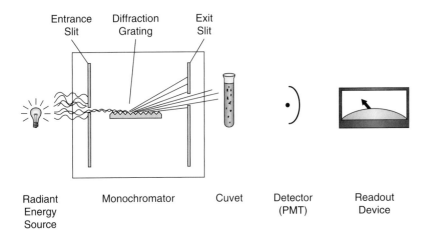

Figure 11–3. Instrument components of a single-beam spectrophotometer.

1. Components of a Single-Beam Spectrophotometer

Light/Energy Source. The light or energy source produces polychromatic light; other components of the spectrophotometer disperse the light and isolate the desired wavelength(s). For wavelengths in the visible region of the electromagnetic spectrum, the tungsten lamp is most often used. The tungsten filament lamp provides a wide range of wavelengths (350 to 950 nm). Measurements in the *ultraviolet* (UV) range (185 to 375 nm) are most often made using deuterium or hydrogen lamps. Laser sources can provide specific, intense, narrow-wavelength light; the wavelength emitted depends on the type of laser. In general, the light source for spectrophotometry must (1) provide intense light, (2) remain relatively cool, and (3) be exactly realigned when replaced. Though lamps are prefocused, their intensities vary; thus, upon replacement, a recalibration must be performed.

▶ *Monochromator.* Beer's law holds true only for **monochromatic light;** therefore, accurate measurements require that the desired wavelength be separated (as much as possible) from other wavelengths in polychromatic light. Wavelength isolation is accomplished

▶ by a monochromator. A **monochromator** is composed of an entrance slit, wavelength dispersing device, and an exit slit. The *entrance slit* focuses a narrow beam of the polychromatic light from the energy source on the dispersing device. The most common dispersing device used in spectrophotometers is the *diffraction grating,* though the prism is often given as an example. Diffraction gratings are composed of a number of parallel lines or grooves that are closely spaced together on a polished metallic surface. Incident light that hits a line or groove is spread out or diffracted; thus, each groove acts as a prism by producing a spectrum of wavelengths so that the polychromatic light from the energy source is separated into dispersed monochromatic wavelengths. The resolution of a diffraction grating depends on the number of lines or grooves; the more grooves, the better the resolution. The final component of the monochromator is the exit slit. The *exit slit* serves to define the range of wavelengths that will ultimately reach the detector. In other words, the exit slit provides wavelength selection by eliminating or blocking unwanted wavelengths from reaching the detector; the width of the

▶ exit slit determines the **spectral bandwidth** or **bandpass**. The spectral isolation or quality of a monochromator is typically defined by its spectral bandpass. Figure 11–4 demonstrates the effect of spectral bandpass on wavelength definition.

Cuvet. Cuvets are tubes or optical cells that hold the absorbing solution. Cuvets can be round or square; the width of the cuvet at the point through which the light beam passes is commonly 1.0 cm. For wavelengths in the visible range, optical plastics or glass cuvets are used. Quartz cuvets are used for wavelengths in the UV range (340 nm and below). The cuvet must not be composed of a substance that itself absorbs light at the measuring wavelength; glass absorbs ultraviolet light, and therefore is not useful for measurements in the UV region of the spectrum. With any cuvet, care must be taken in handling since smudges, scratches, or fingerprints may cause significant error. Cuvets should be cleaned with a mild detergent and wiped with a nonabrasive material before being placed into the spectrophotometer for reading.

Light Sources Used for Spectrophotometry

- Visible light
  —Tungsten filament lamp
- Ultraviolet light
  —Deuterium lamp
  —Hydrogen lamp
- Specific, narrow λ selection
  —Laser sources

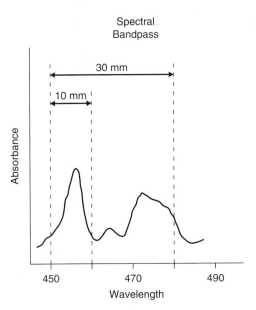

Figure 11–4. Effect of spectral bandpass on wavelength definition. The 30-nm bandpass instrument is unable to resolve the spectral peaks at the wavelengths of 455, 465, and 475 nm. Instead, one broad peak would be detected. In contrast, the 10-nm bandpass instrument is able to resolve the individual peaks.

Detector. **Detectors** measure the amount of light exiting the cuvet by converting the transmitted light to a proportional electrical signal. The most common type of detector used is the **photomultiplier tube** (PMT). In a PMT, a metal layer absorbs light and emits a proportional flow of electrons; the flow of electrons (i.e., the signal) is amplified as electrons pass to subsequent stages in the detector. PMTs generally have good sensitivity and a quick response to changes in light intensity. Theoretically, in the absence of energy striking the detector, no current is produced. Practically speaking, however, a small amount of current is always present. Current that is produced in the absence of light energy is referred to as *dark current*. PMTs exhibiting minimal dark current are preferred.

Photodiode array detectors can be used to monitor light at multiple wavelengths simultaneously. For example, a detector may be composed of an array of diodes that detect wavelengths from 340 to 800 nm at a resolution of 1 nm.

Barrier layer cell detectors convert light to electrical energy (usually without amplification); they are less sensitive and slower to respond than PMTs and are usually used only in inexpensive equipment.

Readout Device. Electrical signals produced by the detector must be converted to a readable form (e.g., %T or absorbance) that relates to concentration. In a readout device, an amplifier is often used to increase the signal for a meter, dial, printout, recorder, or digital readout mechanism. Automated instruments typically have a microprocessor that stores calibration data, makes calculations, and reports analyte concentrations.

Types of Detectors

Photomultiplier tube—a metal layer absorbs light and emits a proportional (and amplified) flow of electrons

Photodiode array—used to monitor light at multiple wavelengths simultaneously

Barrier layer cell—converts light to electrical energy

2. Double-Beam Spectrophotometers. There are two basic designs for double-beam spectrophotometers. The double-beam-in-space unit has one light source that is split into two beams: one directed through the sample and another through a reference cell. These instruments have dual detectors and determine the ratio of the absorbances between the two beams. The second type of double-beam instrument is the double-beam-in-time spectrophotometer. In this unit, the light is split by a chopper to alternately pass through the sample and reference, and transmitted light is measured by one detector. Double-beam spectrophotometers have two important advantages over single-beam instruments: (1) They eliminate any time-related discrepancies between measurement of sample and blank by simultaneously measuring or comparing the measurements of the two, and (2) they compensate for changes or fluctuations in voltage, energy source intensity, or other instrumental drifts.

3. Spectrophotometric Calibration/Maintenance Procedures. Verification of the functional capabilities of a spectrophotometer is imperative to ensure accurate and reliable results. As multianalyzers and spectrophotometric instruments have become more complex with the utilization of lasers, fiber optics, and computers, these functional checks are less visible but nonetheless important. Functional checks for spectrophotometers include wavelength calibration, stray-light corrections, and linearity checks.

Before readings are taken, the spectrophotometer must first be "zeroed" and "blanked" to set the limits of transmittance and absorbance. *Zeroing* refers to blocking the incident light at the cuvet to set the instrument's detection system to 0 %T. *Blanking* refers to setting the instrument's detection system to complete transmittance (i.e., 100 %T or A = 0) while a cuvet that contains water, or none of the substance being measured, is in the sample compartment of the instrument. These steps correct for reflectance or absorbance of light by the cuvet, background or interfering absorbance, and stray light reaching the detection system.

Spectrophotometric methods rely on the ability to accurately select and set a particular wavelength. Wavelength selection is validated through the utilization of a substance that has sharp absorbance peaks at known wavelengths. For example, if it is known that a substance has a sharp absorbance peak at 360 nm, the spectrophotometer should register maximum absorbance when the wavelength is set to 360 nm and the substance is placed in the light path. If maximum absorbance is not observed at the appropriate wavelength, corrective action must be performed as indicated by the manufacturer. The most accurate method used to verify wavelength is based on use of a deuterium lamp, which has strong emission lines at well-defined wavelengths. The most commonly used substance, however, is the holmium oxide glass filter, which exhibits several sharp absorbance peaks over the range of 280 to 650 nm. Filters made with didymium are typically used for wavelength calibration of broad-bandpass instruments.

Photometric linearity is another function that must be checked routinely. The relationship between the amount of energy absorbed and the instrument readout in a properly functioning spectrophotometer is linear. Linearity checks assess the instrument's detector response. One of the simplest methods used to check the linearity is

---

Calibration and Maintenance Procedures for Spectrophotometers

- Setting transmittance/absorbance limits
  —Zeroing: set 0 %T
  —Blanking: set 100 %T (A = 0)
- Wavelength calibration
- Photometric linearity checks
- Stray light corrections

based on the measurement of a series of dilutions of a compound that is known to follow Beer's law; the plot of concentration versus absorbance should be linear. Neutral density filters and commercially prepared, prediluted solutions can also be utilized to check linearity, thus eliminating the possibility of dilution errors.

Stray light within the instrument can also cause errors in spectrophotometric determinations. **Stray light** is defined as energy ◄ reaching the detector through the monochromator and exit slit that is not of the selected wavelength. Filters are often incorporated into the instrument to minimize stray light. Though the effect is generally minimal, stray light can be a factor when absorbance readings are at the high end of the scale (i.e., low transmittance). Stray light can be detected through the use of glass cutoff filters or solutions such as $NaNO_2$, NaBr, acetone, and methylene blue. In either case, the cutoff material is transmissive over one end of the spectrum but essentially opaque below a defined wavelength. Thus, any transmittance detected below the cutoff wavelength is attributable to stray light.

C. REFLECTANCE SPECTROPHOTOMETRY

A variation on standard spectrophotometry is **reflectance spec-** ◄ **trophotometry.** In reflectance methods, a flat, reflective surface (e.g., paper, plastic, or dry film) is illuminated and the amount of light reflected is measured. The analyte or analyte-derived chromogen is adsorbed onto the reflective surface, and the intensity of reflected light from the sample surface is compared to that of a reference that reflects all of the incident light. In general, as concentration increases, reflectance decreases; however, the relationship between reflected light and analyte concentration is nonlinear. Instruments using reflectance photometry apply an appropriate algorithm to calculate concentration.

D. DENSITOMETRY

A **densitometer** is a specialized type of spectrophotometer used to ◄ detect and quantitate the bands produced from electrophoretic or chromatographic separation (see Chapter 3). Densitometers have a motor-driven stage to hold the electrophoretic plate that contains the stained, separated bands. The stage moves passing the electrophoretic patterns through the light beam and the relative absorbances of the bands are measured. Quantitation is achieved by integrating the area under each peak corresponding to the individual bands; peak area is proportional to the sample concentration of the band measured.

E. TURBIDIMETRY

Turbidimetric analyses are done in order to determine the amount of particulate matter in a solution. When light strikes a particle in solution, light may be absorbed, reflected, scattered, or transmitted. Turbidimetry is based on the principle that particles in solution cause a decrease in transmitted light (i.e., incident light is "blocked" by particles in solution). As the number of particles increases, there is proportionately less transmittance of the incident light through the solution. The components of a **turbidimeter** are identical to those of a spectrophotometer; in fact, turbidimetric measurements are typically made on a spectrophotometer. In contrast to spectrophotometry, wavelength selection and bandpass are not critical in turbidimetry since the decrease in light reaching the detector is independent of the wavelength.

F. NEPHELOMETRY

Nephelometry is similar to turbidimetry except that, in nephelometry, light scattered by particles in solution is measured at an angle to the beam of incident light. **Nephelometers** are designed so the detector is at an angle to the incident light passing through the sample. The angle at which the detector is placed often depends on the size and composition of the light-scattering particles. Typically, the detector in a nephelometer is placed at a right angle (90°) to the incident light; however, when analyzing larger particles, such as immune complexes, an angle of 30° may be used to increase sensitivity (Figure 11–5). In nephelometry, both monochromators are set for the same wavelength to measure scattered light.

G. SCINTILLATION COUNTING

Radioisotopes emit energy in the form of particulate or electromagnetic radiation and may be used in assay systems to label or tag components of interest for measurement. *Gamma (γ)-radiation* and *beta (β)-particle* emissions are detected and measured by **scintillation counters.** (Photomultiplier tubes in spectrophotometers are un-

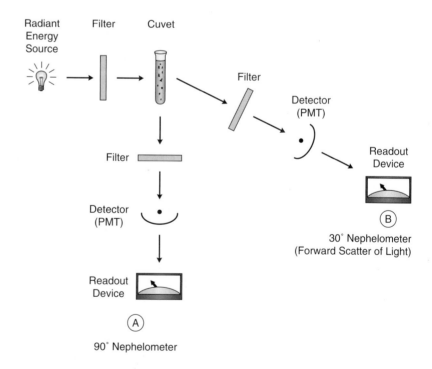

Figure 11–5. Common configurations of a nephelometer. **A.** Detector is placed at a 90° angle from the incident light. **B.** Detector is placed at a 30° angle from the incident light. This configuration measures forward light scatter and is used for larger complexes.

able to directly measure this form of radiant energy.) In scintillation counting, the radiation emitted is converted to light energy. For γ-radiation, this is accomplished using a NaI crystal containing a thallium activator; β-emitters to be measured are mixed into a liquid scintillation cocktail solution. In a crystal scintillation counter, the γ-radiation, a high-energy, short-wavelength form of EM radiation, is converted to flashes of light, which are then detected by a photomultiplier tube. The most commonly used isotope in clinical assays that emit γ-radiation is ^{125}I. The less penetrating β-radiation interacts with the organic molecules in the liquid solution to produce light energy that is measured by a photomultiplier tube. The most commonly used β-emitting isotope used in the laboratory is tritium (^3H). For both γ- and β-emitters, the number of light scintillations produced reflects the number of radioactive decays, and thus the concentration of the radioactive emitter in the sample. A block diagram of a crystal scintillation counter is provided in Figure 11–6.

H. CHEMILUMINESCENCE

Chemiluminescence is the emission of light as a product of a chemical reaction. When certain organic compounds react with oxidizing agents in the presence of an appropriate catalyst, the initial product is in an excited energy state; light is emitted as the product returns to a lower energy level. The light produced by a chemiluminescent reac-

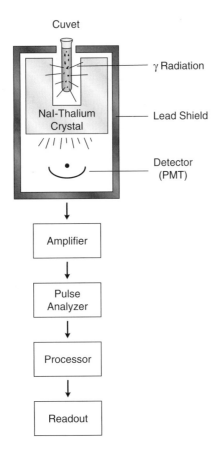

Figure 11–6. Instrument components of a crystal scintillation counter for γ-radiation.

tion can be measured to quantitate one of the reaction components. Often, the enzyme that catalyzes the reaction is used as a label or tag that can be quantitated by chemiluminescence. Enzyme labels are highly sensitive since one enzyme molecule can convert many molecules of substrate.

Examples of chemiluminescent reagent systems include:

- Luminol as substrate and H_2O_2 as oxidizing agent with peroxidase as catalyst
- Adamantyl 1,2-dioxetane aryl phosphate as substrate with alkaline phosphatase as catalyst
- Acridinium ester as substrate and alkaline H_2O_2 as the oxidizing agent

▶ A **luminometer** is the instrument that measures the flashes of light produced by chemiluminescence. A cuvet containing the component to be measured is placed in a light-sealed chamber of the luminometer, and reagents are added to initiate the oxidation reaction. As in a crystal scintillation counter, a photomultiplier tube is used to detect the light flashes produced. The detector then converts the light energy to an electrical signal proportional to the amount of luminescence produced.

Diagnostics manufacturers have designed ultrasensitive, rapid immunoassays using chemiluminescent labels. Refer to Chapter 12 for a more detailed description of these assays.

I. FLUOROMETRY

When molecules absorb light they reach a higher energy level or an excited state. Most molecules lose this excess energy as heat to return to ground state; however, certain light-absorbing molecules will release only a portion of the energy as heat and the remainder by ▶ emitting a photon of light—a phenomenon known as **fluorescence.** Typically, molecules capable of releasing fluorescent energy have a cyclic or ringed structure and/or multiple, conjugated double bonds. The light emitted by a fluorescent compound necessarily has a lower energy (i.e., longer wavelength) than the excitation (i.e., absorbed) light. Each fluorescent compound has characteristic excitation wavelengths (often in the UV region) and emission wavelengths. There is a time delay of about 10^{-8} to 10^{-7} seconds between absorption and emission. For dilute solutions of a fluorophor, the intensity of the fluorescence produced is directly proportional to the concentration of the fluorophor in solution and the intensity of the excitation source, a relationship derived from the Beer–Lambert law.

The fluorescence may also be affected by a number of variables, ▶ including: (1) **quenching** (e.g., the loss of fluorescence due to transfer of energy to the solvent or other solutes), (2) background fluorescence of components (e.g., cuvet material or sample matrix), (3) temperature (fluorescence intensity decreases with increased temperature), and (4) pH. Fluorescence measurements depend on instrument performance and vary between instruments; therefore, fluorescence is expressed in relative intensity units.

Fluorescence is measured using *fluorometers* or *spectrofluorometers* (Figure 11-7), instruments similar to absorption photometers. High-intensity radiation, most often in the UV range, is provided by the excitation source, usually a mercury vapor, halogen, or xenon arc

---

Factors Affecting Fluorescence

- Quenching
- Background fluorescence
- Temperature
- pH

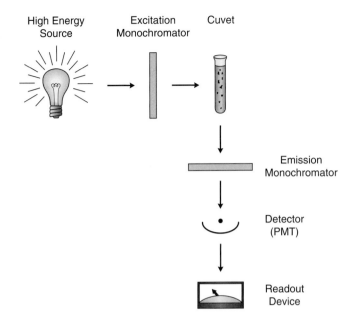

Figure 11–7. Instrument components of a fluorometer. The excitation monochromator is set for a shorter wavelength (higher energy) than the emission monochromator.

lamp. The source light is directed through an excitation monochromator to isolate the excitation wavelength. The excitation beam passes through the sample cuvet, interacting with molecules in the sample solution. A secondary monochromator is usually placed at a 90° angle to the excitation beam, minimizing interference between the excitation light path and emission light path. Light emitted due to fluorescence passes through the secondary monochromator to isolate the emission wavelength (i.e., blocking the scattered light from the excitation source). The emitted energy is detected by a photomultiplier tube connected to a readout device. Fluorometers and spectrofluorometers differ in the types of monochromators used: fluorometers use interference filters or glass filters; spectrofluorometers use gratings or prisms.

Fluorometric measurements are very sensitive and specific. Specificity is derived from the ability to select both the absorbance and emission wavelengths that are characteristic of the fluorophor being measured. Sensitivity may be controlled by altering the intensity of the energy source, thus increasing or decreasing the amount of fluorescence produced. In the clinical laboratory, fluorescence assays are typically used to measure fluorescent tags or labels rather than naturally occurring fluorescent molecules. A description of some of these applications in immunoassay systems may be found in Chapter 12. Still other applications are found in hematofluorometry and flow cytometry.

J. ATOMIC ABSORPTION SPECTROPHOTOMETRY

Free atoms in the ground state absorb light of defined wavelengths (line spectra) characteristic of that element. Atoms in the excited state will emit light of the same wavelengths when returning to ground state. **Atomic absorption spectrophotometry** (AAS) mea- ◀

sures the absorption of a defined wavelength of light by ground-state atoms in a flame. The relationship between absorption and concentration follows Beer's law; thus, the amount of energy absorbed is proportional to the concentration of the element in the flame.

Figure 11–8 is a diagram of the components of an atomic absorption spectrophotometer. Diluted samples are aspirated into an air–acetylene flame. The flame is the atomizer, placing the atoms in their ground state and thereby capable of absorbing energy, and also serves as the cuvet. The energy source utilized in AAS is the *hollow cathode lamp* (HCL). The HCL is lined with the element being measured and generates the line spectrum characteristic of that element, providing monochromatic light of the exact wavelengths needed. As the incident beam passes through the flame, the ground-state atoms in the sample absorb the radiation. A monochromator is located after the flame to remove background energy (from the flame) before the transmitted light reaches the detector. A photomultiplier tube detects the remaining transmitted energy. Most AAS systems are double-beam instruments that use a chopper to send the incident light from the HCL alternately through a reference and a sample pathway. The intensities of the reference and sample beams are compared to generate the readout signal.

Flameless atomic absorption uses a "graphite furnace" to atomize the sample; this technique is more sensitive than the flame methods.

Interferences of several types are potential sources of error in AAS, including chemical, ionization, and matrix interferences. Chemical interference occurs when the cations of interest form complexes with anions that cannot be dissociated by the flame. For example, phosphates present in body fluids form heat-stable complexes with calcium or magnesium cations, preventing their atomization and reducing absorption by the calcium or magnesium in the sample. This interference is reduced or eliminated by incorporating another metal, such as lanthanum or strontium, into the diluent; lanthanum and strontium form tighter bonds with phosphate anions than calcium or magnesium. Ionization interference occurs when metal atoms in the sample become ionized (as opposed to remaining in the ground state) and are unable to absorb the energy from the HCL. This type of interference most often occurs because the flame is too hot; lowering the flame temperature by adjusting the gas mixture prevents significant ionization. Matrix interferences occur when standards and

Types of Interferences in AAS

Chemical—occurs when the cations of interest form complexes with anions that are not dissociated by the flame

Ionization—occurs when metal atoms in the sample become ionized and are unable to absorb the energy from the hollow cathode lamp

Matrix—occurs when standards and samples have dissimilar properties

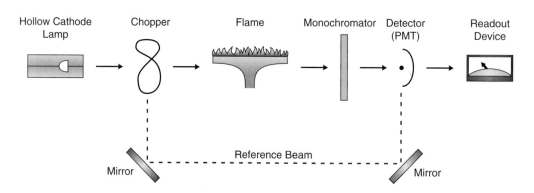

Figure 11–8. Instrument components of an atomic absorption spectrophotometer.

samples have dissimilar properties, such as protein or salt concentration, solvent composition, and viscosity. Dilution of clinical samples for analysis helps to minimize matrix interferences.

Atomic absorption spectrophotometry is accurate and sensitive; it serves as the reference method for some analytes. Though now considered to be a more specialized technique, AAS is used to measure metals, including iron, lead, calcium, magnesium, zinc, and copper, in body fluids. AAS is most often performed in toxicology laboratories. The instrument cost, along with the technical expertise and time required for operation, limit the use of AAS in most clinical laboratories.

III. ELECTROCHEMICAL MEASUREMENTS

Quantitative determinations based on electroanalytical measurements are common in clinical chemistry, second only to photometric determinations. Of the electrochemical methods, potentiometric determinations are the most commonly used in the clinical laboratory. Electrochemical methods also include electrolytic techniques such as amperometry, coulometry, and voltammetry. In electrochemical methods, measurement of electrical potential, current, or resistance is used to quantify clinical analytes.

A. *POTENTIOMETRY AND ION-SELECTIVE ELECTRODES*

Potentiometry is the measurement of the potential difference between two electrodes under equilibrium conditions (i.e., zero current flow); the potential difference is a function of the ion activity (i.e., the effective concentration of the analyte) in the electrochemical cell. A potentiometer requires two electrode components, a *reference electrode* and an *indicator electrode,* each housing a particular half-cell reaction. The reference electrode has a defined half-cell potential, while the half-cell potential of the indicator electrode varies in response to the analyte activity in the sample solution. The difference in potential between the two electrodes is measured by a voltmeter, and the cell potential is related to analyte activity using the *Nernst equation*. The system is calibrated using standard solutions of the analyte.

The **reference electrode,** against which all other electrodes are measured, is the standard hydrogen electrode, having an assigned half-cell potential of 0.000 V at all temperatures. Though it serves as the primary reference electrode, the hydrogen electrode is impractical for general use. In practice, secondary reference electrodes are employed, such as the silver–silver chloride electrode and the saturated calomel electrode (i.e., Hg/Hg_2Cl_2 electrode). Both of these electrodes have a defined reference potential determined by comparison to the standard hydrogen electrode.

The value of potentiometry in the clinical laboratory is the result of the development of **ion-selective electrodes** (ISEs) for important analytes, including sodium, potassium, chloride, bicarbonate, calcium (free), and carbon dioxide. ISEs are indicator electrodes having an analyte-selective membrane; the magnitude of the potential that develops across the selective membrane is a function of analyte activity. Ideally, ISEs would be sensitive only to the ion of interest; however,

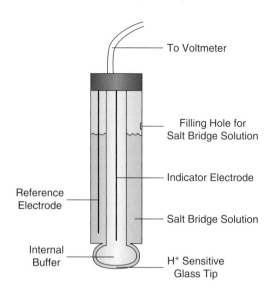

Figure 11–9. Basic configuration of a pH electrode.

in practice, a low degree of interaction with other ions is often present, hence the electrodes demonstrate *selectivity* but are not *specific* for a given ion. The membrane, which may be liquid (water-insoluble), glass, or solid state, determines the selectivity of the electrode. For example, an electrode with a glass membrane of particular composition is used to measure hydrogen ion concentration in a pH meter; the glass pH electrode is shown in Figure 11–9. A membrane incorporating the antibiotic valinomycin is used to select for potassium ions in a potassium ISE; other ionophores have been utilized in developing selective membranes for other ions of interest. CO_2 is measured using a gas-sensing electrode that is a combination of a pH electrode shielded from the sample by a CO_2-permeable membrane; as CO_2 enters the buffer surrounding the pH electrode, the pH changes in relation to the amount of CO_2 in the sample.

In clinical potentiometric systems, proper maintenance and calibration are crucial for reliable measurements. Routine use of electrode cleansing agents to remove protein buildup on the electrode surface is necessary with repeated analysis of biological samples.

B. AMPEROMETRY

▶ **Amperometry**[1] is a modification of voltammetry. In contrast to potentiometry, in which current is maintained at zero and voltage is measured, in amperometry the voltage is held constant and the current is measured; the current is proportional to the concentration of the analyte. The most common clinical application utilizing amperometry is the oxygen electrode (also called the *Clark electrode*) for the determination of the partial pressure of oxygen (PO_2) in the blood. In the O_2 electrode, the indicator electrode is a platinum (or gold) cathode covered with a gas-permeable membrane, while the reference electrode is usually a silver–silver chloride electrode. A voltage sufficient to promote reduction of O_2 at the cathode is applied between the two electrodes. Oxygen from the sample diffuses across the gas-permeable membrane, enters a buffered electrolyte

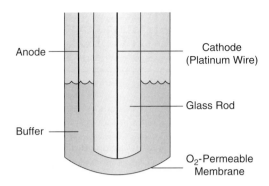

Figure 11–10. Basic configuration of the oxygen electrode for the amperometric determination of PO_2.

solution, and is reduced at the platinum cathode. The current, produced as electrons flow from the anode to the cathode, is proportional to the O_2 present (four electrons are necessary to reduce one molecule of O_2). A diagram of the oxygen electrode is shown in Figure 11–10.

As with potentiometric systems, maintenance of amperometric electrodes also includes periodic application of an electrode cleansing agent to remove protein buildup on the electrode surface.

---

[1] The term *amperometry* is sometimes used interchangeably with polarography, although the term *polarography* is generally reserved for a special type of voltammetry using a dropping-mercury electrode.

SUGGESTED READINGS

Burtis CA, Ashwood ER (eds.). *Tietz fundamentals of clinical chemistry,* 5th ed. Philadelphia: WB Saunders, 2000.

Burtis CA, Ashwood ER (eds.). *Tietz textbook of clinical chemistry,* 3rd ed. Philadelphia: WB Saunders, 1999.

Davis JS. Absorption spectroscopy. In Wandersee JH, Wissing DR, Lange CT (eds.). *Bioinstrumentation tools for understanding life.* Reston, VA: National Association of Biology Teachers, 1996.

Kaplan LA, Pesce AJ. *Clinical chemistry: Theory, analysis, and correlation.* St. Louis: CV Mosby, 1996.

Karselis TC. *The pocket guide to clinical laboratory instrumentation.* Philadelphia: FA Davis, 1994.

Items for Further Consideration

1. Why are most spectrophotometric methods based on the measurement of absorbance as opposed to transmittance?

2. In spectrophotometric determinations, what effect would be observed when measuring a sharp, narrow absorption peak in an instrument with a broad spectral bandpass?

3. A laboratory wishes to implement an immunoassay method for a hormone that is in very low concentration in the blood, even when elevated. The following choice of immunoassay systems is available for this measurement: radioactive, fluorescence, chemiluminescence. Which one would you pick and why?

4. Speculate on why potentiometric methods have replaced the old flame photometry methods for the measurement of electrolytes in body fluids?

Immunoassay

Patrick J. Cummings

Key Terms ◀

AFFINITY

ANTIBODY

ANTIBODY SPECIFICITY

ANTIGEN

ANTIGEN–ANTIBODY COMPLEX

AVIDITY

COMPETITIVE IMMUNOASSAY

DIRECT (NONLABELED) IMMUNOASSAYS

EPITOPE

HAPTEN

HETEROGENEOUS IMMUNOASSAYS

HIGH-DOSE HOOK EFFECT

HOMOGENEOUS IMMUNOASSAYS

IMMUNOGEN

INDIRECT (LABELED) IMMUNOASSAYS

LOG-LOGIT TRANSFORMATION

MONOCLONAL ANTIBODIES

NONCOMPETITIVE IMMUNOASSAYS

POLYCLONAL ANTIBODIES

SANDWICH ASSAYS

Chapter Objectives

Upon completion of this chapter, the reader will be able to:

1. Compare and contrast the terms in the following pairs: polyclonal versus monoclonal antibodies, affinity versus avidity, competitive versus noncompetitive immunoassays, and heterogeneous versus homogeneous immunoassays.
2. Define *antibody specificity.*
3. Identify the types of labels used in radioactive, enzymatic, fluorometric, and chemiluminescent immunoassays.
4. Describe the types of assays in which the "high-dose hook effect" occurs, and corrective steps that will eliminate this source of error.
5. Explain the principles of antigen–antibody interactions, and the role of the law of mass action in antigen–antibody binding.
6. Predict the dose–response curves for the various types of immunoassays.
7. Discuss the various types of separation techniques used in heterogeneous immunoassays.
8. Discuss the general principles of each of the following immunoassays: radial immunodiffusion (RID), radioimmunoassay (RIA), immunoradiometric assay (IRMA), enzyme-linked immunosorbent assay (ELISA), enzyme-multiplied immunoassay technique (EMIT), fluorescence polarization immunoassay (FPIA), microparticle enzyme immunoassay (MEIA), and chemiluminescence.

Instruments/Techniques Included ◀

CHEMILUMINESCENCE	*FPIA*	*MEIA*
ELISA	*IRMA*	*RIA*
EMIT	*NONLABELED AND LABELED IMMUNOASSAYS*	*RID*

Pretest

Directions: Choose the single best answer for each of the following items.

1. The degree to which an antibody binds to its corresponding antigen while not reacting with similar antigens is known as
   a. avidity
   b. sensitivity
   c. specificity
   d. affinity

2. The total strength and stability of an antibody–antigen complex is known as
   a. affinity
   b. avidity
   c. sensitivity
   d. specificity

3. The "high-dose hook effect" seen in immunometric assays for antigen can be eliminated by
   a. diluting the reagent and retesting sample
   b. diluting the sample and assaying the diluted sample
   c. reducing the substrate concentration
   d. decreasing the reaction time

4. A competitive homogeneous immunoassay
   a. requires a separation step and will produce a linear relationship between the signal and analyte concentration
   b. does not require a separation step and will produce an inverse relationship between the signal and analyte concentration
   c. requires a separation step and will produce an inverse relationship between the signal and analyte concentration
   d. does not require a separation step and will produce a linear relationship between the signal and analyte concentration

5. A noncompetitive heterogeneous immunoassay uses

   a. excess reagent and requires a separation step
   b. limited reagent and does not require a separation step
   c. excess reagent and does not require a separation step
   d. limited reagent and requires a separation step

6. The immunoassay that is based on the ability of an antibody to modulate the amount of light emitted from a fluorophore–ligand complex is
   a. immunoradiometric assay (IRMA)
   b. enzyme-linked immunosorbent assay (ELISA)
   c. fluorescence polarization immunoassay (FPIA)
   d. chemiluminescence

7. The ability of immune complexes to be captured and then detected on the surface of a glass fiber matrix is the basis for which of the following immunoassays?
   a. chemiluminescence
   b. FPIA
   c. microparticle enzyme immunoassay (MEIA)
   d. enzyme-multiplied immunoassay technique (EMIT)

8. A low-molecular-weight molecule that is unable to induce an immune response unless it is linked to carrier protein is a(n)
   a. antigen
   b. immunogen
   c. conjugate
   d. hapten

9. An immunoassay based on the direct observation of antigen–antibody reaction within a gel matrix is
   a. MEIA
   b. FPIA
   c. radial immunodiffusion (RID)
   d. ELISA

10. A sandwich ELISA designed to quantitate antigen would result in the following complex (Ab = antibody; Ag = antigen; Ab-enz = antibody conjugated to enzyme; Ag-enz = antigen conjugated to enzyme)
    a. AbAgAg-enz
    b. AgAbAg-enz
    c. AbAgAb-enz
    d. AgAgAb-enz

11. Because of high affinity and specificity for antigen, the isotype most commonly used in immunoassays is immunoglobulin
    a. M
    b. E
    c. A
    d. G

12. The portion of an antibody molecule that determines the binding specificity is determined by the
    a. variable heavy-chain domain
    b. Fc region
    c. variable heavy- plus variable light-chain domains
    d. constant heavy- plus variable light-chain domains

13. The structural region on an antigen that functions as a combining site for an antibody is a(n)
    a. hapten
    b. epitope
    c. parotope
    d. immunogen

14. A measure of the strength of bonding between a single antibody combining site and a single antigenic determinant is
    a. affinity
    b. sensitivity
    c. avidity
    d. specificity

15. The analytical sensitivity of a noncompetitive immunoassay is determined principally by the
    a. affinity of the antibody
    b. lower limit of detection of the label used
    c. cross-reactivity
    d. shelf life of reagents

I. INTRODUCTION

Immunoassays have revolutionized clinical diagnosis since their introduction in the 1970s. Currently, there are more than 600 Food and Drug Administration (FDA)-approved antigen- or antibody-based immunoassays available for infectious disease diagnosis, tumor marker detection, quantitation of specific proteins, and drug monitoring. Immunoassays are procedures that use antibodies for detection and quantitation of analytes. The basis of immunoassays is the specific binding between an **antigen** (or analyte) and a corresponding **antibody,** with subsequent **antigen–antibody complex** formation that allows for quantitation of one of the reactants. Antigen–antibody complexes are detected directly by assaying for immune complex formation (direct immunoassays) or indirectly by detecting labels that have been coupled to one of the reactants (indirect immunoassays). The types of labels that are used in indirect immunoassays include radioisotopes, enzymes, fluorophors, and luminescent substances. Detection of the label allows for identification and quantitation of analytes by various detection systems.

Types of Labels Used in Indirect Immunoassays

- Radioisotopes
- Enzymes
- Fluorophors
- Luminescent substances

II. ANTIBODIES AND ANTIGENS

Antibodies are proteins produced by B lymphocytes in response to exposure to foreign antigens (see Chapter 3). The function of an antibody *in vivo* is to combine with a specific antigen and to facilitate its elimination from the body. The basic unit of an immunoglobulin G (IgG) antibody molecule consists of four polypeptide chains, two heavy chains and two light chains, linked by disulfide bonds to give a "Y"-shaped molecule. The antibody molecule consists of variable and constant regions. The variable heavy- and variable light-chain regions, which contain two binding sites, are responsible for the antibody's binding specificity. The carboxyl-terminal constant region is involved in effector functions, such as complement activation and binding to Fc receptor–bearing leukocytes. The antibody's variable region amino acid sequence is different for each unique antibody and defines the

▶ binding specificity or idiotope. **Antibody specificity** is the degree to which an antibody binds to its corresponding antigen while not reacting with similar substances. Antibodies used in immunoassays must have a high specificity. In contrast, antibodies that react with more than one antigen are referred to as cross-reactive antibodies. These antibodies are not useful for most immunoassay applications. The antibody's constant region determines the immunoglobulin class (isotype) as an IgM, IgG, IgA, IgD, or IgE. Because of high affinity and specificity for antigens, IgG is the most common isotype used in immunoassays.

A foreign substance that is able to induce an immune response
▶ in a host animal by the production of antibodies is known as an **immunogen.** High-molecular-weight foreign proteins are usually the best immunogens. An antigen is a substance that is able to react with an antibody, without necessarily being able to induce antibody formation *in vivo*. The structural region on an antigen that functions as a combining site for an antibody is known as an antigenic determi-
▶ nant or **epitope.** A low-molecular-weight molecule that is unable to induce antibody production, unless it is linked to a carrier protein, is
▶ termed a **hapten** (< 20 kilodaltons). Antihapten antibodies are used to bind and detect free hapten in immunoassays. Haptens that are measured by immunoassays include various drugs, nonpeptide hormones, and vitamins.

Antibodies used in immunoassays can be polyclonal or mono-
▶ clonal. Production of **polyclonal antibodies** involves immunization of an animal host, usually goat or rabbit, with the purified immunogen. Following demonstration of an immune response, serum is collected and the immunoglobulin fraction is purified and used as a specific binding reagent in the assay. Polyclonal serum (antiserum) contains a mixture of antibodies with different specificities to multi-
▶ ple epitopes on the immunogen. In contrast, **monoclonal antibodies** are produced by immunizing mice with purified antigen, followed by *in vitro* fusion of antibody-producing splenic B cells with murine myeloma (B lymphocyte cancer cell line) cells to produce a hybridoma that secretes antibody (monoclonal antibody) to a single epitope on the immunogen. Depending on the design and format of a particular immunoassay, polyclonal or monoclonal antibodies provide the binding specificity required for detection of the analyte.

III. ANTIBODY–ANTIGEN REACTIONS

The noncovalent binding forces between an antibody and an antigen involve electrostatic, van der Waals, hydrogen, and hydrophobic bonds. A measure of the strength of chemical bonding between a single antibody combining site and a single antigenic determinant is known as **affinity.** Because the binding forces between an antigen and an antibody are reversible, the binding affinity is expressed in terms of an equilibrium association constant K_a. According to the law of mass action:

$$K_a = \text{[Ab-Ag complex]}/\text{[Ab][Ag]}$$

where [Ab] is the concentration of unbound antibody at equilibrium (equilibrium concentration), [Ag] is the equilibrium concentration of unbound antigen, and [Ab-Ag] is the equilibrium concentration of the antigen–antibody complex. K_a is defined in reciprocal molar concentrations (M^{-1}) or liters per mole (L/mol), with large K_a values indicating greater affinity of the antibody for an antigen. Because a monoclonal antibody consists of a homogeneous population of binding sites, it will have a uniform binding affinity for the antigen. The monoclonal antibodies that are suitable for most immunoassays have a K_a between 10^8 and 10^{11} L/mol. In contrast, polyclonal antiserum contains a mixture of antibodies with different affinities and binding specificies for the antigen. The summation of the antibody–antigen interactions of a multivalent antibody with a multiepitope antigen is known as **avidity.** Antibody avidity represents the total strength and stability of an antibody–antigen complex, and is often used to measure the binding strength of polyclonal antiserum.

IV. DIRECT IMMUNOASSAY MEASUREMENT

Immunoassays that detect antigen–antibody complex formation directly without labels are known as **direct** or **nonlabeled immunoassays.** One of the most common direct immunoassays is radial immunodiffusion (RID). RID is an immunoassay based on the direct observation of antigen–antibody reaction in the form of a visible precipitate within a gel matrix. RID is a passive diffusion method in which antigen is allowed to diffuse from a well into a gel matrix containing the specific antibody. Antibody–antigen interaction is detected by a ring of precipitation around the well. Calibrators are assayed in parallel with patient samples to allow construction of a calibration or standard curve. In the kinetic method (Fahey–McKelvey method), the concentration varies with the precipitant ring diameter squared (Figure 12–1). The major advantages of RID are its simplicity, accuracy, and adaptability for measuring a variety of proteins in serum, cerebrospinal fluid, and other fluids. However, because of the slow turnaround time, RID methods have been mostly replaced by more rapid and sensitive indirect immunoassays in clinical laboratories.

V. INDIRECT IMMUNOASSAY MEASUREMENT

Assays that detect antibody–antigen complex formation indirectly by detecting labels that have been coupled to one of the reactants are known as **indirect** or **labeled immunoassays.** The types of chemical markers that can be used to label antigens, haptens, or antibodies include radioisotopes, enzymes, fluorophors, and luminescent sub-

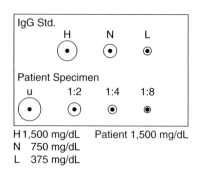

H 1,500 mg/dL   Patient 1,500 mg/dL
N   750 mg/dL
L   375 mg/dL

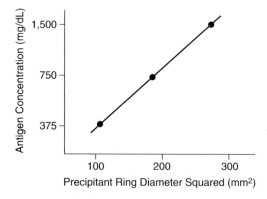

Figure 12–1. Principle of radial immunodiffusion (RID) and a typical dose–response curve.

stances. In general, labeled immunoassays are classified as competitive or noncompetitive based on the concentration of reagents and ▶ the kinetics of the antigen–antibody interactions. In the **competitive** ▶ **immunoassay** format (limited reagent assays), unlabeled test antigen (Ag) competes with a limited quantity of labeled antigen (Ag*) for a limited number of antibody (Ab) binding sites (Ab + Ag + Ag* $\rightleftharpoons$ AbAg + AbAg*).

$$
\begin{array}{ccc}
\text{Ag} & & \nearrow \text{AbAg}^* \\
 & + \quad \text{Ab} & \\
 & \quad \text{(limited)} & \searrow \\
\text{Ag}^* & & \text{AbAg} \\
\text{(limited)} & &
\end{array}
$$

Using these conditions, the probability of antibody binding the labeled antigen is inversely proportional to the concentration of unlabeled antigen present in the patient's sample. Therefore, the amount of bound label (AbAg*) is inversely proportional to the concentration of antigen in the patient specimen, and the *dose–response curve* will have a negative slope. However, if the unbound labeled antigen (Ag*) is measured the dose–response curve will have a positive slope (Figure 12–2A). Because monoclonal antibodies have uniform binding affinity and specificity for antigen, they are used in most competitive binding assays. The overall analytical sensitivity or detection limit of a competitive immunoassay is dependent on both the affinity of the antibody for the antigen and the detection level of the label.

▶   **Noncompetitive immunoassays** or **sandwich assays** for antigen are excess reagent assays, in which the test antigen is allowed to

A

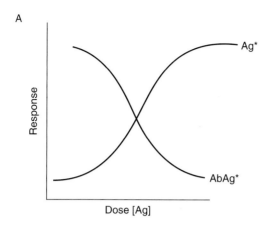

B

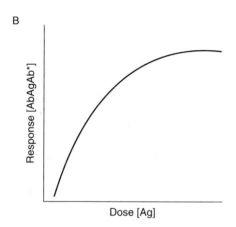

Figure 12–2. **A.** Dose–response curve for a competitive immunoassay where free (Ag*) or bound (AbAg*) label is detected. **B.** Dose–response curve of a noncompetitive immunoassay.

react with an excess of *capture antibody* (Ab) that is attached to a solid support. A labeled antibody (*conjugate,* Ab*) is added that reacts with a different epitope on the bound antigen, forming a "sandwich" complex (AbAgAb*). These assays may be either homogeneous or heterogeneous. **Homogeneous immunoassays** do not require a separation step before determining the extent of antigen–antibody binding. In **heterogeneous immunoassays,** a step is required to separate unbound (free) antigen or *ligand* from bound ligand. Common methods used to separate free from bound ligands in heterogeneous assays include adsorption using charcoal or florisil; precipitation using polyethylene glycol, ammonium sulfate, or secondary antibody; and solid-phase attachment using antibodies, protein A, or avidin–biotin.

The immunological reactions in noncompetitive assays occur in two separate steps. These types of tests are also referred to as *sequential immunometric assays* (see below).

Step 1	⊢Ab	+ Ag	→	⊢AbAg
	(excess)			
Step 2	⊢AbAg	+ Ab*	→	⊢AbAgAb*

Common Methods Used to Separate Free from Bound Ligands in Heterogeneous Assays

- Adsorption using charcoal or florisil
- Precipitation using polyethylene glycol, ammonium sulfate, or secondary antibody
- Solid-phase attachment using antibodies, protein A, or avidin–biotin

Table 12–1. Summary of Immunoassays

TECHNIQUE	END POINT	ASSAY SENSITIVITY	COMMON ANALYTES
RID	Precipitation	50 µg/mL	Serum and cerebrospinal fluid proteins
RIA	Radioisotope counts	10^{-12}–10^{-14} M	Drugs and hormones
IRMA	Radioisotope counts	10^{-11} M	Serum proteins, hormones, cancer antigens
ELISA	Color production	10^{-11}–10^{-14} M	Serum proteins, infectious disease antigens
EMIT	Color production	10^{-10} M	Therapeutic drugs and drugs of abuse
FPIA	Polarized light production	10^{-9} M	Therapeutic drugs and drugs of abuse
MEIA	Fluorescent product	?	Hormones and tumor markers
Chemiluminescence	Light production	?	Hormones and tumor markers

RID, radial immunodiffusion; RIA, radioimmunoassay; IRMA, immunoradiometric assay; ELISA, enzyme-linked immunosorbent assay; EMIT, enzyme multiplied immunoassay technique; FPIA, fluorescence polarization immunoassay; MEIA, microparticle enzyme immunoassay.

Because the amount of bound conjugate is directly proportional to the concentration of test antigen, the dose–response curve will have a positive slope (Figure 12–2B). The analytical sensitivity of a noncompetitive immunoassay is dependent primarily on the detection level of the label.

In immunoassays, antigen–antibody complexes are detected directly by assaying for immune complex formation by precipitation reactions, or indirectly by detecting labels such as radioisotopes, fluorophors, luminescent substrates, or products of enzymatic reactions. The primary factor that determines assay format and method of immune complex detection is the analyte concentration. Because the concentrations of different analytes in biological specimens vary tremendously in both normal and diseased states, various assays are designed to have different analytical sensitivities or lower limits of detection (Table 12–1). Additional factors that are considered in designing immunoassays include cost, test volume, turnaround time, and technical expertise of the laboratory personnel.

VI. RADIOACTIVE-LABELED IMMUNOASSAY SYSTEMS

A. RADIOIMMUNOASSAY (RIA)

Radioimmunoassays (RIAs) are competitive, heterogeneous assays that use radioactive isotopes to detect antigen–antibody reactions. The most common radioisotope used in RIA procedures is ^{125}I. The reaction involves a fixed amount of radiolabeled antigen and unlabeled test antigen competing for a limited number of antibody-binding sites. The antibody is attached to a solid support such as plastic tubes or polystyrene beads, which aids in separation of free from bound antigen by a washing step. The radiolabeled antigen–antibody complex is quantitated using a gamma counter or crystal scintillation counter. Because this is a competitive assay, the amount of radiolabeled bound antigen (ligand) is inversely proportional to the unlabeled test antigen concentration in the sample. The data obtained from a series of standards are used to construct a dose–response curve, which is used to determine test antigen concentration. The standards consist of different concentrations of unlabeled antigen mixed with a standard quantity of labeled antigen. The per-

A

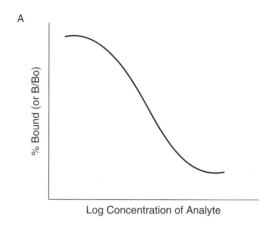

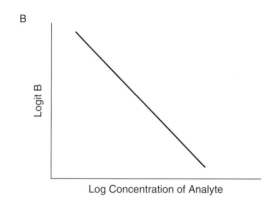

Figure 12–3. Dose–response curves for competitive immunoassays. **A.** Sigmoidal curve, % Bound versus concentration. **B.** Linearized curve, logit B versus concentration.

centage of radiolabeled antigen bound to antibody, expressed as a percentage of the total label in each tube (%B or B/Bo), is plotted on the *y-axis,* and the log concentration of the unlabeled antigen contained in each standard is plotted on the *x-axis* to create a sigmoid curve (Figure 12–3A). To linearize competitive dose–response data, to give additional weight to the middle standards, and to eliminate errors at the ends of the sigmoid dose response curve, the data can be linearized by **log-logit transformation** using the following formula:

$$\text{Logit (B/Bo)} = \ln \text{(B/Bo)/(1-B/Bo)}$$

Logit data conversion is usually performed by computer programs used in most automated assay systems. The logit of the fraction of bound radioactivity (B/Bo) plotted on the *y-axis* and the log of the concentration of the unlabeled antigen in each standard plotted on the *x-axis* generates a linear dose–response curve (Figure 12–3B). By extrapolation, the concentration of the unknown test antigen is determined.

RIA is a very sensitive method that can detect picomolar concentrations of antigen, and can be fully automated for rapid turnaround time. The major disadvantages of RIA involve the inherent danger of working with radioisotopes and the disposal and safety

◀ **Disadvantages of RIA**

- Danger of working with radioisotopes
- Disposal and safety precautions required
- Short shelf life of radiolabeled ligands
- Requirement for a gamma counter
- Tendency of some antigens or haptens to be altered immunologically by labeling

precautions required, the short shelf life of radiolabeled ligands, the requirement for a gamma counter, and the tendency of some antigens or haptens to be altered immunologically by labeling. Because of the numerous disadvantages of RIA, most laboratories have converted to other methodologies. RIA is typically used to quantitate hormones and plasma proteins and to determine therapeutic drug levels.

B. IMMUNORADIOMETRIC ASSAY (IRMA)

Immunoradiometric assays (IRMAs), or immunometric assays, are noncompetitive, heterogeneous assays that use radiolabeled antibody to determine concentration of antigen in biological fluids. IRMA is similar to RIA except that labeled antibody is used to detect antigen in a noncompetitive or sequential/sandwich format. Unlabeled specific antibody, referred to as primary or capture antibody, is adsorbed onto a solid support. The specimen containing the antigen is added to the solid support and allowed to bind to the primary antibody. Following washing to remove unbound antigen, a second labeled antibody (^{125}I) that recognizes a different epitope on the antigen is added to the reaction. Following a second wash step to remove unbound labeled antibody, bound antibody is determined by counting the sample in a scintillation or gamma counter. Using linear graph paper, a dose–response curve is generated by plotting the counts per minute (CPM) for each standard on the y axis and the concentration of each standard on the x axis. Because this is a noncompetitive immunoassay, the amount of bound labeled antibody is directly proportional to the unlabeled test antigen concentration. A smooth curve is drawn through the points, and the concentration of test sample is extrapolated by locating the sample CPM on the standard curve and reading the x axis for concentration (Figure 12–4A). In some specimens with high antigen concentration, there may be a decrease in the amount of radiolabeled antibody bound to the complex, which causes a decrease in the dose response at the higher concentrations (Figure 12–4B). This results in lower CPM values when the concentration of the analyte is actually very high. This misleading and dangerous phenomenon is known as the **high-dose hook effect,** and may be due to saturation of the capture antibody by the excess antigen. Assays must be examined for the hook effect during characterization. In susceptible concentration ranges, the high-dose hook effect can be corrected by diluting the sample and retesting; however, it is often more practical to assay the undiluted and diluted samples in parallel. Because the reactants in this assay are added at two different steps, these assays are referred to as *sequential immunometric assays.* The two antibodies used in immunometric-type assays are usually monoclonal antibodies directed toward different epitopes on the antigen or hapten.

IRMA has a fast turnaround time, has a lower limit of detection than RIA, and produces low background readings. The major disadvantage of IRMA is the problem associated with handling and disposal of radioisotopes, the high-dose hook effect, and the requirement for two monoclonal antibodies directed toward different epitopes on the antigen. Currently, IRMA is used to quantitate analytes such as ferritin, calcitonin, growth hormone, carcinoembryonic antigen, α-fetoprotein, and prostate antigens.

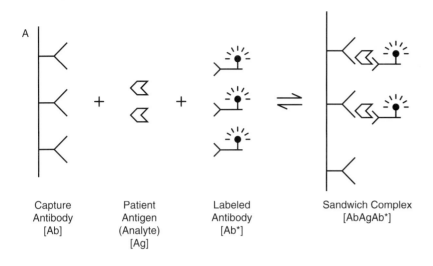

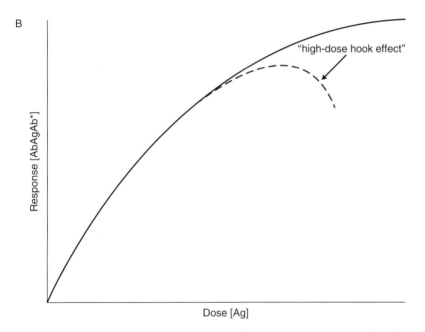

Figure 12–4. Noncompetitive immunoradiometric assay. **A.** Sandwich technique using labeled antibody. **B.** Typical dose–response curve including high-dose hook effect.

VII. ENZYME-LABELED IMMUNOASSAY SYSTEMS

A. *ENZYME-LINKED IMMUNOSORBENT ASSAY (ELISA)*

Enzyme-linked immunosorbent assays (ELISAs) can be competitive or noncompetitive. All ELISAs are heterogeneous assays that use the catalytic properties of enzymes to detect and quantitate antigen–antibody reactions. Depending on the type of immunoassay, enzymes are conjugated to antigen or antibody. Some enzymes used in ELISA include alkaline phosphatase, horseradish peroxidase, glucose-6-dehydrogenase, or β-galactosidase. Depending on the type of enzyme-substrate utilized, the end product may be detected by spectrophotometric or fluorometric methods.

A noncompetitive ELISA for antigen quantitation (sandwich ELISA) is similar to IRMA, in which a primary capture antibody is adsorbed to solid support, usually a microtiter plate. A specimen containing the antigen is added to the well, followed by incubation and wash steps. An enzyme-labeled antibody (Ab-E) is added to react with the bound antigen, forming an AbAgAb-E complex. Following washing to remove unbound labeled antibody, a specific substrate is added to the reaction, and the enzyme converts the substrate to a colored end product that can be measured spectrophotometrically. In the noncompetitive format, the amount of end product produced is directly proportional to the quantity of antigen in the sample; hence, an elevated absorbance reading indicates increased antigen concentration. A standard curve is generated by plotting the average absorbance value for antigen standards on the *y-axis* versus concentration on the *x-axis* to produce a straight line (Figure 12–5B).

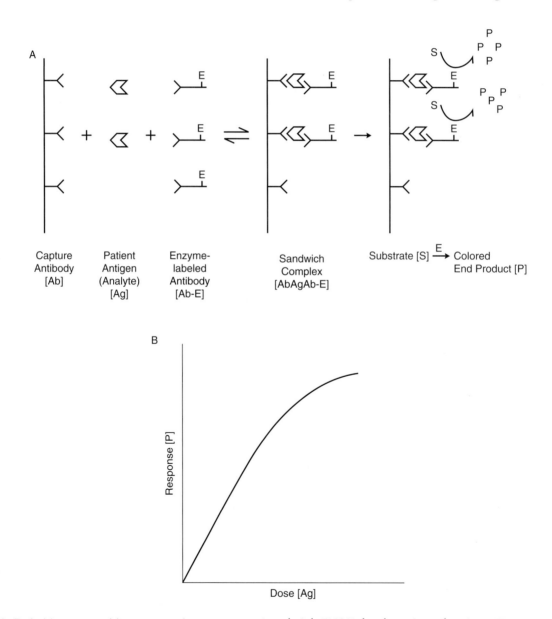

Figure 12–5. **A.** Noncompetitive enzyme immunoassay (sandwich ELISA) for detection of antigen (E = enzyme; S = substrate; P = product). **B.** Typical dose–response curve.

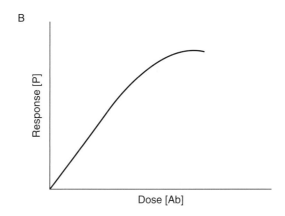

Figure 12–6. **A.** Indirect ELISA (noncompetitive enzyme immunoassay) for detection of antibody. **B.** Typical dose–response curve.

An ELISA for quantitation of antibodies in biological fluids is referred to as an *indirect ELISA*. Indirect ELISA may be used to determine the presence of specific antibody to infectious agents in serum. A specific antigen is adsorbed to a solid support, and diluted serum is added to the plate to allow AgAb complex formation. Following an incubation and wash step, an enzyme labeled antihuman antibody, referred to as the conjugate, is added to the solid support to form an AgAbAb-E complex (Figure 12–6). Following washing to remove unbound conjugate, the substrate is added to produce the end product. The amount of colored end product produced is directly proportional to the quantity of antibody in the sample.

ELISA is a simple, versatile, and sensitive assay, which uses enzyme labels that have a long shelf life and pose minimal risk to the technologist. The major disadvantages of ELISA are the many incubation and washing steps, which introduce potential sources of error.

B. ENZYME-MULTIPLIED IMMUNOASSAY TECHNIQUE (EMIT)

Enzyme-multiplied immunoassay technique (EMIT) is a homogeneous assay that is based on the competitive binding reaction between an enzyme conjugated hapten (E–hapten) and the test hapten, usually a drug or hormone, for a limited number of antibody-combining sites. Binding of antibody to the E–hapten results in inhibition of enzyme activity due to a conformational change, thus blocking the enzyme's active site and preventing its reaction with the substrate. Therefore, the amount of test hapten present in the sample determines the number of antibodies that will be available to bind to the E–hapten conjugate. The higher the test hapten concentration in the sample, the fewer free antibody–binding sites available to inhibit enzyme activity of the E–hapten conjugate, and the higher the signal generated. Therefore, the amount of product generated is directly proportional to the amount of test hapten in the sample (Figure 12–7). A reaction rate is determined by measuring the absorbance at two different times during the assay. The most common enzyme–substrate combination used in the EMIT is glucose-6-phos-

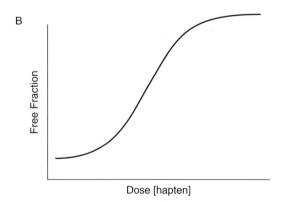

Figure 12–7. **A.** Principle of the homogeneous enzyme immunoassay (enzyme-multiplied immunoassay technique, EMIT). **B.** Typical dose–response curve.

phate dehydrogenase and glucose-6-phosphate plus nicotinamide-adenine dinucleotide (NAD).

The advantages of EMIT include no separation of bound and free antibody (homogenous assay format), full automation, and long reagent shelf life. A limitation is that it is less sensitive than some heterogeneous assays. EMIT is typically used to screen for therapeutic drugs and drugs of abuse and their metabolites.

C. Microparticle Enzyme Immunoassay (MEIA)

Microparticle enzyme immunoassay (MEIA) is a noncompetitive heterogeneous assay used to measure analyte (antigen or antibody) concentration in solution. MEIA is designed to detect antigen–antibody reactions on the surface of latex particles by capturing immune complexes on a glass fiber matrix, and then detecting the immobilized complex with an appropriate conjugate and substrate pair. The test is initiated by mixing a solution of submicron-sized latex particles coated with capture molecules (antigen or antibody), with the test sample to allow immune complex formation. Following this first reaction, the mixture is transferred to a glass fiber matrix that will irreversibly bind the immune complexes, while the unbound reactants flow through the matrix to waste. The immobilized immune complexes are detected by adding an enzyme (alkaline phosphatase)-labeled conjugate followed by a fluorescent substrate (4-methylumbelliferyl phosphate) to the glass matrix. The catalytic activity of the enzyme (dephosphorylation) on the fluorescent substrate results in the generation of a fluorescent end product (methylumbelliferone). The amount of end product is proportional to the concentration of the analyte in the sample (Figure 12–8).

The advantages of MEIA include full automation and rapid turnaround time. MEIA is often a fully automated immunoassay system that is used to screen a variety of antigens and antibodies in biological fluids. MEIA is used primarily to assay for various hormones and tumor markers.

D. Fluorescence Polarization Immunoassay (FPIA)

Fluorescence polarization immunoassay (FPIA) is a homogeneous assay that is based on the competitive binding reaction between a fluorescent conjugated ligand, usually a drug or hormone, with the sample ligand for a limited number of antibody-combining sites. The assay is based on the ability of antibody to modulate the level of polarization of the fluorophore–ligand complex. Specifically, polarization of fluorescence from a fluorophore–ligand conjugate is determined by its rate of rotation in solution. When polarized light is incident on small molecules, such as the fluorophore–ligand conjugate, they undergo rapid rotation in solution emitting light in all directions, and generating a low degree of polarization. However, binding of the fluorophore–ligand conjugate to antibody increases the size of the complex, and thus slows the rate of rotation and increases the degree of polarization. Thus, the amount of ligand present in the sample determines the level of polarized light emitted. The higher the sample ligand concentration, the fewer free antibody-binding sites available to bind with the fluorescein conjugated lig-

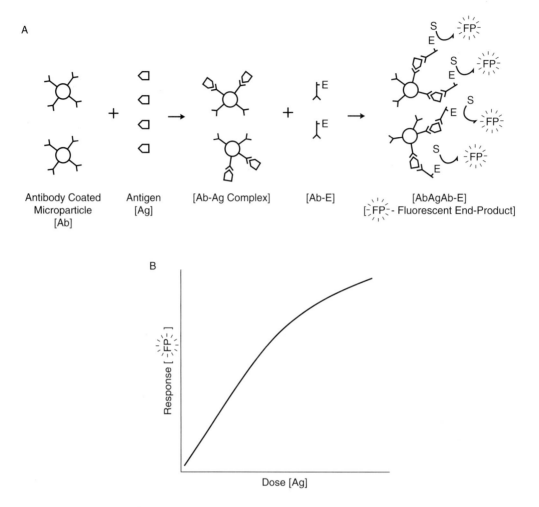

Figure 12–8. **A.** Principle of microparticle enzyme immunoassay. **B.** Typical dose–response curve.

and, and the lower quantity of polarized light emitted. Therefore, the level of polarized light emitted is inversely proportional to the amount of ligand in the sample (Figure 12–9).

FPIA has high sensitivity for a variety of drugs and, because it is a homogenous assay, it does not require a separation step. Some disadvantages of FPIA include the limitation of the technique to relatively small analytes, the requirement for a specialized instrument and presence of interfering fluorescence substances in lipemic and hemolyzed serum specimens that can lead to inaccurate results.

E. CHEMILUMINESCENT IMMUNOASSAYS

Chemiluminescent immunoassays can be designed to be competitive or noncompetitive, and heterogeneous or homogeneous. The most common format is a noncompetitive heterogeneous assay for the quantitation of drugs or hormones in biological fluids. The assay uses a chemiluminescent (acridinium ester or isoluminol)-labeled antibody to detect antibody–antigen reactions. Unlabeled antihapten antibody adsorbed to a solid support is mixed with a specimen con-

0

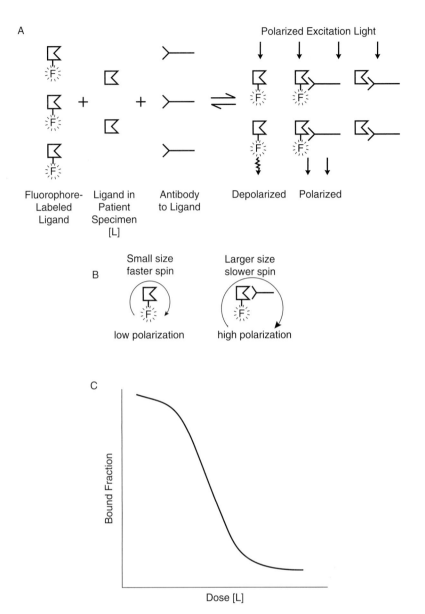

Figure 12–9. **A.** Reaction principle of the fluorescence polarization immunoassay. **B.** Effect of size on degree of polarization. **C.** Typical dose–response curve.

taining a drug or hormone. Following an incubation step, the unbound substances are removed by a wash step. A chemiluminescent-labeled antibody that recognizes a different epitope on the hapten is added to the reaction. Following washing to remove unbound antibody, the bound chemiluminescent label is oxidized by addition of hydrogen peroxide and a catalyst to release chemiluminescent energy in the form of a rapid flash of light, which is detected by a luminometer. The amount of light produced is directly proportional to the amount of hapten in the sample (Figure 12–10).

Chemiluminescent immunoassays are very sensitive assays and are widely used in automated analyzers. The main disadvantage is that a specialized luminometer is required.

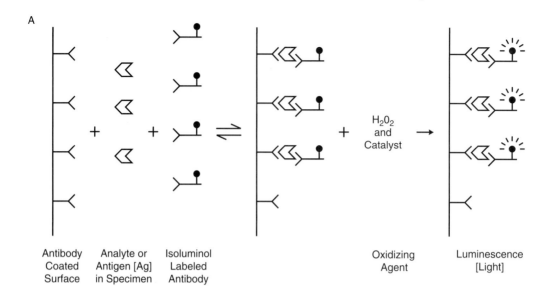

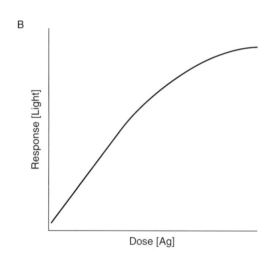

Figure 12–10. **A.** Principle of chemiluminescent immunoassay (sandwich assay). **B.** Typical dose–response curve.

SUGGESTED READINGS

Avrameas S. Amplification systems in immunoenzymatic techniques. *J Immunol Meth* 150:23–32, 1992.

Chard T. *An introduction to radioimmunoassay and related techniques,* 3rd ed. New York: Elsevier Science Publishing, 1987.

Colbert DL, Smith DS, Landon J, Sidki AM. Single-reagent polarization fluoroimmunoassay for barbiturates. *Clin Chem* 30:1765–69, 1984.

Kricka LT. Chemilumescent and bioluminescent techniques. *Clin Chem* 37:1472–81, 1991.

Lazarchik J, Hoyer LW. Immunoradiometric measurements of the factor VIII procoagulant antigen. *J Clin Invest* 62:1048–52, 1978.

Pesce AJ, Michael JG. Artifacts and limitations of enzyme immunoassay. *J Immunol Meth* 150:111–19, 1992.

Pesce MA. Automation. In Kaplan LA, Pesce AJ (eds.). *Clinical chemistry: Theory, analysis, and correlation,* 3rd ed. St. Louis: CV Mosby, 1996.

Rubenstein KE, Schneider RS, Ullman EF. "Homogeneous" enzyme immunoassay: New immunochemical technique. *Biochem Biophys Res Commun* 47:846–51, 1972.

Thompson SG. Principles for competitive-binding assays. In Kaplan LA, Pesce AJ (eds.). *Clinical chemistry,* 3rd ed. St. Louis: CV Mosby, 1996; 251–53.

Items for Further Consideration

1. Discuss the potential factors that must be considered if an immunoassay is designed using polyclonal serum.

2. Design a noncompetitive immunoassay to detect human IgE to ragweed.

3. Draw a typical dose response curve for a competitive immunoassay for a serum analyte where the bound label is measured. What would be the appearance of the curve if the free label is detected?

4. Discuss the factors that determine the detection level or analytical sensitivity of an immunoassay.

5. List the types of immunoassays in which the high-dose hook effect can result in inaccurate results.

6. What precautions can be taken to prevent and detect the high-dose hook effect?

Separation Techniques

Robert H. Christenson and F. Philip Anderson

Key Terms

AMPHOLYTES	INTERNAL STANDARDIZATION	REVERSED PHASE
ANODE	ISOELECTRIC POINT	R_f
CAPACITY FACTOR	KNOX EQUATION	SELECTIVITY COEFFICIENT
CATHODE	LIGAND	STATIONARY PHASE
CHROMATOGRAM	MOBILE PHASE	THEORETICAL PLATE
DIFFUSION	RESOLUTION	VAN DEEMTER EQUATION
EFFICIENCY	RETARDATION FACTOR	WICK FLOW
ELECTROENDOSMOSIS	RETENTION TIME	ZWITTERIONS
ELECTROPHORESIS	RETENTION VOLUME	

Chapter Objectives

Upon completion of this chapter, the reader will be able to:

1. Sketch and label the basic experimental design for electrophoresis.
2. Identify similarities and differences between traditional electrophoresis, isoelectric focusing electrophoresis, polyacrylamide gel electrophoresis, capillary electrophoresis, and two-dimensional electrophoresis.
3. Discuss four factors that may impede resolution in electrophoresis.
4. Compare the properties of the following separation media: cellulose acetate, agarose, and polyacrylamide.
5. Write the equation for the rate of electrophoretic migration and, for each factor, determine whether it is directly or inversely proportional to migration.
6. Name the species separated and identified by Southern blotting, Northern blotting, and Western blotting. Describe the general design of each technique.

7. Name the two phases common to all types of chromatography. Explain the common strategies used to abbreviate the names for different types of chromatography.
8. Name the two types of peak analysis for quantitation of solutes. Name three situations in which use of each would be favored.
9. Describe differences between external calibration and internal standardization.
10. Give the selectivity coefficient value that should be exceeded for good separation.
11. Explain the term *peak resolution* in quantitative terms. Give the value for resolution associated with baseline separation.
12. Define the term *height equivalent to a theoretical plate*. Describe one application for the van Deemter and Knox equations.
13. Describe the stationary and mobile phases for each of the following types of chromatography: thin-layer, adsorption, gel filtration, ion exchange, affinity, gas, and liquid. Name one clinical application for each of these types of chromatography.
14. Explain eddy diffusion. Name and describe the three types of mass transfer that can cause diminished resolution.
15. Identify and describe two differences between liquid chromatography (LC) and high-pressure liquid chromatography (HPLC). Describe what is meant by reversed phase HPLC.
16. Sketch and label the experimental designs for LC, gas chromatography (GC), and thin-layer chromatography (TLC).

Instruments/Techniques Included

AFFINITY CHROMATOGRAPHY

AGAROSE ELECTROPHORESIS

CAPILLARY ELECTROPHORESIS

CELLULOSE ACETATE
ELECTROPHORESIS

CHROMATOGRAPHY

GAS-LIQUID CHROMATOGRAPHY
(GLC)

GEL FILTRATION
CHROMATOGRAPHY

HIGH-PERFORMANCE LIQUID
CHROMATOGRAPHY (HPLC)

ION EXCHANGE
CHROMATOGRAPHY

ISOELECTRIC FOCUSING
ELECTROPHORESIS

MASS SPECTROMETRY (MS)

NORTHERN BLOT

SODIUM DODECYL SULPHATE
(SDS)

SOUTHERN BLOT

THIN LAYER CHROMATOGRAPHY
(TLC)

TWO DIMENSIONAL
ELECTROPHORESIS

WESTERN BLOT

Pretest

Directions: Choose the single best answer for each of the following items.

1. At a buffered pH of 8.6, a protein with an isoelectric point (pI) of 6.0 will
   a. have a net positive charge
   b. have a net negative charge
   c. have no charge
   d. precipitate

2. Proteins that are negatively charged in an electrophoresis experiment will
   a. migrate toward the anode
   b. migrate toward the cathode
   c. precipitate and become fixed in the gel
   d. show no migration

3. Which of the following tends to diminish the resolution of electrophoretic separations?
   a. solute ionic radius
   b. short separation times
   c. the applied electric field
   d. diffusion

4. Which of the following electrophoretic support media requires presoaking in buffer to fill air space?
   a. agarose
   b. polyacrylamide
   c. cellulose acetate
   d. starch

5. For two proteins having the same molecular weight and characteristic isoelectric points (pI) that are 0.1 different, the best method for separation is
   a. traditional electrophoresis
   b. isoelectric focusing electrophoresis
   c. polyacrylamide electrophoresis
   d. electrophoresis on a starch medium

6. In capillary electrophoresis
   a. resolution depends on the length of the capillary column
   b. large sample volumes are still required
   c. electrophoretograms are stained by procedures used for traditional electrophoresis
   d. high voltages may be applied because heat is effectively dissipated

7. Which of the following support media is prepared in multiple gel layers?
   a. agarose
   b. polyacrylamide
   c. cellulose acetate
   d. starch

8. Which of the following types of electrophoresis utilizes ampholytes to establish a pH gradient?
   a. traditional electrophoresis
   b. isoelectric focusing electrophoresis
   c. polyacrylamide electrophoresis
   d. electrophoresis on a starch medium

9. Which of the following is true about increasing the ionic strength of a buffer used in electrophoresis?
   a. increases heat production
   b. decreases heat production
   c. directly increases wick flow
   d. decreases the conductance of the system

10. The rate of migration of a protein in a traditional electrophoresis experiment is directly proportional to
    a. viscosity of the buffer solution
    b. ionic radius of the protein of interest
    c. net charge on the ion
    d. pi (3.1416)

11. Which of the following is a form of planar chromatography?
    a. ion exchange chromatography
    b. high-performance liquid chromatography
    c. thin-layer chromatography
    d. gas chromatography

12. The phase that travels through the chromatography system as a flowing stream of gas or liquid is the
    a. stationary phase
    b. inert phase
    c. enhanced phase
    d. mobile phase

13. Peak area measurement of chromatogram peaks should be considered for quantitation
    a. in separations in which accuracy is the primary concern
    b. in trace analysis in which baseline fluctuations are large compared to the analyte peak
    c. when the response of the detector system is non-linear
    d. in separations in which the possibility of interfering peaks is substantial

14. Which of the following statements is true concerning the calibration of chromatography procedures?
    a. for external calibration, a second compound is added to all calibrators, samples, and controls
    b. the quantity of sample injected onto the column must be carefully controlled when using internal standardization
    c. for internal standardization, an analytical curve is prepared by plotting a ratio of the calibrator-to-internal standard peak heights or areas
    d. calibrators and controls are applied directly to the chromatography system, but samples must be pretreated

15. For a selectivity coefficient (α) of 1.30 for the pair of compounds A and B, which of the following is true?
    a. the compounds are not separated by the chromatography system
    b. the capacity factor for compound A must be increased to have separation
    c. the capacity factor for compound B must be increased to have separation
    d. compounds A and B should be readily separated with this selectivity coefficient

16. Peak resolution for two compounds, A and B
    a. is expressed in terms of time
    b. exceeding 1.25 indicates poor separation
    c. can be calculated from the retention times and flow rate
    d. can be calculated from the volumes of retention and bandwidths of the elution peaks

17. Which of the following tends to increase chromatography efficiency?
    a. mobile-phase transfer
    b. stationary-phase mass transfer
    c. stagnant mobile-phase transfer
    d. increased number of theoretical plates

18. Which of the following statements is false concerning the height equivalent to a theoretical plate (H)?
    a. the calculation for H involves measuring the length of column used for the separation
    b. H is expressed in units of length (e.g., millimeters)
    c. H represents the length of column necessary for 1,000 equilibrations of solute between the mobile and stationary phases
    d. H is dependent on the width of the solute's peak at one-half maximum height

19. The van Deemter and Knox equations are useful for determining
    a. the optimum peak resolution for a chromatography system
    b. the selectivity coefficient for a chromatography system
    c. the optimum mobile-phase flow rate for most efficient separation
    d. migration distance for a solute over a specified time

20. Which of the following statements is true about thin-layer chromatography (TLC)?
    a. TLC is also known as planar chromatography
    b. TLC may be performed in two dimensions
    c. for identification, a ratio of migration of the solute to migration of the mobile-phase solvent is frequently calculated
    d. a, b, and c are true

21. Which of the following describes gel filtration chromatography?
    a. solutes adsorb from the mobile phase onto a solid stationary phase
    b. functional groups on the stationary phase retain solutes by electrostatic attraction
    c. the stationary phase contains crevices or pores that allow passage of certain molecules based on size
    d. the solute may be displaced from the stationary phase by strategies such as use of an inhibitor or changing the pH

22. Which of the following types of detectors can provide definitive identification of a separated solute?
    a. ultraviolet detector
    b. mass spectrometer detector
    c. flame ionization detector
    d. nitrogen/phosphorus detector

23. Which of the following statements is true about gas chromatography (GC) systems?
    a. the stationary phase in GC is a gas
    b. flame ionization detectors may not be used with GC
    c. derivatization of solutes is frequently needed with GC
    d. the Knox equation applies to GC systems

24. Which of the following statements is true about high-performance liquid chromatography (HPLC)?
    a. in "reversed phase" HPLC, the stationary phase is more polar than the mobile phase
    b. either isocratic or gradient elution may be used with HPLC
    c. HPLC systems always involve a liquid stationary phase
    d. the stationary phase for HPLC systems is composed of large particles

I. INTRODUCTION

Analysis of some substances found in a biological matrix may require their prior separation from other molecules in the mixture. A variety of separation techniques are available, including precipitation, dialysis, ultrafiltration, ultracentrifugation, electrophoresis, and chromatography. Among these, electrophoresis and chromatography are by far the most commonly utilized in clinical laboratories. Historically, electrophoresis and chromatography were frequently employed in the first analysis strategies for many substances. However, since electrophoresis and chromatography are not readily adapted to automation, they are relatively laborious, expensive, and slow compared to the majority of routine clinical laboratory techniques. Therefore, electrophoretic and chromatographic techniques are now generally reserved for measurement of substances that cannot be analyzed satisfactorily by traditional wet chemistry or immunoassay methods, preferably using automated analysis systems.

II. ELECTROPHORESIS

Electrophoresis is the separation of charged solutes in a liquid ◀ medium on the basis of their relative migration under the influence of an externally applied electric field. Charged solutes of clinical interest are usually large molecules, including proteins, deoxyribonucleic acid (DNA), and ribonucleic acid (RNA). In the clinical laboratory, electrophoresis is most commonly used for the separation of proteins in biological fluids. Tiselius developed the first electrophoresis method for the study of proteins in 1937, and electrophoresis has since played a significant role in the progress of clinical chemistry.

The general experimental design for electrophoresis is illustrated in Figure 13–1. As shown, the electromotive force is applied via the power supply to generate the charged environment. The **anode** is ◀ the positively charged electrode and, therefore, attracts negatively charged species (hence its name). The **cathode,** the negatively ◀ charged electrode, attracts positively charged solutes. The buffer electrolytes carry the electrical current and maintain the pH. The wicks serve to complete the circuit by bridging the buffer reservoirs to the buffer-saturated support matrix, where separation takes place.

In electrophoresis, the migration and separation of solutes in a mixture depends on the net charge of the individual substances under the pH conditions of the experiment. Proteins are **zwitterions** ◀ (or **ampholytes**)—molecules that may be either positively or nega- ◀

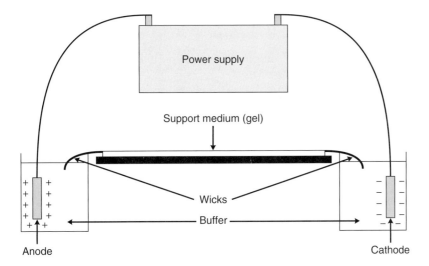

Figure 13–1. Experimental design for the traditional electrophoresis system. The buffer-saturated wicks complete the circuit, and separation occurs on the support medium (gel). Note that the cathode is negatively charged (attracts positively charged solutes) and the anode is positively charged (attracts negatively charged solutes).

tively charged. The amino acid composition of a particular protein species determines the proportions and characteristics of the acidic, basic, and neutral amino acid functional groups, and thus the protein's net charge at a given pH. The **isoelectric point** (pI) is defined as the pH at which the net charge of the molecule is zero. When the pH is higher (more basic) than the pI of the molecule, that molecule acquires a net negative charge as positively charged hydrogen ions (H^+) are donated from the molecule to the buffer. Alternatively, when the buffer pH is lower (more acidic) than the pI, H^+ ions are donated from the buffer and the molecule thus acquires a net positive charge. Generally, the rate of migration of a negatively charged solute toward the anode or a positively charged solute toward the cathode will be proportional to the magnitude of the net charge. In the clinical laboratory, protein electrophoresis is commonly performed at pH 8.6, at which most proteins possess a net negative charge.

The common clinical laboratory electrophoresis method is *zone electrophoresis*, which allows migration of solutes for a fixed time followed by visualization of the pattern of zones of the separated substances (i.e., the electrophoretogram) and analysis of the appropriate fraction(s).

A. GENERAL PRINCIPLES

The forces driving the migration of an ion in an electrophoresis system can be described by the following equation:

$$F = (X)(Q) = (VQ)/d = iRQ/d$$

where

F = force exerted on ion
X = electric field strength (volts/cm)
Q = net charge on ion
V = applied voltage (v)
d = distance across the medium over which the voltage is applied (cm)
R = resistance
i = current

The force opposing migration of the ion (F' or the counterforce) is described by *Stoke's law:*

$$F' = 6\,\pi\,r\,\eta\,v$$

where

π = 3.1416
r = ionic radius of molecule
η = viscosity of buffer solution
v = rate of migration (velocity in cm/s)

The combined effect of the driving force and the counterforce results in a constant velocity such that when $F = F'$, then

$$6\,\pi\,r\,\eta\,v = XQ \quad \text{or} \quad v = \frac{XQ}{6\,\pi\,r\,\eta}$$

By this expression, the rate of migration in the electrophoresis design shown in Figure 13–1 is proportional to the electric field strength and the net charge on the solute, and inversely proportional to the ionic radius of the solute and the viscosity of the support medium. Thus, for separation of two charged species to occur with electrophoresis, the two molecules must have a different net charge and/or a different ionic radius (i.e., a different charge-to-mass ratio).

B. RESOLUTION

The analytical goal of electrophoresis is to optimize the separation of the compounds under analysis into distinct bands. **Resolution** refers to the degree of separation achieved or the ability to identify two species as separate bands. The quality of the resolution depends on (1) selection of experimental conditions appropriate for separating the substances of interest, and (2) controlling experimental phenomena that influence the migration of charged particles in an electric field, including diffusion, electroendosmosis, heat production, pore size, and wick flow.

Diffusion (random molecular motion) leads to zone broadening, which limits resolution. Diffusion is directly proportional to temperature and the square root of time, and inversely proportional to both the viscosity of the medium and to the ionic radius of the charged analyte. Performing the separation in the minimum optimal time helps to limit diffusion; the choice of support media can help limit diffusion caused by solute adsorption or restrictive pore size.

Electroendosmosis is the movement of the buffer solutes and solvent relative to the support matrix when an electric field is applied. Fixed negative charges (e.g., hydroxyl or carboxyl groups) on the surface of solid supports in contact with water induce formation of an ionic "cloud" of buffer cations along the support surface; this positive–negative alignment is termed the *Stern potential*. When voltage is applied, the "cloud" of positively charged ions in the buffer (mobile phase) migrates toward the negatively charged electrode (cathode) carrying water of hydration (i.e., solvent) with it. As a result, uncharged and weakly, negatively charged macromolecules are swept along toward the cathode by this endosmotic flow. In serum protein electrophoresis at pH 8.6, in which proteins carry a net negative charge, gamma globulins are swept behind the line of application during electrophoresis on cellulose acetate, paper, or agar gel (i.e., supports that induce significant endosmotic flow) because of their weak net charge. Use of media that are minimally charged, such as agarose, polyacrylamide gel, or starch gel, can help alleviate this effect.

Heat production occurs during electrophoresis as current flows in opposition to the electrical resistance of the system. Heat can be expressed as

$$Heat = (E)(i)(t)$$

where

$$E = \text{voltage or potential}$$
$$i = \text{current}$$
$$t = \text{time}$$

Heat production decreases the resistance of the system. According to *Ohm's law:*

$$E = (i)(R)$$

where

$$R = \text{resistance; } 1/R = \text{conductance}$$

Factors that Influence Electrophoretic Migration

- Diffusion
- Electroendosmosis
- Heat production
- Pore size
- Wick flow

Therefore, when current is held constant during electrophoresis, voltage decreases as resistance decreases; the heat effect is minimized and migration stays relatively constant. However, when voltage is held constant, current increases as resistance decreases, heat production increases, and migration rate increases. An increase in the system temperature causes evaporation of water from the support media which, in turn, increases the ionic strength of the buffer; high ionic strength decreases the resistance, allowing increased current and thus increased heat production. The increased temperature causes increased diffusion, may denature protein analytes, and causes unpredictable electrophoretic mobility; these effects often result in unsatisfactory, "wavy" electrophoresis patterns.

Although either constant current or constant voltage power supplies are available, most instruments in the clinical laboratory utilize constant voltage since separations can be performed more quickly. In some cases, a cooling system is used to control problematic overheating.

During electrophoresis, moisture evaporates from the surface of the support media, hastened by heat production in the system. In order to replace moisture lost due to evaporation, buffer from the anode and cathode buffer reservoirs flows by wick action up onto the support media (see Figure 13–1)—a phenomenon called **wick flow.** The flow of buffer from each reservoir is equal but opposite in direction. The rate of this flow is directly proportional to the rate of evaporation and is higher at the ends of the support platform, decreasing as it approaches the center. Any condition (i.e., high voltage, high current, etc.) that promotes heating in the system increases evaporation and consequently increases wick flow. The sample application point for serum protein electrophoresis (pH 8.6) is on the cathodic side of center on the support to minimize the effect of wick flow; the migration of negatively charged proteins toward the anode would be opposed by wick flow on the anodic side of the support.

C. GENERAL METHODOLOGY

1. Buffers. In electrophoresis, the applied electric potential drives the migration of solutes through a separation medium in the presence of a buffered system that serves as the electrical environment. As previously discussed, the pH of the buffer affects the net charge on molecules being separated. The ionic strength of the buffer affects the current and the migration velocity of the analyte molecules. Increased ionic strength increases current but decreases migration rate because a thicker ionic cloud hinders movement of the solute(s). High-ionic-strength buffers have the advantage of increasing resolution by producing sharper bands, but have the disadvantage of increased heat production.

2. Support Media. The appropriate support medium used for electrophoresis will depend on the analytical application. Considerations in choosing an electrophoretic medium include potential for acceptable resolution, possible interactions between the analyte molecules and the medium, interactions between the buffer solution and the medium (e.g., causing electroendosmosis), the need for molecular sieving (or not), and stability of the medium upon storage. For most applications in the clinical laboratory, it is desirable to use a medium that is transparent (or able to be made transparent) for analysis by densitometry. Several of the commonly used support media are described below.

Cellulose acetate is formed by using acetic anhydride to acetylate the hydroxyl groups on cellulose. Manufacturers can vary the characteristics of cellulose acetate membranes by varying the extent of acetylation, using various additives and prewashing procedures, and varying the thickness of the membranes. In the dry form, cellulose acetate membranes contain about 80% air space; this space is filled with liquid upon soaking in buffer before electrophoresis. Electrophoretic separation on cellulose acetate can be rapidly accomplished, being completed within about 20 minutes of sample application with most systems. After electrophoresis, cellulose acetate membranes can be rendered transparent by soaking in a solution of 95% methanol and 5% glacial acetic acid. This solvent mixture partially dissolves the cellulose acetate, causing the groups to coalesce, thus eliminating the extensive interfiber space. Cleared (transparent) membranes are stable and can be archived for long periods of time. Cellulose acetate has been replaced by agarose for many applications because agarose does not require soaking prior to electrophoresis or clearing by solvent treatment following electrophoresis.

Agarose is purified from agar by removing agaropectin. Agarose is relatively free of ionizable residues and so does not demonstrate significant electroendosmosis or adsorption (agaropectin is not desirable in separation gels because it contains highly charged sulfate and carboxylic acid groups that contribute to electroendosmosis). Agarose gels are transparent, adsorb proteins minimally, and are suitable for long-term storage after drying. Agarose is the support medium most commonly used for clinical applications, and premade gels are available commercially. It is the medium of choice for routine applications such as serum protein electrophoresis, hemoglobin electrophoresis, and fractionation of alkaline phosphatase isoenzymes.

Polyacrylamide gel electrophoresis (PAGE) is used for high-resolution separation of proteins, small RNA molecules, and small DNA fragments. Polyacrylamide gels are prepared by cross-linking acrylamide with N,N'-methylenebisacrylamide; the amount of cross-linking determines the average pore size in the gel, which can be modified for particular applications. Polyacrylamide can be used in tube gels but is more often used in multisample slab gels. *Disc electrophoresis* (referring to the use of a pH *disc*ontinuity) is a variation of PAGE used to improve resolution. The gels are prepared in a vertical slab or tube with the separation gel on the bottom, a middle spacer (or stacking) gel layer, and a top sample gel layer; the spacer and sample gels have a larger pore size, lower ionic strength, and a different pH than the separating gel. In an electric field, sample molecules migrate rapidly through the upper gels then stack in order of mobility at the boundary before entering the separation gel for fractionation. *SDS-PAGE* is a specialized variation of PAGE that separates proteins on the basis of their molecular weight. Proteins are treated with the detergent **sodium dodecyl sulfate (SDS)**, which binds to the proteins in proportion to their weight; the charge-to-mass ratio is thus determined by the amount of SDS bound. In an electric field, migration rate increases with decreasing molecular weight. Polyacrylamide does not exhibit electroendosmosis; is transparent, allowing analysis by densitometry; and can be stabilized or photographed for storage. PAGE is relatively laborious for routine applications.

Support Media

- Cellulose acetate
- Agarose
- Polyacrylamide gel
- Starch gels

Starch gels have been used for separation of macromolecules based on molecular size and surface charges. The medium is formed by partial hydrolysis of starch, to give the material a gel-like composition. Samples must be applied to starch gels as a thin, fine application; to accomplish this, electric current is frequently required to compact the proteins in a band. While starch gels are of historical importance as a medium first used in protein electrophoresis, they are inconvenient and rarely used in clinical laboratories.

3. Detection of Analyte(s). After electrophoresis has been completed, the support medium must be treated with a wash to remove buffer and other additives that may interfere with staining. Often, the solute molecules must be "fixed" (treated to adsorb) onto the support matrix prior to staining. Some support media also require treatment to render the gel transparent. The separated analytes in the support medium are stained, if necessary, to allow detection. Then the gels or membranes are dried before visual inspection and/or analysis by *densitometry* (see Chapter 3).

Stains used for the general detection of proteins on an electrophoretogram are protein-binding dyes, such as amido black, Coomassie Brilliant Blue, or Ponceau S. Hemoglobins have intrinsic color and may be detected with a visible range densitometer, but are usually stained with Ponceau S. For analysis of proteins in fluids with low analyte concentrations, such as cerebrospinal fluid or urine: (1) samples may be concentrated prior to electrophoresis allowing use of conventional protein stains, or (2) the proteins may be stained after electrophoresis using a highly sensitive staining method such as silver staining. Enzyme activity may be used to detect particular enzymes or isoenzymes. For example, nitroblue tetrazolium is frequently used to locate isoenzymes that produce nicotinamide adenine dinucleotide (phosphate) [NAD(P)H], either directly or via a coupled reaction, where the NAD(P)H reduces the tetrazolium to a purple formazan dye. In some cases, it is possible to detect proteins and nucleic acids by direct ultraviolet (UV) absorption. Lipoproteins can be visualized by use of Fat Red, Oil Red, or Sudan Black. DNA fragments are often stained with ethidium bromide, a fluorescent dye. With most staining methods there are potential problems, including diffusion, inequality of staining between different proteins, and linear response limitations for quantitation.

D. TYPES OF ELECTROPHORESIS

Traditional electrophoresis protocols performed in the clinical laboratory use either agarose or, less frequently, cellulose acetate as the medium for separating proteins. Constant voltage is generally used at a pH of 8.6. The buffers used to maintain this pH include barbitol buffers or a Tris-boric acid–ethylenediaminetetraacetic acid (EDTA) buffer.

Isoelectric focusing electrophoresis (IEF) is a high-resolution technique developed for situations in which the size and net charge of the solutes to be separated are similar (e.g., variants of specific proteins such as α_1-antitrypsin or transferrin). Rather than a constant pH separation as in traditional electrophoresis, IEF separations use a pH gradient formed from a heterogeneous mixture of small, highly charged molecules called ampholytes. The ampholytes migrate in an

---

Types of Electrophoresis

- Traditional electrophoresis
- Isoelectric focusing electrophoresis
- Capillary electrophoresis
- Two-dimensional electrophoresis

electric field to the pH at which they have a net charge of zero (i.e., their individual pI). In this way, the ampholytes create a continuous pH gradient on the gel medium (usually agarose or large-pore polyacrylamide). Mixtures of carrier ampholytes are commercially available to form gradients of various pH ranges. As long as the pH gradient and electric field are present, no further migration occurs. Proteins or other solutes applied to the IEF system also migrate to the point at which pH equals pI; solutes are concentrated into very narrow bands since diffusion is counteracted (i.e., a molecule that diffuses away will acquire a charge and migrate back to its focus point). Proteins with pI differences of as small as 0.02 pH units have been separated by IEF. Staining for visualization depends on the chemistry and concentration of the solutes of interest.

Capillary electrophoresis is performed in a long, thin capillary that may or may not contain a support medium such as a gel. A diagram of a typical experimental setup is shown in Figure 13–2. Typical dimensions for the capillaries are 50 to 100 cm in length by 50 to 150 μm in diameter. Since the volume inside the capillary is quite small, and the sample must be a very small fraction of this, typically only nanoliter quantities are injected. Theoretically, resolution is dependent only on the applied electric field strength and is independent of column length; column length does, however, effect analysis time. Ideally, the highest voltage feasible is applied to a capillary as short in length as possible. This high field strength results in rapid, high-resolution separation. The very high surface/volume ratio of the capillary allows for effective heat dissipation, so very high voltage may be used. Capillary zone electrophoresis is very amenable to automation.

In unmodified, silica glass capillary tubes, the inside surface of the tube is negatively charged and cations align along the inner surface. When the electric field is applied, the cation layers and the associated water migrate toward the cathode, dragging the bulk of the

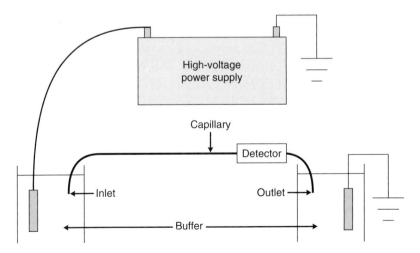

Figure 13–2. The experimental design for capillary electrophoresis. High-voltage power supplies are usually used because the large surface area allows efficient dissipation of heat, thus decreasing diffusion. The detector "window" is integral to the capillary system, and the detector must be very sensitive because the amount of solute applied to the system is small.

solution. Due to the overwhelming electroendosmotic flow, all solutes are dragged to the cathode; positively charged solutes elute first, followed by neutral molecules, with negatively charged solutes eluting last.

The very small (nanoliter) sample volume used in capillary electrophoresis dictates the use of very sensitive detection techniques; however, the on-system concentration of solutes into narrow bands helps to diminish this limitation. Note from Figure 13–2 that capillary electrophoresis does not produce a stable electrophoretogram like that commonly obtained from traditional electrophoresis in the clinical laboratory. Instead, a detector views the solute molecules as they pass a detector "window" in the capillary. The concentration of individual analytes can be determined by calibration or by using an internal standard technique. Traditional UV and visible absorbance detectors are common, but more sensitive laser-induced fluorescence, conductivity, and mass spectroscopy detectors are also commercially available.

Miniaturized capillary electrophoresis systems have been developed that permit very short separation times. In one experimental setup, the capillary was effectively etched into a glass microchip, where the separation column was 15 mm in length; channels allowed introduction and control of a precolumn reaction, buffer, and waste. The fluorescent derivatives of the amino acids arginine and glycine were separated in less than 10 seconds on the microchip system. Further miniature applications will undoubtedly be developed in the future.

► **Two-dimensional (2D) electrophoresis** is a high-resolution technique first developed in the 1970s. Here, separation in one dimension is performed by one type or with one set of electrophoretic conditions, and separation in the second dimension is based on migration using a different type of electrophoresis or under a different set of conditions. In one of the most common variations of 2D electrophoresis, the first dimension separation utilizes IEF (i.e., based on the solute's pI), using either agarose or large-pore polyacrylamide gels. After completion of separation in the first dimension, the IEF gel is linked to a polyacrylamide gel for molecular weight–dependent separation in the second dimension. The gels can be stained with Coomassie blue, amido black, or silver nitrate, depending on the sensitivity needed. The major issue with the two-dimensional technique is interpretation of the many "spots" resulting from separation of macromolecules in biological fluids. Due to the complexity of the fractionation patterns, this technique is rarely used in the clinical laboratory at present; however, advances in computer hardware and software will probably allow increased clinical use of this technology, particularly as the areas of proteomics and pharmacogenetics become further developed.

► **Southern blotting** was developed by Edward Southern in 1975 for use in the identification of DNA fragments. With this technique, DNA fragments are first fractionated by agarose electrophoresis. The agarose gel is then overlaid with either a nitrocellulose or nylon membrane. The separated DNA fragments on the gel are transferred to the overlaid material by capillary action, by electrotransfer, or by pressing the gel to the overlaid material by use of a vacuum. Once transfer is complete, the DNA fragments are identified on the nitrocellulose or nylon membrane by hybridization with complementary DNA probes.

Northern blotting is a method for separation and identification ◀ of RNA. This technique was named by analogy to Southern blotting. First, RNA is separated by agarose electrophoresis, then the gel is overlaid with a nitrocellulose or nylon membrane. Transfer to the membrane is accomplished by capillary action, electrotransfer, or by vacuum. RNA species are detected using specific RNA or single-stranded DNA probes.

Western blotting, also named by analogy to Southern blotting, ◀ is a method used for protein separation and identification. In Western blotting, proteins are separated by agarose electrophoresis and transferred (blotted) onto nitrocellulose or nylon membrane filters. The proteins of interest on the blots are located with specific antibodies.

E. COMMON APPLICATIONS OF ELECTROPHORETIC METHODS

Electrophoresis is used in clinical laboratories for fractionating serum proteins (still the method of choice), identifying abnormal hemoglobin variants, and quantitation of glycated hemoglobin fractions. Electrophoretic methods are also frequently performed for lactate dehydrogenase isoenzyme pattern analysis and separation of bone and liver fractions of alkaline phosphatase isoenzymes; electrophoresis is still occasionally used for creatine kinase (CK) isoenzyme analysis. One manufacturer has developed a dedicated system for performing CK-MB isoform analysis and has recently reported an electrophoretic method for high-density lipoprotein (HDL) cholesterol. In the future, capillary electrophoresis applications may become more prominent in the clinical laboratory due to the ease of automation, development of more sensitive detection systems, and development of miniature technology. Systems for two-dimensional electrophoresis will become more prevalent as proteomics applications are adapted in the clinical laboratory.

III. CHROMATOGRAPHY

A. GENERAL OVERVIEW

Chromatography is defined as the separation of analytes from a ◀ mixture based on their distribution in two different phases termed the **stationary phase** and the **mobile phase.** The stationary phase ◀ may consist of coated solid particles, a gel, or a low-volatility liquid that is contained as a layer on an inert separation support such as a glass plate or within a tube-shaped glass or metal column. The mobile phase moves through or across the stationary phase as a flowing stream of gas or liquid. The fundamental principle is that the chemical and/or physical properties of each analyte will have a different interaction with each of the two phases. Analytes having higher affinity for the mobile phase will interact little with the stationary phase and, therefore, travel faster through the system. Alternatively, passage of solutes having high affinity for the stationary phase will be impeded. Thus, the stationary and mobile phases are selected to enhance separation by exploiting differences in how the analytes interact with the two phases. A variety of interactions may be used as the basis for separation, including molecular size, differences in solubility in the stationary and mobile phases, adsorption of the solute, molecular charge, or various affinity reactions.

Chromatography is fundamentally a very flexible separation technique that can be used for isolation of a wide variety of analytes. Chromatography can be used for qualitative analysis, indicating either the presence or absence of the analyte, or for quantitative analysis, where the solute's actual concentration is measured. Chromatography can also be used for preparation of solutes, with collection of solute at the column's end, or for solute purification.

Chromatography may be classified based on the properties of the mobile and stationary phases used for analysis, or by the type of support used for separation. When a flat surface such as a glass or plastic plate is utilized as the support for the stationary phase, the term *planar chromatography* is used. Most often, the stationary phase on a planar support is a thin layer of sorbent material, so the term **thin-layer chromatography (TLC)** is used. Likewise, *column chromatography* refers to systems in which the stationary phase is contained by a tubelike structure.

A typical column chromatographic system is shown in Figure 13–3. Flow in the mobile phase, whether gas or liquid, is pressure regulated. The sample is introduced onto the column by application (e.g., layering or spotting) or by using an injector appropriate to the type of chromatography or chromatographic apparatus. Separation occurs within the column containing the stationary phase, across which the mobile phase flows. A detector will indicate when the molecules of interest have eluted from the column, and additional data handling equipment can present the data in graphic format and perform necessary data reduction for quantitative analysis. Temperature is a critical variable for the chemistry occurring during chromatography, so a thermal control device is usually incorporated to precisely regulate column temperature.

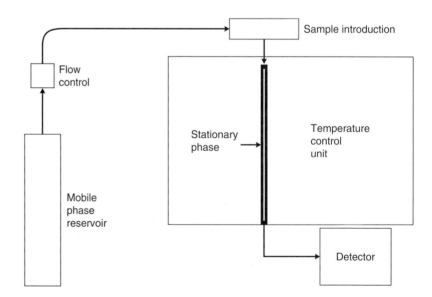

Figure 13–3. Typical system for chromatography separations. In gas chromatography, the mobile phase reservoir is a pressurized gas cylinder; for liquid chromatography, the reservoir consists of a bubble- and particulate-free solvent that enters the system via gravity or by pumps (e.g., for HPLC). See Table 13–1 for a list of detectors.

B. CLASSIFICATION BY PHASE

Systems may also be named for the stationary and/or mobile phases used. For example, gas chromatography (GC) has a gas mobile phase; liquid chromatography (LC) is the term used when the mobile phase is liquid. **High-performance liquid chromatography** ◀ (HPLC; sometimes called *high-pressure liquid chromatography*), a specific type of LC system, uses very fine matrix particles to increase the surface area of the stationary phase and high pressure to maintain flow of the mobile phase, resulting in highly efficient separations. In the original HPLC designs, the stationary phase was more polar than the mobile phase used. When this situation was reversed (i.e., when the mobile phase is the more polar), the term **reversed** ◀ **phase** *HPLC* was coined. Currently in the clinical laboratory, most HPLC applications make use of reversed phase designs. Other types of chromatographic systems include **gas–liquid (GLC),** gas–solid ◀ (GSC), liquid–liquid (LLC), and liquid–solid (LSC) chromatography. Systems may also be referred to by the detection method used; for example, the suffixes MSD and UV refer to mass spectrometry detection and ultraviolet, respectively.

C. DETECTION OF SOLUTES

Detection of solutes after TLC is accomplished by reacting the separated solutes with specific reagents to form spots or bands that can be visualized or scanned with a densitometer for identification. For GC and LC, detection of the separated solutes is accomplished by monitoring the mobile phase as it flows from the column's end. A broad variety of solute detection strategies have been developed, some of which are nonspecific and detect many solutes, while others are extremely selective and specific. Virtually all detectors respond by conversion of the detected solute into an electronic signal, which is sent to a data processor. The data system converts the electronic signal to a graphic display called a **chromatogram.** Chromatograms ◀ (or elution profiles) are usually plotted such that the intensity of the detected signal is expressed versus time, as shown in Figure 13–4.

D. QUANTITATIVE ANALYSIS

Quantification by chromatographic techniques involves analysis by one of two general methods: measurement of the *height* or the *area* of the peak(s) on the chromatogram corresponding to the elution of the solute(s) of interest. The peak height method may be more appropriate (1) in separations in which accuracy is the primary concern, (2) in trace analysis in which baseline fluctuations are large compared to size of the analyte peak, (3) when resolution is not optimal, and (4) in separations in which the possibility of interfering peaks is substantial. Measurement of peak area is generally used for quantitation when (1) peak tailing is a problem, (2) precision must be maximized, and (3) the response of the detector system is nonlinear.

Quantitation of solutes from the height or area of their elution peak on the chromatogram is based on calibration by either external calibration or **internal standardization.** The external calibration ◀ method consists of preparing a set of calibrators with known solute concentrations. The analytical response (i.e., peak height or peak area) of these calibrators is measured, and a plot of the concentra-

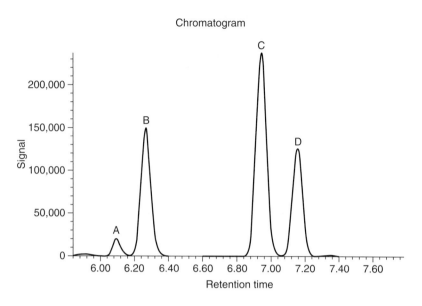

Figure 13–4. A chromatogram from a gas chromatography system showing dihydromorphine (peak A), D3-codeine (peak B), hydromorphine (peak C), and D3-morphine (peak D). The information displayed in chromatograms may be used for solute quantification by assessing either peak height or peak area, with either the external calibration or internal standard techniques.

tion versus analytic response is plotted as a calibration curve. The peak size of an unknown sample is then located on the analytical curve and used to extrapolate the solute concentration of the sample. The internal standard method also uses a set of calibrators of known solute concentrations; however, a second compound, the *internal standard*, is added to each calibrator, control, and sample to be tested, so that the internal standard concentration is equal in each. In theory, the peak size of the internal standard will be constant for calibrators, controls, and samples. An internal standardization curve can be generated by plotting the ratio of the peak size of the solute to the peak size of the internal standard (solute/internal standard) for the calibrators versus the calibrator concentrations. The same peak size ratio (solute/internal standard) can be calculated for samples and used to determine unknown concentrations by extrapolation from the internal standardization curve.

For both the external calibration and internal standardization techniques, calibrators are preferably in the same matrix as the samples. Also, both techniques require that calibrators be treated with the same preparation procedure as samples. It is important that the same volume of calibrators and sample be injected onto the chromatographic system for external calibration. Since the internal standard method is based on a ratio, it has the advantage of being relatively independent of the volume applied onto the column.

E. PEAK RESOLUTION

In a chromatography system, the solute will distribute itself in a highly reproducible manner between the mobile and stationary phases. This partitioning can be mathematically described by a *distribution constant* (K_D) which is defined as the ratio of the concentration of solute in the stationary phase (C_s) to the concentration of solute in the mobile phase (C_m):

$$K_D = C_s/C_m$$

K_D can also be expressed as:

$$K_D = k' \cdot \beta$$

where k' is the **capacity factor** (i.e., the ratio of the number of ◀ solute molecules in the stationary phase to the number of solute molecules in the mobile phase) and β is the ratio of the volume of stationary to volume of mobile phase. This partitioning is one aspect of separation. Another concept important to chromatography is the **retardation factor.** While the capacity factor gives information ◀ about the order of elution for different analytes from a column, it does not address the rate of migration through the column. For compounds eluting from a column, the **retention volume** (V_r) is related ◀ to the capacity factor as follows:

$$V_r = V_m(1 + k')$$

where V_m is the volume of mobile phase required to elute a compound with no affinity for the stationary phase. A large volume of retention (V_r) is necessary for effective separation of solutes. Separation or resolution of two compounds will require differences in their equilibrium constants and, therefore, also in their capacity factors. The separation factor, or **selectivity coefficient,** α, for a pair of ◀ compounds A and B is defined as:

$$\alpha = k'(B)/k'(A)$$

α should be greater than about 1.10 to achieve separation of the two compounds with relative ease.

Resolution is the ability to discriminate, or "resolve," chromato- ◀ graphic peaks for two solutes, A and B. Resolution is a unitless quantity that may be expressed in terms of the bandwidths of each solute, determined from the elution peaks on the chromatogram. The definition of bandwidth is shown in Figure 13–5. Resolution (R_s) can be quantitatively described by the following expression:

$$R_s = \frac{V_r(B) - V_r(A)}{\dfrac{[w(A) + w(B)]}{2}}$$

where $V_r(A)$ and $V_r(B)$ are the respective volumes of retention for solutes A and B; $w(A)$ and $w(B)$ are the bandwidths, in units of volume, at the peak bases for solutes A and B, respectively (Figure 13–5). Alternatively, resolution can be described using time, rather than V_r, in the quantitative expression. Since

$$t_r = V_R / F$$

where **retention time** (t_r) is the time it takes for a solute peak to ◀ migrate from injection to the detector and F is flow rate, then

$$R_s = \frac{t_r(B) - t_r(A)}{\dfrac{[w(A) + w(B)]}{2}}$$

It is important to note that if elution of solutes A and B are expressed in terms of time [e.g., $t_r(A)$ and $t_r(B)$], then $w(A)$ and $w(B)$

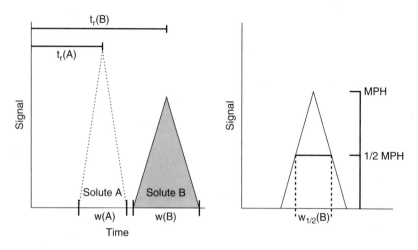

Figure 13–5. *Left panel.* Chromatogram for separation of solute A and solute B. $t_r(A)$ and $t_r(B)$ represent respective retention times for solute A and solute B. $w(A)$ and $w(B)$ represent the respective band widths for solute A and solute B. Resolution is a unitless quantity, so both t_r and $w(A)$ must have the same units (see text). *Right panel.* Display of peak width at ½ maximum peak height for solute B [$w_{1/2}(B)$]. MPH represents the maximum peak height; ½ MPH represents ½ maximum peak height.

(from Figure 13–5) must also be expressed in terms of time because R_s is a unitless quantity.

Figure 13–6 illustrates the effect of various R_s values for two solutes on the quality of resolution. R_s values less than 0.8 are associated with incomplete separations, whereas R_s values exceeding 1.25 result in so-called "baseline" separation.

F. EFFICIENCY

Solute separation by chromatographic systems is achieved through equilibration of solutes between the mobile phase and stationary phase, by either chemical or physical interactions. **Efficiency** refers

$$\text{Resolution} = \frac{V_r(B) - V_r(A)}{\dfrac{[w(A) + w(B)]}{2}}$$

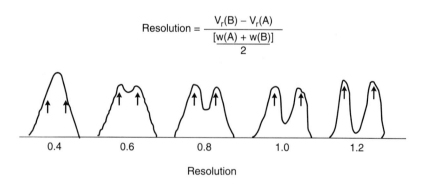

Figure 13–6. Display of the effect of various resolution values on separation quality. In the equation for resolution, the retention volumes for solute A and B are represented by $V_r(A)$ and $V_r(B)$; $w(A)$ and $w(B)$ represent the respective band widths for solute A and solute B (see Figure 13–5).

to the ability of a chromatography system to separate a solute as single peak having a narrow width. This implies that to achieve ideal efficiency, solute molecules must travel through the chromatographic system in a tight or narrow band. In practice, however, there are interactions and processes occurring during chromatography that cause peak broadening, and thus, diminished efficiency.

One process that decreases efficiency is *eddy diffusion,* a phenomenon resulting from the many different possible paths over which solute molecules pass around and between the stationary-phase particles to traverse the length of the column. As illustrated in Figure 13–7, some solute molecules take a tortuous path while others will find a relatively direct route; these differences result in peak broadening and a loss in efficiency. In addition to eddy diffusion, there are three *mass transfer* processes that tend to diminish efficiency. One of these is termed *mobile-phase transfer,* which describes differences in mobile-phase flow within the column. Mobile-phase molecules close to stationary-phase particles flow more slowly than mobile-phase molecules more distant from the particles; this flow rate difference causes peak broadening of the solute. A second mass transfer process that reduces efficiency is termed *stagnant mobile-phase transfer,* which is caused by the delay (or stagnation) of solute molecules in pools of mobile phase between particles of the stationary phase. *Stationary-phase mass transfer,* the third mass transfer process, describes both the movement of a solute into and within the stationary phase, or the association with the stationary phase by diffusion. In other words, solute molecules will spend differing times associated with the stationary phase, resulting in peak broadening and loss of efficiency. All three mass transfer processes are heavily dependent on the column packing used in the chromatography separation.

Individual solute molecules have identical equilibrium characteristics between the mobile and stationary phases, so band broadening must be related to the interaction kinetics between the solute and these phases. As may be visualized in Figure 13–7, stationary phases that minimize eddy diffusion and mass transfer processes will also minimize peak broadening and thus enhance efficiency. Efficiency of chromatographic separations is a trade-off between the equilibration of solute between the mobile phase and stationary phase and the

Column Flow

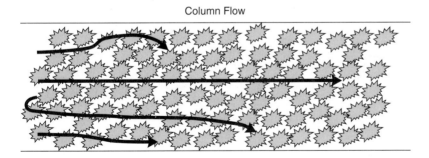

Figure 13–7. Various paths for flow of mobile phase through (across) the stationary phase. Eddy diffusion and the mass transfer processes of mobile-phase transfer, stagnant mobile-phase transfer, and stationary-phase mass transfer affect the efficiency of separation (see text).

▶ forces causing peak broadening. The *number of* **theoretical plates** (N) is used to describe the degree of efficiency exhibited by a chromatographic system and is expressed in the following equation:

$$N = 5.54 \left[\frac{V_r(A)}{w_{1/2}(A)} \right]^2$$

Here, $w_{1/2}(A)$ is the width of solute A's peak at ½ maximum peak height (see Figure 13–5), and $V_r(A)$ is volume of retention, as defined above. In practice, a single theoretical plate can be conceptualized as one equilibration of solute between the stationary and mobile phases. The larger the number of theoretical plates for a solute, A, the better the separation. For example, consider two columns with the same retention volume for solute A; A will elute as a narrower peak from the column with the higher N.

Efficiency can also be expressed as the *height equivalent to a theoretical plate* (H, or sometimes called HETP). H is calculated as:

$$H = \frac{L}{N}$$

where L is the length of the column. In theoretical terms, H is the length of column that is necessary for one equilibration of solute between the mobile and stationary phases. Very efficient columns (small H) can be relatively short, whereas inefficient columns (large H) must be longer. For GC and LC, the various mechanisms that diminish efficiency can be related to H for describing the conditions for optimum efficiency. The relationship with H can be described in an expression that includes flow velocity of the mobile phase (v), eddy diffusion (A), longitudinal diffusion (B), and radial diffusion (C). It would be expected that there are differences in these variables when the mobile phase is a gas versus a liquid. Since this is indeed the case, equations for GC and LC are different. The expression ▶ for GC relating these variables (shown below) is called the **van Deemter equation:**

$$H = A + \frac{B}{v} + Cv$$

For LC, the expression describing the relation of these factors is ▶ termed the **Knox equation:**

$$H = Av^{1/3} + \frac{B}{v} + Cv$$

Plots of H versus the mobile phase flow rate can allow determination of the optimum flow rate for an individual chromatography system. For example, Figure 13–8 shows an example van Deemter (Knox) plot. The nadir for the plot in Figure 13–8 identifies the optimum flow rate that minimizes the H term and, therefore, provides the best efficiency.

G. Mechanisms of Separation

The simplest separation mechanism for chromatography is partitioning between the mobile and stationary phases. This is analogous to an organic extraction technique. In fact, organic solvents are often used as

Separation Mechanisms

- TLC
- Adsorption
- Gel filtration
- Ion exchange
- Affinity
- GLC
- LC/HPLC

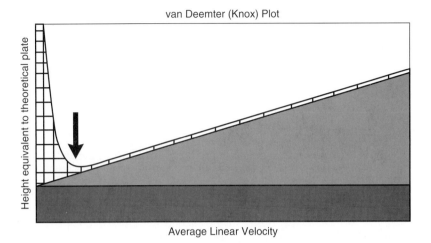

Figure 13–8. A plot of the van Deemter (Knox) equation(s). The arrow indicates the nadir of the height equivalent to a theoretical plate; the average linear velocity at this point is usually the optimum flow rate for the system.

the mobile phase, stationary phase, or both in various chromatography systems. Affinity, ion exchange, and steric exclusion are also physical and chemical mechanisms used for chromatography separations. For all but TLC, chromatography techniques are usually carried out within inert, tube-shaped columns that retain the stationary phase.

1. Thin-Layer Chromatography (TLC). This technique, also referred to as *planar chromatography,* involves a stationary phase which is a thin layer of sorbent, approximately 0.2 mm thick, that is applied to a flat plate made of glass, plastic, or aluminum. The stationary phase (sorbent) is usually composed of silica, alumina, Sephadex, or microcellulose. The mobile phase for TLC is a solvent, which for some applications is a mixture of organic and aqueous liquids. TLC separations are typically carried out in a closed system, often a glass tank sealed with a glass lid. A sufficient volume of mobile-phase solvent is poured into the tank. Samples are applied, or spotted, on the TLC stationary phase at a measured distance from one edge. The TLC plate is then carefully placed into the solvent tank so that the edge with the applied samples is oriented toward the tank's bottom in contact with the mobile phase. (It is critical that the sample application point is above the level of the mobile phase.) The plate is tilted slightly so that the top rests against the side of the tank. The cover is placed on the tank, and this closed system allows the stationary phase to become saturated with vapors from the mobile phase during the run. The mobile phase migrates by capillary action up the stationary phase; flow by capillary action is the balance of the pull of surface tension versus the opposing forces (e.g., viscosity and gravity). After the mobile phase migrates to the desired height on the stationary phase, the separation is stopped by simply removing the plate from the tank. TLC can be carried out in a second dimension, by rotating the plate 90°, and placing it in an appropriate second mobile-phase system.

Run-to-run variability with TLC is high because solvent evaporation, flow rates during the separation, and other variables are difficult to reproducibly control. Solute "spots" on the plate can be qualitatively detected by treating the plate with compounds that will chemically react with the solutes of interest. Quantitative analysis can be performed by densitometry, whereby the ratio of different solutes can be assessed. A common strategy for identification of solutes separated by TLC is to characterize their relative migration in the given system by calculating a ratio of the migration of the solute to the migration distance of the mobile phase solvent. This value is called the ▶ **R$_f$**, which is calculated as follows:

$$R_f = \frac{\text{Migration distance from application point to solute center}}{\text{Migration distance of the mobile phase from application point}}$$

Application of a sample containing the known solute(s) as a reference allows comparison with the unknown spot's R$_f$ for identification.

2. Adsorption Chromatography. Adsorption of solutes from the mobile phase onto a solid stationary phase is a separation mechanism used occasionally in clinical laboratories, mainly for preparation or pretreatment of samples. Figure 13–9 shows a sketch of the principle of adsorption chromatography. Common adsorbents used for stationary phases include silica, which adsorbs basic substances; alumina, which adsorbs acidic substances; and charcoal for adsorption of nonpolar substances. A number of polymers are also available for specialized applications. The adsorption process itself may be relatively nonspecific in nature, in which case the separation quality comes from eluting adsorbed compounds in a selective manner with an appropriate mobile phase. The main disadvantage of adsorption chromatography arises from the difficulty in producing materials having a homogeneous distribution of binding sites. This disadvantage limits its usefulness in the clinical laboratory. Preanalytical separa-

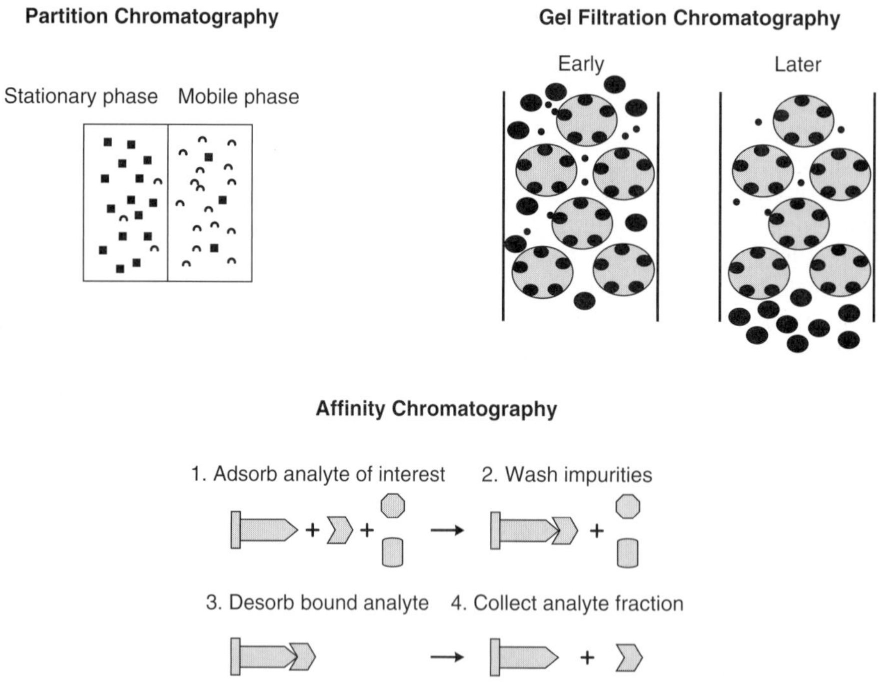

Figure 13–9. Diagrams of various forms of chromatography (see text).

tions to remove interferences before immunoassay for sterols or vitamins are examples of clinical applications.

3. Gel Filtration Chromatography. Gel filtration chromatography, also referred to as *steric exclusion* or *size exclusion chromatography,* is a technique for separating molecules based on their size or molecular weight. As illustrated in Figure 13–9, the stationary phase for this system is a gel that contains crevices or pores that can allow passage of certain molecules based on size. Smaller molecules can penetrate into the pores and their passage is impeded; larger molecules are excluded from the pores and travel through the system more quickly. Examples of gels used include agarose (Sepharose), polyacrylamide, cross-linked dextran, porous glass, and polystyrene–divinylbenzene. Most gel filtration stationary phases have a more or less continuous distribution of pore sizes so that there is a differential migration of molecules smaller than the exclusion limit of the gel. Gel chromatography is mainly used in preparative rather than analytical work.

4. Ion Exchange Chromatography. This technique is useful for the separation of molecules that are or can be charged. The stationary phase contains functional groups attached to a resin that are oppositely charged from the molecules of interest. The pH and composition of the mobile phase can be manipulated so that the analyte is charged, thus causing retention by electrostatic attraction on the stationary phase. Cation exchange resins are prepared with acidic groups that capture positively charged analytes from the mobile phase. In anion exchange, the basic groups on the resin bind negatively charged solutes. Elution of the solute from the column may be accomplished by (1) changing the pH of the mobile phase to alter the charge of the analyte, or (2) by increasing the salt concentration of the mobile phase to provide more charged species that compete with the analyte for sites on the resin. After elution, the solute can be measured by a technique appropriate to detect the analyte (e.g., spectrophotometry). Clinical applications for ion exchange include separation of hemoglobin variants and separation/preparation of amino acids prior to measurement.

5. Affinity Chromatography. As depicted in Figure 13–9, this technique takes advantage of highly selective binding reactions such as antigen–antibody, hormone–receptor, enzyme–substrate, or lectin–carbohydrate attractions. With affinity chromatography, the binding entity is termed the **ligand,** which is the selective component of the stationary phase. Ligands are usually bound to an inert support material such as polyacrylamide, agarose, or cross-linked dextran. The mobile phase is manipulated to optimize binding between the ligand and the solute of interest. The solute is eluted (displaced) from the column using a method appropriate for the specific binding pair; for example, desorption may be accomplished by adding a substrate or inhibitor, changing the pH to disrupt the specific binding, adding a large excess of a solute analog that will compete for binding sites, changing the ionic strength, or using a chaotropic agent such as urea or sulfite that disrupts hydrogen bonding forces. Affinity chromatography is used for preparative separation of proteins and antibodies. In the clinical laboratory, affinity chromatography is used for measurement of low-density lipoproteins and very low-density lipoproteins using heparin columns and for measurement of glycated hemoglobin using a stationary phase of *m*-aminophenylboronic acid cross-linked to beaded agarose.

6. Gas Chromatography (GC). In GC, the mobile phase and the solute(s) to be separated are in the gaseous state for separation and detection. Solutes are separated from each other by differences in vapor pressure as well as interaction with the stationary phase. The stationary phase can be a variety of liquids or solids; the terms GLC (gas–liquid chromatography) and GSC (gas–solid chromatography) are used to describe these systems. The main components of a GC system are diagrammed in Figure 13–3.

Table 13–1 lists the major strategies that have been used for solute detection with GC. The detector output is a chromatogram similar to the example shown in Figure 13–4. Most detection techniques do not definitively identify the solute of interest. For interpretation of the chromatogram, peaks are usually identified by retention times that are consistent for the solute of interest compared to standards and/or with previous experience; the comparative retention times of internal standards can also be used for identification, but this is also not definitive. **Mass spectrometry (MS)** is a detection system that provides definitive information as to the precise identity of the solutes eluting at specific times after injection onto the column; this system is commonly referred to as GC-MS.

Some solutes of interest are appropriate for GC separation without alteration; however, other solutes that are not volatile at temperatures used for GC, or are labile upon heating to volatilize, may be suitable for GC analysis after derivatization. *Derivatization* is the chemical modification of a molecule by replacement of functional groups with other entities. To increase volatility, polar functional groups are usually transformed into more nonpolar groups, such as acyl groups, silyl groups (e.g., trimethylsilyl), monoximes, or esters. In addition, derivatization may enhance selectivity of the separation and/or enhance the detector's response to the solute (Table 13–1).

Table 13–1. Various Detectors Used in Gas or Liquid Chromatography

DETECTOR TYPE	CHROMATOGRAPHY	RANGE OF APPLICATION	OPERATION PRINCIPLE
Flame ionization	Gas	Hydrocarbons	$CHNO + heat \rightarrow CHNO^+ + e^-$; electrons collected for detection
Thermal conductance	Gas	Universal	Change in mobile phase conductivity by solute
Nitrogen/phosphorus	Gas	Nitrogen and phosphorus	Alkali bead selectively ionizes N & P containing compounds
Electron capture	Gas	Electronegative groups (I, Br, F)	$e^- + R + N_2 \rightarrow Re^- + N_2 + e^-$; Excess electrons collected
Mass spectrometer	Gas or liquid	Universal	Monitor charge-to-mass ratio by scanning or single ion monitoring
Ultraviolet photometer	Liquid	Selective if solute absorbs	Absorbance of solute or solute must be derivatized
Diode array	Liquid	Analytes that have spectra	Analyze full spectra of light absorbance of eluted solutes
Fluorometer	Liquid	Selective if solute is fluorescent	Monitor fluorescence of mobile phase, solute must fluoresce or be derivatized
Electrochemical	Liquid	Catecholamines and others	Electrochemically measures oxidized/reduced solutes

7. Liquid Chromatography (LC). Use of a liquid mobile phase characterizes LC. When the stationary phase is comprised of small particles to enhance the efficiency of separation, high pressure is needed to facilitate flow of the mobile phase through the system. Such high pressure is essential for highly effective LC systems referred to as *high-performance liquid chromatography* (HPLC), as mentioned above. The components of HPLC systems can be represented schematically as shown in Figure 13–3.

A few technical issues regarding LC are worthy of mention. It is particularly important to remove particles and bubbles from the mobile-phase solvent because they will negatively affect performance if delivered onto the column. The introduction of bubbles through the sample injector of an HPLC system must also be avoided; HPLC systems often incorporate a "fixed loop" injector, which is easily automated. In LC, solutes may be eluted either by (1) an isocratic elution method in which the composition of the mobile phase used for separation is constant, or (2) gradient elution in which the mobile phase varies (either continuously or stepwise) during the separation. For isocratic separations, a single LC pump is generally used. For gradient systems, two pumps may be programmed to vary the delivery of two different mobile-phase solutions during the separation. Most systems use high-quality pumps to avoid pulsation, and use a damping mechanism to maintain smooth delivery of the mobile phase to the column.

The stationary phase in HPLC columns is usually composed of materials bonded to the surface of silicate esters. For normal phase HPLC, the stationary phase is frequently charged silanol, amino, or nitrile groups. For reversed phase HPLC, nonpolar C18-type octadecylsilane molecules are bonded to the particulate support. Polymeric column packings, including graphite carbon, polymers, or mixed copolymers, may be derivatized on the stationary phases with C4, C8, or C18 entities; these packings are particularly stable over a wide range of pH. Chiral packings, for separation of mirror image forms of the same compounds (i.e., enantiomers), are also available. Detector systems available for HPLC systems are listed in Table 13–1. It is of note that technology for MS detectors has advanced quickly in recent years and will undoubtedly be an increasingly important detection strategy for combination with LC and HPLC in the future.

LC, and particularly HPLC, systems have been used for many separations in the clinical laboratory. These include analysis of many drugs, catecholamines, homocysteine, amino acids, and bone markers such as deoxypyridinoline, to mention a few.

IV. SUMMARY

Electrophoresis and chromatography are often viewed as separation and analysis methods to be avoided in the clinical laboratory. However, due to the excellent flexibility and separation power of these techniques, they will continue to play an important role, particularly when coupled to superior detection systems such as MS. The areas of drug analysis, molecular diagnostics, and genomics applications will continue to depend on these principles of separation and analysis for the foreseeable future. Also, futuristic miniaturization technologies will include components utilizing these techniques.

SUGGESTED READINGS

Burtis CA, Ashwood ER (eds.). *Tietz textbook of clinical chemistry,* 3rd ed. Philadelphia: WB Saunders, 1999; Chapters 7 and 8.

Ewing AG, Wallingford RA, Olefirowicx TM. Capillary electrophoresis. *Anal Chem* 61:292A–303A, 1989.

Jacobson SC, Hergenroder R, Moore AW Jr., Ramsey JM. Precolumn reactions with electrophoretic analysis integrated on a microchip. *Anal Chem* 66:4127–32, 1994.

Kaplan LA, Pesce AJ (eds.). *Clinical chemistry: Theory, analysis, correlation,* 3rd ed. St. Louis: CV Mosby, 1996; Chapters 5, 6, 7, 8, and 10.

Rao ML, Staberock U, Baumann P, et al. Monitoring tricyclic antidepressant concentrations in serum by fluorescence polarization immunoassay compared to gas chromatography and HPLC. *Clin Chem* 40:929–33, 1994.

Items for Further Consideration

1. Describe the various effects that temperature can have on chromatography and electrophoresis separations. Are there any benefits to performing testing at elevated temperatures?

2. Two-dimensional electrophoresis is a very high-resolution technique; however, run-to-run variability in the position of spots can be rather large, making interpretation of patterns difficult. Articulate a strategy that could be used to compensate for run-to-run variability.

3. List and explain the characteristics necessary for chromatographic separation of a polar solute in a polar solvent, a nonpolar solute in a polar solvent, and a nonpolar solute in a nonpolar solvent.

4. In one blanking strategy used for HPLC detection, the elution stream is split after the column and a reagent is added to facilitate measurement of a low-concentration solute. Sketch and briefly describe your idea of the experimental design for such a system.

5. Suggest possible systems for three-dimensional electrophoresis and describe them briefly.

Molecular Diagnostics

Hassan M.E. Azzazy

Key Terms ◀

AMPLICON

AMPLIFICATION

ANNEAL/HYBRIDIZE

BASE PAIRING

COMPLEMENTARY DNA (cDNA)

COMPLEMENTARY SEQUENCES

DNA POLYMERASE

LIGASE

NUCLEOTIDE

NUCLEIC ACIDS

PRIMERS

PROBE

REPLICATION

REVERSE TRANSCRIPTASE

RNA POLYMERASE

STRINGENCY

TARGET DNA

TRANSCRIPTION

TRANSLATION

Chapter Objectives

Upon completion of this chapter, the reader will be able to:

1. Describe the structure and function of nucleic acid molecules (DNA, RNA).
2. Explain the terms: hybridize, anneal, amplify, ligation, primer, probe.
3. Briefly describe the principles and different formats of hybridization assays.
4. State the principles of the following amplification techniques: polymerase chain reaction (PCR); variations of PCR: RT-PCR, nested PCR, and multiplex PCR; ligase chain reaction (LCR); and gap–ligase chain reaction.
5. Discuss the advantages and disadvantages of utilizing molecular diagnostics in the clinical laboratory.
6. Discuss different decontamination strategies that are applied in the molecular diagnostic laboratory.
7. Suggest future applications of molecular diagnostics.

Instruments/Techniques Included ◀

GAP-LCR	**MULTIPLEX PCR**	**REVERSE TRANSCRIPTASE (RT)-PCR**
HYBRIDIZATION	**NESTED PCR**	
LIGASE CHAIN REACTION (LCR)	**POLYMERASE CHAIN REACTION (PCR)**	**THERMOCYCLER**

Pretest

1. Nucleotides, the building blocks of nucleic acids, are linked together by
   a. peptide bonds
   b. glycosidic bonds
   c. phosphodiester bonds
   d. hydrogen bonds

2. Eukaryotic mRNA can be purified by using an
   a. oligo dT column
   b. oligo dA column
   c. oligo dG column
   d. oligo dC column

3. For chemical decontamination of deoxyribonucleic acid (DNA), bleach inactivation of amplicons is dependent on amplicon length:
   a. longer products can be inactivated within 30 seconds
   b. shorter products (< 100 base pairs) require 30 seconds
   c. longer amplicons can be inactivated within 30 to 60 minutes
   d. shorter products (< 100 base pairs) require 30 to 60 minutes

4. Which of the following enzymes catalyzes the synthesis of complementary DNA (cDNA) from ribonucleic acid (RNA)?
   a. RNA polymerase
   b. reverse transcriptase
   c. *Taq* polymerase
   d. DNA ligase

5. Which of the following represent stringent hybridization conditions?
   a. low salt concentration, high temperature
   b. low salt concentration, low temperature
   c. high salt concentration, high temperature
   d. high salt concentration, low temperature

6. Each cycle of a polymerase chain reaction (PCR) consists of three steps that proceed in the following order:
   a. DNA denaturation, primer annealing, extension
   b. primer annealing, extension, DNA denaturation
   c. DNA denaturation, extension, primer annealing
   d. extension, DNA denaturation, primer annealing

7. Which of the following amplification reactions utilizes two sets of primer pairs specific for one target sequence?
   a. PCR
   b. nested PCR
   c. RT-PCR
   d. multiplex PCR

8. Which of the following amplification reactions utilizes two or more sets of primer pairs specific for different target sequences?
   a. PCR
   b. nested PCR
   c. RT-PCR
   d. multiplex PCR

9. Which of the following enzymes is widely used for pre-PCR decontamination procedures?
   a. DNA glycosylase
   b. DNA polymerase
   c. DNA ligase
   d. reverse transcriptase

10. DNA ligase is an enzyme that catalyzes
   a. the removal of nucleotides one at a time from the end of a DNA strand
   b. the formation of phosphodiester bonds between nucleotides
   c. the synthesis of an RNA strand complementary to its DNA template
   d. the cleavage of DNA at a specific site in the strand

11. RNA polymerase is a specific enzyme that catalyzes
    a. the removal of nucleotides one at a time from the end of a DNA strand
    b. the formation of phosphodiester bonds between nucleotides
    c. the synthesis of an RNA strand complementary to its DNA template
    d. the cleavage of DNA at a specific site in the strand

12. The process by which a single nucleic acid strand "attaches" to a complementary target DNA or RNA sequence is called
    a. amplification
    b. annealing
    c. ligation
    d. polymerization

13. If the thymine content of DNA from an organism is 31%, what is the percentage of its cytosine content?
    a. 19%
    b. 31%
    c. 38%
    d. 69%

14. Amplification of DNA by means of PCR requires
    a. reverse transcriptase, heat-stable DNA polymerase, and two primers
    b. DNA ligase, heat-stable DNA polymerase, and two primers
    c. repeated cycles of hybridization and ligation
    d. at least one molecule of DNA as a starting template, heat-stable DNA polymerase, and two primers

15. Nucleic acid probes are best described as
    a. short nucleotide sequences that are labeled with a reporter molecule
    b. nucleotide sequences that hybridize to a variety of target DNA sequences

    c. nucleotide sequences that can be targeted only to DNA sequences
    d. short double-stranded DNA sequence that is labeled with a reporter molecule

16. The double strands of DNA can be reversibly separated by
    a. acid
    b. alkali
    c. phenol
    d. ethidium bromide

17. Which of the following techniques is used for DNA detection?
    a. Western blot
    b. Northern blot
    c. Southern blot
    d. Eastern blot

18. Which of the following techniques is used for RNA detection?
    a. Western blot
    b. Northern blot
    c. Southern blot
    d. Eastern blot

19. Amplification primers and hybridization probes are usually
    a. > 100 nucleotides
    b. > 1,000 nucleotides
    c. < 50 nucleotides
    d. between 100 and 200 nucleotides

20. Molecular diagnostic tests can be used to
    a. detect tumor antigens
    b. monitor levels of therapeutic drugs
    c. detect infectious diseases
    d. monitor hormone levels

I. INTRODUCTION

Nucleic acid technology has assumed an essential role in different areas of *in vitro* diagnosis, including human genetics, clinical chemistry, and microbiology. Molecular diagnostics encompasses the arsenal of molecular biology techniques that can be employed as diagnostic tests where DNA and RNA are the analytes. Molecular diagnostic tests can be applied to differential diagnosis, monitoring and follow-up, and prognosis. Analysis of genetic information may even be utilized to detect disease prior to clinical manifestations or to assess a person's probability of acquiring a disease. The high sensitivity of molecular diagnostics has the potential to detect a single organism of an infectious agent or a single tumor cell.

While molecular testing has augmented diagnostic capabilities in some areas of the medical laboratory, it has opened entirely new avenues of investigation in others. Applications for molecular diagnostics testing fall into several major categories:

1. Microbiology: quick and reliable detection of infectious diseases
2. Cancer diagnosis
3. Genetic diseases: detection and prediction of genetic disorders to help counseling and early treatment
4. Prenatal screening
5. Forensic science: identification of tissue samples in crime scenes

II. BIOCHEMISTRY

A. CHEMICAL COMPOSITION OF NUCLEIC ACIDS

▶ **Nucleic acids** (DNA and RNA) are polymers of nucleotides. Each
▶ **nucleotide** is composed of:

1. A nitrogenous base (Figure 14–1A):
   • Four bases occur naturally in DNA: adenine (A), guanine (G), cytosine (C), and thymine (T)
   • RNA contains: adenine (A), guanine (G), cytosine (C), and uracil (U)
2. A pentose sugar: deoxyribose in DNA and ribose in RNA
3. A phosphate group

The sugar and phosphate alternate in the polynucleotide chain. Nucleotides are linked together via phosphodiester linkages: the 5′ position of the ribose residue of one nucleotide is linked, via a phosphate group, to the 3′ position of the pentose sugar of the next nucleotide (Figure 14–1B).

B. STRUCTURE OF DNA

DNA is a double-stranded molecule (double helix) in which the sugar-phosphate backbone is on the outside. Bases from opposite strands are oriented toward the inside of the helix and associate by hydrogen bonding: G associates with C via 3 hydrogen bonds; T associates with A via two hydrogen bonds (Figure 14–1C). This is
▶ termed **base pairing,** also known as *complementarity* of the strands.

C. OVERVIEW OF NUCLEIC ACID PROCESSING

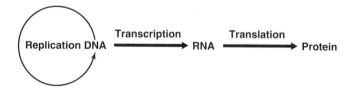

1. ▶ **Replication:** A process by which parental DNA is duplicated by the action of **DNA polymerase.**
2. ▶ **Transcription:** One strand of DNA is "transcribed" by a specific enzyme, **RNA polymerase,** to produce a complementary strand of RNA.

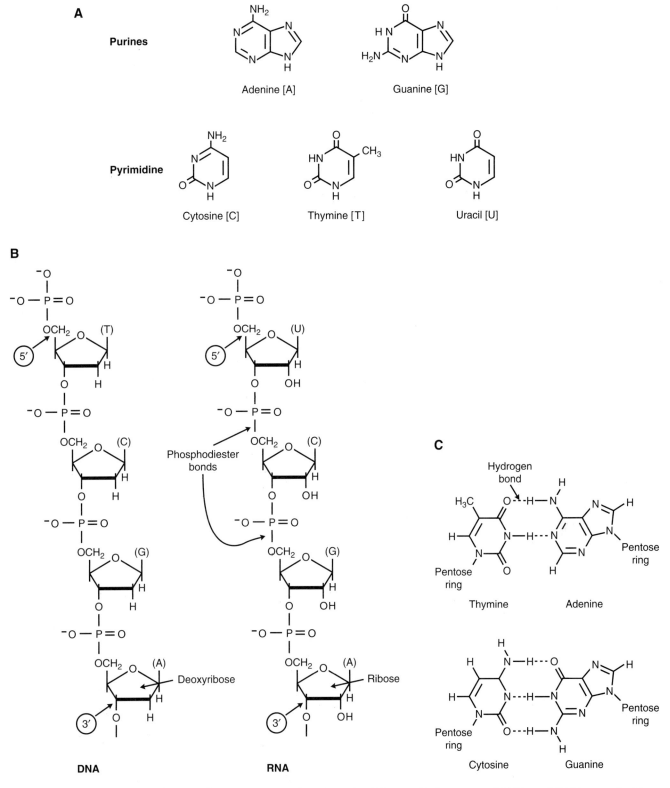

Figure 14–1. **A.** Structure of pyrimidine (C, T, U) and purine (A, G) bases. **B.** Structure of RNA and DNA strands. Nucleotide units are linked by phosphodiester bonds that link the 5′ carbon of the pentose sugar of one nucleotide to the 3′ carbon of the pentose sugar of the next nucleotide. **C.** Base-pair matching between A:T and G:C. Dashed lines indicate hydrogen bond formation.

▶ 3. **Translation:** Messenger RNA (mRNA) is "translated" by ribosomes into proteins.

The nucleotide sequence of DNA dictates the nucleotide sequence of mRNA. Similarly, the sequence of mRNA dictates the amino acid sequence of the protein (each amino acid is encoded by three nucleotides).

III. HYBRIDIZATION

A. PRINCIPLE

The double strands of DNA or DNA–RNA hybrid can be reversibly separated by heat (or alkali), which disrupts the hydrogen bonds between the complementary bases. When temperature is reduced, free
▶ (dissociated) DNA strands will spontaneously **anneal (hybridize)** with complementary strands, thus reforming a double helix.

B. PROBES

▶ A **probe** is a short, single-stranded DNA (or RNA) sequence that is labeled with a reporter molecule. It binds to, and specifically identi-
▶ fies, **complementary sequences** of nucleotides in another single-stranded nucleic acid sequence (target).

C. HYBRIDIZATION FORMATS

1. Solid Support Hybridization. The target nucleic acid is immobilized, then denatured on a nitrocellulose or nylon membrane (Figure 14–2). A labeled probe is added, and hybridization will occur if the nucleic acid target contains a sequence complementary to the probe. Types of solid support hybridization include Southern blot (for detection of DNA), Northern blot (for detection of RNA), and dot blots.

2. In-Solution Hybridization. A small amount of target nucleic acid is required. The speed of the reaction is increased by (1) adding a high concentration of labeled probe and (2) the free probe–target interaction. Applications include:

- Culture confirmation assays
- Microorganism identification in clinical samples
- To confirm the identity of amplification products

3. *In Situ* Hybridization. The probe is applied to formalin-fixed or paraffin-embedded whole cells or tissue sections (on microscope slides) that may contain the target DNA.

D. FACTORS AFFECTING HYBRIDIZATION ASSAYS

Too low a temperature, too high an ionic strength, or both will re-
▶ duce the **stringency** of hybridization and may negatively affect the specificity of the detected signal. In contrast, raising the temperature, reducing the ionic strength, or both, will increase the hybridization

---

Characteristics of Nucleic Acid Probes

- Composed of nucleotide sequences that are labeled with a reporter molecule.
- Hybridize (bind) to their complementary sequences with high specificity.
- Can be targeted to either RNA or DNA sequences.
- Oligonucleotide probes (< 50 nucleotides) can be chemically synthesized.
- Under stringent conditions (such as high temperature, low salt concentration), oligonucleotide probes can detect the change of a single nucleotide within a specific nucleic acid sequence.

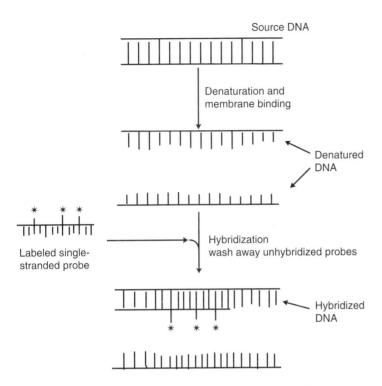

Figure 14–2. DNA hybridization. The source DNA is denatured, and the two strands are kept apart by binding them to a nitrocellulose or nylon membrane. Labeled single-stranded DNA probe is added to the bound source DNA strands. If source DNA contains a sequence complementary to the probe, the probe will hybridize to the source DNA. The membrane is then washed to remove unhybridized probes. If probe hybridizes with the source DNA, it can be detected with an assay that identifies its labeled tag. Asterisks indicate the label of the probe.

Factors Affecting Hybridization Assays

- Temperature
- Ionic strength
- Length and base composition of probe
- pH

stringency. Therefore, a tight control of temperature and of reagents is necessary to avoid false-positive as well as false-negative results. Other factors that influence hybridization include length and base composition of the probe and pH.

Variables in selecting nucleic acid probes include:

1. Specificity: probes must be highly specific for the target sequence.
2. Length of the probe: Depending on the test protocol, probes may be long (> 100 nucleotides) or short (< 50 nucleotides).
3. Guanine:cytosine (G:C) content of the probe: A probe with a high G:C content will hybridize to its target more strongly than a probe with a low G:C content (and consequently high adenine:thymine [A:T] content), due to the formation of three hydrogen bonds between G:C base pairs versus two hydrogen bonds between A:T base pairs.
4. DNA (stable) versus RNA (less stable).
5. Labeling method used: isotopic, nonisotopic.

Reporter molecules include:

1. Isotopic labels:
   - ^3H, ^{32}P, ^{33}P, ^{35}S, ^{125}I
   - Photographic exposure: radiation to x-ray film
   - Quantification: scintillation counter, digital imaging

2. Nonisotopic labels (enzymes, fluorophores, lumiphores):
   - Enzyme reactions:
     - Substrates: colored or luminescent substrates
     - Enzymes: horseradish peroxidase, alkaline phosphatase
   - Fluorescent molecules (e.g., fluorescein)
   - Luminescence: dioxetene analogs, luminol, acridinium esters

IV. AMPLIFICATION TECHNIQUES

▶ Without **amplification,** nucleic acid analysis often has the disadvantage of low sensitivity. Amplification techniques, producing millions of copies of a specific sequence in a short period of time, increase sensitivity dramatically while still retaining a high specificity.

A. POLYMERASE CHAIN REACTION (PCR)

PCR is a technique for oligonucleotide primer–directed enzymatic amplification of a specific DNA sequence. PCR is capable of amplifying a single copy of template DNA by a factor of $\geq 10^6$ from a mixture of irrelevant sequences. Amplification by PCR requires knowledge of unique sequences that are adjacent to the DNA segment to
▶ be amplified so that specific **primers** can be designed that will anneal to the sequences flanking a specific DNA segment. Applications of PCR include detection of genetic mutations, chromosomal translocations, infectious diseases, and forensic testing.

The PCR method includes repeated cycles that amplify selected nucleic acid sequences. Each cycle consists of three steps (Figure 14–3):

1. DNA denaturation by heating to ~94°C (the double strands of DNA are separated)
2. Primer annealing, at low temperature (~54°C), in which primers anneal to their complementary target sequences
3. Extension: DNA *Taq (Thermus aquaticus)* polymerase extends the newly annealed primers (at ~72°C)

At the end of each cycle, the quantities of PCR products are theoretically doubled. The whole procedure is carried out in a programma-
▶ ble **thermocycler.** PCR is characterized by high sensitivity that allows detection of signals from nonviable microorganisms, degraded DNA samples, and sometimes from individual cells.

1. Modifications of the Standard PCR

Reverse Transcriptase (RT)-PCR. RT-PCR was developed to amplify
▶ RNA target sequences. **Reverse transcriptase** converts target RNA
▶ sequences (either messenger or ribosomal) into **complementary DNA (cDNA)** by means of DNA primers. The cDNA is replicated to produce double-stranded DNA by the action of DNA polymerase. The PCR steps can then proceed normally to amplify the target cDNA.

Applications of PCR

- Detection of genetic mutations
- Chromosomal translocations
- Infectious diseases
- Forensic testing

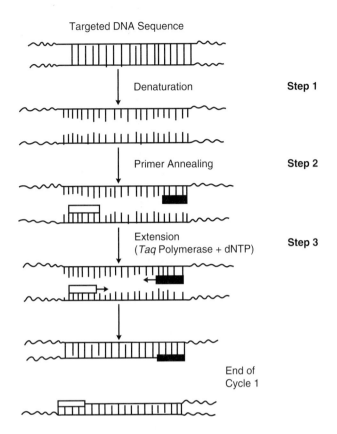

Figure 14–3. Polymerase chain reaction (PCR). PCR is used to amplify specific DNA sequences. Each PCR cycle is composed of three steps: (1) denaturation: double-stranded DNA is heated to separate it into individual strands, (2) primer annealing: specific primers that define the segment to be amplified anneal to distinct sequences of the single strands, and (3) extension: DNA *Taq* polymerase extends the primers in each direction and synthesizes two strands complementary to the original two strands. This cycle is repeated multiple times to produce an amplified product of defined sequence.

Enzymes that can perform both the RT and PCR stages of RT-PCR in one tube are now available. RT-PCR is used in diagnosing infections caused by RNA viruses, detecting viable *Mycobacterium* species, and monitoring antimicrobial therapy.

Nested PCR. This technique, designed mainly to increase the sensitivity and specificity of PCR, utilizes two sets of amplification primers. One set of primers is used for the first round of amplification (15 to 30 cycles). The products of the first amplification round are then subjected to a second round of amplification with another set of primers that hybridize within the **target DNA** amplified by the first primer set. In this method, even if nonspecific amplification products are generated in the first round of amplification, the second pair of primers acts to select only the specific targets for further amplification. Therefore, the amplification by the second primer set verifies the specificity of the product of the first amplification round.

Multiplex PCR. In this amplification reaction, two or more sets of primer pairs specific for different targets are introduced in the same tube. Therefore, simultaneous amplification of more than one unique target DNA sequence can be carried out at the same time. It is essential that primers used in multiplex reactions be designed to have very similar annealing temperatures. This technique can be used to detect multiple pathogens in the same specimen.

2. Selected Methods for Detection of PCR Products

▶ *Direct Detection by Ethidium Bromide.* Detection of a PCR product **(amplicon)** can be simply performed by adding ethidium bromide, a dye that becomes fluorescent after intercalating with DNA strands, to the reaction mixture at the end of the amplification process. Upon examination of the tubes under an ultraviolet lamp, only positive reactions (tubes containing the amplicon) will emit light.

Detection by Gel Electrophoresis. PCR amplification products can be separated according to their size by agarose or polyacrylamide gel electrophoresis. The separated DNA fragments can be visualized by staining the gel with ethidium bromide or silver. Use of electrophoresis allows confirmation of the size of the PCR products, thus allowing the specific product to be distinguished from other nonspecific amplification products.

Blotting and Hybridization with Specific Probes. This method is particularly utilized when screening for known point mutations in the amplified sequence. Following agarose gel electrophoresis, PCR amplicons can be transferred (blotted) to nitrocellulose or nylon membranes and analyzed by hybridization with a labeled oligonucleotide probe specific for the mutation. After hybridization, the membrane is washed at a critical temperature to remove probes bound to sequences that were not a perfect match.

Sequencing of PCR Products. The nucleotide sequence of the PCR product can be determined by sequencing the amplified stretch of DNA. This can be particularly important in case of infection with either human immunodeficiency virus (HIV) or hepatitis C virus (HCV) in order to determine the sequence variability in genes coding for viral envelopes.

B. LIGASE CHAIN REACTION (LCR)

LCR relies on repeated cycles of oligonucleotide hybridization and ligation to generate multiple copies of a nucleic acid sequence of interest. This procedure requires at least two complementary pairs of oligonucleotide probes (probes A and A′, and probes B and B′). The probes (probes A and B, and probes A′ and B′) hybridize at adjacent positions on each strand of heat-denatured target DNA leaving a "nick" in between (Figure 14–4). The probes, bound to target DNA, provide a substrate for joining by the "nick-sealing" activity of DNA ▶ **ligase.** Following ligation, the target-ligation product duplex is separated by thermal denaturation. Both the single-stranded target DNA and the ligated probe products are then available to act as templates for the hybridization and ligation of more probes. Theoretically, each cycle of oligonucleotide hybridization and ligation results in doubling of the number of template sequences. The recent development of a thermostable DNA ligase greatly simplified this technique.

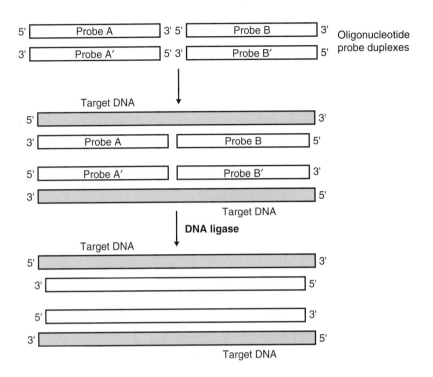

Figure 14–4. Ligase chain reaction (LCR). LCR involves repeated cycles of oligonucleotide hybridization and ligation to amplify target nucleic acid sequence. Probes (A and B, and A′ and B′) hybridize at adjacent positions on each strand of the target DNA. The target-bound probes are then joined by DNA ligase. Following ligation, the target-ligation product duplex is denatured by heating and the single-stranded target DNA and the ligated products act as templates for hybridization and ligation of additional probes.

C. GAP–LIGASE CHAIN REACTION (GAP-LCR)

Gap-LCR is a modified version of LCR in which oligonucleotide probes are modified so that they cannot be joined in the absence of a target-specific step. The probes are designed so that when they are hybridized to a target DNA, they are not immediately adjacent but are separated by a gap. In Gap-LCR (Figure 14–5), the specimen is first heated by the thermal cycler to denature the double-stranded target DNA and the probes. The temperature is then lowered, and the two complimentary probes (probes A and B, and probes A′ and B′) hybridize at adjacent positions on each strand of heat-denatured target DNA, leaving a gap between them ranging in length from one to several bases. The gap between the probes is filled using a thermostable polymerase and deoxyribonucleotides. The probes, now adjacent and bound to target DNA, provide a substrate for ligation by the nick-sealing activity of DNA ligase. Following ligation, the target-ligation product duplex is separated by thermal denaturation, and both the single-stranded target DNA and the ligated probe products become available to act as templates for the hybridization and ligation of more probes.

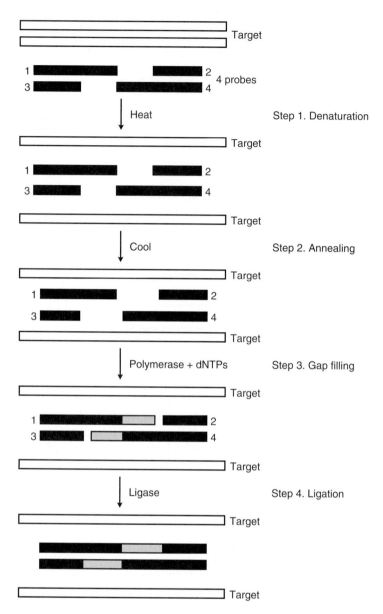

Figure 14–5. Gap–ligase chain reaction (LCR). Each cycle of gap-LCR involves 4 steps: (1) denaturation: heat is applied to denature target DNA and the probes, (2) annealing: temperature is lowered to allow hybridization of the probes to their target sequences, (3) gap filling: the gap between the probes is filled by the action of DNA polymerase, and (4) ligation: the adjacent probes are ligated by the action of a thermostable DNA ligase.

V. CONTAMINATION IN THE MOLECULAR DIAGNOSTICS LABORATORY

The development of reliable diagnostic tests based on nucleic acid amplification relies on the ability to control contamination. This is because the greatest advantage of these techniques—the ability to amplify minute amounts of target DNA—also renders them susceptible to contamination. Because of the very high sensitivity of amplification-based assays, any minute amounts of DNA from exogenous sources can generate a false-positive and thus incorrect diagnosis. Carryover contamination, contaminating DNA originating from previously amplified products (amplicons), is the most frequent cause of false-positive results.

A. SOURCES OF CONTAMINATION

Contamination during the preanalytical phase is due to the presence of nucleic acids that originate from a source other than the patient sample. Potential sources of contamination during the analytical phase can occur due to compromised reagents, commercially available enzyme preparations, consumables and laboratory equipment, and spreading of aerosols from a positive test sample to an originally negative one.

B. STRATEGIES TO AVOID AMPLICON CONTAMINATION

1. Preamplification Decontamination

UDG-dUTP. The combined use of the enzyme uracil DNA glycosylase (UDG) and deoxyuridine triphosphate (dUTP) is the most widely used pre-PCR decontamination procedure. UDG is an enzyme that removes the uracil residues from DNA, thus creating abasic sites from which polymerase cannot extend. Because these abasic sites are susceptible to hydrolysis by heat or alkali, the DNA is cleaved and rendered unamplifiable in subsequent reactions. This decontamination procedure requires a pre-test protocol in which PCR is performed with dUTP instead of deoxythymidine triphosphate (dTTP); amplicons containing deoxyuridine instead of dTTP are generated. Consequently, the DNA products become distinct from the true target DNA, which contains deoxythymidine in place of deoxyuridine. Newly assembled PCR mixtures can then be incubated with UDG prior to temperature cycling. If any deoxyuridine-containing amplicons were inadvertently introduced into the fresh PCR mixtures, they would be rendered unamplifiable by UDG. Prior to temperature cycling, UDG is inactivated at 95°C for 10 minutes.

2. Postamplification Decontamination

Photochemical Decontamination. This sterilization method is based on psoralen photochemistry. Psoralens are linear furocoumarins that intercalate between the base pairs of DNA. Upon excitation with light of wavelength between 320 and 400 nm, psoralens react with pyrimidines forming monoadducts and inter- and intra-strand crosslinks that block polymerase extension. However, if hybridization of amplicon products is required after amplification, isopsoralens should be used. Isopsoralens are angular furocoumarins that form only monoadducts, leaving the modified amplicons single stranded and ready for hybridization. The isopsoralen derivatives can be added to the reaction mixture prior to amplification and the reaction tube is irradiated for 15 minutes with ultraviolet (UV) light after PCR. This prevents reamplification of contaminating DNA amplicon in new reaction mixtures.

Sources of Contamination

Preanalytical phase:
- Nucleic acids that originate from a source other than the patient sample

Analytical phase:
- Reagents
- Commercially available enzyme preparations
- Consumables and laboratory equipment
- Spreading of aerosols from a positive sample to an originally negative one

▶ *Chemical Decontamination.* Ten percent sodium hypochlorite (bleach) was found to eliminate all ethidium bromide stainable DNA and to prevent PCR amplification of DNA within 1 minute of treatment. By contrast, 2 M HCl did not destroy DNA detectable by PCR within 5 minutes. Bleach inactivation is dependent on amplicon length. Longer products can be inactivated within 30 seconds, shorter products (< 100 base pairs) require ~5 minutes.

3. Standard Decontamination Procedures. These include chemical cleaning of surfaces with 10% sodium hypochlorite, permanent UV irradiation (254 nm) of laboratory surfaces and other surfaces after use, autoclaving of laboratory equipment, the use of positive displacement pipettors or regular pipettors in conjunction with aerosol-proof pipette tips, dedication of lab coats and pipettors to a specific area, and routine analysis of reagents for contamination. The molecular diagnostics laboratory should be designed so that (Figure 14–6) work flow is unidirectional from areas free of amplification products to areas where amplicons are analyzed.

Standard Decontamination Procedures

- Cleaning of surfaces with 10% sodium hypochlorite
- Permanent UV radiation of laboratory surfaces
- Autoclaving of laboratory equipment
- Use of positive displacement pipettors or regular pipettors in conjunction with aerosol-proof tips
- Dedication of labcoats and pipettors to specified areas
- Routine analysis of reagents for contamination
- Unidirectional work flow

VI. ADVANTAGES AND DISADVANTAGES OF MOLECULAR DIAGNOSTIC TECHNIQUES

Advantages of molecular diagnostics are listed in Table 14–1. In microbiology, molecular diagnostic techniques are ideal in the following situations: (1) to facilitate detection of unculturable microorganisms such as *Tropheryma whippelii* (causative agent of Whipple disease) and microorganisms that should be grown under fastidious and specialized conditions; (2) for viral load determinations—these methods are currently used to evaluate progression of HIV and HCV diseases; and (3) identification of nonviable infections. Molecular diagnostic techniques enhance laboratory safety. Despite the fact that organisms such as *Coxiella burnetii, Mycobacterium tuberculosis,*

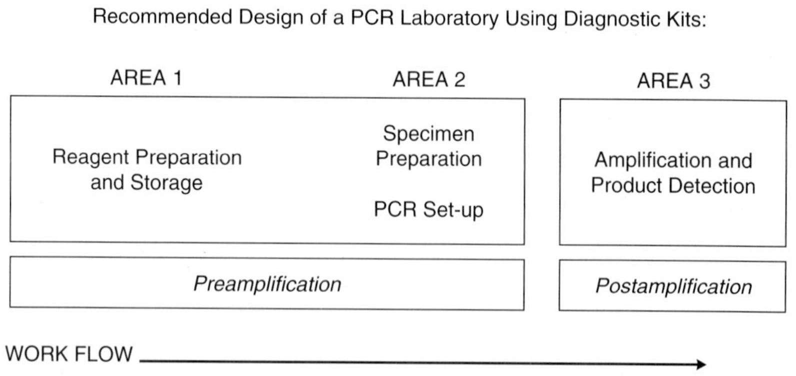

Recommended Design of a PCR Laboratory Using Diagnostic Kits:

AREA 1	AREA 2	AREA 3
Reagent Preparation and Storage	Specimen Preparation PCR Set-up	Amplification and Product Detection
Preamplification		*Postamplification*

WORK FLOW ⟶

Figure 14–6. Design for a molecular diagnostics laboratory. Work flow should follow a unidirectional path from preamplification areas free of amplicons to areas where amplicons are produced and detected.

Table 14–1. Advantages of Molecular Diagnostic Techniques

High sensitivity of the amplification techniques
High specificity
Decrease time required for diagnosis
Identification of genetic disorders
Epidemiological studies

and *Coccidioides immitis,* and other viruses are easily grown, they are considered laboratory hazards. These organisms may infect laboratory workers and cause serious illness or death. Molecular detection of microbes is considered a potentially effective replacement for traditional culture. The high sensitivity of molecular diagnostic methods has permitted detection of microorganisms from as few as 10 organisms in patient specimens. Some of the molecular diagnostic tests used for detection of microorganisms are listed in Table 14–2.

Disadvantages of molecular diagnostics include (1) the high cost and limited number of commercially available kits, (2) application of special procedures to avoid contamination, and (3) potential of false-negative results due to the presence of inhibitors of amplification in some patient samples.

Future applications of molecular diagnostics include:

- Molecular screening of particular at-risk populations for a group of possible pathogens or for the most probable etiologic agents of disease.
- The use of nucleic acid "chip" technologies for molecular screening. DNA chips are made by bonding numerous specific DNA probes on a stationary support. Hybridization will occur within a particular well if the appropriate sequence is present in the DNA analyte. This technology can detect several pathogens in one test and can be easily automated.

Disadvantages of Molecular Diagnostics

- High cost
- Limited number of commercially available kits
- Application of special procedures to avoid contamination
- Potential of false-negative results due to presence of inhibitors of amplification in some patient samples

---

Table 14–2. Some Molecular Diagnostic Assays Available for Detection of Microbial Pathogens

TARGET ORGANISM	BASIC TECHNIQUE	DETECTION SYSTEM	PRIMARY APPLICATION
HIV-1	PCR	EIA	Confirmatory testing
HCV	bDNA	EIA	Quantitation during therapy
Mycobacterium tuberculosis	TMA	HPA-ECL	Smear-positive and untreated patients
Chlamydia trachomatis	LCR	EIA	Monitoring and confirmation
Hepatitis B virus	bDNA	EIA	Quantitation and monitoring
Enterovirus	PCR	EIA	Monitoring and confirmation
Herpes simplex virus	PCR	WB-ECL	Monitoring and confirmation
Neisseria gonorrheae	PCR	EIA	Monitoring and confirmation

TMA, Transcription-mediated amplification; HPA, Hybridization protection assay; EIA, enzyme immunoassay; WB, Western blot; ECL, electrochemiluminescence.

SUGGESTED READINGS

Association for Molecular Pathology Statement. Recommendations for in-house development and operation of molecular diagnostics tests. *Am J Clin Path* 111:449–463, 1999.

Farakas DH. Establishing a clinical molecular biology laboratory. In Farakas DH (ed.). *Molecular biology and pathology: A guidebook for quality control.* San Diego, CA: Academic Press, 1993.

Farakas DH. Molecular techniques and their applications. *Lab Med* 25:633, 1994.

Glick BR, Pasternak JJ. *Molecular biotechnology: Principles and applications of recombinant DNA.* Washington, DC: American Society for Microbiology Press, 1994.

Hill CS. Molecular diagnostics for infectious diseases. *J Clin Ligand Assay* 19:43–52, 1996.

Macario AJL, De Macario EC (eds.). *Gene probes for bacteria.* New York: Academic Press, 1990.

Mulcahy GM. The integration of molecular diagnostic methods into the clinical laboratory. *Ann Clin Lab Sci* 29:43–54, 1999.

Mullis KB, Faloona FA. Specific synthesis of DNA in vitro via a polymerase-catalyzed reaction. *Meth Enzymol* 155:335–50, 1987.

Persing DH, Smith TF, Tenover FC, White TJ. *Diagnostic molecular microbiology: Principles and applications.* Washington, DC: American Society for Microbiology Press, 1993.

Rapley R, Walker MR (eds.). *Molecular diagnostics.* Oxford: Blackwell Scientific Publications, 1993.

Ross JS. Financial determinants of outcomes in molecular testing. *Arch Pathol Lab Med* 123:1071–1075, 1999.

Items for Further Consideration

1. Predict pitfalls during the establishment of a molecular diagnostics laboratory.

2. Discuss strategies for integrating molecular diagnostics methods into the clinical laboratory.

3. Develop a plan to train laboratory workers in molecular techniques.

Automation and Computerization

Wendy R. Sanhai

Key Terms ◄

AUTOMATION

BATCH ANALYSIS

CARRYOVER

CLOSED SYSTEM

CONTINUOUS-FLOW

DISCRETE ANALYZERS

ELECTRONIC PATIENT RECORD

HOSPITAL INFORMATION SYSTEM (HIS)

LABORATORY INFORMATION SYSTEMS (LIS)

MODULAR CONFIGURATION

MULTICHANNEL ANALYSIS

OPEN SYSTEM

QUALITY ASSURANCE

RANDOM-ACCESS ANALYSIS

ROBOT

SAMPLE ANALYSIS

SEQUENTIAL ANALYSIS

SPECIMEN IDENTIFICATION

THROUGHPUT

UNIT OPERATIONS

Chapter Objectives

Upon completion of this chapter, the reader will be able to:

1. Define and discuss the following terms as they relate to clinical laboratory automation: robot, open and closed systems, modular configuration, specimen throughput, batch analysis, random-access analysis, discrete analysis, multichannel analysis, sequential analysis, and single-channel analysis.
2. Discuss the following steps required to complete an analysis of a patient's sample: specimen identification, specimen preparation, specimen delivery, specimen handling and transport, specimen processing, sample transport and delivery (within analyzer), reagent handling and storage, reagent delivery, chemical reaction phase, measurement approaches, signal processing, data handling, and process control.
3. List major benefits of automation in the clinical laboratory.
4. State the major advantage of a laboratory information system (LIS).
5. Define the basic unit of information in an LIS.
6. Sketch the relationship between LIS, hospital information system (HIS), and electronic patient record system.

7. Define the following quality assurance terms: pending lists, turnaround time (TAT) reports, error correction reports, test utilization reports, delta check limits, global lists of testing, comment list.
8. Describe the type of information involved in data transactions within a health care organization.
9. Sketch the flow of laboratory data transactions.
10. Describe each of the following data transactions: patient admission transfer, discharge; test ordering; specimen collecting; test analysis; test result reporting/interpreting; patient/guarantor billing.
11. Describe some of the potential errors that could occur with the following laboratory data transactions: test ordering, specimen collection, specimen analysis, test reporting, test interpretation.
12. Predict future adaptations of LIS in clinical laboratories.

Instruments/Techniques Included ◄

BATCH ANALYSIS

HOSPITAL INFORMATION SYSTEM

LABORATORY INFORMATION SYSTEMS

MODULAR CONFIGURATION

MULTICHANNEL ANALYSIS

RANDOM-ACCESS ANALYSIS

SEQUENTIAL ANALYSIS

UNIT OPERATIONS

Pretest

Directions: Choose the single best answer for each of the following items.

1. In batch analysis, specimens are processed such that
   a. all specimens requiring a particular test are processed in the same analytical session
   b. all specimens requiring a particular test are processed in subsequent analytical sessions
   c. all tests on a sample are completed before analysis on another sample begins
   d. samples are analyzed in random order

2. Specimen throughput is
   a. the rate at which specimens are collected for automated analysis
   b. the rate at which an analyzer processes specimens
   c. the rate at which results are reported
   d. independent of incubation and reaction times

3. Random-access analysis is a type of analysis in which a specimen

   a. can be analyzed only in continuous sequence with other specimens
   b. is subjected to a single process that produces results for a single analyte
   c. can be analyzed by any available process without regard to the initial order of specimens
   d. can be analyzed by any available process but only in the initial order of specimens

4. In single-channel analysis, each specimen
   a. is subjected to multiple processes
   b. has different results associated with it
   c. is subjected to a single process and the result for only one analyte is obtained
   d. is subjected to a single process and results for multiple analytes are obtained

5. In an open system, the
   a. operator cannot modify assay parameters
   b. operator cannot purchase reagents from multiple vendors
   c. reagents exist only as a unique format
   d. operator can modify assay parameters and use custom reagents

6. Multichannel analysis
   a. is synonymous with discrete analysis
   b. has each specimen being subjected to several analytical processes
   c. has a single test result for each specimen
   d. has result order independent of specimen order

7. Modular instrument configuration
   a. provides flexibility and expandability to meet changing needs
   b. does not allow different technologies to be used for all incorporated assays
   c. is used for batch analysis testing
   d. requires complete redesign of the system to meet changing demands

8. The most commonly used technology for specimen identification is
   a. radiofrequency identification
   b. touch screens
   c. handprint tablets
   d. bar coding

9. Which of the following methods of specimen delivery is considered a batch process?
   a. courier service
   b. pneumatic tube system
   c. electric track vehicles
   d. mobile robots

10. With regard to reagents
    a. small analyzers typically use liquid reagents for analysis
    b. small analyzers typically use reagents in dry tablet form for analysis
    c. larger analyzers commonly have compartments maintained at 4 to 10°C for reagent storage
    d. larger analyzers typically use reagent-impregnated slides and strips for analysis

11. Reagent delivery in automated systems
    a. has a better than ±1% reproducibility and an accuracy within ±1%
    b. has a better than ±0.1% reproducibility and an accuracy within ±10%
    c. is often performed by negative-pressure syringe devices
    d. is commonly accomplished through multiple platinum probes

12. Signal processing involves
    a. analog-to-digital conversion of signals that is quite slow
    b. analog-to-digital conversion of signals that is quite rapid
    c. mainframe computers but not desktop and laptop computers
    d. very simple algorithms to produce meaningful output to health care providers

13. According to the flow of data transactions in an LIS, at which step can technologists verify results and apply coded comments to delta checks and critical values?
    a. test interpretation
    b. test ordering
    c. specimen analysis
    d. test reporting

I. INTRODUCTION

Rapid changes in health care coupled with parallel advances in technology have stimulated the evolution of new approaches for laboratory automation. Increasing cost-containment pressures make the application of this technology extremely attractive for improving clinical laboratory efficiency. In an effort to reduce medical care costs, many smaller laboratories are integrating into larger, more efficient entities. These systems are equipped with "front-end" sample handling and transport systems resulting in almost total automation in such laboratories. In addition, previously nonautomated areas may be evaluated for subsequent automation. The successful implementation of large-scale automation in mega-laboratories is likely to be key in the future prosperity of the clinical laboratory.

One of the benefits of automation is a reduction in the variability of results and errors in analysis by eliminating tasks that are repetitive and monotonous for technologists. Robots can now perform many of these repetitive tasks. By definition, a **robot** is any machine that can be programmed to perform any task with human-like skill. Robots are often considered simply a mechanical extension of a computer. The greatest asset of a robot is that it can be configured to perform a multiplicity of tasks with the same level of "skill."

The term **automation** describes the process by which robotic systems and/or instruments perform many tests with minimal involvement from an analyst. Automation, defined by the International Union of Pure and Applied Chemistry (IUPAC) as the replacement of human manipulative effort, facilitates the performance of a given process by mechanical and instrumental devices. Many automated instruments are regulated by feedback information so that the system is self-monitoring or self-adjusting. This assumes a degree of intelligence to be present in automated devices that have the capacity to respond appropriately to changing conditions.

Another benefit of automation is the higher level of safety in regard to the handling of potenially biohazardous samples by technologists. In addition, there is less risk of contamination of samples due to human error.

Equipped with a computer, analytical hardware, and supporting software, automated analyzers have reduced cost and enable laboratories to process a larger workload without the need for increased staff. Although the initial cost of automated equipment is high, substantial cost savings usually accrue because of increased throughput and productivity.

II. METHODOLOGY

The evolution of automation in the clinical laboratory has paralleled that in the manufacturing industry, progressing from fixed automation, whereby an instrument performs repetitive tasks by itself, to programmable automation, which allows performance of a variety of different tasks. This section addresses each of these steps and identifies potential shortcomings. Figure 15–1 shows the individual automated steps, also known as **unit operations,** required for completion of an analysis.

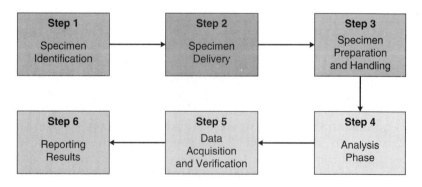

Figure 15–1. Unit operations in automated clinical analysis.

A. STEP 1: SPECIMEN IDENTIFICATION

Typically, the identifying link between patient and specimen is made at the patient's bedside. This connection must be maintained during transport of the specimen to the laboratory. Some of the mechanisms currently used for **specimen identification** include automated bar coding, optical character recognition, magnetic strip and magnetic ink character recognition, voice, radiofrequency identification, touch screens, light pens, hand-printed labels, optical mark readers, and smart cards. Bar coding is the most widely used technology. The risks of misidentification begin at the bedside and are compounded with every processing step that the specimen undergoes.

B. STEP 2: SPECIMEN DELIVERY

Unless analysis is done at the patient's bedside, as in point-of-care testing, samples such as blood, urine, sputum, and the like are collected and routed to a central or "core" laboratory. Delivery of these samples can be accomplished by a variety of methods, which include courier, pneumatic tube systems, electric track vehicles, and mobile robots. Courier service, a batch process, has the potential problems of damage or loss of specimens due to manual handling. Though mechanical problems may arise and cause the tubes or carriers to be misrouted, many health care institutions have adopted pneumatic tube systems and have found them to be quite reliable. These systems have evolved to minimize specimen damage due to rapid acceleration or deceleration of the tube in transport.

C. STEP 3: SPECIMEN PREPARATION AND HANDLING

Upon arrival at the laboratory, many specimens require pretreatment or initial preparation steps before they may be analyzed. Such steps include, but are not limited to, clotting of blood in specimen collection tubes, centrifugation, dilution, and transfer of sample to secondary reaction tubes. In some cases, due to the time required to complete these steps, delays in analysis may become costly or produce inaccuracies in results. When performed manually, delayed or rough handling of specimens before analysis may cause hemolysis or result in changes in analyte concentration due to continued cellular metabolism. Evaporation of specimens left standing for extended periods of time (> 4 hours) will vary depending on temperature, relative humidity, and air flow, and may cause analytical errors as great as 50%. To eliminate such problems associated with specimen preparation, systems are being developed to automate these steps.

Some assays may require an initial separation step(s) to ensure consistency of assay performance. For example, in heterogeneous immunoassays, free and bound fractions must be separated. (See Chapter 12 for further review of this technique.) In other systems, proteins and other interferents must first be removed before testing may be performed. Several automated immunoassay analyzers use bound antibodies or proteins in a solid-phase format to accomplish this preanalytical preparation. In this approach, antigens bind to a solid surface to which the antibodies or other reactive proteins have been adsorbed or chemically bonded. Polystyrene tubes coated with "sticky" proteins that bind analytes, such as streptavidin, are also used as effective separation devices.

Mechanisms for Specimen Identification

- Automated bar coding
- Optical character recognition
- Magnetic strip and magnetic ink character recognition
- Voice
- Radiofrequency identification
- Touch screens
- Light pens
- Hand-printed labels
- Optical mark readers
- Smart cards

Methods of Specimen Delivery

- Courier services
- Pneumatic tube systems
- Electric track vehicles
- Mobile robots

D. STEP 4: SAMPLE ANALYSIS

▶ **Sample analysis,** which may involve many steps, begins after initial preparation steps are completed. The source of reagent in an ana-
▶ lyzer may be referred to as being *open* or *closed*. In an **open system,** the operator can modify assay parameters and purchase
▶ reagents from a variety of vendors. In a **closed system,** assay parameters are set by the manufacturer, who is also the sole provider of reagent(s) for the system in a unique container or format.

Many automated systems use liquid reagents stored in plastic or glass containers. The volumes of reagents stored depend on the number of tests to be performed. Some analyzers use reagents in dry tablet form or strips. In analyzers in which specimens are not processed continuously, reagents are stored in refrigerators and introduced into the instruments as required. In larger systems, sections of the reagent storage compartments are maintained at 4 to 10°C. Before reaction, these reagents may be brought to room ($\cong25°C$) or body ($\cong37°C$) temperature in an incubation station using warm air or a water bath.

Nonliquid reagent systems, in which reagents are dispersed in multilayered slides, are also commonly used. Here, individual layers within the slide contain reagents necessary for the assay. In this dry-slide technology, the patient's specimen provides the fluid that reconstitutes the dry reagents.

The method of sample delivery and transport into and within the analyzer is often different among automated systems. For exam-
▶ ple, in **continuous-flow** systems, the sample is aspirated through
▶ the sample probe into a continuous reagent stream, whereas in **discrete analyzers,** the sample is aspirated into the sample probe and then delivered, often with reagent, through the same orifice into a separate reaction cup or other container.

Syringe devices for both reagents and sample delivery are common to many automated systems. These liquids are taken up and delivered to mixing and reaction chambers either by pumps through tubes or by positive-displacement syringe devices, where the volumes of reagent delivered can be changed. Reproducibility of delivered volume is frequently better than ±1% with an accuracy of within ±1%. In analyzers in which more than one reagent is acquired and dispensed by the same syringe, washing or flushing of the probe is essential to prevent reagent carryover.

▶ The effect of **carryover** between specimens (cross-contamination) is always of concern in automated clinical analysis. The extent of carryover can be determined by a procedure published by the National Committee for Clinical Laboratory Standards (NCCLS) (1993). Most manufacturers of discrete systems reduce carryover by setting an adequate flush-to-specimen ratio (as much as 4:1), by incorporating wash stations for probes and/or by using disposable tips. Combined with appropriate choices of inert sample probe material, geometry, and surface conditions, these methods minimize imprecision and inaccuracy.

Concerns related to the chemical reaction phase and product measurement are also addressed in the design of every automated analyzer. The time needed to complete the reaction phase depends on a number of factors such as rate of transport through the system, mixing and transport of reactants, thermal conditioning of fluids, time of incubation, and time for detection or measurement. Most automated chemistry analyzers have traditionally relied on photometers and spectrophotometers for measurement of analyte concentration. Alternative approaches utilize reflectance photometry, densitometry, fluorometry, and chemiluminescence. Ion-selective electrodes and other electrochemical techniques are also widely used. (See Chapter 11 for further information on these systems.)

An additional term relating to the format of analyzer design is *analyzer configuration.* In a **modular configuration** (Figure 15–2), an ◀ analyzer is assembled from individual components or modules either by the manufacturer or the user. Modular laboratory automation involves a stepwise integration strategy that does not require the vast initial investment of total laboratory automation. It also provides flexibility and expandability to meet changing needs and increase operational efficiency, and enables different technologies to be used for all of the incorporated assays. Furthermore, a modular approach benefits manufacturers by reducing the need to entirely redesign a system to satisfy changing demands. Typically, a single computer coordinates activities of all modules. A common example of an add-on module is an ion-selective electrode unit. Such add-on modules increase the throughput capabilities of these systems. Some configurations combine two or more modules to improve function as well as increase throughput.

Specimen **throughput** is the rate at which an analytical system ◀ processes specimens. In most multichannel random-access systems, this rate is independent of reaction times; instead, it depends on a fixed sequence of events involving mechanical motions and optical measurement cycles. This is in contrast to batch analysis, in which throughput rate is affected by the assays performed. The number of tests per hour is dependent on the test combinations selected in which assays requiring longer times will result in a slower specimen throughput.

Another type of analyzer configuration is **batch analysis.** Batch ◀ analyzers process or analyze many specimens in the same session or "run," as shown in Figure 15–3. In batch analysis, the same test is performed on all samples in a particular session. Subsequent tests are performed in the same way until all the programmed tests for

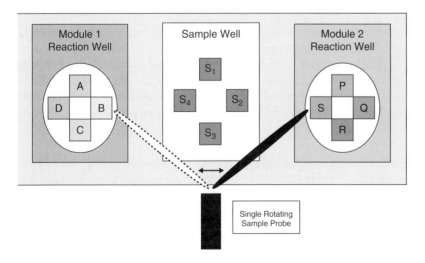

Figure 15–2. Modular instrument configuration. (A, B, C, D) and (P, Q, R, S) are different assay systems available in modules 1 and 2, respectively. S_1, S_2, S_3, and S_4 represent different samples.

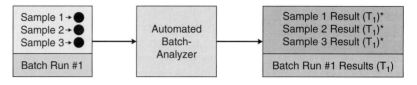

*Where T_1 (time) is the same in all cases.

Figure 15–3. Automated batch analysis.

that session have been completed. For example, in a session containing 20 samples for sodium- and potassium-ion analysis, all sodium tests are completed on the 20 samples before all potassium tests are done on the same 20 samples.

▶ In contrast to batch analysis, **random-access analysis** is the process by which each specimen or sample, by a command to the processing system, is analyzed by any available test in or out of sequence with other specimens and without regard to their initial order. This system is demonstrated in Figure 15–4.

Additional concepts relating to chemical analyzer configuration ▶ are multichannel and sequential analysis. **Multichannel analysis,** depicted in Figure 15–5, is also known as *multitest analysis.* Here, each specimen is subjected to multiple analytical processes so that a set of test results is obtained on a single specimen.

▶ In **sequential analysis,** demonstrated in Figure 15–6, each specimen enters the analytical process one after the other and a requested test or group of tests is performed for each specimen. In this discrete analysis, each specimen has its own physical and chemical space separate from every other specimen. A result or set of results for a particular specimen emerges in the same order as the specimens are entered.

E. STEP 5: DATA ACQUISITION AND VERIFICATION

A spectrum of computers is now available for the acquisition and processing of analytical data. These include mainframe computers; workstations; desktop, laptop, and notebook personal computers; and microprocessors. The interfacing and integration of such computers into automated analytical systems has had a major impact on the speed with which results are reported.

Analog signals from detectors are rapidly converted to digital form by analog-to-digital converters (10^{-3} to 10^{-5} sec). The computer then processes the digital data by means of algorithms into immediately useful and meaningful output. Result presentation may include

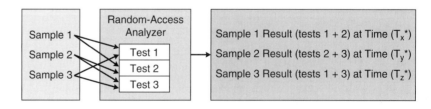

*Where T_x, T_y, and T_z represent different times that are independent of each other and may not occur in succession.

Figure 15–4. Random-access analysis.

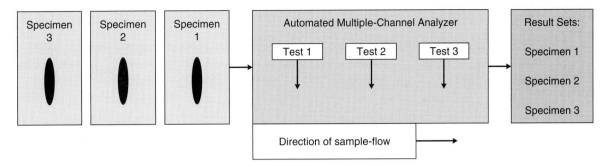

Figure 15–5. Multichannel analysis.

critical values, i.e., test-values that require immediate caregiver notification, and delta checks. Delta checks compare results of the same analyte with previous results for the same patient over a time interval. Delta checks are used for detecting specimen identification errors. All results are verified by laboratory technologists. Footnotes can be added, and the health care provider can be contacted before results are downloaded into laboratory or hospital information systems.

F. STEP 6: REPORTING RESULTS

Laboratory personnel and health care providers may access results from laboratory or hospital information systems or from the **electronic patient record.** Computer-assisted interpretative reports generated from these results may also be available. The College of American Pathologists has developed an extensive coding system called SNOMED (Systematized Nomenclature of Medicine), which has been widely recognized as the most complete standard nomenclature system for communicating clinical information.

III. LABORATORY INFORMATION SYSTEMS

Laboratory information systems (LIS) have facilitated the establishment of standard laboratory nomenclature and information processing procedures. The LIS integrates the clinical laboratory with the medical staff and various departments of a health care organization. The basic unit of information in an LIS file is referred to as a data element. Some examples of data elements include patient's name, sex, and individual test results.

The choice of an LIS begins with a systematic study of the existing system at the institution. If the decision is to purchase a new LIS,

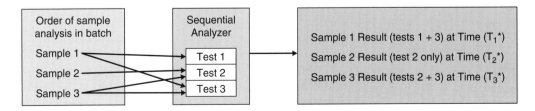

*Where the order of sample analysis: sample 1 → sample 2 → sample 3 corresponds to the order of results for these samples at times T_1 → T_2 → T_3, respectively.

Figure 15–6. Sequential analysis.

specifications for the system must be derived. These specifications must be detailed and specific and address the laboratory personnel and those of the health care organization and medical staff. If a **hospital information system (HIS)** exists or is soon to be incorporated, an LIS–HIS interface specification is included. The health care organization may choose to develop its own LIS or, more often, purchase one from an LIS vendor.

Data such as patient identification code; date(s) of admission, transfer, and discharge; and specimen collection information are often entered by nonlaboratory personnel. With an HIS, this information is automatically transmitted to the LIS through either a shared database or an interface. Several different strategies exist for inquiry and reporting data. Billing information is transferred to the hospital's financial management system electronically or by magnetic tape.

A. INTEGRATION OF LIS, HIS, AND PATIENT RECORD SYSTEMS

Each step in the flow of clinical laboratory information is referred to as a data transaction, in which there exists a unidirectional transfer of information from a data provider (e.g., patient, laboratory instrument, or clinician), to a data receiver (e.g., nurse, emergency room clerk, or clinician). Although laboratory personnel are involved at each step, the patient, medical staff, and other services participate in the overall process. The relationship between LIS, HIS, and electronic patient record systems is demonstrated in Figure 15–7.

B. QUALITY ASSURANCE

LIS and HIS are crucial components in an institution's **quality assurance** program. Table 15–1 defines key terms that relate to a quality management system. In regard to the role of an LIS in institutional quality assurance, there are eight principal steps that describe data transactions. Figure 15–8 demonstrates the order in which these transactions occur in an LIS.

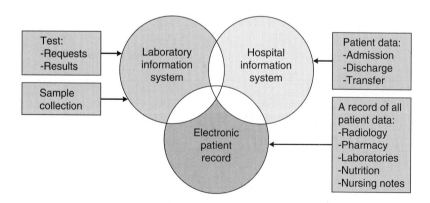

Figure 15–7. Integration of information systems. Examples of data elements are listed in boxes and connected to each information system into which they are downloaded, by arrows. Overlap areas represent the interconnection of the three information systems.

Table 15–1. Quality Assurance Terms and Definitions

TERMS	DEFINITIONS
Pending list	List of tests ordered but not resulted or verified
Turnaround time (TAT) report	Reports of actual times of specimen collection, test accessioning in the laboratory, and reporting/verification of results
Error correction reports	Reports of changes made to previously released test results (includes a record of technologist's ID, date and time of change)
Test utilization reports	Reports of tests ordered by health care providers and the suitability of their results in making clinical decisions
Delta check limits	Percent differences allowed in consecutive test results from different specimens, for the same patient
Global lists of testing	Lists of all tests available for ordering in clinical laboratories
Comment list	List of comments that can be accessed with a laboratory code

1. Step 1: Patient Identification. All hospital services for each patient are related to a unique patient identification number. Upon presentation, registration or admission demographics (data elements), such as the patient's name, birth date, ethnicity, and sex, are recorded. Once a unique patient identifier or number is assigned by the HIS, the data elements are transmitted to the LIS.

2. Step 2: Patient Admission, Discharge, and/or Transfer. Data elements included in these processes include:

- Admission date
- Admission status (e.g., outpatient, inpatient, preadmission, ambulatory, and surgery)
- Location (ward, service, or outpatient site)
- Room or bed
- Health care provider
- Diagnosis or diagnoses
- Patient transfer(s)
- Guarantor(s)
- Billing or finance designation

Such data elements for patients may differ from one health care encounter to the other.

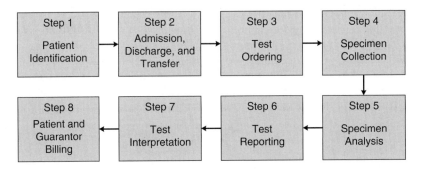

Figure 15–8. Data transactions in an LIS.

3. **Step 3: Test Requisition.** When tests are requested by a health care provider, nursing and laboratory services ensure that all specimens are collected. This process includes the following data elements: date, time, test name, test code, special collection containers, and so on.

4. **Step 4: Specimen Collection.** A variety of health care personnel may collect samples for laboratory analysis. An accession number may be assigned either when the test is ordered or upon receipt in the lab. Time of receipt is recorded in the LIS. In most cases, a barcoded sticker containing the patient identifier information and the accessioning number is printed and attached to the specimen. Any additional test requests made in reference to a previously collected specimen are typically added to the accession number already in existence for that specimen. This number is used by both the HIS and LIS for reporting and inquiry. Other data elements include date, time, technologist's ID, test number, test name, test code, source, and turnaround time. A record of cancellation in the LIS is maintained via the specimen number, test identification, date cancelled, reason for cancellation, and user identification.

5. **Step 5: Specimen Analysis.** The LIS sorts tests into groupings or work lists. The format for work lists is designed specifically for each work area. Examples of work lists include a loading list for an automated instrument or a format for recording data for manually performed tests. Tests that are pending are automatically advanced to the next work list. In this way, work lists serve as "conveyor belts" for work in progress in the laboratory. Data elements from specimen analysis include results, units of measure, and verification of delta checks. These data elements are entered in the LIS automatically through the instrument interface, or manually by the technologist performing the analyses. Technologists are responsible for ensuring that results are accurate and valid before verification and reporting.

6. **Step 6: Test Reporting.** A test report is the "product" of the laboratory. These reports include test results with measurement units, free text, comments, and reference intervals. The LIS retains the identification of the technologist performing the test and verifying the results. Through the LIS, records of all patient results that are abnormal or out of delta check limits can be accessed by the laboratory supervisor or director.

7. **Step 7: Test Interpretation.** Ideally, interpretative reports assist clinicians in making informed medical decisions. Abnormal results and pathophysiological relationships may be highlighted. Physiologically related analytes are typically grouped together, and abnormal and critical values are flagged or highlighted. These reports may also list possible causes of a single abnormal result or differential possibilities offered by multiple abnormal results. This type of report has been useful for analytes such as calcium, thyroid hormones, and serum proteins for which the pathophysiology is defined.

8. **Step 8: Patient and Guarantor Billing.** The business office is responsible for processing laboratory charges. Together with the laboratory, health care administration develops a charge list with the

following data elements: name, charge code, and charge for each test from the hospital's chart of accounts.

C. PITFALLS ASSOCIATED WITH LABORATORY INFORMATION SYSTEMS

The responsibility for accurate and timely test reports lies with the laboratory. In order to monitor and control errors, one must be familiar with the potential errors associated with each step in laboratory testing. Table 15–2 presents potential errors that may occur in laboratory data transactions.

D. CLINICAL APPLICATIONS AND THE FUTURE OF LIS

Most LISs today were developed for the clinical laboratory in the 1980s. In the late 1990s, clinical laboratories were cost centers within integrated health care organizations. In order to meet the challenge of reducing costs through economies of scale, LISs will have to find better ways to link dispersed geographical sites and instrumentation. With the high cost of software development, it is possible that individual LIS vendors will become closely associated with specific health care organizations.

Computer technology is also addressing the challenge of interpretative reporting and data utilization. However, clinical laboratory data are only a portion of the total database required for making accurate patient-specific diagnostic, therapeutic, and prognostic decisions. Addressing this challenge will require computer access to clinical information and the development of rational models of medical decision-making processes. Another "layer" of software, the *decision support system*, is currently emerging. This software utilizes data from all the computers in an organization to provide cost- and resource-utilization information for management planning. Decision support systems will also assist laboratory directors in managing organizational compliance with health plan accreditation requirements and clinical pathways for patient care.

Table 15–2. Factors That Affect Laboratory Data Transactions

DATA TRANSACTION	POTENTIAL ERRORS
Test ordering	Inappropriate test, illegible writing, wrong patient identification, special requirements not specified, delayed order
Specimen collection	Incorrect tube or container, incorrect patient identification, indadequate volume, invalid specimen (hemolyzed or too dilute), incorrect collection time, improper transport conditions
Specimen analysis	Incorrect instrument calibration, specimen mix-up, incorrect volume, interfering substances present, improper dilution, instrument imprecision
Test reporting	Incorrect patient identification, transcription error, results not verified, report delayed
Test interpretation	Interfering substances not recognized, specificity of test not understood, precision limitation not understood, analytical sensitivity not appropriate, previous values not available for comparison

The stage is set for extraordinary advances in computerized information processing. Health care organizations, medical groups, and computer vendors will develop enterprise-wide information systems, enhanced by color graphics and sound, connecting a central data repository with multiple sites over broad geographical areas. The LIS will be a module within this system that is essential for the control of quality and cost of patient care. Generally accepted standards for data exchange among computer systems will be required for these developments to occur.

REFERENCE

National Committee for Clinical Laboratory Standards, document 10-T2. *Preliminary evaluation of quantitative clinical laboratory methods,* 2nd ed. Tentative Guideline, 1993.

SUGGESTED READINGS

Bazzoli F. Laboratory systems evolve to meet data demands. *Health Data Management* 66–71, 1999.

Boyd JC et al. Robotics and the changing face of the clinical laboratory. *Clin Chem* 42(12):1901–10, 1996.

Burtis CA, Ashwood ER (eds.). *Tietz textbook of clinical chemistry,* 3rd ed. Philadelphia: WB Saunders, 1999; Chapters 10 and 15.

Felder RA et al. Robotics in the medical laboratory. *Clin Chem* 36(9):1534–43, 1990.

International Union of Pure and Applied Chemistry: Nomenclature for automated and mechanized analysis (recommendation 1989). *Pure Appl Chem* 61:1657–64, 1989.

Sainato D. Laboratory automation: Coming of age in the 21st century. *Clin Lab News* 26(1), 2000.

Items for Further Consideration

1. Compare and contrast possible analyzer configurations for a large reference laboratory, a hospital laboratory for a 500-bed institution, a dedicated trauma center laboratory, and a physician's office laboratory.

2. How has laboratory automation reduced risk of exposure to infectious agents?

3. List similarities and differences in the requirements for information systems used in a large reference laboratory, a hospital laboratory for a 500-bed institution, a dedicated trauma center laboratory, and a physician's office laboratory.

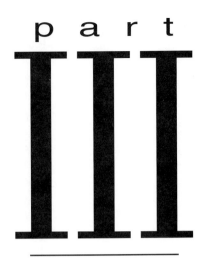

Laboratory Operations

Calculations and Statistical Analyses

Glenn C. Flodstrom

Key Terms

BEER'S LAW

COEFFICIENT OF VARIATION (CV)

CONSTANT ERROR

CORRELATION COEFFICIENT (r)

DILUENT

DILUTION FACTOR

F-TEST

HYDRATE

LINEAR REGRESSION

MEAN

MOLARITY (M)

NORMALITY (N)

NULL HYPOTHESIS

PERCENT VOLUME SOLUTIONS (%v/v)

PERCENT WEIGHT/VOLUME SOLUTIONS (%w/v)

PRECISION

PROPORTIONAL ERROR

RANDOM ERROR

SERIAL DILUTIONS

SIGNIFICANT FIGURES

SIMPLE DILUTION

STANDARD DEVIATION (SD)

STANDARD ERROR OF THE ESTIMATE (s_{Y/X})

SYSTEMATIC ERROR

t-TEST

UNIT CONVERSION

Chapter Objectives

Upon completion of this chapter, the reader will be able to:

1. Apply rules for rounding off and significant figures to all calculations and final results.
2. Convert units for laboratory values.
3. Determine the weight of a chemical (or volume of concentrated acid/base) needed to prepare a solution of a given molarity.
4. Perform calculations using molarity (*M*), normality (*N*), percent weight/volume (%w/v), and percent volume/volume (%v/v).
5. Determine the quantity of the hydrated form of a chemical needed to prepare a solution of known molarity.

6. Calculate the dilution factor(s) for simple and serial dilution schemes.
7. Use the appropriate formula to calculate volumes needed to prepare a dilute (working) solution from a concentrated (stock) solution.
8. For a spectrophotometric analysis, calculate the concentration of an unknown solution, given the concentration and absorbance of a known standard of a linear method.
9. From a given set of data, perform statistical calculations of the mean, standard deviation, and coefficient of variation.
10. Perform linear regression calculations; give an interpretation of the data based on the values for the slope, intercept, correlation coefficient, and standard error of the estimate ($s_{y/x}$).
11. Evaluate the statistical significance of accuracy and precision using the *t*-test and *F*-test.
12. Identify the appropriate statistical test to use when examining sets of laboratory data.

Calculations and Statistics Covered

CONCENTRATION OF AND PREPARATION OF SOLUTIONS:

MOLARITY (M)

NORMALITY (N)

% w/v

% v/v

CONCENTRATED ACIDS

HYDRATED SALTS

CHANGING THE CONCENTRATION OF SOLUTIONS

UNIT CONVERSIONS

DILUTIONS: SIMPLE AND SERIAL

PHOTOMETRIC MEASUREMENTS:

BEER'S LAW

STATISTICS:

MEAN ($\bar{x}$)

STANDARD DEVIATION (SD)

COEFFICIENT OF VARIATION (CV)

LINEAR REGRESSION ANALYSIS (y = mx + b)

–CORRELATION COEFFICIENT (r)

–STANDARD ERROR OF THE ESTIMATE ($s_{Y/X}$)

t-TEST

F-TEST

Pretest

Directions: Choose the single best answer for each of the following items.

1. The number 4.515 rounded to three significant figures is
   a. 4.5
   b. 4.51
   c. 4.52
   d. 4.510

2. Express the answer to the sum of 1.119 g plus 0.0035 g to the appropriate number of significant figures.
   a. 1.122
   b. 1.123

   c. 1.1225
   d. 1.12250

3. Express the concentration (in milligrams per deciliter) to the appropriate number of significant figures for a solution that contains 550 mg in 10.000 dL.
   a. 55 mg/dL
   b. 60 mg/dL
   c. 55.0 mg/dL
   d. 55.00 mg/dL

4. Convert the glucose result, 255 mg/dL, to millimoles per liter (MW glucose = 180.2 g/mole):
   a. 1.42 mmol/L
   b. 14.2 mmol/L
   c. 18.0 mmol/L
   d. 25.5 mmol/L

5. Determine the weight of NaCl necessary to prepare 250 mL of a 4.0% (%w/v) NaCl solution:
   a. 4.00 g
   b. 6.00 g
   c. 8.00 g
   d. 10.00 g

6. Express the concentration in millimoles per liter of a solution that contains 7.80 g of NaCl (MW = 58.44 g/mole) in 500.0 mL of solution.
   a. 266.9 mmol/L
   b. 267 mmol/L
   c. 270 mmol/L
   d. 300 mmol/L

7. Determine the molarity (M) of a 7.50%(w/v) KNO_3 solution (MW KNO_3 = 101.1 g/mole)
   a. 0.736 M
   b. 0.742 M
   c. 0.758 M
   d. 0.774 M

8. Calculate the amount of calcium chloride ($CaCl_2$) needed to prepare 500 mL of a 0.750 N solution if the molecular weight of $CaCl_2$ is 110.99 g.
   a. 10.4 g
   b. 15.6 g
   c. 20.8 g
   d. 41.6 g

9. Determine the number of grams of NaOH needed to prepare 250.0 mL of a 0.35 M solution of NaOH (MW = 40.00 g/mole):
   a. 2.2 g
   b. 3.5 g
   c. 4.0 g
   d. 4.5 g

10. A previous calculation has determined that in order to prepare 100.0 mL of a 0.75 M solution of $CuSO_4$, 12.0 g of the anhydrous form is needed (MW $CuSO_4$ = 159.6 g). If only the hydrate, $CuSO_4$ • 5 H_2O (MW = 249.7 g/mole) is available, calculate the amount of hydrate to be weighed.
    a. 7.67 g
    b. 12.5 g
    c. 18.8 g
    d. 38.4 g

11. Given that the concentrated HNO_3 available has a specific gravity of 1.418 g/mL and a percent assay (%w/w) of 69.90%, determine the molarity of the concentrated HNO_3 (MW HNO_3 = 63.02 g/mole).
    a. 15.66 M
    b. 15.7 M
    c. 16 M
    d. 61.7 M

12. Calculate the necessary volume of 1.50 M HCl needed to prepare 500.0 mL of 0.050 M HCl.
    a. 16.7 mL
    b. 17 mL
    c. 37 mL
    d. 37.5 mL

13. A serum for alkaline phosphatase (ALP) determination exceeded the limit of linearity for the assay. The sample was diluted with 0.200 mL of saline and 0.100 mL of serum and reanalyzed. If the diluted sample result was 543 U/L, determine the activity of the original undiluted sample.
    a. 362 U/L
    b. 814 U/L
    c. 1090 U/L
    d. 1630 U/L

14. A serial dilution was performed in which 0.200 mL of diluent was added to all tubes in the series followed by the addition of 0.100 mL of sample to tube 1. After mixing, 0.100 mL of the solution in tube 1 was added to tube 2. This process was repeated to tube 8, where 0.100 was discarded after mixing. The dilution factor of tube 4 is
    a. 32.0
    b. 64.0
    c. 81.0
    d. 243

15. Calculate the molarity of an NADH solution (ϵ = 6.22 × 10^3 M^{-1} cm^{-1}), in which the path length was 1.00 cm and the absorbance of the solution was 0.894 (under the specified conditions for ϵ).
    a. 1.44 × 10^{-4} M
    b. 5.56 × 10^{-3} M
    c. 1.22 × 10^3 M
    d. 6.21 × 10^3 M

16. A spectrophotometric procedure for glucose yields the following data:
    Absorbance of 500 mg/dL standard = 1.274
    Absorbance of unknown sample = 0.675
    Calculate the concentration of the unknown sample.
    a. 192 mg/dL
    b. 265 mg/dL
    c. 338 mg/dL
    d. 943 mg/dL

17. Standard deviation results on level I quality control material were obtained for the months of January and February for a calcium assay. These results were 0.32 mg/dL for January and 0.66 mg/dL for February. Choose the correct statement below regarding the relative accuracy or precision of this assay during these two months.
    a. quality control results for calcium were more accurate in January than in February
    b. quality control results for calcium were more accurate in February than in January
    c. quality control results for calcium were more precise in January than in February
    d. quality control results for calcium were more precise in February than in January

Use Table 16–1 to answer questions 19 and 20:

19. With respect to the *t*-test results, it can be correctly stated that
    a. there is not a statistically significant difference between January and February
    b. there is a statistically significant difference between January and February
    c. the *t*-test does not provide enough information to answer this question
    d. the January results are more precise than the February results

20. With respect to the *F*-test results, it can be correctly stated that
    a. there is not a statistically significant difference between January and February
    b. there is a statistically significant difference between January and February
    c. the February results are more accurate than the January results
    d. the data from January and February are statistically identical

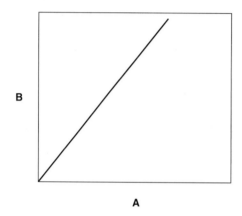

B

A

Figure 16–1.

18. Interpret the linear regression data from two calcium methods (Figure 16–1 above) (reference method = A and candidate method = B) given the data below:
    Regression equation: $y = 1.10\,x + 0.0$
    Correlation coefficient (r) = 0.999
    $s_{y/x}$ (residuals) = 1.2
    a. proportional error in the candidate method is evident
    b. random error in the candidate method is evident
    c. proportional error in the reference method is evident
    d. random error in the reference method is evident

Table 16–1. Quality Control (Glucose, Level I)

MONTH	NO. OF SAMPLES	MEAN (mg/dL)	SD (mg/dL)	$t_{calculated}$	$t_{tabular}$	$F_{calculated}$	$F_{tabular}$
January	30	87.2	3.1				
				1.25	2.01	1.067	1.84
February	24	86.6	3.0				

The need for laboratory mathematics skills at the bench remains important for maintaining quality in the clinical laboratory despite the extensive use of computers in many calculations. In addition to the usual quality control procedures for a run, a laboratory scientist should critically review all results, particularly with respect to whether such a result is physiologically credible or suggestive of a calculation error. By regulation, all computer calculations must be verified twice a year, with documentation.

I. SIGNIFICANT FIGURES/ROUNDING RULES

It is often necessary to round off numbers when performing calculations in the laboratory, particularly if hand calculators are used, in order to produce an answer with the appropriate number of digits. Throughout this section, the following rounding rules will be used: If the number to be dropped is less than 5, the preceding number remains the same. If the number to be dropped is greater than 5, then the preceding number is raised by 1 (rounded up). If the number to be dropped is equal to 5, then the preceding number will be raised by 1 if it is an ODD number (rounded up), but will be unchanged if it is an EVEN number.

There is uncertainty or imprecision inherent in any value, whether obtained by direct measurement or indirectly through calculation. A review of the rules determining the number of **significant figures** will assist in the performance of calculations performed later in this chapter.

In general, any nonzero digit is considered significant. Zero is considered significant if its removal would affect the value of the number (e.g., 5,085, 1.609), but not if it is there to establish the position of the decimal point (e.g., 0.0055, 12,300). If a zero is not necessary to establish the position of the decimal point, but was recorded in a value, then that zero (and all of the intervening zeros) are significant; thus, in the examples 2,500.00 and 1.00700, all of the zeros are significant, for a total of six significant figures in each number.

> ## Rules for Rounding Off
>
> - If the number to be dropped is < 5, the preceding number remains the same.
> - If the number to be dropped is > 5, the preceding number is raised by one.
> - If the number to be dropped equals 5
>   —and the preceding number is ODD, the preceding number is raised by one.
>   —and the preceding number is EVEN, the preceding number remains the same

Example 1
Round off the following numbers to three significant figures:

    a. 4,561
    b. 24.05
    c. 134.8
    d. 1.355

Answers
    a. 4,560 (The last zero is not significant, and the tens digit remains unchanged.)
    b. 24.0 (Since the tenths digit is even, it remains unchanged.)
    c. 135 (The ones place is rounded up, since 8 is more than 5.)
    d. 1.36 (The tenths digit is rounded up, since it is odd and the hundredths digit is equal to 5.)

Example 2
Determine the number of significant figures in the following values:

    a. 4.567 g
    b. 100.0 mL

c. 450.7 IU/mL

d. 0.0034 mg

Answers

a. 4.567 g   = four significant figures

b. 100.0 mL   = four significant figures

c. 450.7 IU/mL = four significant figures

d. 0.0034 mg   = two significant figures

It is important during any calculation that the imprecision of the original measurement(s) be retained. Performing a calculation will *not* increase the precision of a measurement. In the process of *adding* or *subtracting* two or more values of differing significant figures, the least precise value determines the correct number of significant figures in the final answer, as shown in the following examples:

Example 3

Determine the correct number of significant figures in the following problems:

a. 7.45 g + 5.113 g + 25 g = ?

b. 1,000.0 mL − 75 mL − 140 mL = ?

Answers

a. 38 g (The least precise measurement, 25 g, determines the number of significant figures in the final answer. Note that the rounding rule was also applied here, since 37.5 is rounded up to the next higher digit, whereas if the "ones" digit were even, it would remain unchanged.)

b. 780 mL (Unless a measuring device of known imprecision [e.g., ± 0.5 mL] is used, assume that zero is a "place holder" in 140 mL and that the imprecision of this measurement is one-half of the last significant digit (±5 mL); thus, 785 is rounded to 780 mL [two significant figures]).

Example 4

In the process of *multiplying* or *dividing* two or more values of differing significant figures, the general rule is that the value that has the least number of significant figures determines the number of significant figures in the final answer, as shown in the following examples:

a. 4.56 g ÷ 2.500 dL = ?

b. 315 U/mL × [(0.400 mL + 0.100 mL) ÷ 0.100 mL] = ?

Answers

a. 1.82 g/dL (three significant figures). Note that the two zeros are significant, and therefore 4.56 g has the least number of significant figures (3).

b. 1580 U/mL (three significant figures). This example uses a ("X5") dilution scheme (dilution schemes will be discussed later in this chapter), such as might be employed in an enzyme assay that is out of range. Note that the precision of the pipetted volumes (0.400, 0.100 mL) determines the number of significant figures in the final answer. The dilution factor (5) should not be considered "exact."

It should be noted that unit conversions or mathematical constants can be entered into a calculation with as many places as required, such that the conversion has more significant values than the measurement with the least number. This will minimize rounding errors.

II. SOLUTION PREPARATION

A. GENERAL CONSIDERATIONS

Although few clinical laboratories prepare their own reagents, the knowledge to do so has value in particular situations, such as when the necessary reagents are not available in kit form, or when working in research settings. Use of volumetric glassware and balances that are capable of weighing ±0.0001-g quantities will ensure that the reagent will be within its expected concentration range as calculated. The use of the appropriate significant figures will yield a realistic estimate of the expected concentration range.

B. MOLAR SOLUTIONS

Molarity (*M*) is used to express the concentration of solutions in ◄ terms of the number of moles per liter of solution. To prepare a reagent of a given molarity, the molecular weight (mass) of the chemical must be known; this information is usually provided by the chemical manufacturer on the bottle label. A convenient volume is chosen that will prepare a sufficient quantity of the reagent without excessive waste. The weight of the chemical is most often the critical variable. Care should be taken in the calculation as well as the actual weighing process, since errors in either can lead to reagent failure. The calculation can be done in two steps: (1) multiplication of the necessary volume times the molarity (yielding moles needed after volume units cancel) and (2) conversion of moles to grams.

Example 5
   a. Prepare 500 mL of a 0.500 *M* NaCl solution (MW NaCl = 58.44 g/mole).
   b. Prepare 50 mL of a 0.785 mg/dL $Pb(NO_3)_2$ solution.

Answers
   a. Select a 500.0-mL volumetric flask; then calculate the weight of chemical needed:

   500.0 mL × 0.500 moles/1000 mL × (58.44 g/1 mole) = 14.6 g NaCl

   b. Select a 50.00-mL volumetric flask; then calculate the weight of chemical needed:

   50.00 mL × 0.785 mg/100 mL = 0.392 mg $Pb(NO_3)_2$

C. NORMAL SOLUTIONS

Normality (*N*) expresses the concentration of solutions as the num- ◄ ber of equivalent weights of a substance per liter of solution; the *equivalent weight* of an element or compound is equal to its molecular weight divided by the *valence*. Preparation of a solution of a given normality is similar to the preparation of molar solutions except the valence of the chemical must be known. If an element (such as iron [Fe] or copper [Cu]) has more than one oxidation state, the valence can often be determined from the compound's anion ($NO_3 = -1$, $SO_4 = -2$, etc.). The formula for conversion to normality from molarity is $N = nM$, where N is equal to the normality, n is equal to the valence, and M is equal to the molarity.

Example 6

    a. Prepare 1.00 L of a 0.40-N solution of Na_2SO_4 (MW = 142.0 g/mole).

    b. Convert a 5.34-M solution of H_2SO_4 to normality.

Answers

    a. Since the valence is 2 ($SO_4 = -2$), the equivalent weight of Na_2SO_4 is 142.0/2 = 71.0 g; to prepare a 0.4-N solution, 71.0 g $\times$ 0.4 = 28.4 g would be diluted to one liter.

    b. 5.34 M H_2SO_4 $\times$ 2 = 10.68 N H_2SO_4

D. PERCENT SOLUTIONS

Solutions can also be prepared by adding a particular weight to a given volume solution; these are most commonly expressed as **percent weight/volume solutions (%w/v).** A weight/volume solution is defined as the number of grams (or other unit of mass) contained within 100 mL of solution. For example, a 5.0% NaCl solution would have 5 grams in 100 mL solution. These can be converted to molarity or normality if given the molecular weight. **Percent volume solutions (%v/v)** are often used when mixing solvents together such as in thin-layer chromatography solutions. A solution that contains 5 mL methanol, 10 mL chloroform, and 85 mL water would be an example of a percent volume solution, where the concentration of methanol is 5 %v/v and the concentration of chloroform is 10 %v/v.

Example 7

    a. Prepare 100 mL of a 2% w/v solution of KCl.

    b. Determine the volume percent for benzene of a solution that contains 10 mL benzene, 20 mL n-propanol, and 60 mL isopropanol.

Answers

    a. Weigh 2.0 g of KCl and dilute to 100 mL.

    b. %v/v benzene = 100% $\times$ 10 mL / (10 + 20 + 60) = 11.1% v/v benzene

E. SOLUTIONS OF HYDRATES

It is occasionally necessary to use a hydrated chemical or salt in the preparation of a solution if the anhydrous form is not available. This is acceptable, since once water is added to the dry chemical or salt, the water of hydration (in the hydrate) becomes part of the solution. However, the weight of the water in the hydrated form must be taken into account when determining the amount needed to be weighed for a solution of a given concentration.

If the chemical or salt being used is a **hydrate** ($X \cdot yH_2O$), where X is the salt itself, and y is some integer, an additional calculation is required prior to the weighing of that salt. As with nonhydrated salts, the formula (molecular) weight can be obtained from the manufacturer's label. It is necessary to weigh a larger amount of chemical (compared to the anhydrous form) in order to compensate for the water molecules of the hydrate. The exact amount needed is calculated by multiplying the ratio of the hydrated (or most hydrated) to the anhydrous (or least hydrated) forms.

Example 8

Prepare 250.0 mL of a 0.750-M solution of $CuSO_4$, given that only the hydrate form of the chemical is available (MW $CuSO_4$ [anhydrous] = 159.6 g/mole, MW $CuSO_4 \cdot 5\ H_2O$ [hydrate] = 249.7 g/mole).

Answer

Amount of anhydrous $CuSO_4$ needed = 0.2500 L × (0.750 moles/L)

$$= 0.1875\ moles$$

0.1875 moles × 159.6 g/mole = 29.925 g $CuSO_4$ (anhydrous) needed

$$29.925\ g\ (anhydrous) \times \frac{249.7\ g\ (hydrate)}{159.6\ g\ (anhydrous)} = 46.82\ g\ (hydrate)$$

To spot check the calculation: when using the hydrated form, a larger mass will be required than if the anhydrous form were used, and vice versa. The desired weight should be quantitatively transferred to a 250-mL volumetric flask containing some deionized water, and then additional deionized water is added to the mark after allowing the chemical to dissolve. [Of course, the weight of the hydrate needed can be calculated directly using the molecular weight of the hydrated form (i.e., 0.1875 moles × 249.7 g/mol of hydrate = 46.82 g of hydrate needed).]

Example 9

Determine the amount of sodium phosphate monobasic, dihydrate ($NaH_2PO_4 \cdot 2H_2O$), MW = 155.99 g/mole that is equivalent to 13.7 g of sodium phosphate monobasic, monohydrate ($NaH_2PO_4 \cdot H_2O$), MW = 137.99 g/mole.

Answer

$$13.7\ g\ (monohydrate) \times \frac{155.99\ g\ (Dihydrate)}{137.99\ g\ (Monohydrate)} = 15.5\ g\ (Dihydrate)$$

F. CONCENTRATED ACIDS

If solutions of dilute acids or bases are required in the clinical laboratory, they can be easily prepared from stock solutions of concentrated acids or bases, which can be conveniently stored for an extended period of time (see Chapter 19 concerning occupational exposure to hazardous chemicals). On the label of the bottle, concentrated acids or bases provide the *specific gravity* and *percent assay* or purity of the solution rather than the molarity, because many of these chemicals may vary slightly. However, the molarity (or normality) can be calculated from the specific gravity and percent assay by the following formula:

$$M = \frac{(\text{Specific gravity}) \times (\text{Percent assay}) \times 10}{\text{Molecular weight of the acid or base}}$$

Example 10

Determine the molarity of a bottle of glacial acetic acid (CH_3COOH, MW = 60.05), which has a specific gravity of 1.046 g/mL and a percent assay of 99.2%.

Answer

$$M\,(CH_3COOH) = \frac{1.046 \times 99.2 \times 10}{60.05} = 17.3\ M$$

G. CHANGING CONCENTRATIONS

The formula $C_1V_1 = C_2V_2$ is used when changing the concentration of an existing solution (of known concentration) to prepare a solution of lesser concentration, where C_1 = concentration of the stock solution, V_1 = volume of the stock solution (usually in milliliters), C_2 = concentration of the solution to be prepared, and V_2 = volume of the new solution required (in the same units as V_1). Any appropriate units of concentration may be used (e.g., molarity, normality, w/v, etc.) as long as C_1 and C_2 are expressed in the same units. If the molarity is calculated for each bottle of concentrated acid or base, and the container labeled appropriately, only the $C_1V_1 = C_2V_2$ portion of the formula would be needed for subsequent dilutions. This formula is also highly useful in the preparation of more dilute solutions from "stock" (more concentrated) solutions, many of which may have an acceptable shelf life if refrigerated. This practice precludes repeated weighing and solution mixing, saving time and expense. An exercise in the use of stock solutions is shown in Example 11.

Example 11

Prepare 250.0 mL of a 0.435-M solution of HNO_3 by diluting a stock concentrated HNO_3 solution that has a molarity of 15.7 M.

Answer

$$15.7\ M \times V_1 = 0.435\ M \times 250.0\ \text{mL}$$
$$V_1 = 6.93\ \text{mL}$$

Therefore, 69.4 mL of concentrated HNO_3 would be added to approximately 200 mL deionized water contained in a 250-mL volumetric flask and additional deionized water used to bring the total volume to the calibrated mark.

Example 12

Prepare 100.0 mL of a 0.35-M NaOH solution using a previously prepared NaOH stock solution that has a concentration of 2.50 M.

Answer

$$2.50\ M \times V_1 = 0.35\ M \times 100.0\ \text{mL}$$
$$V_1 = 14\ \text{mL NaOH (stock)}$$

Therefore, the 14 mL NaOH would be added to a 100.0-mL volumetric flask and diluted to the mark with deionized water.

III. UNIT CONVERSIONS

It is often necessary in the laboratory to convert a measurement or calculation from the given units of expression to more desirable units of expression. This **unit conversion** does not change the measurement involved, but often will convert the measurement to a unit that is easier to interpret or more appropriate for additional calculations. Any unit conversion factor should always be equal to 1; therefore, multiplication of the original measurement by that conversion factor will not change that measurement, except to convert it to the desired units. Unit conversions are often "exact"; for example, 1 gram is exactly equal to 1,000 milligrams; therefore, significant figures should be ignored (or considered as infinitely large). Similarly, with other conversion factors, such as *Système International* (SI) to conventional, care must be paid to have at least as many significant figures as decimal places in the conversion factor to conserve the correct number of significant figures in the final answer. Calculations may require that more than one conversion factor be used to reach the final desired unit.

Example 13

    a. 0.346 g/dL   = ? mmol/L (MW = 245.1 g/mole)

    b. 1.58 L/24 hours = ? mL/min

Answers

    a. $\dfrac{0.346 \text{ g}}{\text{dL}} \times \dfrac{10 \text{ dL}}{1 \text{ L}} \times \dfrac{1 \text{ mole}}{245.1 \text{ g}} \times \dfrac{1000 \text{ mmol}}{1 \text{ mole}} = 14.1 \text{ mmol/L}$

This method has also been called the *unit cancellation* method, because the original unit will appear in both numerator and denominator, and thus is "canceled out." The unit dL is only useful as it is part of the frequent clinical chemistry unit, mg/dL.

    b. $\dfrac{1.58 \text{ L}}{24 \text{ hr}} \times \dfrac{1000 \text{ mL}}{1 \text{ L}} \times \dfrac{1 \text{ hr}}{60 \text{ min}} = 1.10 \text{ mL/min}$

Conversions of different units that express solute concentration (e.g., molarity, normality, percent solution, and milligrams per deciliter) are commonly seen in the clinical laboratory. Most calculation errors involve the inversion of one or more factors, such that the units do not cancel. Therefore, it is recommended that one write out the entire equation with the units as well so it can be observed that the desired units will "cancel out." Recall the definitions of these common units :

    1 molar (1 *M*) solution = 1 mole solute per liter solution
    1 normal (1 *N*) solution = 1 equivalent weight per liter solution
    1 percent (1% w/v) solution = 1 gram solute per 100 milliliters
        solution
    1 mg/dL solution = 1 milligram solute per 1 deciliter solution

Example 14

Convert the following quantities to the desired units:

    a. 0.90 *M* NaCl solution to milligrams per deciliter (MW NaCl = 58.44 g/mole)

    b. 155 mg/dL glucose solution to millimoles per liter (mmol/L) (MW glucose = 180.2 g/mole)

c. 2.50 M H_2SO_4 solution to percent w/v (MW H_2SO_4 = 98.07 g/mole)

d. 3.4 N K_2SO_4 to molar (M) (MW K_2SO_4 = 101.1 g/mole)

e. 0.77 M $CaCl_2$ solution to millimoles per liter (MW $CaCl_2$ = 111.0 g/mole)

Answers

a. 0.90 moles/L NaCl × (58.44 g/1 mole) × (1,000 mg/1 g) × (0.1 L/1 dL) = 5259.6 mg/dL or 5300 mg/dL when rounded off to two significant figures

b. 155 mg/dL glucose × (1 mole/180.2 g) × (1 g/1,000 mg) × (1000 millimole/1 mole) × (1 dL/0.1 L) = 8.601 mM or 8.60 mM when rounded off to three significant figures

c. 2.50 moles/L H_2SO_4 × (98.07 g/1 mole) × (0.1 L/100 mL) = 24.5 %w/v

d. 3.4 N K_2SO_4 × (1 MW/2 equivalent weights) = 1.7 M K_2SO_4

e. 0.77 moles/L $CaCl_2$ × (1,000 millimoles/1 mole) = 770 mmol/L

IV. DILUTIONS

A. SAMPLE DILUTIONS

Although highly automated instruments in the clinical laboratory may have "autodilute" programs that will dilute and reanalyze samples with concentrations outside the analytical range for that method, it is necessary for technologists to be able to perform dilutions as part of their daily work. A **simple dilution** consists of the addition of a measured volume of sample (usually serum) to a measured volume of diluent (often deionized water or saline, but will vary according to the analyte). In most laboratories, automated pipettes are used for both volume measurements.

After the instrument has analyzed the diluted sample, it is necessary to multiply the result by the **dilution factor** (ratio of total volume divided by sample volume). Use of the terms x2, x5, and so on, indicating the dilution factor, should be encouraged in order to minimize the use of the notations 1:1, 1/2, and so on, which may be misinterpreted. Technically speaking, the expression "one to two" is denoted as 1:2 and represents the ratio of one part sample to two parts **diluent.** The notation of 1/2 or "one in two" represents one part sample to two total parts. Hence, these two notations or expressions are suggestive of two different dilution factors. The dilution factor of 1:2 is 3, while the dilution factor of 1/2 is 2. Therefore, when expressing a dilution ratio, the ratio should always be expressed in terms of sample volume and *total* volume (i.e., the latter) rather than sample volume versus the amount of diluent added. When expressed in this way, the inverse of the dilution ratio is the dilution factor (i.e., for the dilution ratio 1/10, the dilution factor is 10), which then can be multiplied by the result of the diluted sample to determine the original analyte concentration or activity in the undiluted sample.

Ways to Express Dilutions

- By dilution factor:
  x2, x5, etc.
- By dilution ratio as a fraction:
  ½ ⇒ 1 part sample *in* 2 parts total solution; dilution factor = 2
- By dilution ratio as a proportion:
  1:2 ⇒ 1 part sample *to* 2 parts diluent; dilution factor = 3

Example 15

Patient sample 23 from the morning glucose run was out of the linear range of the assay (> 600 mg/dL). A 0.100-mL aliquot of this sample was added to 0.200 mL of saline and the diluted sample reanalyzed. The result obtained for the diluted sample was 356 mg/dL. What is the calculated glucose value of the original patient sample?

Answer

356 mg/dL × (0.200 mL + 0.100 mL)/ 0.100 mL = 1068 mg/dL

B. SERIAL DILUTIONS

Though multiple **serial dilutions** are more often performed in other ◀ laboratory areas (e.g., immunology, microbiology), the principle is identical to simple dilutions. Each dilution step requires the calculation of a dilution factor, and all dilution factors multiplied together yield the dilution factor for a particular tube in a sequence.

Example 16

A X2 serial dilution was performed in a series of five test tubes as follows: each tube received 0.500 mL saline initially, then 0.500 mL of serum was added to tube 1. The contents of tube 1 were mixed, and then 0.500 mL of tube 1 was removed and added to tube 2. The same sequence was followed, until tube 5, from which 0.500 mL was discarded after mixing. What is the dilution factor for tube 3?

Answer

Dilution factor (tube 1) = (0.500 mL + 0.500 mL) / 0.500 mL = 2.00
Dilution factor (tube 2) = dil. fac. tube 1 × (0.500 mL + 0.500 mL) / 0.500 mL
   = 4.00
Dilution factor (tube 3) = dil. fac. tube 2 × (0.500 mL + 0.500 mL) / 0.500 mL
   = 8.00

It is a convenient short-cut to recall that the dilution factor in a serial dilution that is prepared with equal volumes throughout can be determined by the number X^y, where X is the initial dilution factor and y is the tube number. This formula is not applicable if, for example, the initial dilution was x10, and subsequent dilutions were x2, which is often done if the highly dilute solutions yield the most important data.

Example 17

A serial dilution was performed as diagrammed in Figure 16–2 by first adding 0.5 mL diluent to all six tubes. 0.1 mL of serum was

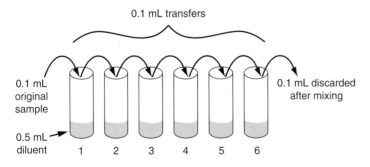

Figure 16–2. Dilution scheme for Example 17.

added to tube 1, then 0.1 was removed after mixing and added to tube 2. This sequence was continued until tube 6, where after mixing, 0.1 mL was removed and discarded. What is the dilution factor in tube 4?

Answer

The initial dilution factor in tube 1 is 0.1/0.6 (which is equal to "X6"). In subsequent tubes, a 6X dilution factor also exists. Therefore, the dilution factor in tube 4 is (6 X 6 X 6 X 6) = 1,296.

V. BEER'S LAW CALCULATIONS (ABSORPTION SPECTROSCOPY)

Manual spectrophotometric measurements are rare in today's clinical laboratory because they are usually performed using highly automated equipment. However, an understanding of **Beer's law** for a particular method is useful in troubleshooting either the method or the instrument, since a great deal of chemistry assays are still based on absorbance principles. In those rare cases in which it is required that a calibration curve be plotted manually, or in a research setting, the calculations below are essential (also see Chapter 11). Detection of deviations from Beer's law (A = abc), such as the presence of interfering substances, may be detected by observing the raw data from spectrophotometric measurements, by dilution of the analyte or by multiple wavelength readings.

The most practical form of Beer's law is the calculation of an unknown concentration if one or more standards have been measured within the linear range of the assay. The general form of the equation is:

$$A_1/C_1 = A_2/C_2$$

where A_1 = absorbance of the standard, C_1 = concentration of the standard, A_2 = absorbance of the unknown, and C_2 = concentration of the unknown. Typically, C_2 (concentration of the unknown) will be the variable we wish to solve for. In the rearranged form, $C_2 = A_2 \times (C_1/A_1)$, (C_1/A_1) is often referred to as a "factor" by which the unknown absorbance is multiplied to obtain the unknown concentration.

Example 18

Determine the concentration of an unknown uric acid sample whose absorbance is 0.342 at the wavelength of the assay. A 10-mg/dL standard was also run, which had an absorbance of 0.673 under the same conditions.

Answer

$C_2 = 0.342 \times (10.0$ mg/dL $/ 0.673) = 5.082$ mg/dL or 5.1 mg/dL when
rounded off to two significant figures

Example 19

If a 5.0-mg/dL creatinine standard has an absorbance of 0.875, and an unknown sample has an absorbance of 0.432, what is the concentration of creatinine in the unknown sample?

Answer

$C_2 = 0.432 \times (5.0 \text{ mg/dL} / 0.875) = 2.47$ mg/dL or 2.5 mg/dL after rounding
off to two significant figures

VI. STATISTICS

A. *Mean/Standard Deviation/Coefficient of Variation*

The **mean, standard deviation (SD),** and **coefficient of variation** ◀
(CV) are frequently used in quality control (see Chapter 18) and
method evaluation (see Chapter 17). The use of a hand calculator
will facilitate these and other statistical calculations, as will other
commonly available software. Technologists should be able to interpret quality control results from these statistics, given sufficient data.

The mean, or arithmetic average, of a set of data is an indication
of the *central tendency* of those data. If used in the context of quality control results, the mean can reflect the accuracy of those measurements on replicate samples. The standard deviation is a measure
of dispersion, and reflects the **precision** (or reproducibility) of those ◀
measurements. The formula for standard deviation is:

$$\text{Standard deviation} = \sqrt{\frac{\Sigma (\overline{X} - X_i)^2}{n-1}}$$

In the formula for standard deviation, each individual result (X_i)
is subtracted from the mean ($\overline{X}$) and the difference is squared. These
squares are summed and divided by the number of samples (n) minus one. Finally, the square root is taken. Samples having the same
mean can reflect differences in precision by the comparative magni-
tudes of the standard deviations; the larger the standard deviation,
the less precision.

Example 20
Using Figure 16–3, determine the more precise method (A or B) from
the two sets of data obtained from repeated assays of quality control
material (Level I, Sodium)

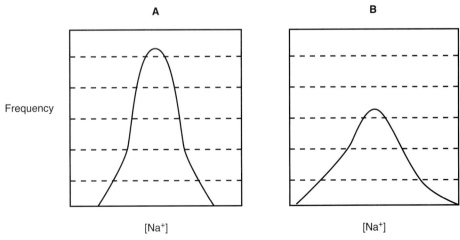

Figure 16–3. Frequency diagrams for two sets of data (A and B) for Example 20.

Statistical Parameters of Precision and Accuracy

- Mean ($\bar{x}$): an indicator of central tendency; reflects accuracy
- Standard deviation (SD): an indicator of reproducibility or precision
- Coefficient of variation (CV): a relative indicator of precision; expresses the SD as a percentage of the mean

Linear Regression Analysis

- Least squares analysis for "best fit" line, $y = mx + b$
    - m = slope of regression line
    - b = y-intercept of regression line
- Correlation coefficient (r)
- Standard error ($s_{y/x}$)

Answer

Figure 16–3 represents a smaller standard deviation (less dispersion) for method A and, therefore, higher precision than method B.

If the means of two samples are not equal, but they have been run on the same method (for example, two levels of control for the same analyte), it is still possible to evaluate the relative precision at the two concentrations using the coefficient of variation. The coefficient of variation, which is the standard deviation expressed as a percentage of the mean,

$$CV = \frac{SD}{mean} \times 100$$

allows comparison of analytical precision over a wide range of measurements. In practice, it is not unusual for samples lower in analyte concentration to have higher CVs than samples high in analyte concentration. This increased imprecision at low concentrations may be acceptable depending on the method.

The CV is also useful to compare precision of different methods for the same analyte.

Example 21

Calculate the mean, standard deviation, and coefficient of variation from the following glucose results for quality control Level I (all in milligrams per deciliter):

86, 85, 80, 80, 83, 81, 84, 81, 85, 87

Answer

Mean = 83.2 mg/dL; SD = 2.6 mg/dL; CV = 3.1%

Example 22

Calculate the mean, standard deviation, and coefficient of variation from the following glucose results for quality control Level II (all in milligrams per deciliter):

245, 252, 251, 238, 257, 255, 248, 255, 249, 244

Answer

Mean = 249 mg/dL; SD = 5.9 mg/dL; CV = 2.4%

Compare the results in Examples 21 and 22, noting that the fluctuation that occurs around the mean is somewhat greater for Level II (as reflected in the standard deviation). However, this is offset by the denominator (mean) in the CV formula.

B. LINEAR REGRESSION

Linear regression, or regression analysis, is often performed as part of a method comparison to evaluate the performance of a new method when compared to an existing or reference method (see Chapter 17). Linear regression is a statistical method to determine the best linear relationship between two variables. Results from the reference or existing method are represented by the independant variable (x) and are plotted on the x-axis; results from the evaluated method are represented by the dependent variable (y) and are plotted on the y-axis. The determination of the "best fit" line is often performed by computer or calculator and is based on the *least squares* analysis. The calculated line will be in the form **y = mx + b,** where m is the slope, and b is y intercept. The **correlation coefficient (r),** and **standard** ◀ **error of the estimate ($s_{y/x}$)** also referred to as *the standard deviation of the residuals,* are also often calculated in linear regression analysis.

Consider the three plots of a reference method (x axis) and a candidate method (y axis) shown in Figures 16–4 A, B, and C. Ideally, the two methods would show perfect correlation, which would be reflected by a positive slope of 1, a y-intercept of 0, a correlation coefficient of 1.0, and $s_{y/x}$ equal to 0 as demonstrated by Figure 16–4 A. If the slope is either less than or greater than one, this is considered **proportional error,** as shown in Figure 16–4B. Proportional ◀ error is present when the difference between two data sets is proportional to (i.e., a percentage of) the concentration of the analyte being measured. If the y-intercept is greater or less than zero, this is **constant error** as shown in Figure 16–4C. Constant error, a bias as ◀ indicated by the name, is constant over the concentration range tested. Both constant and proportional error are considered **system-** ◀ **atic errors** because they will be present in every measurement.

Random error in a method is best shown by the correlation ◀ coefficient (r) and the *residuals.* When the data are plotted as shown in Figures 16–4 D and E, random error or precision is reflected by the scatter of points around the regression line and by the relative magnitude of the residuals.

> **Ideal Values for Parameters of Regression Analysis**
> ----
> - Slope (m) = 1.0
> - y-intercept (b) = 0.0
> - Correlation coefficient (r) = 1.0
> - Standard error of the estimate $(s_{y/x})$ = 0.0

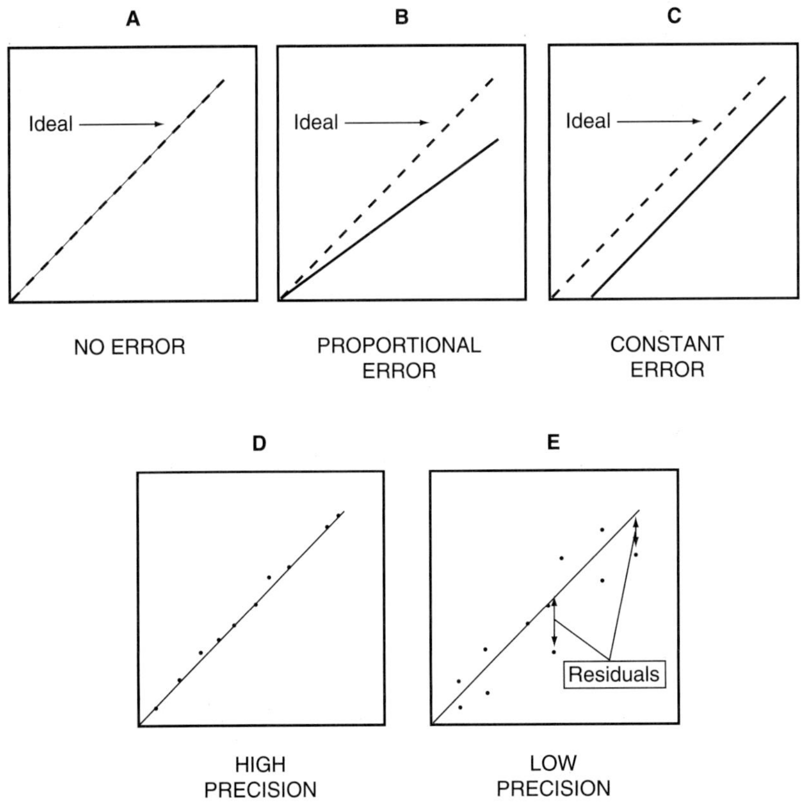

A

Ideal ⟶

NO ERROR

B

Ideal ⟶

PROPORTIONAL ERROR

C

Ideal ⟶

CONSTANT ERROR

D

HIGH PRECISION

E

Residuals

LOW PRECISION

Figure 16–4. Regression analysis plots.

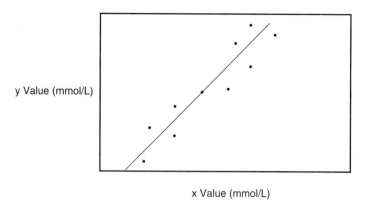

y Value (mmol/L)

x Value (mmol/L)

Figure 16–5. Plot of data for method comparison of sodium analyses for Example 23.

The ideal correlation coefficient is 1.0, and the ideal standard deviation of the residuals ($s_{y/x}$) is 0.0. To calculate residuals, the square of the distance between each data point and the regression line is summed, then divided by (n – 2) and the square root taken of that result. The $s_{y/x}$ is zero for a comparison of two methods when all data points lie on the regression line. To summarize, the slope (m) is sensitive to proportional error, the y-intercept (b) is sensitive to constant error, and both $s_{y/x}$ and the correlation coefficient (r) are sensitive to random error.

Example 23
Determine if a new method for sodium (y-axis) versus a reference method (x-axis) has systematic or random error (or both). The data are plotted in Figure 16–5 with the "best fit" regression line.

In addition, comment on the acceptability of the new method based on the analysis of accuracy and precision. The linear regression information for these data are: slope (m) = 1.13; y-intercept = –6.5; correlation coefficient (r) = 0.933; $s_{y/x}$ = 14.8 mmol/L.

Answer
Two types of systematic error are present. The slope (1.13) indicates proportional error and the y-intercept value (–6.5) indicates a negative constant error. The candidate method values for sodium will differ from the reference value according to the relationship y = 1.13x – 6.5. Although the correlation coefficient appears to be acceptable, an inspection of the data plot (Figure 16–5) and determination of $s_{y/x}$ suggests that this method also has excessive random error. Although the decision would be based on analytical goals, it is likely that the candidate method would not be acceptable.

C. TESTS OF STATISTICAL SIGNIFICANCE (t-TEST, F-TEST)

▶ Both the **t-test** and **F-test** are used in the clinical laboratory to determine if the differences between two sets of data are statistically significant. The t-test can be used to test for significant differences between two means, for example in a comparison of the mean from a candidate method versus the mean from a reference method, or to

---

Regression Parameters As Indicators of Error

- Slope (m) is sensitive to proportional error
- y-intercept (b) is sensitive to constant error
- Correlation coefficient (r) is sensitive to random error
- Standard error of the estimate ($s_{y/x}$) is sensitive to random error

compare the mean of a data set to a given (e.g., true) value; thus, the t-test can help evaluate accuracy or bias. The F-test is used to test for statistically significant differences between the *variances* (dispersion) of two sets of data; thus the F-test can help evaluate precision. Both the t- and F-test calculations require the use of a table and a specialized program or other software. Use of the t- and F-test tables also require determination of the *degrees of freedom* (df) which is the number of independent observations minus the number of restrictions on the data set. For a t-test comparison of means, df = n_1 + $n_2 - 2$, where n_1 and n_2 are the respective sample sizes. The F-test table requires degrees of freedom for both numerator ($n_1 - 1$) and denominator ($n_2 - 1$), where n_1 is the sample with the largest standard deviation. These values are necessary to determine the appropriate cutoff of significance (critical value) using t- and F-tables.

If a calculated t value or F value (using appropriate formulas) is greater than the "critical" value from the t- or F- table, then the t- or F-test demonstrates a statistically significant difference between the compared values at a chosen confidence level. The **null hypothesis** ◄ states that there is not a statistically significant difference (or one not due to chance). If the results of the t- or F-test are statistically significant, then the null hypothesis is rejected, indicating that the data differences cannot be explained by chance alone.

The probability that this conclusion is incorrect is established by a probability value, alpha (α). In general, the smaller the value for alpha, the less likely that the result for the t- or F-test will be incorrect. The commonly chosen limit for alpha is $p < 0.05$, which indicates the probability allowed for the conclusion to be incorrect (< 1 in 20 chances). In other words, using a $p < 0.05$ probability limit, if the null hypothesis is rejected, the compared values can be assumed to be statistically different with 95% confidence.

The t-test may be applied in different forms depending on the type of data; the form of t-test being used must be known to choose the correct t-table. There are two designs of the t-test: paired versus unpaired, and one-tailed versus two-tailed. Data such as repeated measurements on the same individual utilize the *paired* formula, whereas those not expected to be identical (e.g., from different individuals) would utilize the *unpaired* formula. Normally, the two-tailed formula would be used to test for differences in either direction (higher or lower). However, if the outcome was expected to be in one particular direction, the one-tailed formula is appropriate (e.g., a new treatment is being evaluated and there are preliminary data to indicate that the new therapy is better than the control therapy). Figure 16–6 shows the t distribution indicating the one-tailed and two-tailed limits when $p < 0.05$.

Example 24

A t-test is performed on 2 months' quality control (QC) data (January, February) to see if the mean had shifted significantly during that time. The results of the t calculations were as follows:

> Calculated t value = 1.451
> Degrees of freedom = 56
> Critical ($p < 0.05$) t-table value = 2.134 (for the given df)

Interpret these results and accept or reject the null hypothesis (that there is no difference in means between January and February QC results).

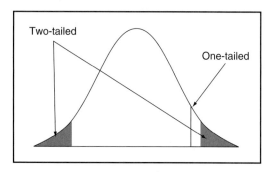

Figure 16–6. The distribution when p > 0.05.

Answer
The calculated *t* value is less than the critical *t* value, and therefore the null hypothesis is not rejected. According to this *t*-test, the results that were obtained could be explained by chance variation alone; they were not statistically significant. With respect to the comparison of means, we can say that there is no difference between the means that cannot be explained by chance variation (i.e., the observed difference between the means is likely due to chance variation).

Example 25
The same data from Example 20 can also be used to compare variances, using the formula for the *F*-test and an *F* table. Data from these calculations are below:

Calculated *F* value = 1.751
Degrees of freedom = 27 (numerator)
29 (denominator)
Critical (p < 0.05) *F*-table value = 1.534 (for the given df)

Interpret these results and accept or reject the null hypothesis (that there is no difference in variances between January and February QC results).

Answer
Because the calculated *F* value is larger than the critical (table) *F* value, one can state with 95% confidence that there is a statistically significant difference between variances. Therefore, the null hypothesis is rejected, and the difference between the variances would not be expected to occur by chance alone.

SUGGESTED READINGS

Byrne AE. Basic laboratory principles and techniques—calculations in clinical chemistry. In Kaplan LA, Pesce AJ (eds.). *Clinical chemistry: Theory, analysis and correlation,* 3rd ed. St. Louis: CV Mosby, 1996; 34–44.

Campbell JM, Campbell JB. *Laboratory mathematics: Medical and biological applications,* 5th ed. St. Louis: CV Mosby, 1997.

Doucette LJ. *Mathematics for the clinical laboratory.* Philadelphia: WB Saunders, 1997.

PRACTICE EXERCISES

1. Determine the number of significant figures in 1,350.

2. How many significant figures are in 1,500 U/L?

Describe how the following solutions (items 3 through 12) would be prepared:

3. 500 mL of 6.2 M Na_2SO_4 (MW = 142.0 g/mole)

4. 100 mL of 5.0 N NaH_2PO_4 (MW = 120.0 g/mole)

5. 100 mL of 2.0 M HCl (MW = 36.46 g/mole) from concentrated HCl (specific gravity = 1.078, % assay = 96.5%)

6. 500 mL of 0.50 M NaOH (MW = 40.0 g/mole) using 6.0 M NaOH

7. 50.0 mL of a 5% (w/v) solution of K_2SO_4 (MW = 174.2 g/mole)

8. 100 mL of 0.2 N NaOH (MW = 40.0 g/mole) from 10% (w/v) NaOH

9. 50 mL of a 5% (v/v) aqueous solution of methanol (CH_3OH, MW = 32.04)

10. 100 mL of 0.45 M $CuSO_4$ (MW = 159.6 g/mole) from $CuSO_4 \bullet 5\ H_2O$ (MW = 249.7 g/mole)

11. 100 mL of 0.3 M KCl (MW = 74.53 g/mole) from 10% (w/v) solution

12. 50.0 mL of 6 M NaOH (MW = 40.0 g/mole) from 2.5 M NaOH

13. A serum sample for AST analysis exceeded the limit of linearity for the method (400 U/L). A dilution was prepared by adding 0.2 mL serum to 1.8 mL diluent. If the AST result of the diluted sample was 63, determine the AST activity in the undiluted sample.

14. A serial dilution is prepared by adding 0.2 mL of serum to 0.8 mL diluent in tube 1, followed by transfer of 0.5 mL of this mixture to tube 2, which contains 0.5 mL of diluent (as do all subsequent tubes). 0.5 mL of the mixture from tube 2 is transferred to tube 3, and this process is continued through tube 8. Determine the dilution factor in tube 7.

15. Convert 350 mg/dL glucose (MW = 180.0 g/mole) to mmol/L.

16. Convert 5.0 %(w/v) NaCl (MW = 58.5 g/mole) to mmol/L.

17. Calculate the creatinine concentration in an unknown sample if a 10.0 mg/dL creatinine standard has an absorbance of 0.952 and the unknown had an absorbance of 0.347.

18. Calculate the glucose concentration in an unknown sample if a 500 mg/dL glucose standard has an absorbance of 1.252 and the unknown had an absorbance of 0.191.

19. Calculate the mean, standard deviation, and coefficient of variation for the following calcium QC samples (all in milligrams per deciliter): 9.0, 9.3, 9.1, 9.0, 9.2, 9.1, 9.3, 9.0, 9.3, 9.1.

20. Describe the linear regression parameters (slope [m], y-intercept [b], and correlation coefficient [r]) expected if a method comparison is performed on a candidate method that has a positive proportional error and increased random error.

21. If the calculated t value for a comparison of means is 3.652, and the critical value ($p < 0.05$) at the appropriate number of degrees of freedom is 2.753, is this result statistically significant?

22. What is the importance of a calculated F value that is less than one?

PRACTICE EXERCISES (ANSWERS)

1. Three.

2. Two.

3. Weigh 440.2 g of Na_2SO_4 and dilute to 500.0 mL.

4. Weigh 20.0 g of NaH_2PO_4 and dilute to 100.0 mL.

5. Add 7.02 mL concentrated HCl to water and dilute to 100.0 mL.

6. Measure 41.7 mL of 6.0 M NaOH and dilute to 500 mL.

7. Weigh 2.5 g of K2SO4 and dilute to 50.0 mL.

8. Measure 8.0 mL of 10% w/v NaOH and dilute to 100 mL.

9. Measure 2.5 mL of methanol and add it to 47.5 mL water.

10. Weigh 11.23 grams of $CuSO_4 \bullet 5 H_2O$ and dilute to 100 mL.

11. Measure 22.4 mL of 10% KCl and dilute to 100 mL.

12. It is not possible to make a concentrated solution from a more dilute one.

13. 630 U/L.

14. The dilution factor is X320 (1 to 320 dilution).

15. 19.4 mmol/L.

16. 854.7 mmol/L or 860 mmol/L when rounded off to two significant figures.

17. 3.6 mg/dL.

18. 76 mg/dL.

19. Mean = 9.14 mg/dL; SD = 0.13 mg/dL; CV = 1.4%.

20. A method that has a positive proportional error will have a slope value of greater than one, but the y intercept should not be affected (near zero). Random error would be shown by a widely distributed pattern of data points around the regression line, and this may or may not be reflected in a correlation coefficient value that is somewhat less than 1.0.

21. Yes, if the calculated value for t exceeds the critical value, then the result is significant, and the null hypothesis is rejected.

22. The equation for the F-test indicates that the larger variance should be in the numerator and the smaller in the denominator, so a mathematical error is likely.

Items for Further Consideration

1. How might the use of too many or too few significant figures in a test result sent to a physician affect the interpretation of that test result?

2. List some possible sources of error in the preparation of solutions.

3. When performing a dilution, how does the selection of the appropriate measuring device(s) affect the accuracy of the final result?

4. Why is the determination of a correlation coefficient less sensitive than the calculation of $s_{y/x}$ in the evaluation of the precision of a method?

5. Given the linear regression plot shown in Figure 16–7, evaluate the new method under consideration. Include in the discussion the potential types of error present and which linear regression parameters lead to the conclusions drawn.

Slope (m) = 1.000

y-intercept (b) = –5.5

$s_{y/x}$ = 12.5

Correlation coefficient = 0.945

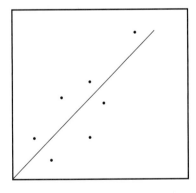

Figure 16–7. Linear regression plot.

Method Evaluation

Janine Denis Cook

Key Terms ◀

ACCURACY	HOOK EFFECT	PRECISION
ANALYTICAL GOALS	INTERFERENCE	PROPORTIONAL ERROR
ANALYTICAL SENSITIVITY	LIMIT OF QUANTIFICATION	RANDOM ERROR
ANALYTICAL SPECIFICITY	LINEAR RANGE	RECOVERY
BETWEEN-DAY PRECISION	MEDICAL DECISION LEVELS	REFERENCE RANGE
BETWEEN-RUN PRECISION	METHOD COMPARISON	REPORTABLE RANGE
BIAS	METHOD EVALUATION	SYSTEMATIC ERROR
CARRYOVER	MINIMUM DETECTABLE	WITHIN-DAY PRECISION
CONSTANT ERROR	CONCENTRATION (MDC)	WITHIN-RUN PRECISION
CORRELATION COEFFICIENT (r)	OUTLIERS	

Chapter Objectives

Upon completion of this chapter, the reader will be able to:

1. Discuss five issues that affect method choice.
2. Describe the evaluation experiments that must be performed as dictated by the Clinical Laboratory Improvement Act (CLIA) for the establishment and verification of method performance specifications of a Food and Drug Administration (FDA)-approved moderate- to high-complexity method.
3. Outline the additional experiments that must be performed per CLIA for methods that are modified, "home brew," or not FDA-approved.
4. Discuss one experiment that provides an estimate of a method's precision.
5. Discuss one experiment that provides an estimate of a method's accuracy.

Method Evaluation Studies Covered ◀

PRECISION:	WITHIN-RUN	METHOD COMPARISON:	MINIMUM DETECTABLE
	WITHIN-DAY	REGRESSION ANALYSIS	CONCENTRATION (MDC)
	BETWEEN-RUN	CORRELATION COEFFICIENT (r)	
	BETWEEN-DAY	STANDARD ERROR (S_Y/X)	RECOVERY

PRECISION: WITHIN-RUN WITHIN-DAY BETWEEN-RUN BETWEEN-DAY

INTERFERENCE TESTING (ANALYTE SPECIFICITY)

METHOD COMPARISON: REGRESSION ANALYSIS CORRELATION COEFFICIENT (r) STANDARD ERROR ($S_{Y/X}$) BIAS

MINIMUM DETECTABLE CONCENTRATION (MDC)

RECOVERY

REFERENCE RANGE VERIFICATION

REPORTABLE RANGE (LINEARITY)

Pretest

Directions: Choose the single best answer for each of the following items.

1. In choosing a new method, it is important that this test
   a. have medically unacceptable errors that result in misdiagnosis
   b. satisfy the medical applications required by the physician
   c. have less quality than the method being replaced
   d. increase the cost and labor requirements needed to perform the assay

2. The four experiments required by CLIA to evaluate a moderate- to high-complexity FDA-approved test are
   a. accuracy, precision, reportable range, and analytical specificity
   b. accuracy, precision, analytical specificity, and analytical sensitivity
   c. reportable range, reference range, analytical specificity, and analytical sensitivity
   d. accuracy, precision, reportable range, and reference range

3. The error specifications for the analytical coefficient of variation (CV) calculated from biological variability is defined as
   a. CV < 20%
   b. two to three times the standard deviation (SD) of the blank average
   c. one-half of the within-individual CV
   d. one-fourth of the reference range

4. The imprecision of a method is expressed as the
   a. mean
   b. coefficient of variation
   c. linear regression line
   d. reportable range

5. A method that is inaccurate but precise would be represented on a bull's-eye target as
   a. loose scatter around the eye center
   b. loose scatter off-center
   c. tight scatter around the eye center
   d. tight scatter off-center

6. With error specifications defined as ±2 SD from the mean for a normally distributed reference range, how many nondiseased patients would statistically be excluded?
   a. 1/5
   b. 1/20
   c. 1/100
   d. 1/500

7. A hemolysis interference study performed by subjecting an unspun specimen to a freeze–thaw cycle is invalid for serum potassium because the
   a. potassium concentration within the red blood cell is many-fold higher than found in serum
   b. freeze–thaw cycle would not hemolyze the specimen
   c. specimen should be centrifuged prior to the freeze–thaw cycle
   d. experiment as designed would test for icterus interference

8. For the limit of detection and limit of quantification, it is true that the
   a. limit of detection and the limit of quantification are different aspects of analytical specificity
   b. limit of quantification specifies a minimum concentration with imprecision at a specified limit of error
   c. detection limit is calculated as one times the SD from the measured blank average
   d. customary CV limit associated with the limit of quantification is 4%

9. The total error is calculated as the
   a. sum of the constant and proportional errors
   b. sum of the within-run and within-day precisions
   c. sum of both the random and systematic errors
   d. CV plus the SD

10. In order to determine analytical specificity, which of the following experiments is often performed?
    a. within-day precision
    b. statistical mean
    c. lipemia interference
    d. reference range verification

11. An estimate of the accuracy of a method can be performed by a
    a. limit of detection experiment
    b. reference range study
    c. method comparison study
    d. between-day precision study

12. Verification of a method's reportable range must be performed per CLIA with at least
    a. five linearly related concentrations spanning the analytical range
    b. 20 data points
    c. 20 days' worth of runs
    d. three concentrations—lower, midpoint, and high

13. To ensure that a testing material will not inject matrix effects into a study, the material should
    a. be a mixture of the analyte of interest dissolved in water
    b. contain an organic solvent
    c. be stabilized by the addition of numerous preservatives
    d. be a patient specimen

14. The additional experiments required by CLIA for non-FDA approved tests are
    a. analytical sensitivity and analytical specificity
    b. accuracy and analytical sensitivity
    c. reportable range and reference range
    d. accuracy and precision

15. A preanalytical variable that might affect the results of a method evaluation is
    a. the quality control data
    b. intravenous (IV) fluid contamination during specimen collection
    c. the calibration curve
    d. failure to perform daily maintenance

16. The precision study with the shortest analysis time frame is
    a. within-day precision
    b. within-run precision
    c. between-run precision
    d. between-day precision

17. The experiment that establishes the reportable range is
    a. linearity
    b. least detectable dose
    c. hemolysis interference
    d. precision

18. The statistical analysis used in a method comparison study that assesses both random and systematic error is
    a. mean
    b. CV
    c. SD
    d. linear regression

19. One cause of proportional error is
    a. operator variability
    b. instrument temperature variation
    c. incorrect calibrator assignment
    d. pipetting errors

20. Recovery is
    a. the ratio of the amount of analyte recovered to the amount of analyte added (expressed as a percentage)
    b. ideally approaches 0%
    c. an estimate of a method's precision
    d. the ability of a method to measure results that can be reliably reported

I. INTRODUCTION

The quality of service offered by the clinical laboratory is influenced by its personnel, instrumentation, and test methods. Therefore, method selection and evaluation are an integral aspect of ensuring quality laboratory results. The wrong method choice is costly, both in financial terms and in the relationships with clients. It is therefore important to select the new method with care.

The ultimate clients to be satisfied are the end users: the physician and the patient. The chosen method must satisfy the medical applications and produce results within medically acceptable error; excessive error could result in misdiagnosis.

▶ The trend over the last few years in the clinical laboratory setting has been toward **method evaluation** as opposed to method development. Most laboratorians today use commercially prepared test methods that have been approved by the Food and Drug Administration (FDA).

When submitting a method to the FDA for clearance, a manufacturer must make certain that the required analytical performance claims are supported by data. As a user, it is not necessary to repeat all of the evaluation testing performed by the manufacturer for FDA submission. Rather, a limited evaluation to *verify* the manufacturer's claims and to understand the performance of the method *in your laboratory* is conducted. The goal of the evaluation must be to achieve results that are *at least* as good as those claimed by the manufacturer.

Simply stated, a laboratory must at least perform the "CLIA Four" as part of its method evaluation protocol—an examination of four important characteristics of method performance. As mandated by the Clinical Laboratory Improvement Act (CLIA) [493.1213 (b)(1) and (2)], the following is the standard for the establishment and verification of method performance specifications: ". . . a laboratory that introduces a new procedure for patient testing using a moderate to high complexity method (instrument, kit or test system) cleared by the FDA as meeting the CLIA requirements for general quality control, must demonstrate that, prior to reporting patient test results, it can obtain the performance specifications for accuracy, precision, and reportable range of patient test results, comparable to those established by the manufacturer. The laboratory must also verify that the manufacturer's reference range is appropriate for the laboratory's patient population."

Moreover, if "a laboratory that introduces a new procedure for patient testing using a method developed in-house; a modification of the manufacturer's test procedure; or a method (instrument kit, or test system) that has not been cleared by the FDA as meeting the CLIA requirements for general quality control, [it] must, prior to reporting patient test results—verify or establish for each method the performance specifications for the following performance characteristics, as applicable: accuracy; precision; analytical sensitivity; analytical specificity to include interfering substances; reportable range of patient test results; reference range(s); and any other performance characteristic required for test performance."

Table 17–1 summarizes the CLIA requirements for the evaluation of a moderate- to high-complexity method.

When a thorough and labor-intensive method evaluation is undertaken, benefits are realized. Ideally, the new method provides some improvement, such as improved accuracy and precision, enhanced automation, a reduction of reagent and/or labor cost, increased safety, improved client satisfaction, or state-of-the-art technology.

Table 17–1. CLIA Requirements for the Evaluation of Moderate-
to High-Complexity Tests

FDA-Cleared Methods
Accuracy
Precision
Reportable range
Reference range

Other Methods[a]
Accuracy
Precision
Reportable range
Reference range
Analytical sensitivity
Analytical specificity (interferences)

[a] "Other Methods" refers to "home brew" methods, FDA-approved methods that are modified from the manufacturer's recommended protocol, and nonapproved methods.

II. METHOD EVALUATION PROTOCOL

The topics covered under the discussion of method evaluation are found in Table 17–2. The approach is a healthy blend of practicality and regulatory requirements. The emphasis of this chapter is on carefully thought-out initial planning to ensure an efficient method evaluation.

A. PRACTICAL CONSIDERATIONS

Properly preparing for an effective method evaluation involves defining the problem in as much detail as possible, enlisting the input of all necessary and interested parties and identifying all issues that could affect the choice of methods. Listed in Table 17–3 are some factors to consider as part of this initial assessment.

B. PAPER COMPARISON

The method evaluation must be well organized or the process will be costly and tedious. Beginning with a "paper" comparison ensures

Table 17–2. Protocol for Method Evaluation

1. List and prioritize method/instrument issues.
2. Perform a paper comparison of available methods/instruments versus performance on identified issues.
3. Define analytical goals.
4. Gather the necessary materials for the evaluation.
5. Perform the initial experiments.
6. Evaluate data from initial experiments and determine if method warrants further evaluation.
7. Perform the final experiments.
8. Evaluate data from final experiments and determine if method acceptable for adoption and implementation
9. Implement the chosen method/instrument.

Table 17–3. Important Practical Considerations

- **Analytical issues** (accuracy, appropriate reference ranges [no bias with established methods], analysis time, data handling, easily scales up to anticipated testing volume, interferences, expected analytical performance, linear range, methodology, method protocol, performs well with proficiency challenges, personnel skill level required, precision, quality control, reagents [composition, quantity supplied, storage requirements, stability, shelf life], recovery, run size, minimum detectable concentration, analyte specificity, standardization, suitable for diagnostic need, turnaround time)
- **Cost analysis** (reimbursement, consolidation of testing, hands-on time, labor, instrument cost [purchase, lease or reagent rental], reagent cost, testing volume)
- **Instrument capabilities** (stats, open system, bar code reading, interfacing, primary tube, random access, on-line QC, self-diagnostics, test menu, throughput, error flags, etc.)
- **Instrument issues** (ease of operation, carryover, maintenance, reliability, service, dead volume, run size, service contracts, consumables, service support, size "footprint," technical support, training, calibration, and stability)
- **Logistical issues** (automation, space needs [including reagent storage], special instrument considerations [electrical, plumbing, temperature, humidity, ventilation, access, waste], specialized ancillary equipment [hoods], training, workload)
- **Safety considerations** (environmental hazards)
- **Specimen requirements** (patient preparation, sample size, species specificity, specimen handling, specimen preanalytical issues, sample preparation)

that all issues are identified and will limit to a reasonable number those methods that warrant further research.

Once determined to be worthy of more intensive investigation, the analyte to be measured, as well as its diagnostic need and medical usefulness, are identified. For certain diagnoses, several analyte options may be available. Input from physicians can be vital at this stage to ensure that the correct analyte will be chosen to provide the greatest diagnostic utility. Identification of the goals to be accomplished as a result of this investigation are clearly stated and understood, such as the required accuracy and precision. Tied into the utility of an analyte as a diagnostic tool is the determination of the method's analytical performance, such as the minimum detectable concentration and analyte specificity. Service issues are also considered such as specimen requirements, setup times, and turnaround time.

The paper comparison begins with a technical literature review. Professional journals, proficiency surveys such as those offered by the College of American Pathologists (CAP), "throw-away" references such as the Medical Laboratory Observer (MLO) Clinical Laboratory Reference, and manufacturers' information such as package inserts and Web sites are important resources that are often consulted to identify methods that are available for a specific analyte. In addition, the value of conferring with colleagues and enlisting testimonials from current method users cannot be overstated.

The paper comparison can be accomplished in several steps: (1) identify the requirements that are applicable to the current evaluation using the list in Table 17–3 as a guide; (2) create a table of important issues versus available methods; (3) gather information on the various methods' applications, methodologies, and performances; (4) use an objective-rating scheme with weighting factors for the most important characteristics to help identify the leading candidate methods. This paper assessment will narrow the field of methods that require actual research to a manageable number and begin to rank the remaining methods in order of quality.

C. ANALYTICAL GOALS

The issue of **analytical goals** must also be considered before embarking upon the actual method evaluation procedures. For each experiment performed, acceptability criteria are established beforehand. These are the allowable errors that can be tolerated without invalidating the medical usefulness of the result. Ideally, acceptable error limits are centered on the medical decision levels for that particular analyte.

For the more common analytes, performance standards already exist. For those without established standards or guidelines, use of professional judgment and input from colleagues, clinicians, and clients is required.

General guidelines exist for establishing performance goals. For example, one guideline suggests estimating the maximum error upper limit such that there is only a 5% chance that the actual error would exceed the upper limit (the 95% limit of the analytical error; stated differently, one out of every 20 samples will exceed the error limit). References for identifying error specifications and establishing quality goals include:

- **Medical decision levels**
- Physician surveys
- *Tonk's rule* (allowable error = one-fourth of reference range or 10%, whichever is less)
- *Inter- and intraindividual variation* (CAP 1996 Aspen Conference established an analytical CV goal of one-half the within-individual CV)
- Health Care Financing Administration (HCFA)/CLIA and CAP/Centers for Disease Control and Prevention (CDC)/other proficiency testing requirements
- State-of-the-art performance as indicated on proficiency summary reports

D. RESEARCH PREPARATION

Again, to have the evaluation be as efficient as possible, it is important to be prepared by ensuring that the necessary instruments, reagents, and supplies are readily available. Before proceeding with the initial experiments, ensure that all applicable instrumentation is functioning properly and method calibration is acceptable. Determine which quality control material will be used for the study; a multilevel control with concentrations in the normal range and at the medical decision points is ideal. Analysts performing the method evaluation must be competent and adequately trained in the new procedure to guarantee valid results. The research experience must mimic the real-life laboratory as much as possible. If the evaluated analyte is uncommon, specimens are collected and stored in advance. Also, specimens containing the analyte of interest with possible interferences such as hemloysis, icterus, and lipemia are procured ahead of time.

If known, it is important to understand the effects of preanalytical variables on the various methods. These effects include, for example:

- Demographic factors (age, sex, race)
- Physiological changes (diet [food ingestion, prolonged fasting])
- Rhythmic changes [diurnal variation, menstrual cycle, other cyclic changes (i.e., seasonal)]
- Stress-related changes (posture, exercise, acute and chronic illness, mental stress, pregnancy, drug effects)
- Changes occurring during collection (sample source [arterial, venous, capillary], tourniquet effect, IV fluid contamination, hemolysis, anticoagulants and preservatives, mislabeling)
- Changes occurring after collection and before testing (analyte deterioration prior to testing, sample storage conditions, excessive exposure to the clot, and changes due to hematological disorders)

In some instances, the effects of these preanalytical variables can be minimized by carefully selecting the patient specimens to be used in the evaluation.

Part of the preparation involves understanding the experiments to be performed. Detailed discussion of most of these experiments is provided in the various National Committee for Clinical Laboratory Standards (NCCLS) documents mentioned in the chapter text.

E. INITIAL EXPERIMENTS

The method evaluation must be efficient and designed to detect unacceptable performance with a minimum number of experiments. The experiments recommended and/or required for a complete evaluation are listed in Table 17–4.

If possible, determine an easy *"acid test"* that will eliminate some of the candidate methods immediately and thus decrease additional time and expense in the remaining evaluation. An acid test is related to a specific characteristic of the method, to the needs of

Table 17–4. Method Evaluation Experiments

Initial
1. "Acid test" (evaluation of a method-characteristic required by the laboratory and/or medical staff)
2. Within-run precision
3. Method comparison
   - Other accuracy experiments
   - Analysis of proficiency materials, NIST standards, etc.
   - Recovery study
4. Reference range verification

Final
1. Between-day precision
2. Reportable range

the laboratory or to the needs of the physicians. For instance, the laboratory may perform testing on both human and veterinary specimens. A new thyroxine method is needed and the goal is to find one method with the dynamic range to encompass human, canine, and feline reference ranges. An acid test in this scenario involves testing a human, feline, and canine specimen once with each of the candidate methods. Another example of an acid test involves identifying a human chorionic gonadotropin (hCG) method that is not affected by hook effect (see Chapter 12: Immunoassay). Serum from a woman with a molar pregnancy is analyzed by each of the candidate methods. Any method yielding a hooked result when this specimen with an exceedingly high hCG concentration is analyzed neat (i.e., undiluted) is excluded from further investigations. A final example of an acid test involves the *in vitro* fertilization laboratory that needs an estradiol method with quick turnaround to meet the needs of the physicians it services; those methods with excessive analysis times are removed from further consideration.

The first required experiment to be included in the initial evaluation is within-run precision. **Precision** is defined as agreement between sample replicate measurements. It is an estimate of random analytical error—error that can be either positive or negative, whose direction and exact magnitude cannot be predicted. The calculated standard deviation (SD) and coefficient of variation (CV) (see Chapter 16) of the results provide an estimation of a method's imprecision. Precision can be further stratified with experiments that estimate **within-run precision** (replicates analyzed within the same analytical run), **between-run precision** (replicates analyzed between different runs), **within-day precision** (replicates analyzed on several runs on same day) and **between-day precision** (replicates analyzed repeatedly on different days). A detailed description of all the types of precision studies can be found in NCCLS Document EP-5, "Evaluation of Precision Performance of Clinical Chemistry Devices." Within-run precision provides an assessment of precision over a very short time frame, thus judging a method as acceptable or not before proceeding to longer-term studies. The within-run precision experiment, involving the analysis of the same material multiple times, is conducted at concentrations near the analyte's medical decision levels. Precision is performed with samples that have a matrix similar to patient specimens. Usually, the quality control material that will ultimately be employed after method implementation to assess the acceptability of results is used to determine precision. The control material is analyzed about 20 times in the same run to achieve a statistically significant number. A mean, SD, and CV are calculated and the results compared to the analytical goals established previously. Those methods with within-run precision failing to meet the established analytical goals are removed from further evaluation. The determination of an F-test of significance (refer to Chapter 16) will decide whether there is a difference between the measured variance and that determined as the analytical goal.

The next experiment assesses **accuracy,** defined as the agreement between the measured quantity and its true value. The **method comparison** study functions as an assessment of the accuracy of the method. Accuracy can either be *established* by comparing results from the evaluation method with those from a definitive or reference method or *verified* by comparing results from the evalua-

Specimen Analysis for a Method Comparison Study

- Analyze 40–100 patient specimens over a 5-day period
- Analyze samples in duplicate by both methods (ideally at the same time)
- Selected specimens should be representative of the patient mix in the lab and representative of a variety of disease states
- Analyte concentration should span the entire analytical range

Types of Error

- Random (varies from sample to sample; scatter)
- Systematic (always in one direction; bias)
  —Constant (same magnitude over concentration range)
  —Proportional (magnitude is a percentage of analyte concentration)
- Total Error (random + systematic error)

Causes of Random Error

- Instability of the instrument
- Temperature variations
- Variations in reagents
- Variations in handling techniques
- Operator variabilities

tion method with those from an established comparative method. A detailed description of accuracy studies can be found in NCCLS Document EP-9, "Method Comparison and Bias Estimation using Patient Samples." The chosen reference or comparative method is accurate and precise and functions as the benchmark. Typically, 40 to 100 *patient* specimens are analyzed in duplicate by both the evaluation and the reference or comparative methods, ideally at the same time, over a period of 5 days. Specimens for this study are selected to be representative of the patient mix in the laboratory. Patient analyte concentrations are chosen to span the entire analytical range of the method. Inclusion of samples from patients with a variety of disease states normally encountered is also important. Graphically plotting the paired results from the evaluated method against those from the reference or comparative method by performing a linear regression analysis allows depiction of the type of error inherent in the evaluated method (see Chapter 16). If two methods correlate perfectly with one another, then the data pairs plotted as concentration values from the reference method (x) versus the respective evaluation method (y) result will produce a straight line ($y = mx + b$), with an ideal slope of 1.0, an ideal y-intercept of 0.0, and a correlation coefficient of 1.0 (see Chapter 16). Errors that cause a deviation from the ideal one-on-one agreement with a benchmark method are **systematic error** (error that is always in one direction—bias) or **random error** (varies from sample to sample and manifested as scatter about the line of best fit). Systematic error can be further subdivided as **constant error** (error always in the same direction and of the same magnitude even as the concentration of analyte changes and manifested as a shift of the best fit line in one direction and a nonzero intercept) and **proportional error** (error that is always in one direction and whose magnitude is a percentage of the concentration of the measured analyte and manifested as a nonideal slope in the best fit line) (see Chapter 16). Causes of random error include instability of the instrument, temperature variations, variations in reagents, variations in handling techniques (pipetting, mixing, etc.), and operator variabilities. A constant error is independent of analyte concentration and can be caused by interferent that gives rise to a false signal. A proportional error may be due to such things as incorrect calibrator assignment, erroneous calibration, or a side reaction. More recently, the trend has been toward the estimate of total method error, which encompasses both random and systematic errors.

Systematic error differences that exist between the evaluated and the comparison methods are best detected with **bias,** the difference between the average results from the evaluated and comparison methods. Whether the noted bias is statistically significant is determined by a *t-test* (see Chapter 16). Systematic error can be estimated by substituting values into the regression equation and determining the difference between the substituted and calculated values.

Random error is demonstrated by the **correlation coefficient, (r),** the ratio of covariance to the total variance. This statistic indicates how well changes in the test method track changes in the reference method. One drawback to evaluating the significance of the correlation coefficient is that the r value is extremely sensitive to the analyte concentration range. The *standard error of the estimate, $s_{y/x}$,* also estimates the random error between the compared methods (i.e., $s_{y/x}$ is the standard deviation of the differences between the measured values of y and the linear regression values of y).

The plotted data are inspected to identify any **outliers.** An outlier can be readily identified as a point on the graph that is too far removed from the linear regression line. All data discrepancies between the two methods are reanalyzed as soon as possible for resolution. The presence of any true outliers will disproportionately affect the linear regression analysis because the line is "pulled" towards these outliers. The data are also visually inspected for nonlinearity. If nonlinearity is found, the data are condensed into the acceptable linear range. All statistics are recalculated if any data are rejected for nonlinearity or as outliers.

A method comparison study can also be performed with qualitative methods. The qualitative interpretations (positive, negative; present, absent) from each method are compared and, if in agreement, are considered acceptable.

If the control material used for precision studies has established values, the mean value from this precision experiment when compared to assigned mean also serves to assess method accuracy.

Proficiency materials, National Institute of Standards and Technology (NIST) standards, other certified material, or split samples from another laboratory can serve as estimates of accuracy when analyzed with the evaluation method and the answers compared to refereed results.

Another way to determine a method's accuracy is by analyzing **recovery,** defined as the ability of a method to correctly measure pure analyte when added to the samples routinely measured. A known amount of analyte is added to an aliquot of sample to achieve a final concentration that ideally is near the medical decision levels. This experiment may be difficult to perform if pure analyte material is not readily available. To avoid excessively altering the specimen matrix, the spiked aliquot amount comprises < 10% of the total sample volume. Pipetting accuracy is crucial when such small volumes are manipulated. A comparison baseline sample is prepared by adding a similar amount of only diluent to the specimen. Prepared samples are analyzed two to four times each to reduce the effects of assay imprecision on results. The result for the spiked sample is compared with the result for the baseline sample to calculate recovery: the difference between results from the spiked sample and the baseline sample is the amount recovered; the amount recovered divided by the amount added is the recovery and should be expressed as a percentage. Ideally, recovery approaches 100% and any deviation is taken as an estimate of the proportional error. A recovery experiment is particularly critical if the method has a pretreatment step in which analyte can be lost. Though recovery evaluates matrix effects and tests for competitive interferences, the comparability of results may be limited since the spiked samples are contrived and may not reflect *in vivo* conditions.

If some of the patients for the correlation study are verified as "normal," this patient subset will also serve as the reference range verification study. Otherwise, a separate reference range study is designed. The **reference range** or interval for an analyte is defined as the limits of laboratory results that define a patient population without

---

Data Analysis for a Method Comparison Study

- Plot data: evaluated method values on the y-axis versus comparison method values on the x-axis
- Linear regression analysis $(y = mx + b)$
  - m = slope of line (ideally 1.0)
  - b = y-intercept (ideally 0.0)
  - correlation coefficient (ideally 1.0)
- Evaluation of linear regression
  - Bias (test for significance with t-test)
  - Standard error of the estimate $(s_{y/x})$
- Inspect for outliers (discard any value deemed an outlier and recalculate all statistics)
- Inspect for linearity

disease—in other words, healthy. Usually, a reference range is further qualified by demographic factors such as age, sex, race, genetics, and so on. What constitutes "normal" can be defined several ways, including the spread of values around a statistical mean, those values found most often in healthy individuals, values excluding those associated with disease, values that are unlikely to cause harm, and a consensus of "approved" values or an "ideal" value. Any of these definitions can be employed to describe a reference range. A detailed description of reference range studies can be found in NCCLS Document C28-A, "How to Define and Determine Reference Intervals in the Clinical Laboratory." For some difficult analytes (e.g., therapeutic drugs and cerebrospinal fluid total protein), the manufacturer's suggested reference range or literature values may be appropriate. Choose the "normal" population representative samples with care, taking into consideration variables that may affect the reference range such as age, sex, race, genetics, diet, and the like. After analysis, the data are inspected to identify outliers. There are several methods to test for outliers; one method is outlined below:

- Sort the data from highest to lowest.
- Calculate the mean and SD for the data.
- For suspected outliers at the high end of the range, calculate the ratio value (d) as:

$$\frac{\text{Outlier value} - \text{Mean}}{\text{SD}} = d$$

- For suspected outliers at the low end of the range, calculate the d value as:

$$\frac{\text{Mean} - \text{Outlier value}}{\text{SD}} = d$$

- If d > 2.75, the suspected outlier is indeed a true outlier. Discard that value and recalculate the SD.

Example

For the following reference range data, determine if outliers exist:

9.0	9.8	10.6
9.1	9.9	10.7
9.2	10.0	10.8
9.3	10.1	10.9
9.4	10.2	11.0
9.5	10.3	11.5
9.6	10.4	
9.7	10.5	

Answer

First, the mean (10.07) and standard deviation (0.68) are calculated for the data. Visual inspection of the data identifies the point 11.5 as a suspected high outlier. The formula for a suspected high outlier is applied:

$$d = \frac{11.5 - 10.07}{0.68} = 2.1$$

Since the ratio value is < 2.75, 11.5 is not an outlier.

As a rule of thumb, no more than one result per every 40 patients analyzed should be excluded. The cause of the outlier should be investigated before rejecting that data point. Verification of a patient as abnormal is also important before excluding that patient's result. Finally, it is important to recalculate the statistics for all remaining data after rejection of outliers.

The remaining data are plotted as a *histogram* with concentration versus the frequency of that result occurring in this patient reference range study. The resultant curve is identified as a normal bell-shaped curve (Gaussian) or non-Gaussian distribution. The data from a Gaussian curve are analyzed with *parametric statistics* to calculate a patient mean, SD, and the limits for the central 95% of the results. Likewise, a non-Gaussian curve is analyzed with *nonparametric statistics* and the mean, SD, and 2.5 and 97.5 percentiles calculated similarly. Determination of the reference range can be based on several different methods, including the reference interval within a specified percentage (usually 95%), limiting reference values (usually 0.025 and 0.975), and the analysis of clinical outcomes, a risk-based procedure. Typically, for a normal distribution, the reference range is calculated as the mean ±2 SD. Because the purpose of this experiment is to *verify* the manufacturer's stated reference range, if all the "normal" patient values or the derived statistical range fall within the manufacturer's recommended reference interval, the stated range is verified as acceptable.

It is very informative to perform a linear regression plot of the values from the "normal" study population derived from the method comparison study. Any bias present between the two methods will be evident from the graph.

F. FINAL EXPERIMENTS

Those methods that have successfully survived the initial experiments are subjected to the final tests. The between-day precision study and the reportable range study are the last experiments to be completed.

The between-day precision study ideally is started at the beginning of the evaluation and continued forward. This experiment is usually performed over a period of 20 days and with at least 20 data points being collected. Material, known to be stable over the course of the study, is analyzed over multiple runs, employing multiple lots, multiple calibrations, and multiple analysts to achieve an accurate estimate of true variability. Concentrations are chosen at medical decision levels. Practically speaking, the control material that will ultimately be used to assess run acceptability is used for the precision study. Statistics, including mean, SD, and CV, are calculated and the results are compared to the established analytical goals. An F-test of significance will determine whether there is a difference between the measured variance and that determined as the analytical goal. These data may be used in the future as the quality control parameters for Levy–Jennings plots once the laboratory has gone "live" with the new method.

The last experiment to be performed on a manufacturer's FDA cleared method used without modification is the **reportable range** ◀ study. The method's reportable range is a CLIA term that includes "all results that may be reliably reported. The reportable range is the range of values that a laboratory establishes as providing accurate laboratory results for the intended clinical use. The reportable range is established by demonstrating linearity of the results of the method system. The limits of the reportable range are based on meeting accuracy and precision requirements such as the minimal limit of quantification or sensitivity, when applicable."

A *linearity experiment* is typically used to establish the reportable range. **Linear range,** or analytical range, represents concentrations in which a linear relationship exists between the assay readout and changing analyte concentration. Within this linear range, the method is usable without modification. Ideally, the linear range encompasses at least 95% of clinical specimens without dilution. A method may be linear, yet nothing is implied about its accuracy. A detailed description of an ideal linearity study is provided in NCCLS Document EP-6, "Evaluation of the Linearity of Quantitative Analytical Methods." A linearity study is performed, employing a least five linearly related but different concentrations measured at least in duplicate (quadruplicate is recommended) in a single run and evenly spaced over the entire range recommended by the manufacturer. In some instances, it may be necessary to spike samples with an aliquot of a stock solution of the analyte to be measured to achieve the desired concentration levels. The matrix of the linearity materials must match that of the patient specimens. To evaluate the possibility of matrix effects, the linearity experiments can be done initially with aqueous samples and then repeated with biological specimens. Preference is given to linearity materials prepared by mixing varying amounts of a patient specimen or pool containing a high concentration of the analyte with a patient specimen or pool containing little or no analyte of interest to produce samples with constant final volumes. The data are inspected for outliers. A plot of the assigned versus measured concentrations ideally fits a linear model and passes through the origin. Recovery for each concentration level ideally approximates 100% over the defined linear range. Bias is determined at each concentration level and judged to be acceptable by comparison with analytical goals. If the manufacturer's reportable range cannot be verified, the laboratory will have to modify its stated working reportable range. Patient specimens having results that exceed the calculated reportable range or linear range are diluted and reanalyzed, and the appropriate dilution factor applied or the results reported as "greater than the upper linearity limit."

For methods that are not cleared by the FDA, are "home brew," use analyte-specific reagents, or are modified from the manufacturer's approved protocol, additional experiments are required, including analytical sensitivity and analytical specificity.

Analytical sensitivity, an aspect of precision, is the ability of a method to detect small quantities of the measured analyte. Two aspects of the analytical sensitivity are the least detectable dose and the limit of quantification.

The **minimum detectable concentration (MDC),** or least detectable dose, is the smallest analyte concentration that can be distinguished from a suitable blank or zero with stated level of confidence, usually 95%. Note that this value may not necessarily mean the result measured is clinically useful, only that it can be statistically distinguished from zero. The least detectable dose is determined from blank readings, the responses observed due to reagent and sample constituents, not including the desired analyte. Ideally, blank readings should be negligible. Typically, detection limit is calculated as three times the SD of the measured blank average. The blank or zero calibrator is measured 20 times within a run. The MDC of the method is calculated as either two or three times the SD from the measured blank average. At the mean plus two times the SD limit, there is a 2.5% chance that a result above this limit is a zero sample. Alternatively, sensitivity has been calculated as the **limit of quantification,** which is the minimum concentration whose imprecision is within some required or specified limit of error, typically expressed as a CV < 20%. Material containing analyte at several low concentration values is measured repeatedly over several runs spanning several days to accumulate a total of 20 replicates. The mean, SD, and CV are calculated for this data set. The calculated CV versus concentration is plotted. The analyte concentration at which the CV reaches 20% is identified.

    Analytical specificity is the ability of a method to measure only the analyte for which it is designed and is an indicator of a method's freedom from interferences. This term is related to accuracy and provides an estimate of the method's constant error. Aspects of specificity include freedom from interference due to hemolysis, icterus, lipemia, and cross-reactivity. **Interference** is defined as the effect of a component, which does not by itself produce a reading, on the accuracy of the measurement of the desired analyte. Analytical specificity is determined by experiments investigating interferences and cross-reactivities. A detailed description of an ideal interference study is provided in NCCLS Document EP-7, "Interference Testing in Clinical Chemistry." CLIA requires that a laboratory be aware of common interferences by performing studies or having information available from interference studies performed elsewhere, either by the manufacturer or reported in literature, for example. A patient specimen, with an analyte concentration near the medical decision level, is spiked with known amounts of the suspected interfering substance. The interfering material is added to produce a final concentration that approaches physiological limits. The amount of interferent added is small to avoid altering the specimen matrix. Pipetting accuracy and precision become crucial when adding such a small volume. A baseline sample is also prepared by spiking the specimen with the same volume of interferent solvent only. Both the spiked and baseline samples are analyzed at least in duplicate. Any difference in the results between the spiked and baseline samples is attributable to the interferent and considered bias. Differences are compared to the analytical goals to determine if the effect is acceptable.

    Common interferences that are routinely investigated include *hemolysis, icterus,* and *lipemia* (turbidity). The recovery of analyte in the presence of interferent when compared to the blank ideally approximates 100%. The predetermined analytical goals dictate the acceptability levels for each investigated interferent.

    Interferographs are plots of the concentration of interferent added versus the ratio of the final to original result expressed as a percent; they are excellent depictions of the effect of interferences for common chemistry analytes measured on the major automated

Common Interferences

- Hemolysis
- Icterus
- Lipemia
- Drugs
- Collection tube additves and preservatives
- Specimen collection devices
- Heterophilic antibodies

analyzers. Alternatively, the interferograph depicts the concentration of interferent versus the analyte concentration by instrument.

The experiment to determine the effect of hemolysis on a specific method is begun by collecting paired samples and centrifuging and analyzing one of the samples directly. The other specimen is subjected to physical trauma to effect hemolysis. Note that this approach is not valid if red blood cells contain the analyte of interest in high concentrations.

Lipemia effects are initiated by dividing a lipemic sample in half. One half is analyzed directly while the other half is analyzed after it has been subjected to ultracentrifugation to clarify the specimen. If a suitably lipemic specimen cannot be located, a specimen is spiked with the drug Intralipid® to mimic lipemia. The comparison baseline sample is an aliquot of the same specimen that has been spiked with water.

Spiking samples with increasing amounts of unconjugated bilirubin from a stock solution assesses the effect of icterus. A baseline sample is prepared by adding an identical amount of the solvent to sample.

Other interferences that could be considered include those from drugs, collection tube additives and preservatives, specimen collection devices, and heterophilic antibodies.

The experiments previously discussed are by no means comprehensive. Other experiments may be performed as dictated by the needs of the laboratory and its clientele. Specific characteristics that are critical to a particular test can be investigated further, such as ▶ **carryover** and **hook effect** (see Chapter 12). Other experiments include specimen, reagent, calibrator, and control stabilities; calibration; QC schedule; and matrix effects. The performance of a method when scaled up to the laboratory's large patient specimen volume may be an important variable to explore; often, a method is not robust or is unsuitable in providing the necessary efficiency to process a large number of specimens within a certain time constraint.

III. METHOD IMPLEMENTATION

Once all the evaluation experiments are complete, a summary report of the evaluation findings is prepared. This report contains experimental results, all conclusions, and pertinent statistics. Following the report is the actual raw data. Per CLIA, "the laboratory must have documentation of the verification or establishment of all applicable test performance specifications." Store this report and the associated data in a secure place; the documentation is subject to regulatory inspection.

It is hoped, of course, that the method evaluation experiments pinpoint a clear winner for the new method to be adopted by the laboratory. The actual choice represents a balance of quality and cost to yield the best method to meet the needs of the laboratory and the physicians that it services.

Since so much effort goes into the initial preparation for the evaluation, a similar amount of energy needs to be expended to prepare for method/instrument implementation. Items for consideration before going "live" are listed in Table 17–5.

Table 17–5. Implementation Considerations

- **Communication** (notify clients of method change, notify programmers if an interface is required, notify staff/clients of any special specimen handling or transportation needs, prepare educational/marketing materials for clients/staff)
- **Documentation** (file research data appropriately, prepare maintenance/QC forms, update lab literature [e.g., service manual], update reference ranges, verify all the necessary changes were made, write the instrument operating procedure, write the maintenance procedure, write the testing protocol in NCCLS format)
- **Logistics** (assess staffing needs for the new test, arrange for any special instrument requirements [plumbing, electrical, ventilation], determine if patient rebaselining is needed between the old and new method, determine the result's reporting format, identify an implementation date, identify the location for reagent storage, identify QC material, identify the location in the lab for the instrument, test all aspects of the new test before going live, train analysts in method/instrument operation and document)
- **Purchasing** (alert Purchasing about the arrival of the reagents/equipment; decide on service contracts; negotiate reagent cost; negotiate the instrument cost; order QC material; order reagents and supplies; procure any ancillary equipment; procure necessary reagents, instruments, etc.; purchase/reagent rent instrumentation; use up reagents of test to be replaced)
- **Safety** (identify any reagents/instrument safety hazards, obtain reagent MSDS)

MSDS = material data safety sheets

Once the method has been carefully evaluated and implemented, by no means is it time to rest. Ensuring that the method continues to produce quality results is an ongoing process. The criteria that are used to judge whether a method is performing acceptably may be laboratory specific, such as adequate precision, or regulated. The regulated requirements, dictated either by the CLIA or an accrediting agency such as the CAP, are necessary to ensure continuing quality laboratory results from a test method. Some of these requirements and their references are listed in Table 17–6.

Table 17–6. Regulatory Requirements

Requirement	CAP	CLIA
Enroll in proficiency testing	42 CFR 493.801	
Actively review proficiency testing results		.1407(e)(4)(iii)
Document corrective action to unacceptable proficiency results		(iv)
If no graded proficiency testing available, have another procedure to validate performance at least semiannually		.1709
Develop a quality improvement program	03:2000	
Have defined goals for monitoring analytical performance, procedures, policies, tolerance limits, corrective action, etc.	01.3000	
Have specimen collection, preservation, transport, and storage procedure	03:2005	
Actively review QC, surveys, maintenance, and function checks	01:3010	
Detect unusual lab results (absurd values), significant analytical errors, and clerical errors	03:2020	
Write procedure manual in NCCLS GP2-A3 format	.1211	
Review procedures at least annually		.1407(e)(13)
Ensure analysts are knowledgeable about contents and changes in the procedure manual	03:2115	
Maintain copy of discontinued procedures for 2 years		.12119(g)
Write procedure to verify specimen integrity	03:2200	
Write criteria for unacceptable or suboptimal specimens		.1703(c)
Establish or verify reference range		.1213(2)(i)(F)
Establish or validate reportable patient range		.1213
Define upper and lower limits of all reportable parameters		.1213
Define turnaround time		.1219(c)
Ensure analysts are familiar with critical limits and reporting		.1109(f)
Properly label all reagents		.1205(d)
Perform reagent checks	03C:2414	
Provide information on common interferences available or evaluated	03C:ZXVQ	
Perform calibration		.1217
Initially verify linearity and when calibration verification fails	03C:2435	
Comply with method calibration verification procedures		.1217
For quantitative assays, analyze controls of more than one concentration at least once daily		.1218(b)(2)
For qualitative tests, analyze positive and negative controls with each batch of patient specimens		.1218(b)(1)
Analyze controls like patient specimens		.1218(f)(1)
Verify QC acceptable before releasing patient results		.1218(e)
Provide maintenance schedule and parameter tolerance limits	03C:2695,2710	
Write instructions for maintenance checks	03C:2700	
Write instructions for minor troubleshooting and instrument repairs	03C:2715	
Check for carryover with automatic pipetting	03C:3040	
At least semiannually, verify compatibility of tests performed by different methodologies or instruments or at different sites.		.1709
Evaluate specimen containers to ensure they do not contribute any analytical interference	01:4052	
Identify the testing analyst		.1107(d)
Verify analytical accuracy and precision		.1449
Identify analytical interferences		.1213(b)(2)(i)(D)
Verify the reportable range		
Verify the reference intervals		
Periodically evaluate the appropriateness of the reference intervals	01:4211	
Provide a list of current test methods and performance specifications to clients		.1109(g)
Explain method changes to clients		.1109(g)
Periodically verify the transmission of patient results	01:4335	
Verify the autoverification process quarterly	01:4338	
Verify the accuracy and consistency of patient results across all interfaces	01:4340	
Periodically verify patient data calculations performed by computer	01:4345	
Annual review of content and format of computer-printed patient reports	01:4350	
Analysts meet requirements of CLIA 88 for moderate- to high-complexity testing		.1489
Assess competency of analysts to perform duties		.1713
Have a departmental continuing education program	01.5420	
Identify and control electrical, chemical, toxic, radioactive, and biological hazards	01:7035,7040,7055,7085,7090	
Have protective equipment available appropriate to the test's hazards	01:7105	

CAP citations are in a 01: or 03: format; CLIA citations all begin with 42 CFR 493.

SUGGESTED READINGS

Boyd JC. Reference limits in the clinical laboratory. In Dufour DR (ed.). *Professional practice in clinical chemistry: A companion text.* Washington, DC: American Association for Clinical Chemistry, 1999; 2–1 through 2–7.

Carey RN, Garber CC. Concepts and practices in the evaluation of laboratory methods. Presented at the AACC Capital Section meeting, April 24, 1993.

Carey RN, Garber CC. Evaluation of methods. In Kaplan LA, Pesce AJ (eds.). *Clinical chemistry: Theory, analysis, and correlation,* 3rd ed. St. Louis: CV Mosby, 1996; 402–423.

Dufour DR. Preanalytical variation, or: what causes "abnormal" results (besides disease). In Dufour DR (ed.). *Professional practice in clinical chemistry: A companion text.* Washington, DC: American Association for Clinical Chemistry, 1999; 1–1 to 1–11.

Glick MR, Ryder KW, Glick SJ. In *Interferographs: User's guide to interferences in clinical chemistry instruments,* 2nd ed. Indianapolis, IN. Science Enterprises, 1991.

Koch DD, Peters T. Selection and evaluation of methods. In Burtis CA, Ashwood ER (eds.) *Textbook of Clinical Chemistry.* Philadelphia: WB Saunders, 1999; 320–335.

National Committee for Clinical Laboratory Standards, Villanova, PA. (Refer to document numbers mentioned in the text.)

Items for Further Consideration

1. Attempt a "dry run" method evaluation to practice the concepts presented in this chapter. Assume that you are the laboratory manager of a midsize hospital with an acute coronary care unit and an active emergency room. The laboratory director has charged you with selecting a cardiac marker method for the clinical chemistry laboratory.

   - Identify the best analyte to test for that disease.

   - List all issues to be taken into consideration concerning method selection.

   - Identify five methods that are commercially available to test for that analyte.

   - Perform a paper comparison of available methods. Include in the table at least 10 criteria.

   - Decide what analytical goals will be used to judge method acceptability.

   - Narrow the method candidates to two. List your rationale for this decision.

   - Design the evaluation protocol experiments.

Quality Assurance and Quality Control

Mary Kay O'Connor

Key Terms ◀

ASSAYED CONTROLS

CLINICAL LABORATORY
   IMPROVEMENT ACT (CLIA
   '88)

CONTROL SAMPLES

CRITICAL VALUES

DELTA CHECKS

LEVEY–JENNINGS CHART

MEAN

95% CONFIDENCE LIMIT

PROFICIENCY TESTING (PT)

QUALITY ASSURANCE (QA)

QUALITY CONTROL (QC)

SHIFT

STANDARD DEVIATION (SD)

TREND

UNASSAYED CONTROLS

WESTGARD MULTIRULE SYSTEM

Chapter Objectives

Upon completion of this chapter, the reader will be able to:

1. Outline and describe the general components of a quality assurance program.
2. Relate preanalytic, analytic, and postanalytic variables to patient test result outcomes.
3. Explain how quality control programs are used to verify the accuracy of patient test results.
4. Discuss the following concepts as they relate to quality control: matrix effects, lot number, assayed/unassayed material, control level, Gaussian distribution, and 95% confidence limit.
5. Explain the principle behind the use of the ±2 standard deviation control range for the evaluation of quality control data.
6. Compare and contrast quality control data patterns demonstrating a shift and a trend.
7. Apply the Westgard Multirule System in regard to the evaluation of quality control data.
8. Differentiate between a "warning" rule and a "rejection" rule in the Westgard System.
9. List control rules from the Westgard System that are sensitive to random and systematic error.
10. Describe how delta checks are used to ensure accuracy and reliability of patient test results.
11. Describe how external quality control programs are used to monitor the accuracy of laboratory results.

Pretest

Directions: Choose the single best answer for each of the following items.

1. An example of a preanalytic error that might affect the accuracy of patient test results is
   a. incorrect dilution of patient sample
   b. improper sample collection
   c. improper reconstitution of test reagents
   d. failure to perform instrument maintenance

2. In a population with a Gaussian distribution, 95.5% of all measurement values will
   a. lead to a high false rejection rate
   b. be precise
   c. fall within the ±2.0 standard deviation (SD) range of the mean
   d. be accurate

3. The mean and SD values published in manufacturers' package inserts for quality control materials are derived from
   a. repetitive testing of the control material
   b. statistical probability tables
   c. the average values from at least five different lot numbers
   d. calibration data

4. For quantitative testing of most analytes, how many levels of control must be analyzed?
   a. none
   b. one
   c. two
   d. three

5. In regard to quality control programs, it is desirable to have a high probability rate for
   a. false rejection
   b. error detection
   c. true negatives
   d. false positives

6. The method of recording quality control data that produces a plot of the results by graphing analyte concentration versus observation point is referred to as a _____ plot.
   a. Shewhart
   b. CUSUM
   c. Levey–Jennings
   d. delta average

7. A slow, gradual increase in the values obtained for quality control analysis is indicative of a
   a. shift
   b. calibration error
   c. miscalculation
   d. trend

8. Which of the following Westgard Rules is primarily sensitive to systematic error in the test system?
   a. 1_{2s}
   b. 1_{3s}
   c. 2_{2s}
   d. R_{4s}

9. Which of the following Westgard Rules is considered first, before any other rule, when evaluating quality control data?
   a. 1_{2s}
   b. 1_{3s}
   c. 2_{2s}
   d. 4_{1s}

10. Which of the following Westgard Rules requires evaluation of data from previous control measurements?
    a. 1_{2s}
    b. 1_{3s}
    c. 4_{1s}
    d. R_{4s}

11. Which of the following QC monitoring methods is useful in detecting the mislabeling of patient samples?
    a. Westgard Multirule System
    b. Shewhart analysis
    c. delta check
    d. anion gap

12. An example of an external quality control program is
    a. standard deviation index
    b. t-test
    c. youden analysis
    d. proficiency testing

For items 13 through 16, use the Levey–Jennings charts shown in Figure 18–1 to choose the appropriate Westgard Rule that best corresponds to the QC observation points.
    a. 10_x rule
    b. R_{4s} rule
    c. 1_{2s} rule
    d. 2_{2s} rule

13. Observation point 4

14. Observation point 7

15. Observation points 8 through 17

16. Observation point 26

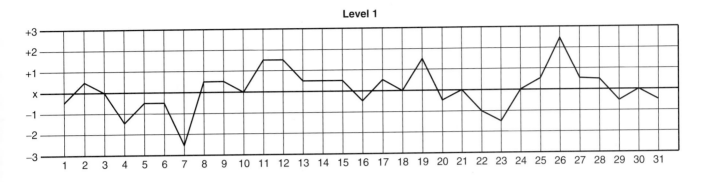

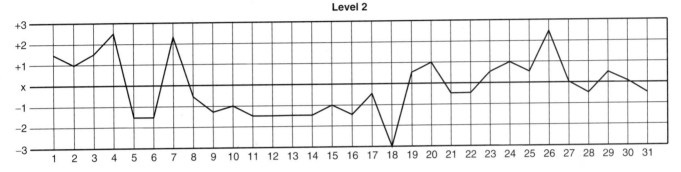

Figure 18–1. Levey–Jennings charts for questions 13 through 16 in the Pretest.

I. INTRODUCTION

An effective **quality assurance (QA)** program monitors and evaluates the quality of the entire testing process, including the preanalytic, analytic, and postanalytic phases, and is a process used by laboratories to monitor and assess quality on a continuous basis. The total testing process is a cascade of activities that affects the outcome of test results. With quality assurance, problems are identified and corrected and changes are implemented to minimize the possibility of problem recurrence. A good QA plan should be periodically reviewed for effectiveness. The laboratory director is responsible for overseeing the implementation of the QA plan and is ultimately responsible for all outcomes. As described in Chapter 19, the **Clinical Laboratory Improvement Act (CLIA '88)** requires each laboratory to develop a QA program that includes policies and procedures regarding the following:

- Personnel
- Specimen collection and handling
- Patient test management/record keeping
- Patient test results
- Safety
- Communications
- Quality control
- Proficiency testing

Quality control (QC) is just one component of a global quality assurance program. Quality control monitors the testing procedures during the analytic phase and will be discussed in greater detail later in this chapter.

II. COMPONENTS OF A QUALITY ASSURANCE PROGRAM

A. PERSONNEL

Minimum personnel requirements for clinical laboratories have been established by CLIA '88, and these standards are based on the complexity of testing that the laboratory performs (i.e., waived, provider-performed microscopy [PPM], moderately complex, or highly complex). The laboratory director oversees the entire operation and administration of the laboratory. The technical consultant is responsible for the technical and scientific management of the laboratory. This includes selecting test methods, establishing and monitoring a QC program, resolving technical problems, and evaluating the competency of all testing personnel. A clinical consultant is responsible for determining test appropriateness and interpretation of results and also serves as a liaison between the laboratory and its patients. The laboratory director or patient physician often fills this position. A general supervisor is responsible for the day-to-day supervision of testing personnel and the reporting of results. Testing personnel are responsible for processing specimens, performing tests, and reporting results. Personnel files should include, but not necessarily be limited to:

- Job application materials
- Copies of diplomas, certifications, or registrations necessary for job qualification
- Job description
- Orientation/training worklists
- Continuing education records
- Performance evaluations
- Competency assessments
- Occupational Safety and Health Administration (OSHA) blood-borne pathogens training record
- Hepatitis B virus immunization record or declination form
- OSHA exposure incident reports, if appropriate
- Any other personnel forms necessary for accreditation purposes

Preanalytic Variables

- Diet and state of fasting
- Postural changes
- Level of emotional stress
- Exercise
- Alcohol ingestion
- Dehydration
- Drugs
- Pregnancy
- Diurnal rhythms

B. PREANALYTIC VARIABLES

1. Specimen Collection and Handling. Proper patient preparation and specimen collection are crucial to guaranteeing accurate test results. To ensure that both of these issues are handled appropriately, a specimen collection procedure manual should be developed that follows the specific guidelines established by the National Committee for Clinical Laboratory Standards (NCCLS). Preanalytic variables that might affect test result outcomes include diet and state of fasting, postural changes (i.e., supine vs. ambulatory), level of emotional stress, exercise, alcohol ingestion, dehydration, drugs, pregnancy, and diurnal rhythms. Specimen requirements will determine how and in what container a sample is collected. Chemistry testing is usually performed on serum samples collected in a red top or serum separator tube (SST), hematology in an ethylenediaminetetraacetic acid (EDTA)-preserved whole blood (lavender top) tube, and coagulation studies in citrate-preserved plasma (blue top) tube. Urine collections are usually a clean-catch midstream collection for urinalysis and urine culture and sensitivity. Timed urine collections, with or without preservatives, are collected for quantitation of specific chemical analytes. Collection of a specimen in an incorrect container or in an improper manner may cause erroneous results.

Before sample collection, patients must be questioned to ascertain their correct identification. In some clinical situations, a patient may be unable to answer verbally with his or her name. In this type of situation, a patient identification band or like means of positive identification must be verified before phlebotomy is initiated. Immediately after collection, each and every specimen collected on a patient must be properly labeled. Computer-generated labels are convenient for this purpose and should include pertinent patient information and the tests ordered by the physician. In the event that labels are not available, each specimen container should be labeled with the patient's first and last name, birth date, and a unique identifier such as the patient's Social Security or medical record number. Date and time of collection should also be included, as well as initials of the person collecting the specimen. (*Note:* Blood banking specimens have their own unique patient identification system.) There should be a stated policy for dealing with any unlabeled or mislabeled specimens received in the laboratory. In most cases, unlabeled and mislabeled specimen policies call for the sample to be rejected.

Once a specimen is collected, certain analytes may be affected by time, temperature, and/or light. Glucose, lipids, phosphate, potassium, albumin, white blood cells, and cortisol are examples of analytes that change significantly over time. In the case of Food and Drug Administration (FDA)-cleared procedures, specimens should be collected and handled according to the assay manufacturer's instructions for each test. It is important to provide a specimen collection procedure manual containing all necessary information including instructions for transporting samples to the laboratory. The evaluation and rejection of unsatisfactory specimens is an important step in specimen handling. Some conditions that may deem a specimen unsatisfactory include:

- Inadequate quantity of blood or urine collected
- Inadequate blood-to-anticoagulant ratio
- Improper collection container
- Improperly labeled containers
- A specimen that has clotted (inappropriately)
- Specimens not collected in appropriate sterile or anaerobic transport containers
- Specimens that are hemolyzed, lipemic or grossly contaminated
- Too much time has elapsed since collection
- Samples that have been transported at inappropriate temperatures

Sometimes it is necessary to collect blood samples at a specific time of day (e.g., cortisol) or at a specific time interval (e.g., therapeutic drugs, serial cardiac markers, and postprandial blood sugar).

Specimen rejection logs are kept to describe which samples were rejected, why they were rejected, and who was informed.

Once a specimen is collected from the patient, a number of changes begin to occur. Bacterial growth and the action of enzymes in the sample are two factors that can drastically alter results. When testing is to be performed on serum or plasma samples, the cells must be separated promptly by centrifugation at an appropriate temperature, g force, and specified length of time. When a specimen cannot be analyzed immediately after collection, it is important to handle the specimen so that it is properly preserved for future testing according to the test manufacturer's recommendations.

2. Patient Test Management/Record Keeping. It is extremely important that a specimen remain positively identified throughout the entire testing process. A *test requisition*, written by an authorized person or submitted electronically through a laboratory information system (LIS), should accompany the correctly labeled specimen to the laboratory. All oral requests must be followed by a written request within a specified time frame. The test requisition should include the patient's name and unique identifier, the name and address of the physician requesting the test(s), the tests to be performed, the date and time of specimen collection, and other relevant information needed to ensure accurate and timely testing and reporting of test results. The requisition's information must correspond to the labeled specimen's information.

A specimen accession log, either computerized or paper, should be used to record when specimens are received in the laboratory. Normally, a unique sample identifier, such as an accession number, is assigned to the specimen to track the sample throughout the testing process. The specimen log should include all tests ordered on a particular specimen. Again, this is usually handled by an LIS (see Chapter 15).

C. ANALYTIC VARIABLES

After a specimen has been properly processed and readied for testing, it is important to make sure that analytic checks are performed on laboratory and instrument conditions to ensure that the testing system is monitored and "in control". Water quality is an extremely important item for consideration when preparing and reconstituting standards, control materials, and reagents. Water quality can also be an important factor in the making of sample dilutions. Generally, Type I water meeting NCCLS specifications is used. All scheduled and *preventative maintenance* and function checks must be performed according to the instrument manufacturer's guidelines. Electronic, mechanical, and operational checks help to assure accurate and reliable results. Surge protectors are an important precaution against current fluctuations that might interfere with instrument electronics. All temperature-dependent equipment (including refrigerators, freezers, water baths, and incubators) must be checked daily or just prior to use, or continuously monitored with a recording system. Some testing systems require a humidity check that is performed with a hygrometer. Daily recording of temperature and humidity is good standard laboratory practice to ensure optimal conditions exist for testing. Balances and mechanical pipettes must be kept clean and

---

Information Included on the Laboratory Test Requisition

- Patient name and unique identifier
- Name and address of physician requesting the test(s)
- Tests to be performed
- Date and time of specimen collection

Checks on Laboratory and Instrument Conditions

- Water quality
- Preventive maintenance on equipment and instruments
- Function checks on equipment and instruments
- Surge protectors
- Monitoring of temperature in temperature-regulated compartments of equipment
- Proper cleaning of glassware and equipment
- Daily record of laboratory temperature and humidity

properly maintained, which includes checking for accuracy and precision on a specified routine basis. Balances must be checked against weights from the National Institute of Standards and Technology (NIST). Glassware should be clean and free from chips, cracks, or soap residue. Volumetric "Class A" glassware is customarily required for precise and accurate pipetting. This is considered essential when reconstituting calibrators, standards, and control materials.

The dependability of a method to give accurate and reliable results depends primarily on the quality of standards used for instrument calibration or method standardization. A laboratory should only use the purest possible calibration material or calibrators recommended by the manufacturer for an instrument or test kit. For FDA-cleared commercial kits, calibration must be performed at least as often as required by the manufacturer. Recalibration or calibration verification may also be necessary after major repairs or when certain parts, such as lamps or syringes, are replaced. Calibration may be necessary when controls are out of range or demonstrate an unexplained shift or trending pattern. Calibration verification is a procedure used to determine whether an instrument is still properly calibrated, and is usually less costly than a full recalibration. To verify that a calibration has not changed, calibration material must yield values that agree within manufacturers' specifications for low-, mid-, and high-level calibrator values. If a calibration cannot be verified, recalibration of the instrument is indicated. Calibration or calibration verification must be performed at least every 6 months or more often if recommended by the manufacturer.

Procedure manuals serve as a reference, enabling testing personnel to have access to essential elements of testing needed for each test method. Each laboratory's manual will vary depending on a laboratory's specific needs. Essential elements of the manual, as required under CLIA '88 are reviewed in Chapter 19.

1. Internal (Intralaboratory) Quality Control. Quality control can be described as a process or system used by laboratorians to ensure the quality of their laboratory's overall analytical performance. It should realistically assess a laboratory's normal or usual performance by identifying any significant problem as soon as it occurs, allowing for correction of the problem as promptly as possible. This includes the documentation of the problem and how it was solved. Quality control sets limits and incorporates all aspects of testing that might affect the integrity of the specimen and the reporting of results. Each laboratory must establish a written quality control procedure program that includes verification and assessment of accuracy, measurement of precision, and detection of error for all procedures performed in the laboratory.

Control samples must be treated and tested in the same manner as patient specimens. The type of control used is determined by whether the test will have a qualitative, semiquantitative, or quantitative result. A qualitative test determines only whether a substance is present or absent and is usually reported as either positive/negative or detected/not detected. A pregnancy test is an example of this kind of test. In this case, only positive and negative controls are necessary. With semiquantitative immunological testing (e.g., rheumatoid factor), titered controls are run indicating a positive or negative result up to a specified dilution point.

Quantitative testing measures the concentration of analyte present in a sample. The accuracy of quantitative methods must be checked using control samples containing the analyte present at a normal level, an abnormally high level and, where relevant, an abnormally low level. Thus, a minimum of two (and in some cases three) control levels covering different analyte concentration ranges is used. When testing therapeutic drug levels, therapeutic, subtherapeutic, and toxic levels of controls should be employed. With blood gas analyzers, controls in the acidotic, normal, and alkalotic ranges must be used. In regard to the composition or base of the control material, controls are designed to consist of the same matrix as the analyte being measured (i.e., serum, urine, whole blood, or cerebrospinal fluid). To be cost effective, the purchasing of control material should be in a quantity sufficient to last as long as possible (usually a minimum of a year) and should exhibit good vial-to-vial stability. Control material is available in lyophilized (powdered) form that must be reconstituted with either deionized water or a specific reconstituting solution before use, or in liquid form that is ready to use.

Assayed and unassayed controls are readily available for the standard analytes. There are a number of manufacturers that produce both ▶ types of these controls. **Assayed controls** have been analyzed repeat-▶ edly by the manufacturer and thus have an established target **mean** ▶ and **standard deviation (SD)** for each analyte. These values are provided in the manufacturer's package insert for the control materials and should be considered only as a target range for each analyte. It is good laboratory practice to verify the package insert values for each analyzer (and each method of analysis) before the control is placed into use. A laboratory may find that the mean value of an analyte is slightly shifted from the stated value. The SD for a particular analyte within a laboratory should be less than or equal to the range stated in the package insert. The SD used in the laboratory should not exceed that listed in the package insert. Over time, a laboratory will acquire a historic standard deviation for each analyte performed on a particular instrument.

When an unassayed control material is purchased, a mean and standard deviation must be determined in the laboratory prior to ▶ placing the control material into use. **Unassayed controls** must first be analyzed over a period of time to collect data for computation by statistical methods. These determinations should be made while running in parallel with the existing lot number of control. A reasonable overlap is at least 20 values from 20 or more separate runs. For low-volume testing, a provisional target value from 20 determinations on fewer than 20 runs may be used. However, no more than three data points should be acquired from any one run.

If a frequency distribution of the data were plotted, as demonstrated in Figure 18–2, the distribution of points in relation to the mean or average result may be assumed to be a normal continuous distribution, termed *Gaussian distribution*. This assumption implies that the majority of the data will cluster evenly about the mean. Based on statistical probabilities for repetitive measurements and the inherent fluctuation of data for such measurements, it is also presumed that, for Gaussian data distribution, 68.2% of all values measured will fall within plus or minus ($\pm$) 1 SD of the mean, 95.5% of all measurements will fall within ±2 SD of the mean, and 99.7% of all values will fall within ±3 SD of the mean. This leaves a 0.3% chance

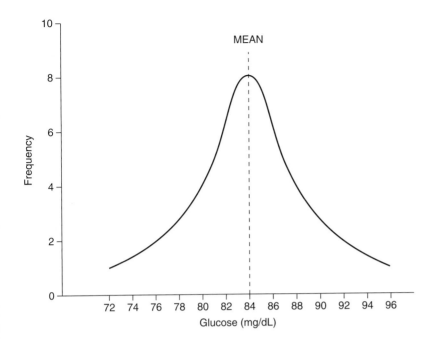

Figure 18–2. Frequency distribution of glucose measurement data.

that a value will occur outside the ±3 SD limit due to normal fluctuation of data points. Moreover, points may be expected to fall outside of the ±2 SD limits 4.5% of the time due to random fluctuation. This fluctuation does not automatically indicate that an error is present within the test system; instead, it demonstrates the expected variability of data obtained from repetitive measurements. Thus, in the case of the ±2 SD range, 4.5% of the time, a measurement point will fall outside of the range due to the expected random fluctuation of data. For a range encompassing ±3 SD from the mean, one may expect that a measurement point will fall outside of this range only 0.3% of the time. Since the probability of measurements falling within the range is so high, points falling outside of the range are highly suspect in regard to the possibility of an error being present in the test system. These statistical probabilities provide the underlying foundation for our evaluation and interpretation of quality control data.

A confidence level of p = 0.05, or **95% confidence limit** is the most prevalently accepted limit for control ranges. Since ±2 SD in a Gaussian distribution curve incorporates 95.5% of all values obtained during control data collection, control ranges are set at the calculated mean ±2 SD from that mean. A 95% confidence level balances out the probability of rejection, called the probability for false rejection, and the probability of detecting certain analytical errors, called the probability for error detection. It is desirable to have a low probability for false rejection, (less than 5%) and a high probability for error detection, (approaching 100%).

Quality control monitors several basic aspects of analyte determination, including accuracy, precision, systematic error, and random error. See Chapters 16 and 17 for a more detailed discussion of these concepts.

2. Methods of Recording Quality Control Data. Many laboratory instruments have the capacity for on-board storage of QC data and allow for data to be visualized in lists and/or charts while also handling data reduction (i.e., calculations of mean, SD, and coefficient of variation). Some laboratories store QC results on a laboratory computer. If neither of these cases are an option, QC sheets can be prepared manually.

The most commonly used QC recording system is the ▶ **Levey–Jennings chart** shown in Figure 18–3. For each analyte and each control level, a chart is constructed to display the analyte concentrations on the *y-axis* and the day/event on the *x-axis*. Horizontal lines are drawn for the mean and the upper and lower control limits (the mean +2SD and the mean −2SD concentrations, respectively). Each time QC material is analyzed, the result is plotted at the corresponding concentration in sequence after results previously plotted. After several points are plotted, a pattern emerges showing the fluctuation of the data points in the repeated measurements. In the normal variation of measurements, points are expected to randomly fall above and below the mean in no set pattern with 95% of the data falling within ±2 SD of the mean.

3. Evaluation of Quality Control Data. The first step involved in the evaluation of QC data is visual inspection. Visual inspection of the data is crucial in regard to detecting and troubleshooting errors in the test system. Developing patterns such as shifts and trends indi-
▶ cate the presence of error. A **shift** is a sudden or abrupt change in the data, which establishes a new distribution pattern above or be-
▶ low the established mean. A **trend** is a more gradual deviation in the data above or below the mean. Recognition of such patterns is helpful in determining the most effective steps to be taken for identification and correction of the problem.

The second step in the evaluation of QC data involves a more detailed inspection of the individual data points. In 1981, James

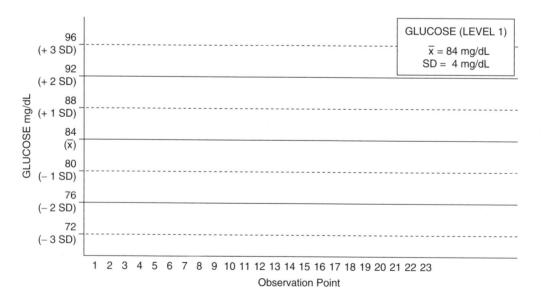

Figure 18–3. Levey–Jennings quality control record.

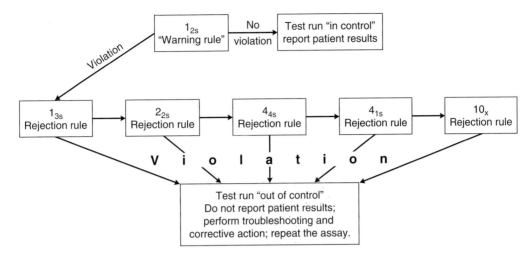

Figure 18–4. Application of the Westgard Multirule System for the evaluation of quality control data.

Westgard and his associates developed a logic diagram, shown in Figure 18–4, for applying a set of decision criteria to data plotted on the Levey–Jennings chart. The most basic application of Westgard's system is composed of six rules for interpreting QC data (summarized in Table 18–1). These rules are evaluated in sequential order, beginning with the 1_{2s} "warning" rule. If there is no violation of this rule, no further investigation is necessary. A stepwise progression of this **Westgard Multirule System** follows.   ◄

- *1_{2s}: "Warning" rule.* One control result exceeds the ±2 SD range. Violation of this rule requires that the remaining rules be evaluated in sequence to determine if the test is "in control" and if patient results may be signed out. If no other rule violations exist, patient results may be reported.
- *1_{3s}: "Rejection" rule.* One control result exceeds the ±3 SD range. This rule is sensitive to random error. Patient results should not be reported if this rule is violated.
- *2_{2s}: "Rejection" rule.* This rule can be applied to either two consecutive control results of the same control level exceeding the ±2 SD limit, or, it can be applied to control results from two different control levels exceeding their respective ±2 SD limits. This rule is sensitive to systematic error. Patient results should not be reported if this rule is violated.
- *R_{4s}: "Rejection" rule.* According to this rule, the run is rejected when the range or difference between two control observations within the run exceeds 4 SD. One control result will exceed the mean by plus 2 SD while the other control exceeds the mean by minus 2 SD, making a total of 4 SD difference between them. This rule is sensitive to random error. Violation of this rule indicates that patient results should not be reported.
- *4_{1s}: "Rejection" rule.* This rule is violated when (a) four consecutive control observations from the same control level exceed the same 1 SD limit or (b) consecutive control observations from two different control levels exceed the ±1 SD limit in the same direction. Evaluation of this rule requires checking the controls for both the present run and the previous runs. This rule is sensitive to systematic error. Violation of this rule indicates that patient results should not be reported.

Table 18–1. Westgard Multirules

1$_{2s}$	*"WARNING RULE"* Violated when one control result exceeds +2 standard deviations from the mean OR −2 standard deviations from the mean
1$_{3s}$	Violated when one control result exceeds +3 standard deviations from the mean OR −3 standard deviations from the mean *Violation indicates presence of random error
2$_{2s}$	May be violated in two different ways: 1. Two consecutive results from one level of control exceed +2 standard deviations from the mean OR −2 standard deviations from the mean 2. Results from both levels of control exceed +2 standard deviations from the mean OR −2 standard deviations from the mean *Violation indicates presence of systematic error
R$_{4s}$	Violated when the difference between two consecutive control results, having assay values in opposite directions from one another, exceeds 4 standard deviations. *Violation indicates presence of random error
4$_{1s}$	May be violated in two different ways: 1. Four consecutive control results for one level of control exceed +1 standard deviation from the mean OR −1 standard deviation from the mean 2. Two levels of control have two consecutive results that are +1 standard deviation from the mean OR −1 standard deviation from the mean *Violation indicates presence of systematic error
10$_x$	May be violated in two different ways: 1. Ten consecutive control results for one level of control occur on the same side of the mean 2. Two levels of control have consecutive results that are on the same side of the mean *Violation indicates presence of systematic error

- *10$_x$: "Rejection" rule.* This rule is applied in a similar manner to the 4$_{1s}$ rule in that one or two levels of control may be involved. It is violated when 10 consecutive control observations from the same level of control occur on the same side of the mean or when consecutive control observations from two control levels demonstrate a pattern of 10 data points falling on the same side of their respective mean. Assessment of this rule may require checking control values from the present run and previous runs. This rule is sensitive to systematic error. Patient results should not be reported if this rule is violated.

If any of these "Rejection" rules are violated, troubleshooting and corrective measures ensue. After performing corrective action, all samples, patient and control, should be reanalyzed. Once the error condition has been rectified, the QC data obtained during the period of instability should not be incorporated into the control data pool.

It is not necessary to incorporate all of the rules into a QC program, but at least two rules should be selected. One should be sensitive to random analytical error and the other sensitive to systematic analytical error. Based on the rule violation, the technologist has some indication as to the source of the error and thus the approach to troubleshoot the problem most effectively. When control rules point to systematic error, the most probable causes may be out-of-date or contaminated standards, an instrument out of calibration, temperature problems, or a problem with reagent blanking. Mechanical pipettes that have not been checked and are out of calibration may also cause a bias error. These errors affect all results in the same direction. Violation of random error rules can point to reagent instability, unstable instrument conditions (worn syringes, motor/board errors, etc.), or too much variability between individual technique in timing or pipetting.

A laboratory must document and maintain records of all QC activities and retain records for at least 2 years.

4. Other Quality Control Monitoring Systems.

Other types of QC monitoring systems may also be utilized to monitor the integrity of the results being reported.

Delta checks are a comparison of a patient's result for a given analyte with the same patient's previous result. A limit of variability is established for each analyte (most commonly a percent limit) and is commonly set in the LIS. Specimens that have differences greater than the established limit are flagged and must be reviewed by laboratory personnel. A flag may not necessarily be associated with an analytical error but instead may demonstrate a true change in the patient's condition. Conversely, an error potentially detected by the delta check system is the mislabeling of specimens.

Another monitoring method involves the use of alert checks or technical flags that are designed to detect unlikely values. These checks are often incorporated into the analyzer computer system or LIS and may point to a result that has exceeded the linearity limits of the method and requires dilution, or falls below the sensitivity limit of the analytical system. "Absurd" results or results that are "incompatible with life" must also be recognized. An example of this is a potassium result of 12.0 mmol/L.

Pattern recognition may also help to detect unlikely test combinations. Certain analytes occur in clinical specimens in very distinctive combinations. An example of this is blood urea nitrogen (BUN), creatinine, and uric acid values.

Arithmetic checks have also been used to monitor the integrity of patient results. Here, two or more related results from the same patient specimen are applied to a calculation. The most commonly recognized mathematical approximation used in this manner is the anion gap. When anion gap results consistently run high or low for a series or set of patient specimens, an error in one or more components of the electrolyte test panel may be suspected. See Chapter 6 for a clinically focused discussion of anion gap.

5. External (Interlaboratory) Quality Control

Regional Quality Control Programs. Many of the major manufacturers of QC material and various professional societies make these QC programs available to laboratories. A laboratory subscribes to such a program and purchases a quantity of QC material that is analyzed over a period of time (usually 1 year). The results from the analysis of these materials are returned to the supplier of the program and are analyzed against peers using the same lot number of control material, the same reagents, and the same instrumentation. A summary report from the supplier is generated and includes statistical evaluation parameters such as the mean, SD, and coefficient of variation, as well as a standard deviation index (SDI) for the laboratory's results. The SDI correlates an individual laboratory's performance to its peers. The SDI is calculated by subtracting the group mean from the laboratory's mean then dividing by the group SD. An SDI greater than 2.0 indicates that the laboratory's mean varies more than ±2 SD from the group. Often, this indicates a systematic error in the laboratory's measurement for that analyte.

▶ ***Proficiency Testing.*** **Proficiency testing (PT)** is a means by which a laboratory compares its performance of analytes with other laboratories using similar reagents and instrumentation. The difference between proficiency (survey) testing and daily QC is that challenges of survey sample material are sent to the laboratory three times a year and are analyzed without knowing target values or ranges. In essence, these materials mimic a patient sample. Several professional organizations distribute such surveys and analyze results for a yearly fee. The most well-known organization is the College of American Pathologists (CAP). All laboratories with a CLIA certificate must participate in an accredited proficiency testing program. After a specific cutoff date, all data is analyzed and each participating laboratory receives an individualized survey evaluation comparing it to its peer group. A copy of the report is also sent to the Health Care Financing Administration (HCFA), as well as relevant state and local authorities, for all regulated analytes. Laboratories must achieve a score of 80% or higher to pass the "challenge" event. A failure of 60% or lower, on two consecutive proficiencies, places the laboratory in jeopardy (inability to bill for the test) for that analyte. A laboratory must then pass two consecutive proficiencies in order to reinstate the reporting (and associated billing) of that particular analyte. Although PT programs are available for a wide variety of analytes, not all testing services at an institution may be covered. For those analytes for which no PT is available, the laboratory must develop an acceptable evaluation strategy.

D. POSTANALYTIC FACTORS

1. Calculations. Calculations are used in the clinical laboratory for a variety of reasons. The most common is in the preparation of simple dilutions to bring the concentration of some body fluid component into the linear range of an assay. Dilutions can also be used to prepare reagents or standard solutions. Once a result is obtained on a diluted specimen, it is necessary to calculate the actual concentration of the analyte in the specimen using the appropriate dilution factor. Other calculations commonly performed include creatinine clearance, anion gap, estimation of osmolality, and low-density lipoprotein (LDL) cholesterol concentration.

Commonly Performed Calculations

- Sample dilutions
- Concentration of analyte in specimen
- Creatinine clearance
- Anion gap
- Osmolality
- LDL cholesterol concentration

Whatever the calculation, it is essential that it be performed accurately. If an LIS is in use in the laboratory, it is possible to create software programs to perform these types of calculations. Regulatory agencies require that the accuracy of computer-performed calculations be verified periodically, with appropriate documentation.

2. Patient Test Report. The test report submitted to the ordering physician must include the following: patient name and unique identifier, name and address of the laboratory performing the test, test(s) performed, test result(s), unit(s) of measure, reference interval(s), and information regarding specimen integrity. The date/time of specimen collection and when the report was generated can also be useful in managing quality assurance.

Certain analytes have value limits that, when exceeded, indicate that the patient has a possible life-threatening condition. Consequently, these results, referred to as **critical values,** require immediate action on the part of the laboratory. Analyte values that exceed these limits should be repeated for verification and then immediately reported. If the result is given verbally, the name of the person receiving the result and time of the report must be documented and a hard copy of the test report sent to the ordering physician as promptly as possible.

The laboratory must keep all records and logs for a minimum time period, which may vary due to local regulation. The minimum time required by most regulatory authorities is 2 years. Documents included in this requirement are test requisition forms, preliminary and final test report forms, all testing records that include patient logs, quality control logs, instrument printouts, and instrument maintenance.

Information Included in Patient Test Report

- Patient name and unique identifier
- Name and address of lab
- Test(s) performed
- Test result(s)
- Unit(s) of measure
- Reference interval(s)
- Information regarding specimen integrity
- Date/time of specimen collection
- When the report was generated

III. THE QUALITY ASSURANCE PERSPECTIVE

Quality assurance programs, which include the laboratory's quality control program, are designed to maintain the highest standards of patient care. Cooperation between the various departments within the health care institution is of utmost importance. Implementation of the protocols outlined in this chapter requires interaction between the many departments in order to achieve high-quality patient care. QA programs are "living" documents in that they must be constantly evaluated and modified to meet the changing needs in today's health care environment.

SUGGESTED READINGS

Burtis CA, Ashwood ER (eds.). *Tietz fundamentals of clinical chemistry,* 5th ed. Philadelphia: WB Saunders, 2000.

Henry JB (ed.). *Clinical diagnosis & management by laboratory methods,* 19th ed. Philadelphia: WB Saunders, 1996.

Kaplan L, Pesce A (eds.). *Clinical chemistry: Theory, analysis and correlation,* 3rd ed. St. Louis: CV Mosby, 1996.

Matthews D, Farewell V. *Using and understanding medical statistics.* Basel: S Karger AG, 1985.

NCCLS Standard C24-A, "Internal Quality Control Testing: Principles and Definitions," Approved Guideline 1991.

NCCLS document C3-A3, "Preparation and Testing of Reagent Water in the Clinical Laboratory," 2nd ed., Approved Guideline 1997.

NCCLS documents—H1-A4, "Evacuated Tubes and Additives for Blood Specimen Collection," 4th ed., Approved Standard 1996; H3-A3, "Procedures for the Collection of Diagnostic Blood Specimens by Venipuncture," 3rd ed., Approved Standard 1991; H5-A3, "Procedures for Handling and Transfer of Diagnostic Specimens and Etiologic Agents," 3rd ed., Approved Standard 1994.

Westgard J et al. A multi-rule Shewhart chart for quality control in clinical chemistry. *Clin Chem* 27(3):493–501, 1981.

Westgard J, Quam E, Barry T. *Basic QC practices—Training in statistical quality control for healthcare laboratories.* Westgard Quality Corp., 1998; *www.westgard.com.*

Items for Further Consideration

1. A specimen collected with EDTA as an additive was received in the laboratory for *serum* protein electrophoresis. What are the ramifications associated with this situation? How should the laboratory respond?

2. If you were a CAP inspector, what preanalytic, analytic, and postanalytic variables would you be checking to make sure a laboratory was performing daily quality assurance?

3. Discuss the steps that should be taken prior to putting a new lot number of control into use. How is this transition most effectively made?

4. Contrast laboratory situations that produce a shift and a trend in QC data and indicate appropriate troubleshooting strategies for each.

Regulatory Issues for the Clinical Laboratory

Audrey E. Hentzen

Key Terms ◀

CERTIFICATE OF ACCREDITATION

CERTIFICATE OF COMPLIANCE

CERTIFICATE OF PROVIDER-PERFORMED MICROSCOPY

CERTIFICATE OF REGISTRATION

CERTIFICATE OF WAIVER

CLINICAL LABORATORY IMPROVEMENT ACT OF 1988 (CLIA)

EXPOSURE CONTROL PLAN

HIGH-COMPLEXITY TESTING

JOINT COMMISSION ON ACCREDITATION OF HEALTHCARE ORGANIZATIONS (JCAHO)

MATERIAL SAFETY DATA SHEETS (MSDS)

MODERATE-COMPLEXITY TESTING

OCCUPATIONAL SAFETY AND HEALTH ADMINISTRATION (OSHA)

PERSONAL PROTECTIVE EQUIPMENT (PPE)

PROFICIENCY TESTING

PROVIDER-PERFORMED MICROSCOPY (PPM)

QUALITY ASSURANCE (QA)

RIGHT-TO-KNOW LAW

UNIVERSAL PRECAUTIONS

Chapter Objectives

Upon completion of this chapter, the reader will be able to:

1. Define and relate the following terms to laboratory organization and function: waived test, provider-performed microscopy (PPM), moderate complexity, high complexity, Department of Health and Human Services (HHS), Health Care Financing Administration (HCFA), Federal Register, Clinical Laboratory Improvement Act (CLIA), Joint Commission on Accreditation of Healthcare Organizations (JCAHO), Occupational Safety and Health Administration (OSHA), Chemical Hygiene Plan, Exposure Control Plan, Material Safety Data Sheets (MSDS), and Universal Precautions.
2. Describe components of quality control and quality assurance and their interrelationship.

3. Interpret and explain the Federal Register 42 CFR Part 493 laboratory requirements for Laboratory Certification, Participation in Proficiency Testing, Patient Test Management, Quality Control, Personnel for Moderate-Complexity (Including the Subcategory) and High-Complexity Testing, Quality Assurance, and Inspections for Compliance.
4. Translate OSHA regulations for chemical hygiene and exposure control plans into laboratory policies and procedures.
5. Explain the role HHS-designated accreditation programs play in CLIA regulation of laboratories.

Regulations and Agencies Covered ◄

BLOOD-BORNE PATHOGEN EXPOSURE CONTROL PLAN

CHEMICAL HYGIENE PLAN

CLINICAL LABORATORY IMPROVEMENT ACT (CLIA)
- *CLIA-67*
- *CLIA-88*

FOOD AND DRUG ADMINISTRATION (FDA)

FORMALDEHYDE AND ETHYLENE OXIDE EXPOSURE CONTROL PLAN

HEALTH CARE FINANCING ADMINISTRATION (HCFA)

HEALTH AND HUMAN SERVICES (HHS)

HEPATITIS B VIRUS (HBV) VACCINATION PROGRAM

MYCOBACTERIUM TUBERCULOSIS EXPOSURE CONTROL PLAN

OCCUPATIONAL SAFETY AND HEALTH ADMINISTRATION (OSHA)

Pretest

Directions: Choose the single best answer for each of the following items.

1. All of the following laboratories are exempt from CLIA regulations except
   a. forensic laboratories
   b. physician office laboratories
   c. laboratories certified for drugs of abuse testing (SAMHSA)
   d. federal laboratories

2. In a laboratory certified for provider-performed microscopy, all of the following are allowed to perform the tests, *except:*
   a. registered nurses and physicians
   b. physicians and registered nurse-practitioners
   c. dentists and registered nurse-practitioners
   d. registered nurse-practitioners

3. Which type of CLIA certificate is needed in order to perform laboratory tests of moderate or high complexity after an inspection has found the laboratory meets CLIA regulations?
   a. Certificate of Registration
   b. Certificate of Accreditation

   c. Certificate of Waiver
   d. a and b

4. The agency that set the standard for proficiency testing is
   a. American Association of Blood Banks (AABB)
   b. HCFA
   c. College of American Pathologists (CAP)
   d. JCAHO

5. Proficiency testing is required for laboratories performing
   a. only waived testing
   b. moderate-complexity testing
   c. high-complexity testing
   d. both b and c

6. Which of the following does not have to be included in the written procedure for a test system?
   a. step-by-step instructions
   b. distribution of the kit
   c. panic values
   d. reference range

7. A proper test report must include all of the following except
   a. name and address of laboratory where test was performed
   b. name of laboratory director
   c. test methodology used
   d. proper reference range

8. Immunohematology records must be retained and made available upon request for a minimum of how many years?
   a. 2
   b. 5
   c. 7
   d. 10

9. For a laboratory that performs moderate complexity testing, which of the following personnel positions is *not* required?
   a. clinical consultant
   b. technical consultant
   c. technical supervisor
   d. laboratory director

10. Quality assurance measures include all of the following except
    a. patient test management assessment
    b. complaint investigations
    c. proficiency testing
    d. parking policies

11. Which of the following is considered to be a waived test procedure?
    a. spun hematocrit
    b. enzyme-linked immunosorbent assay (ELISA)
    c. blood typing
    d. KOH prep

12. Employees may decline hepatitis B virus (HBV) vaccination for any of the following reasons except
    a. HBV antibody titer already present
    b. they have an allergic reaction to the vaccination
    c. they refuse and won't sign the declination statement
    d. they just received their booster HBV vaccination

13. OSHA mandates that MSDS
    a. be readily available and accessible at the workplace
    b. be written in English and Spanish for easy use
    c. direct employees to sources of first aid
    d. instruct employees to have safety personal protective equipment (PPE) available

14. OSHA directives are all of the following except
    a. optional
    b. provided to encourage safe working conditions
    c. the result of public laws
    d. subject to unannounced inspections

I. INTRODUCTION

As clinical laboratories advanced in complexity and technology, personnel skills and decision-making abilities became more rigorous. It became clear to federal agencies such as the **Department of Health and Human Services (HHS)** and the **Occupational Safety and Health Administration (OSHA)** that there needed to be clear guidelines and directives to provide reliable quality laboratory results and safe working environments. Initially, Medicare, Medicaid, and the **Clinical Laboratory Improvement Act of 1967 (CLIA)** set standards to regulate policies and performance requirements in clinical laboratories. The **Clinical Laboratory Improvement Act of 1988 (CLIA '88)** replaced the Medicare, Medicaid, and CLIA '67 standards with a single set of requirements that were to be generally effective September 1, 1992. Amendments to CLIA '88 are ongoing as they are recommended by the Clinical Laboratory Improvement Advisory Committee (CLIAC), and amendments are published in the *Federal Register*, a weekly publication of the United States Congress. Some regulations of CLIA '88 required a timeline for implementation, and thus laboratories were given effective dates later than September 1, 1992, for compliance. For instance, proficiency testing for newly regulated laboratories were given until January 1, 1994, to comply, whereas laboratories that were regulated (currently in 1988) were to continue their proficiency testing and meet the earlier CLIA effective dates. Additionally, laboratories screening or interpreting gynecologic cytology preparations had to enroll in proficiency testing and meet CLIA regulations for quality control and personnel requirements by January 1, 1994.

OSHA has been active in establishing regulations that protect health care workers while performing their job. Regulations for occupational exposure to hazardous chemicals in laboratories were published January 31, 1990, in the *Federal Register*. Specific directives for occupational exposure to ionizing radiation, formaldehyde, benzene, and ethylene oxide were added as these chemicals became commonplace in laboratories. OSHA identified other areas of employee health risks and biohazards, such as blood-borne pathogens and *Mycobacterium tuberculosis* (TB), providing regulations and directives (December 2, 1991, and October 28, 1994) to protect health care workers.

II. CLINICAL LABORATORY IMPROVEMENT AMENDMENTS OF 1988

All laboratories must meet certain regulations codified as 42 CFR 493 (Table 19–1) for certification to perform testing on human specimens under the Clinical Laboratory Improvement Act of 1988. CLIA implements statutes described by the Social Security Act and the Public

Table 19–1. CLIA '88 regulations as stated in Federal Regulations (CFR Part 493)

SUBPART	CFR NUMBER	SUBJECT AREA
A	§493.1–25	General provisions and definitions
B	§493.35–39	Certificate of Waiver
C	§493.43–53	Registration Certificate, Certificate of Provider-Performed Microscopy Procedures (PPM), and Certificate of Compliance
D	§493.55–63	Certificate of Accreditation
E	§493.501–521	Accreditation by a private, nonprofit accreditation organization or exemption under an approved state laboratory program
F	§493.602–649	General administration
G		Reserved [in document]
H	§493.801–865	Participation in proficiency testing for laboratories performing tests of moderate complexity (including the subcategory), high complexity, or any combination of these tests
I	§493.901–959	Proficiency testing programs for tests of moderate complexity (including the subcategory), high complexity, or any combination of these tests
J	§493.1101–1111	Patient test management for moderate complexity (including the subcategory), high complexity, or any combination of these tests
K	§493.1201–1285	Quality control for tests of moderate complexity (including the subcategory), high complexity, or any combination of these tests
L		Reserved [in document]
M	§493.1351–1495	Personnel for moderate-complexity (including PPM subcategory) and high-complexity testing
N-O		Reserved [in document]
P	§493.1701–1721	Quality assurance for moderate-complexity (including the subcategory) or high-complexity testing, or any combination of these tests
Q	§493.1775–1780	Inspection
R	§493.11800–1850	Enforcement procedures
S		Reserved [in document]
T	§493.2001	Consultations

Health Service Act (Public Law 100-578), which applies to all laboratories seeking payment under the Medicare and Medicaid programs. CLIA is administered by the **Health Care Finance Administration (HCFA),** part of HHS.

The regulations set forth by CLIA were established for all laboratories that performed clinical testing for the purpose of diagnosis, monitoring, or treatment of patients used in medical decisions. Exceptions to these rules include laboratories that perform forensic testing, research laboratories that test human specimens but do not report patient-specific results for diagnostic or treatment purposes, laboratories certified for drugs of abuse testing, and federal laboratories under the jurisdiction of the federal government (Table 19–2).

All other laboratories must be either CLIA-exempt or possess one of the following CLIA certificates: (Table 19–3)

- Certificate of Registration
- Certificate of Waiver
- Certificate of Provider-Performed Microscopy (PPM) Procedures
- Certificate of Compliance
- Certificate of Accreditation (Table 19–3)

The certificate(s) required depend on the category(ies) of testing performed by the laboratory: waived tests, PPM procedures, **moderately-complex** tests, and/or **highly-complex** tests. A **Certificate of Waiver** is awarded to laboratories that perform only waived tests. Tests for Certificate of Waiver must meet descriptive criteria as evaluated by the Food and Drug Administration (FDA) or Centers for Disease Control and Prevention (CDC) for non–commercially available tests. Tests and their classification are published in the *Federal Register.* Laboratories awarded a Certificate of Waiver must follow manufacturers' instructions for performing each test categorized as waived. If a laboratory is certified to conduct waived tests and it initiates any moderate- (including subcategory PPM) or high-complexity testing, it must notify the HCFA within 30 days and apply for Certificates of Registration, PPM, and Compliance and meet those CLIA requirements. A Certificate of PPM will be awarded to laboratories that perform only waived tests and test procedures categorized as PPM. A **Certificate of Registration** will be issued to all laboratories that

---

Exceptions to CLIA Regulations

- Laboratories that perform forensic testing
- Research laboratories that test human specimens but do not report patient-specific results for diagnostic or treatment purposes
- Laboratories certified for drug abuse testing
- Federal laboratories under the jurisdiction of the federal government

---

Table 19–2. CLIA-Exempt Laboratories

State exempt laboratory: A laboratory that has been licensed or approved by a state in which the HCFA has determined that the state has enacted laws relating to laboratory requirements that are equal to or more stringent than CLIA requirements and has been approved by HCFA.

Forensic laboratories: Testing on human specimens is performed for forensic purposes.

Research laboratories: May test human specimens but do not report patient-specific results.

Substance Abuse and Mental Health Services Administration (SAMHSA): Approved drug testing laboratories which must adhere to established guidelines and regulations (see Chapter 9).

Federal laboratories: Under the jurisdiction of an agency of the federal government and are subject to the rules of CLIA, except that the secretary may modify the application of such requirements as appropriate.

Table 19–3. CLIA Certification

Types of CLIA Certificates

Certificate of Waiver: A certificate is issued to a laboratory to perform only tests categorized as waived tests.

Registration Certificate: A certificate is issued that enables the entity (laboratory) to conduct moderate- or high-complexity laboratory testing or both until the entity is determined to be in compliance through a survey by HCFA or by an approved accreditation organization.

Certificate of Provider-Performed Microscopy: A certificate issued to a laboratory in which a physician, midlevel practitioner, or dentist performs no tests other than provider-performed microscopy procedures and, if desired, tests categorized as waived tests.

Certificate of Compliance: A certificate issued to a laboratory performing tests of moderate (including subcategory) or high complexity, after an inspection that finds the laboratory to be in compliance with all applicable condition level requirements.

Certificate of Accreditation: A certificate is issued based on the laboratory's accreditation by an accreditation organization approved by HCFA, indicating the laboratory is deemed to meet applicable CLIA requirements.

Private, Nonprofit Accreditation Organizations with Deemed Status
American Association of Blood Banks
American Osteopathic Association
American Society of Histocompatibility and Immunogenetics
College of American Pathologists
Commission on Office Laboratory Accreditation
Joint Commission on Accreditation of Healthcare Organizations

perform tests other than waived. Certificates of Registration are required initially when a laboratory applies and intends to perform test procedures of moderate complexity, high complexity, or both. A new laboratory or a laboratory that adds new levels of testing to its test menu must apply for a Certificate of Registration as it works toward its **Certificate of Compliance.** Laboratories that perform PPM and moderate- and high-complexity testing must meet the standards and requirements set forth by CLIA and obtain a Certificate of Compliance once inspection has verified all CLIA regulations have been met. If a laboratory chooses to be accredited by a private nonprofit accreditation program having deemed status (as approved by the HCFA; see Table 19–3), which has standards equal to or more stringent than the CLIA, then the laboratory may be granted a **Certificate of Accreditation.**

A. WAIVED-TEST CATEGORY

Laboratories that perform only waived testing must apply and meet the requirements (subpart B) for a Certificate of Waiver. Waived test criteria include that test systems are simple laboratory examinations or procedures that employ methodologies that are easy to perform, accurate, and render little likelihood of erroneous results or pose no risk of harm to the patient if the test is performed incorrectly. Table 19–4 lists many tests currently identified as meeting the criteria for waived tests. Revisions to the list of waived tests approved by HHS

Table 19–4. Tests Designated as Waived or PPM

CDC-Approved Waived Tests
Dipstick or tablet reagent urinalysis
    Bilirubin
    Glucose
    Hemoglobin
    Ketone
    Leukocytes
    Nitrite
    pH
    Protein
    Specific Gravity
    Urobilinogen
Fecal occult blood
Ovulation test (color comparison)
Urine pregnancy tests
Erythrocyte sedimentation rate
Hemoglobin (copper sulfate)
Blood glucose by monitor (FDA approved)
Spun microhematocrit
Hemoglobin by single analyte instrument
(See CFR Doc. 96-17097 for additional test systems listed by test and manufacturer)

Provider-Performed Microscopy Criteria
The primary instrument for performing the test is the microscope
    Direct wet mount
    KOH preparations
    Pinworm examinations
    Fern tests
    Postcoital qualitative examinations of vaginal or cervical mucus
    Urine sediment examinations
    Nasal smears for granulocytes
    Fecal leukocyte examinations
    Qualitative semen analysis

will be published in the *Federal Register* as new tests are developed and evaluated by the CDC.

B. Provider-Performed Microscopy Category

PPM procedures are a subcategory of moderate-complexity testing, which requires a Certificate of **Provider-Performed Microscopy** ◀ and a Certificate of Compliance. Laboratories that are eligible to perform PPM examinations must meet the applicable requirements in CLIA subpart C or D and subparts F, H, J, K, M, and P and be subject to inspection as specified under subpart Q. The examination must be performed by a physician, midlevel practitioner (nurse midwife, nurse practitioner, or physician assistant), or dentist. The procedure, categorized by the CDC, is in the subcategory of moderately complex, and the primary instrument for performing the test is the microscope (Table 19–4). The PPM approved test list is subject to revisions as the CLIAC and HHS determine necessary. Laboratories that perform PPM testing may perform waived testing as well without seeking other certification. If a laboratory begins to offer a test that is categorized as moderate (or high) complexity, it must meet CLIA standards and conditions for a Certificate of Compliance.

C. MODERATE- AND HIGH-COMPLEXITY CATEGORY

Test evaluation and categorization (waived, moderate or high complexity) is performed by the FDA (manufactured kits) and the CDC. Each specific test system, assay, or examination is evaluated for complexity using specific criteria. All tests are then listed by complexity and published in the *Federal Register*. Seven criteria are used to evaluate and categorize a test:

1. Personnel knowledge, training, and experience
2. Reagent and material preparation
3. Characteristics of operational steps
4. Calibration
5. Quality control and proficiency testing materials
6. Test system troubleshooting and equipment maintenance
7. Result interpretation and judgment

Each of these criteria are graded based on the required level of skill or expertise required to perform the test system or assay, with level 1 having minimal requirements, level 2 having limited requirements, and level 3 having substantial requirements. Test systems or assays receiving scores of 12 or less will be categorized as **moderate complexity,** while those receiving scores above 12 are categorized as **high complexity.**

Laboratories performing tests of moderate or high complexity must meet the requirements in subpart C or D and subparts F, H, J, K M, P, and Q, and apply for a Certificate of Registration and Certificate of Compliance or Accreditation. The requirements in the CLIA regulations are very specific and rigorous for laboratories performing moderate- and high-complexity testing. Subpart C specifies the requirements for Certificates of Registration, PPM, and Compliance, while subpart D gives detailed requirements for compliance by accreditation through an approved private, nonprofit accreditation organization. Subparts F, H, J, K M, P, and Q describe the standards and conditions for proficiency testing, patient test management, quality control, personnel assignments, quality assurance, and inspections by CLIA-designated officials. Deficiencies in any one of these subparts may lead to enforcement procedures (sanctions and/or suspensions) detailed in subpart R and nonpayment for services by Medicare and Medicaid agencies.

1. Subpart F: General Administration. This subpart sets forth the methodology for determining the fee amount for issuing the appropriate certificates and determining compliance (inspection) validation for accredited laboratories and CLIA-exempt laboratories. Laboratories must pay a fee for the issuance of Certificate of Registration, Certificate of PPM, Certificate of Waiver, Certificate of Accreditation, or Certificate of Compliance. These costs include issuing the certificate; collecting the fees; evaluating and monitoring proficiency testing and programs; procedure, test, or examination categorizations; and implementation of CLIA regulations. Fee amount is set annually by HHS and is based on the category of the test complexity, laboratory test volume, specialties tested, inspection, and general administration of CLIA. Upon receipt of an application for certificate, HHS or its designee will notify the laboratory of the required fee. Instructions for submitting the fee are accompanied by this notice, and no certification will be issued until the fee has been paid.

2. Subpart H: Participation in Proficiency Testing Programs.

Each laboratory must enroll in a **proficiency testing** (PT) program ◀
that meets the criteria specified in subpart I (Table 19–5). The laboratory must meet all requirements of the PT program, and failure to comply will result in sanctions or penalties. A proficiency program sends biologic samples to each enrolled laboratory for assay as an unknown sample, as though it were an actual patient sample. The laboratory must enroll in a PT program for all analytes of specialties or subspecialties for which it performs moderate- or high-complexity testing. CLIA regulations include specific requirements for cytology services, including proficiency testing of personnel involved in screening and interpreting gynecologic preparations and workload limits per day. The results are sent back to the PT agency, and results from all the laboratories are compared to the true value as well as to each other. The laboratory's measurement is compared with other laboratories using similar methods. Target values are posted and acceptable proficiency testing performance ranges (standard deviation, coefficient of variation, or percent of target value) are provided. Results generated by each laboratory are evaluated and, based on the accrediting agency's criteria, graded as acceptable or unacceptable performance. PT results are reported to HHS and the accrediting agency. CLIA regulations require specific standards of performance within each of the specialties or subspecialties for which the laboratory provides moderate- or high-complexity testing. Unsuccessful PT scores carry sanctions based on the severity of the failure and potential risk to patients or the general public. Sanctions include immediate corrective actions, monetary penalties (loss of Medicare or Medicaid reimbursement), suspension, limitation, or revocation of the laboratory's CLIA certificate.

Sanctions for Unsuccessful PT Scores

- Immediate corrective actions
- Monetary penalties (loss of Medicare or Medicaid reimbursement)
- Suspension
- Limitation
- Revocation of CLIA certificate

Table 19–5. Proficiency Testing (PT) for Laboratories Performing Moderate-Complexity (Including the Subcategory), High-Complexity, or Any Combination of These Tests

Laboratory Participation

Laboratories must enroll in approved program(s) for each specialty and subspecialty

Samples must be examined or tested with regular patient workload

Test samples the same number of times as routine patient samples

No interlaboratory communications pertaining to proficiency testing sample(s)

PT samples or portions of samples must not be sent to another laboratory

Meet the criteria for acceptable proficiency testing performance

Criteria for PT Program Approval

Administered by a private, nonprofit organization, federal or state agency

Ensure the quality of test samples

Appropriately evaluate and score the testing results

Identify performance problems in a timely manner

Demonstrate ability to prepare and distribute samples

Provide sufficient annual challenge

Determine the correct results for each challenge

Provide statewide or nationwide reports to regulatory agencies

Provide shipping schedule

Resolve administrative and technical problems

Maintain records for a period of 5 years

3. Subpart I: Proficiency Testing Programs. HHS-approved PT programs must be administered by a private, nonprofit organization, a federal or state agency, or an entity acting as a designated agent for the state. PT organizations must submit an application to HHS and show ability to provide technical assistance to laboratories seeking to qualify under the program for each specialty, subspecialty, or analyte for which it provides testing. CLIA has established specific criteria that the PT program must meet to be an approved CLIA PT program (Table 19–5). A PT program must submit its application to HHS by July 1 of the current year, one year prior to the year it plans to offer PT services. The PT program must meet the requirements of subpart I, which includes requirements for providing PT samples, replacement samples, result analysis, and PT reporting. The PT agency must provide HHS or its designated agent with an electronic or hard copy of PT reports for its subscribers. These reports include summaries of its PT history, so that each laboratory's compliance can be ascertained. The PT agency must maintain records of its subscribers' PT performance for 5 years, and these must be made available upon request by HHS or its designated agent.

4. Subpart J: Patient Test Management. CLIA regulations set forth in subpart J require laboratories to have in place a system to maintain the integrity, quality, and identification of all patient samples during the preanalytic, analytic, and postanalytic processes. Patient test management systems for moderate- or high-complexity testing must include procedures for specimen submission, handling, test requisitions, test records, reporting, and referral (Figure 19–1). Each of the components of the system must be evaluated for quality assurance and, where possible, quality control materials or protocols be implemented. Any one component of the system affects the outcome of the overall testing process and quality. Procedures for specimen submission and handling require the laboratory to have and follow a written policy and procedure manual for patient preparation, specimen collection, specimen preservation, conditions for transport, processing, and storage. Quality control measures could include monitoring the number of hemolyzed or rejected specimens collected

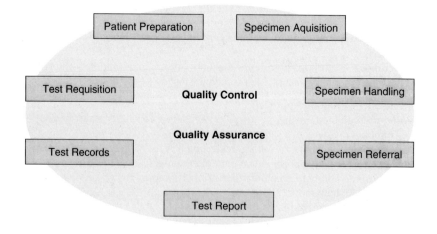

Figure 19–1. Patient test management requirements for moderate (including subcategory) and high complexity.

over a period of time, or specimens lost during transit due to breakage or spillage. The laboratory must maintain test requisition records for at least 2 years. Written authorization for oral requests for laboratory tests must be requested within 30 days. Test requisitions must include patient name, address, tests to be performed, date of specimen collection, and for Pap smear testing, the patient's last menstrual period, date of birth, and any history of previously abnormal results. Laboratory information systems have been a valuable addition for management of laboratory data, patient processing, result reporting, and record keeping. The recent addition of bar coding patient samples and instrumentation has greatly improved the efficiency of test requisition, processing, result reporting, and record keeping. The laboratory must maintain and be able to retrieve records of patient testing, including instrument printouts for 2 years. Immunohematology results must be maintained for 5 years. The laboratory results must be sent to the ordering individual, and a copy of this report must be retained for a period of 2 years or, if immunohematology, 5 years. Quality control indicators for patient results reporting may include the use of turnaround times and number of results reported with transposition errors, missing reference ranges, or misdirected results. The laboratory must develop and follow written procedures for identifying and reporting life-threatening results or critical values to appropriate individuals. In the event that laboratory testing cannot be performed, the laboratory must notify the ordering individual of the delay and expected time of completion. Any referring laboratory used for testing must also be CLIA certified.

5. Subpart K: Quality Control. Subpart K is divided into two sections, general quality control and quality control for specialties and subspecialties including cytology. These requirements are to be followed unless an alternative procedure is specified in the manufacturer's protocol. This protocol must be cleared by the FDA as meeting certain CLIA requirements or HHS may approve an equivalent procedure that is specified in Appendix C of the State Operations Manual (HCFA Pub. 7). Procedures for quality control are general (Table 19–6) and apply to all areas and procedures of the laboratory (preanalytic, analytic, and postanalytic). These general guidelines include requirements for test system calibrations, quality control material requirements, testing frequency, remedial action policies, and documentation. Specific instructions for calibration and calibration verification procedures are also included in subpart K (Table 19–7).

Quality control for laboratories performing moderate-complexity (including the subcategory PPM), high-complexity, or any combination of these tests must establish and follow written quality control procedures for monitoring and evaluating the quality of the analytical testing process of each test. For each test of high complexity performed, all applicable standards of subpart K must be met. For each test of moderate complexity performed, the laboratory must follow the manufacturer's instructions for instrument and test system operation and performance. A written procedure manual for the performance of all analytical methods used by the laboratory must be readily available and followed by laboratory personnel. The written procedure manual must contain all required information (Table 19–8) and must be updated, maintained, and reviewed annually by the laboratory director. Testing personnel must perform and docu-

Table 19–6. General Procedures for Quality Control

For FDA-Approved Kits or Test Systems
Follow the manufacturer's instructions for control procedures, testing requirements, and calibration.

For Moderate- or High-Complexity (or Both) Testing:
Evaluate instrument and reagent stability and operator variance.
Determining the number, type, and frequency of calibration or control materials:
   Qualitative tests must include a positive and negative control.
   Quantitative tests must include at least two samples of different
     concentrations.
   For electrophoretic determinations at least one control sample must be
     used.
   Direct antigen systems must be tested each day using a positive and
     negative control.
   If no calibration or control materials are available an alternative mecha-
     nism to ensure the validity of patient test results must be developed.
Control samples must be tested in the same manner as patient specimens.
Calibration or control materials must have statistically derived parameters.
Criteria for control acceptability must be evaluated prior to reporting patient
   test results.
Notify the appropriate individual of delayed testing or any errors in reported
   patient results.
The laboratory must check each batch or shipment of reagents for reactivity.
Remedial action policies and procedures must be established and applied as
   necessary.
Remedial action must be documented.
Document and maintain records of quality control activities.

---

ment calibration procedures or calibration verification at least once every 6 months (or as described in Table 19–7). Testing personnel must perform and document control procedures using at least two levels of control materials as specified by the manufacturer, each run or at least each day of testing (§493.1218 for control procedures). Appropriate testing personnel must perform and document applicable specialty and subspecialty control procedures (§493.1223-85 specialty and subspecialty standards). Testing personnel must perform and document remedial action taken when problems or errors are identified (§493.1219 for remedial actions policies and procedures). Records of all quality control activities must be maintained (§493.1221 for specific retention periods). The laboratory must utilize test methods, equipment, instrumentation, reagents, materials, and supplies that provide accurate and reliable test results and reports. For each method, the laboratory must verify or establish and document the accuracy, precision, analytic sensitivity, specificity, reportable reference range, and performance specifications prior to patient testing. Testing or designated personnel must perform and document maintenance and function checks of equipment, instruments, and test systems.

6. Subpart M: Personnel for Moderate- and High-Complexity Testing. This subpart describes the personnel requirements that ▶ must be met for laboratories performing **moderate-complexity** ▶ **testing,** PPM procedures, **high-complexity testing,** or any combi-

Table 19–7. Calibration and Calibration Verification Procedures

Definitions

Analyte—a substance or constituent for which the laboratory conducts testing.

Calibration—testing and adjusting an instrument, kit, or test system to provide a known relationship between the measurement response and the value of the analyte being measured.

Calibration verification—assaying of calibration materials as patient samples to confirm the calibration accuracy of the instrument, kit, or test system.

Reportable range—range of patient test result values over which the laboratory can establish or verify the accuracy of the instrument, kit, or test system.

Standard

FDA-approved kits or test systems—the laboratory must follow the manufacturer's instructions for calibration and calibration verification using calibration material.

For all other methods, calibration procedures must be performed, according to the manufacturer's instructions, or in accordance with criteria established by the laboratory.

The calibration materials must be appropriate for the methodology and traceable to a reference method or reference material of known value.

Calibration verification must be performed at least once every 6 months and whenever any of the following occur:

   A complete change of reagents for a procedure is introduced (unless no change in reportable range is demonstrated).

   Major preventive maintenance or replacement of a critical part has occurred.

   Controls reflect an unusual trend or shift or are outside of the laboratory's acceptable limits.

   Laboratory's established schedule requires more frequent calibration verification.

Table 19–8. Written Procedure Manual (§493.1211)

The written procedure manual must include, when applicable:

1. Specimen collection requirements, processing, and rejections
2. Procedures for microscopic examinations
3. Step-by-step performance of the procedure
4. Description of any items that need to be prepared
5. Calibration and calibration verification procedures
6. Reportable range for patient results
7. Control procedures
8. Out-of-control protocols
9. Limitations in methodologies, including interfering substances
10. Reference range (normal values)
11. Imminent life-threatening laboratory results or "critical values"
12. Pertinent literature references
13. Description of the course of action to be taken in the event that a test system becomes inoperable

The laboratory must maintain a copy of each procedure with the dates of initial use and discontinuance. These records must be kept for 2 years. Manufacturer inserts or operator manuals may be used but the procedures must be approved, signed, and dated by the director.

nation of these tests. The only requirement of personnel that perform waived testing is to have basic education and sufficient training to perform and report the tests. Laboratories certified to perform PPM procedures, a test of subcategory of moderate complexity, must meet requirements for the positions of laboratory director and testing personnel, which includes midlevel practitioners (Figure 19–2). Laboratories certified to perform tests of moderate complexity must meet requirements for the positions of laboratory director, clinical consultant, technical consultant, and testing personnel (Figure 19–2). Laboratories certified to perform tests of high complexity must meet requirements for the positions of laboratory director, clinical consultant, technical supervisor, general supervisor, cytology general supervisor, cytotechnologist, and testing personnel (Figure 19–2). All testing personnel must meet competency criteria for each specialty or subspecialty in which they perform tests. A competency test program must be performed annually and documented.

7. Subpart P: Quality Assurance. Each laboratory must establish and follow written policies and procedures for a comprehensive ▶ **quality assurance (QA)** program that is designed to monitor and evaluate the quality of all phases of the testing process (Table 19-9). Procedures must be in place to evaluate the effectiveness of the policies; identify and correct problems; ensure accurate, reliable, and prompt reporting of test results; and ensure adequacy and compe-

TEST CATEGORY

PERSONNEL	PPM	Moderate Complexity	High Complexity
Laboratory Director	State License, MD, DO, DDS, PhD w/Board Certification	State License, MD, DO, PhD w/Board Certification, MS + 1 yr training or BS + 2 yr, training and 2 yr supervisor experience	State License, MD, DO, PhD w/ Board Certification
Clinical Consultant	Not Required	State License, MD, DO, PhD w/Board Certification	State License, MD, DO, PhD w/Board Certification
Technical Consultant	Not Required	State License, MD, DO, PhD w/Board Certification, MS + 1 yr training or BS + 2 yr, training and 2 yr supervisor experience	Not Required
Technical Supervisor	Not Required	Not Required	State License, MD, DO, PhD w/Board Certification, MS + 2 yr training or BS + 2 yr, training and 2 yr supervisor experience
General Supervisor	Not Required	Not Required	State License, MD, DO, PhD w/Board Certification, Technical Supervisor, AA + 2 yr training, Cytotechnologist
Testing Personnel	State License, MD, DO, DDS, PhD w/Board Certification Nurse-midwife/practitioner or Physician Assistant	State License, MD, DO,PhD w/Board Certification, MS, BS, AA, High School Diploma w/training	State License, MD, DO, PhD w/Board Certification, MS, + 2 yr experience, BS + 2 yr experience and 2 yr training, Cytologist

Figure 19–2. Personnel requirements for laboratories performing PPM procedures, moderate-, or high-complexity testing.

Table 19–9. Quality Assurance for Moderate-Complexity (Including Subcategory) or High-Complexity Testing or Any Combination of These Tests

General requirements: Establish and follow written policies and procedures for a quality assurance program to monitor and evaluate the ongoing and overall quality of the total testing process. Evaluate effectiveness, document problems, and make revisions.

Patient test management assessment: The laboratory must monitor, evaluate, and revise, patient preparation, specimen collection, labeling, preservation, transportation, test requisition, specimen rejection, test report information, timeliness, and accuracy and reliability of test reporting systems.

Quality control assessment: Evaluate and review the effectiveness of corrective actions for problems identified during calibration and control runs and patient testing.

Proficiency testing assessment: Unsuccessful proficiency testing must be evaluated, corrective actions taken, and evaluated for effectiveness.

Comparison of test results: The laboratory must have a system that twice a year evaluates and defines the relationship between test results from different methodologies (i.e., cross checks).

Relationship of patient information to patient test results: Identify and evaluate patient test results with relevant criteria, such as age, sex, and diagnosis.

Personnel assessment: Mechanism to evaluate employee or consultant competence.

Communications: A system must be in place to document problems in communication between the laboratory and ordering individual.

Complaint investigations: All complaints and problems are investigated and reviewed, and corrective action documented.

Quality assurance review with staff: Document and assess problems, review, and communicate to staff.

Quality assurance records: Maintain documentation of all quality assurance activities.

tence of staff. Complaints must be documented and investigated, and appropriate corrective actions instituted. Remediation may require revision of policies and procedures, changes in work flow, and reassignment of personnel. Quality assurance must be reviewed and discussed with all staff as part of the ongoing effectiveness of the quality assurance program.

8. Subpart Q: Inspection. HHS, or its designee, has the right to make announced or unannounced inspections to see if CLIA regulations, standards, and conditions are being met. The laboratory may be required to permit HHS or it designee to interview all employees, provide access to all areas of the facility including specimen procurement, processing, storage, testing, and reporting areas. Inspectors may want to observe employees while performing tests, data analysis, and reporting. Upon request, the laboratory must allow inspectors access to review of all information, data, records, procedure manuals, documentation of QA, and remediation performed. Failure to comply can result in suspension of Medicare and Medicaid payments or revocation of the laboratory's CLIA certificate. Laboratories certified to perform PPM or waived testing and CLIA-exempt laboratories are also inspected to assess compliance and verify exemption status.

III. OCCUPATIONAL SAFETY AND HEALTH ADMINISTRATION (OSHA)

Congress finds a substantial burden is placed on commerce in terms of personal injury and illnesses arising out of work-related conditions, leading to lost wages, medical expenses, and disability compensation. Congress has declared that through its powers, it shall regulate commerce through direct policies, to reduce work-related expenses by providing safe environmental working conditions. Public Law 91-596, enacted December 29, 1970, amended by Public Law 101-552, Section 3101, and Public Law 105-198 (November 5, 1990, and July 16, 1998, respectively) put into effect regulations to ensure safe and healthful working conditions for men and women. Standards were developed that provide opportunity for research, gathering information, personnel education, and training in occupational safety and health to strive for safer working conditions. The federal regulations affect all working environments, not just the laboratory, and under this legislation, OSHA is authorized to inspect work environments to determine if the employer is complying with mandatory safety standards. OSHA was authorized to enforce, through fines and penalties, the standards for safe work environments.

The employer is ultimately responsible for providing a safe work environment. Through laboratory policies, procedures, supervision, training, and education of employees, the employer provides the opportunity for a safe working environment. The employee is responsible for following all safety rules, and failure to do so may limit their safety and the safety of fellow employees. Failure to follow safety rules and policies is often grounds for dismissal because the employer can be fined and penalized for allowing unsafe practices.

OSHA regulations are published in the *Federal Register,* and directives provide detailed information and specific standards the employer must meet for potential hazards. The laboratory must meet general safety practices for hazards, such as electrical, fire, broken glass, and provide safety equipment, such as fire extinguishers, fire blankets, safety showers, and eyewash stations. The National Fire Protection Association (NFPA) has developed a standard hazards identification system that is used to label working reagent bottles, containers, and storage cabinets. A diamond-shaped symbol, divided into color-coded quadrants, with a number in each quadrant designating the magnitude of severity, is used as an easily identifiable label so employees can be aware of potential hazards (Figure 19–3). Some manufacturers have provided reagent bottles and containers that have permanently imprinted diamonds for easy labeling and use.

Clinical laboratory personnel are exposed to a variety of potential hazards in the performance of their daily tasks. The laboratory is filled with instruments that contain moving mechanical parts, chemicals, radiant heat, electrical components, water baths, toxic vapors, flammable liquids, corrosive or irritant substances, carcinogens, and biological hazards. Some of these hazards have specific OSHA directives (Table 19–10) that regulate the allowable exposure levels and how the employer is to provide a safe work environment. The first steps toward OSHA compliance and providing a safe environment starts with identifying the potential hazards in the laboratory environment. Each laboratory is charged with this task followed by implementation of a safety plan that employs engineering controls and safe work practice policies.

Potential Hazards for Laboratory Personnel

- Instruments that contain moving mechanical parts
- Chemicals
- Radiant heat
- Electrical components
- Water baths
- Toxic vapors
- Flammable liquids
- Corrosive or irritant substances
- Carcinogens
- Biological hazards
- Needle sticks
- Aerosols

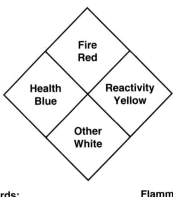

Health Hazards:
0 No hazard
1 Can irritate
2 Can injure
3 Can seriously injure
4 Can cause death

Flammability Hazards:
0 Will not burn
1 Ignites with considerable preheating
2 Ignites with moderate heat
3 Ignited at all normal temperatures
4 Very flammable

Other:
OX Oxidizer
ACID Acid
ALK Alkali
COR Corrosive
-W- Use no water

Reactivity:
0 Stable
1 Stable unless exposed to high
  temperature or pressure
2 Normally unstable, but will not detonate
3 Can detonate or explode, but requires
  initiating force or pressure
4 Readily detonates or explodes

Figure 19–3. National Fire Protection Agency hazard rating diamond.

A. OCCUPATIONAL EXPOSURE TO HAZARDOUS CHEMICALS IN LABORATORIES

The **right-to-know law** was published August 1987 in the *Federal Register* and was developed for employees who may be exposed to hazardous chemicals while performing their duties. In May 1988, this law was expanded to include clinical laboratories and ultimately led to the requirement for chemical hygiene plans for every laboratory.

Table 19–10. OSHA Environmental and Occupational Safety Directives

OSHA STANDARD	SUBJECT
29 CFR 1910.1001–1016	Specified carcinogens
29 CFR 1910.1028	Benzene
29 CFR 1910.1030	Blood-borne pathogens
29 CFR 1910.1047	Ethylene oxide
29 CFR 1910.1048	Formaldehyde
29 CFR 1910.120	Hazardous waste operations and emergency response
29 CFR 1910.139	Respiratory protection for *Mycobacterium tuberculosis*
29 CFR 1910.1450	Occupational exposure to hazardous chemicals in the laboratory
29 CFR 1910.1904	Record keeping and reporting of occupational injuries and illness
29 CFR 1910.95	Exposure to occupational noise
10 CFR 1910.96 (Part 20)	Ionizing radiation

Final OSHA regulations for occupational exposure to hazardous chemicals in the laboratory was published January 31, 1990, in the *Federal Register* (29 CFR 1910.1450). These regulations direct the employer of a clinical laboratory to develop a systematic approach to identify significant risks to employees, monitor permissible exposure limits, and have in place a **chemical hygiene plan.**

Risk assessment is based on epidemiological information, level of exposure (chemical, carcinogenic, toxins), and reported adverse health effects from exposure (research). Safe work practices (prevention and policy considerations) are evaluated for effectiveness and documented. Engineering controls and safe work practices must be put into practice to reduce exposure to potential hazards that cannot be removed from the work environment. Monitoring permissible exposure limits for personnel exposure to hazards (formaldehyde, benzene, ionizing radiation) must be performed if exposure cannot be reduced below OSHA-specified action levels. A written chemical hygiene plan must be formulated and implemented by the employer. Under the right-to-know law, any employee who works where hazardous chemicals are present must be informed of the specific chemical hazards (Table 19–11). **Material Safety Data Sheets (MSDS)** must be obtained from the manufacturer for each hazardous chemical in the workplace. MSDS are to be readily available and accessible to employees. MSDS provide information about the chemical hazards, safety precautions, and first aid treatment instructions for accidental exposure. Employees must be educated (annually) on how to interpret safety labels and MSDS, and respond correctly if accidental exposure occurs. A designated chemical hygiene officer maintains the written hazardous communication plan and undertakes development, implementation, monitoring, and administration of the chemical hygiene plan (Table 19–12).

Exposure to toxic substances has the potential to cause acute or chronic effects that act through localized injury or they may act systemically, creating physiological dysfunction throughout the body. The severity of the effects may be related to the duration and extent of the exposure. Some substances are toxic at very low levels and may require very little exposure to cause poor health or illness. Exposure to hazardous chemicals may be through inhalation, direct contact, inoculation, or injection. In the laboratory, personnel should be sensitive to the risk in their environment and wear appropriate **personal protective equipment (PPE).** When engineering controls and safe work practices are unable to reduce potential hazard exposure below OSHA action limits, then PPE may require additional

---

Table 19–11. OSHA-Mandated Communication of Hazards to Employees

All employees must receive free biosafety training during working hours.

Training sessions must be presented in a clear and easy-to-understand way.

Training schedule and sessions are designed for particular hazards and are site specific.

Instructors must be knowledgeable.

Instructors must demonstrate use and location of safety devices.

Employers and instructors must document all training.

Instructors must check employee understanding to ensure that employees understand how to cope with hazards they may encounter during their daily tasks.

Table 19–12. Chemical Hygiene Officer Duties

Develop standard operating procedures for safety and health of persons who work in the presence of hazardous chemicals.

Set criteria and implement measures to control and reduce employee exposures (engineered controls and safe work practices).

Describe and document the employee information and training programs.

Act as the laboratory liaison to OSHA-required medical examination and consultations.

Designate specific laboratory operations or activities that cannot be performed without prior approval.

Provide additional protection for work with particularly hazardous materials (carcinogens, teratogens, and highly toxic materials).

Ensure that all chemicals are properly labeled (NFPA) and that label reflects degree of hazard involved in its use.

Maintain a written chemical hygiene plan, annually reviewed, updated with changes in regulations or chemical inventory.

equipment, such as high-efficiency particulate air (HEPA) filter respirators, physical barriers or shields, and biosafety cabinets.

B. FORMALDEHYDE AND ETHYLENE OXIDE EXPOSURE CONTROL

Poisonous vapors may or may not give warnings (olfactory sensitivity). Toxic vapors from organic solvents have little odor, while bromide, ammonia, and formaldehyde have distinct odors that serve as a warning to laboratory personnel. OSHA has provided mandatory directives for occupational exposure to formaldehyde (29 CFR 1910.1048) and ethylene oxide (29 CFR 1910.1047) for employers. Exposure risk must first be determined through testing and then engineering controls and safe work practices put into place to reduce exposure. Each laboratory must meet OSHA action level limits or provide appropriate HEPA filter respirators (certified HEPA filters, types N, R, and P). The health risks for these toxic vapors are defined at very low concentrations, and employee awareness and demand for appropriate personal protective equipment has dramatically increased over the last 7 to 9 years. A recent study showed that complaints from employees to OSHA regarding employer noncompliance (formaldehyde, ethylene oxide, and *Mycobacterium tuberculosis* exposure) had increased 15 times more than previously documented (1990 to 1991).

C. BLOOD-BORNE PATHOGEN EXPOSURE CONTROL

Clinical specimens are inherently high-risk sources for infections and therefore pose a significant risk for exposure to employees. Specimen collection, processing, handling, and testing all provide opportunity for exposure. The CDC recommends preventive universal measures **(universal precautions)** and OSHA has written regulations for occupational exposure to blood-borne pathogens (*Federal Register,* December 2, 1991, 29 CFR 1910.1030). OSHA regulations became effective March 6, 1992. These regulations limit occupational exposure to blood and other potentially infectious body fluids or materials that can transmit blood-borne pathogens such as hepatitis B virus (HBV), human immunodeficiency virus (HIV), and others.

▶ OSHA directs each laboratory to have in place a written **exposure control plan,** PPE education and training program, and employer-supplied PPE. Engineering controls need to be in place to minimize exposure. Safe work practices must be developed, implemented, and evaluated to reduce or eliminate employee exposure to blood or other potentially infectious materials (Table 19–13). Communication of these potential hazards to the employee is required by OSHA. Housekeeping procedures must be in place to maintain a clean and sanitary work environment, and work practices must be in place to protect housekeeping staff (labs must have sharps containers, designated biohazard waste receptacles, and procedures for biohazard waste disposal).

An HBV vaccination program must be provided to all employees who work with potentially infectious materials or who are assigned to work in an area where exposure to blood-borne pathogens is possible. HBV vaccine must be offered within 10 days of employee assignment to these areas and must be at no cost to the employee. The HBV vaccination program must be established as outlined in the *Federal Register* (Table 19–14). Employees may decline the HBV vaccination if they have been previously vaccinated and have immunity or if the HBV vaccination is contraindicated for medical reasons. These employees must sign the declination statement published in Appendix A of the *Federal Register* (29 CFR 1910.1030, App A). Exposure incidents must be documented and postexposure or follow-up evaluations provided at no cost to the employee. A record-keeping system must be in place so that training and exposure incidents are kept as part of each employee's per-

Table 19–13. Elements of Blood-Borne Pathogen Exposure Control Plan

Identification of job classification, tasks, and procedures where there is potential for exposure.
Explanation and use of personal protective equipment (PPE) must be available:
  PPE must be supplied at no cost to employee.
  Equipment is appropriate and it does not permit blood or other potentially infectious materials to pass through to an employee's clothes, skin, mouth, or mucous membranes.
  Type of PPE includes: gloves, gowns, laboratory coats, face shields or masks, and eye protection.
Engineering controls must be in place to minimize exposure:
  These are devices that isolate or remove the pathogen hazard from the workplace.
  Engineering controls must be made available to all employees.
  Examples of engineering controls include sharps containers, splash guards, biosafety cabinets and fume hoods, mechanical pipetting devices, handwashing facilities, eyewash stations, and showers.
Develop work practices to reduce employee exposure to blood or potentially infectious materials:
  Employees must fully understand work practice procedures.
  The procedures must be implemented when appropriate.
  Work practices include handwashing, no recapping of contaminated needles, no food or drink in the laboratory, no mouth pipetting, warning labels affixed where needed, biohazard symbols placed appropriately, and a dress code.

Table 19–14. HBV Program Protocol (*Federal Register* 29 CFR 1910.1030)

The employer shall not make prescreening a prerequisite for receiving the
  HBV vaccination.
HBV vaccine series must be made available to all employees who are at risk
  for exposure to blood or potentially infectious materials.
The HBV vaccine must be made available to the employee within 10 days of
  being assigned to a work area where risk of exposure is present.
Employees who have previously received the HBV vaccine may be verified
  through antibody testing.
An employee may initially decline an HBV vaccination but at a later date
  may accept the vaccination at no cost.
An employee may decline an HBV vaccination but must sign the statement of
  declination found in Appendix A (29 CFR 1910.1030, App A).
A routine booster dose of HBV vaccine is recommended by the U.S. Public
  Health Service and shall be made available.
The vaccination series must be made available at no cost to the employee.
Postexposure evaluation and follow-up of an exposure incident shall be doc-
  umented and include at least the following:
  Route of exposure and circumstances
  Identification and documentation of source
  Determine HBV and HIV infectivity of the source
  Results made available to employee
  Employee HBV and HIV serological status (90-day grace period for HIV
    testing)
  Postexposure prophylaxis, as recommended by the U.S. Public Health
    Service
  Counseling
  Evaluation of reported illnesses
  All findings or diagnoses shall remain confidential

manent record. This record must be kept for 30 years, remain confidential, and be transferred to a new attending physician when the
employee leaves.

D. Mycobacterium tuberculosis *Exposure Control*

Clinical laboratory employees who work with *Mycobacterium tuber-culosis,* the causative agent of tuberculosis (TB), have the additional
potential biological hazard of exposure with this highly infectious
agent. Recent resurgence of TB infections has led OSHA to publish
guidelines for preventing transmission of *M. tuberculosis* in health
care facilities (29 CFR 1910.139). Each laboratory is responsible for
having a policy for *M. tuberculosis* exposure control based on their
laboratory testing and personnel exposure to tuberculosis. The policy must include risk assessment, engineering controls, safe work
practices, employee TB screening program, employee TB education,
PPE training, and program evaluation for effectiveness. OSHA directs
health care facilities to provide three levels of transmission prevention.

- Level 1: Develop and implement policies for rapid identification and isolation of TB-infected patients.
- Level 2: Use PPE and respirators (HEPA filters) to protect
  health care workers from exposure.
- Level 3: Coordinate activities (report cases, provide patient
  care and therapy) with local public health departments.

IV. PRIVATE, NONPROFIT ORGANIZATIONS WITH DEEMED STATUS

HCFA may deem a laboratory to meet all applicable CLIA program requirements if the laboratory meets requirements of an approved accreditation organization. Private, nonprofit organizations (Table 19–5) may apply for deemed status and receive approval from HCFA once they provide assurance that their organization meets all requirements equivalent to CLIA condition-level requirements. Each organization's application is reviewed for specialties and subspecialties, requirement stringency, inspector qualifications, frequency of inspection, ability to provide electronic data and timely reports to the HCFA, PT result evaluation, and ability to manage and administer the accreditation processes. HCFA will validate the accreditation agency's process through inspection of an accredited clinical laboratory to validate that it has met CLIA requirements. HCFA will review and maintain oversight of those agencies with deemed status.

▶   The **Joint Commission on Accreditation of Healthcare Organizations (JCAHO)** is recognized as the highest level of accreditation throughout the United States and many other countries. The JCAHO provides accreditation services for health care organizations, encompassing all health service disciplines, setting the most rigorous quality performance standards. JCAHO standards represent an industry consensus, with a focus on quality patient care. JCAHO standards are divided into two major functions: organizational and technical. Within each of these functions are subgroups that apply to all areas of the hospital or health care, which are scrutinized for quality. The JCAHO Comprehensive Accreditation Manual for Pathology and Clinical Laboratory Services focuses on laboratory standards of performance. This manual incorporates OSHA directives and CLIA regulations as part of each performance subgroup standard. JCAHO accreditation decisions are made by applying decision rules to scored grid elements for all areas of the health care organization. Decision rules are based on what is considered accurate and fair. *Accreditation with Commendation* is the highest accreditation decision possible. *Accreditation without Type I recommendations* results when a hospital has demonstrated acceptable compliance with JCAHO standards. Accreditation with Type I recommendations results when a hospital receives at least one recommendation addressing insufficient or unsatisfactory standards of compliance in a specific performance area. Resolution of the insufficiency must be made within a specified time. *Provisional accreditation* results initially when a hospital has demonstrated satisfactory compliance with selected JCAHO standards of performance. These types of accreditation are followed up with a full institution survey 6 months later. *Conditional accreditation* results when a hospital is not in substantial compliance with JCAHO standards but is capable of achieving acceptable standards within a stipulated time period. *Preliminary nonaccreditation* is assigned to a hospital when it is found to be in significant noncompliance with JCAHO standards or may be applied as a penalty for other reasons (falsification of documents) until a final decision is made. *Not accredited* status results when a hospital is denied accreditation because of significant noncompliance with JCAHO standards or accreditation is withdrawn for other reasons or the hospital withdraws from the accreditation process. Based on the accreditation report to HCFA, sanctions may be imposed which suspend a laboratory's testing and Medicare reimbursement.

SUGGESTED READINGS

42 CFR Part 493, CLIA Laboratory Requirements, *Federal Register,* October 1, 1997.

CLIAS 19th National Meeting, *Complying with CLIA '88* (Spring 1993) Convention Issue (Official Publication of the Clinical Ligand Assay Society, *Journal of Clinical Immunoassay*).

Comprehensive Accreditation Manual for Pathology and Clinical Laboratory Services, Joint Commission on Accreditation of Healthcare Organizations, Oakbrook Terrace, IL, 1998–1999.

Krishnan J, Usha T, Janicak CA (1999) Compliance with OSHA's respiratory protection standard in hospitals. *American Industrial Hygiene Association Journal* 60(2): 228–234.

Items for Further Consideration

1. Your laboratory just expanded to include a same-day surgical center draw station. Develop a quality assurance plan that meets CLIA regulations, to evaluate patient test management.

2. Design a competency program for a department of the clinical laboratory. This laboratory performs moderate- and high-complexity testing and is seeking a Certificate of Compliance.

3. Use of a new hazardous chemical was added to the laboratory when a new automated instrument was purchased. Describe steps that need to be taken to meet OSHA standards and directives regarding hazardous chemicals in the workplace.

Management and Supervision

Janine Denis Cook and Paul A. Griffey

Key Terms ◀

ADMINISTRATOR

BALANCE SHEET

BREAK-EVEN ANALYSIS

BUDGETING

CAPITAL ACQUISITIONS

CAPITAL BUDGET

COST ACCOUNTING

COST PER TEST

DIRECT COSTS

DIRECTOR

FINANCIAL ACCOUNTING

FINANCIAL STATEMENTS

FIXED COSTS

INCOME STATEMENT

INDIRECT COSTS

INVENTORY MANAGEMENT

LABORATORY MANAGEMENT

LEADERSHIP STYLE

MANAGER

MATERIALS MANAGEMENT

OPERATING BUDGET

PERFORMANCE APPRAISAL

POLICY MANUAL

PROFITABILITY

QUALITY IMPROVEMENT

SEMIVARIABLE COSTS

SENSITIVITY ANALYSIS

STATEMENT OF CASH FLOWS

SUPERVISOR

TECHNICAL PROCEDURE MANUAL

UNIT COSTS

VARIABLE COSTS

Chapter Objectives

Upon completion of this chapter, the reader will be able to:

1. Describe the financial impact of the health care industry on the U.S. economy.
2. Explain how the goals and objectives of the laboratory are derived from its mission statement.
3. Differentiate between the laboratory responsibilities of the director, administrator, manager, and supervisor.
4. Explain how project management and problem-solving skills contribute to a manager's success.
5. Identify factors that invalidate the economic law of supply and demand for the clinical laboratory.
6. Define *cost accounting* and give an example to illustrate its importance and function in the clinical laboratory.
7. Provide an example for each of the following: fixed, semivariable, and variable costs.

8. Given a balance sheet, explain each of the financial categories located on the statement.
9. Justify a purchase option for an instrument.
10. Outline an inventory management scheme for a clinical laboratory department.
11. Identify and discuss the important considerations for a laboratory information system (LIS).
12. List five factors to consider when designing a laboratory.
13. Discuss the manager's role in motivating employees and delegating responsibilities.
14. Prepare a mock disciplinary action discussion for an unsatisfactory employee.

Techniques/Practices/Methods/Systems Included ◀

BREAK-EVEN ANALYSIS

BUDGETING

CLIENT RELATIONS/MARKETING MANAGEMENT

COST ACCOUNTING

FINANCIAL ACCOUNTING

FINANCIAL MANAGEMENT

INVENTORY MANAGEMENT

HUMAN RESOURCE MANAGEMENT

MATERIALS MANAGEMENT

SENSITIVITY ANALYSIS

TECHNOLOGY MANAGEMENT

Pretest

Directions: Choose the single best answer for each of the following items.

1. The *primary* goal of financial management is to
   a. focus on the cash flow and financial position of the institution
   b. assure the profitability and survival of the organization
   c. produce the operating budget of the organization
   d. provide a means for payment of liabilities

2. The goals of an organization are
   a. created without taking into consideration the future plans of the organization
   b. inconsistent with the organization's management style
   c. congruous with the mission
   d. quantifiable work plans for implementation within a designated time frame

3. Cost accounting is
   a. a system that provides a forecast of information to be used in financial decision making
   b. a system for formalizing in writing a quantitative financial plan
   c. a document of financial history
   d. the financial position of the organization at a given time

4. A financial statement that documents the financial history of the institution is
   a. inventory
   b. a balance sheet
   c. an operating budget
   d. the capital budget

5. An example of a capital budget item is
   a. reagents
   b. analyst salaries
   c. a building addition
   d. stationery supplies

6. The largest component of the day to day operating budget is
   a. calibrators
   b. office supplies
   c. salaries and benefits
   d. water and sewer

7. A tool of financial analysis that provides a quantitative rationale and is used to evaluate profitability for funding of a capital project is
   a. flexible budgeting
   b. inventory holding cost
   c. cost allocation
   d. payback analysis

8. The systematic processes of managing and controlling the acquisition and use of supplies to ensure both availability and cost effectiveness in the day to day operations of a clinical laboratory is referred to as the process of
   a. inventory
   b. usage
   c. cost of ordering
   d. materials management

9. An inventory technique that requires delivery of supplies at the moment of need is called
   a. standing order
   b. perpetual
   c. minimum/maximum
   d. just-in-time

10. A variable in a break-even analysis is
    a. testing volume
    b. utilities
    c. inventory storage costs
    d. office supplies

11. An example of an intangible asset is
    a. building structures
    b. client base
    c. instrumentation
    d. office furniture

12. In the laboratory design or redesign process, regulatory requirements must be addressed for
    a. location of the restrooms
    b. placement of cabinetry
    c. site of instrumentation
    d. safety and ergonomics

13. Marketing management provides quality service to
    a. lenders
    b. staff
    c. customers
    d. vendors

14. Getting things done through people is the definition of
    a. human resource management
    b. motivational management

c. leadership styles
d. scheduling

15. A variable in the cost per test calculation is the
    a. cost of biological waste disposal
    b. cost of result reporting
    c. cost of reagent storage
    d. number of controls per run

16. A personnel management task is
    a. marketing
    b. hiring
    c. laboratory information system (LIS) implementation
    d. cost per test calculations

17. An example of a human resource managerial task is
    a. inventory
    b. cost accounting
    c. marketing
    d. scheduling

18. After the first year of hire, analyst technical competency is assessed per the Clinical Laboratory Improvement Act of 1967 (CLIA)
    a. semiannually
    b. monthly
    c. annually
    d. quarterly

19. The performance appraisal is a document for
    a. the formal process of communicating job performance feedback to employees
    b. the formal process of administering discipline to employees
    c. establishing the chain of command to employees
    d. providing a means of feedback to the manager about employee attitudes

20. An example of a variable cost is
    a. analyst salaries
    b. utilities
    c. instrument purchase price
    d. reagent cost

I. INTRODUCTION

The health care industry is a major component of the U.S. economy. This industry has experienced tremendous growth, climbing from 3.4% of the gross national product (GNP) in 1964 to 13.6% in 1996.

The clinical laboratory is an evolving and very dynamic aspect of health care; approximately 70% of patient diagnoses are derived from clinical laboratory diagnostic testing results. Cost containment has changed the clinical laboratory from the highly profitable enterprise of the 1980s into an industry striving for survival under the weight of government regulations and accreditation standards (as discussed in Chapter 19).

Health care, which impacts so heavily the financial well-being of this country, must be carefully managed. **Laboratory management** can be defined as the process of *planning, organizing, directing,* and *controlling* resources (Figure 20–1). The resources available to the laboratory manager are *finances, people,* and *technology.* The proper allocation of these resources and the application of continuous quality improvement processes (as discussed in Chapter 18) ensure quality results flowing from efficient, cost-effective operations.

II. GOALS AND OBJECTIVES

A clinical laboratory, just like any other business, must develop a mission statement that embodies the vision, purpose, and the values of the organization. The primary mission of the clinical laboratory is to provide quality care to the patient. Secondary statements may include the applicable services offered by the laboratory, the laboratory's managerial philosophy, and the rationale for cost-effective operations. Finally, statements relating to the environment in the laboratory, including interpersonal relations, working conditions, and environmental safeguarding, can be included. For success, every employee must "buy into" the organization's mission. Each employee must understand his or her role in communicating and implementing the laboratory's mission.

From the mission statement, *goals* are developed that outline the processes which, when implemented, will achieve the specific statements presented in the mission. Goals must be congruous with the laboratory's mission, consistent with one another, and reflect basic human values. Since these goals, if achieved, will influence the future activities of the laboratory, they are created with the vision of the short- and long-term destiny of the laboratory. To be effective, these goals are derived with input from personnel at each level of the organization and from all laboratory customers. Goals are communicated to every individual in the laboratory and the organization as well as caregivers. To be successful, the goals must be consistent

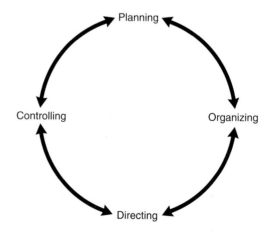

Figure 20–1. The management process.

with the management style of the organizational officers and achievable with the available resources, including facilities, equipment, and trained staff. Goals should be reviewed annually and modified as necessary to reflect any change in the mission of the laboratory.

Objectives are quantifiable work plans designed to implement the goals and are achievable within a designated period of time. In the formulation of objectives, input and commitment from all personnel involved ensure the success of the plan and encourage discussion, interaction, and consensus. Objectives must be communicated to every employee to ensure shared values. Furthermore, they are prioritized to mirror the organizational goals. Periodic review of these priorities permits efficient utilization of time and resources, builds flexibility to adapt to changing needs and circumstances, and keeps the mission in focus.

It is critical for the manager to thoroughly understand the mission, goals, and objectives of the clinical laboratory organization, in addition to believing and supporting the mission. Many managers fail, not from lack of skill, but from not adapting to the culture or sharing the values of the organization.

"A leader does the right things; a manager does things right." The long-range goals and objectives of the clinical laboratory are communicated from the laboratory's leadership. A manager's responsibility is to implement the mission effectively by creating short-term goals and objectives that support the long-term plans.

III. MANAGEMENT TITLES

The responsibilities of managing the laboratory are divided among the individuals who work together as a management team. The cooperation and interaction of the management team is essential for successful performance of the clinical laboratory. Management responsibilities are typically defined by title.

Management Titles
• Director • Administrator • Manager • Supervisor

- A **director** leads the affairs of an organization by establishing the vision, mission, and values that determine the direction of the organization. Vision is that part of the imagination that thinks "outside the box" to navigate into new uncharted territory. This can be referred to as "doing the right things."
- An **administrator** runs an organization within the framework of the various policies provided. He or she must efficiently move the organization to achieve its purpose.
- A **manager** takes charge of and oversees the functioning of an activity to achieve a set of goals. The manager's strength comes from the ability to allocate resources to get things done properly. This can be referred to as "doing things right."
- A **supervisor** oversees the activities of others to help them accomplish specific tasks or perform specific activities as efficiently as possible.

IV. MANAGEMENT SKILLS

Change, which is a given in today's health care industry, is difficult for everyone. The process of change cannot be forced but needs to be nurtured to guarantee success. The manager should have an arsenal of skills at his or her disposal in order to effectively implement change.

Successful management includes the creation of well-planned systems and processes that produce an efficient operation. Most encountered problems arise from imperfect or outdated processes. ▶ **Quality improvement** involves bettering troublesome processes to prevent future problems and stopping the "fire-fighting" management mentality. Teamwork involving staff at all levels is important to solve problems and implement solutions that are the integral aspects of quality improvement.

Appropriate project management skills achieve success by organizing and completing delegated tasks through effective use of time and resources. Projects are divided into manageable sections with established objectives, responsibilities, and timelines. Regular updates ensure that the project is proceeding as designed.

One of a manager's duties involves attending meetings. A properly conducted session is an invaluable tool for successful business. If done poorly, the cost, time, and energy of committee members is wasted. To be effective, each meeting has a designated chair to facilitate the efficiency of the session. An agenda is prepared and distributed. Agenda items are discussed and decisions are made. Meeting minutes are prepared afterward to outline the proceedings and detail decisions, assignments, and timelines.

A manager is often required or requested to serve on committees. The purpose of the committee is developed with all members in agreement. From this purpose, objectives are created with timelines and work plans and responsibilities are assigned.

Problem-solving skills are vital if a manager is to successfully maneuver the intricacies of the laboratory. The laboratorians utilize the manager as a technical resource. Clients expect the manager to resolve service issues satisfactorily. Logical thinking, critical analysis of the problem, a clear understanding of technical issues, and the development of creative solutions are steps involved in the problem-solving process.

Managers must continually make decisions concerning the operation of the laboratory. The ability to collect and assimilate pertinent information to formulate timely decisions is the key to success. They must be able to put personal feelings aside and deal objectively with all personnel issues.

Interpersonal skills are essential! A manager must respect all coworkers and strive for mutual understanding.

Steps in the Problem-Solving Process

- Logical thinking
- Critical analysis of the problem
- Clear understanding of technical issues
- Development of creative solutions

V. FINANCIAL MANAGEMENT

▶ The primary goal of **financial management** is to assure the profitability and survival of the organization. The bottom line or **prof-** ▶ **itability** has a direct relationship to this survival. The entire management team must understand basic accounting, economic, and financial principles and apply them to daily laboratory operations to maintain and achieve success.

A. ECONOMICS OF HEALTH CARE

The economic law of supply and demand states that an equilibrium price will be established where the amount of the product produced equals the desire for consumption. However, while patients consume laboratory services, it is the caregivers that utilize the services and the insurers that pay for them. In contradiction to this law, insurers such as Blue Cross and Blue Shield, Medicare and Medicaid, and other third-party payers reimburse providers a given fee for the services rendered to their members without regard for the demand for those services. Further disruption of the law is caused by such things as improvements in technology that dramatically increase test costs and prospective payments (diagnostic-related groups) while managed care (health maintenance organizations [HMOs], preferred provider organizations [PPOs]) decreases reimbursement.

The laboratory manager must analyze the external fiscal environment and its impact on the laboratory operations, including such issues as the laboratory's market share, its product line, and the relationship of the laboratory's fiscal status to that of the parent organization. In today's environment of managed care and governmental regulations, the relationships of costs, volume, and revenue are complex but remain paramount in the financial decision-making processes of the laboratory manager.

B. Cost Accounting

Cost accounting is a system for providing a forecast or analysis of ◀ information to be used in financial decision making. In today's managed care environment and with industry competitiveness, correct cost information ensures participation in those contracts, guaranteeing profitability for the laboratory. The total costs associated with services must be identified and responsibility for those costs assigned.

Total costs consist of three component costs :

1. **Fixed costs** are expenses that do not change over a ◀ given period of time regardless of testing volume; an example of fixed costs is instrument leases (Figure 20–2).
2. **Semivariable costs** are expenses that will change in in- ◀ crements based on workload volume change; an example of a semivariable cost is technologist salary.
3. **Variable costs** are expenses that are attributable to work- ◀ load volume and change in a direct relationship to that volume. Reagents and supplies required to perform a specific test are variable costs (Figure 20–2).

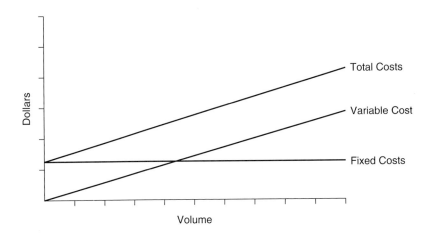

Figure 20–2. Laboratory cost behavior. (Total cost here assumes no semivariable component.)

Identifying and allocating the various costs to the services being rendered will yield the **cost per test.** An example cost per test analysis is presented (Table 20–1) for the fictitious test, serum rhubarb, performed by radioimmunoassay (RIA). The total number of patients per run is 13, which, when analyzed in duplicate, is 26 tubes. Two patients per run require dilution, adding four additional tubes. The three controls per run are analyzed in duplicate to yield six control tubes. The seven calibrators in duplicate require 14 tubes. Fifty tubes (26 + 4 + 6 + 14) are needed per run, which equates to 200 tubes per month. The one failed run per month increases the total to 250 tubes (200 + 50) per month. The $500 in reagents per month divided by the number of tubes per month (250 tubes) yields the reagent cost per tube. The analyst time is 130 minutes per run with 13 patients per run, so the time per patient specimen is 10 minutes. The rate for 10 minutes of analyst time at $18 per hour is $3 per patient specimen. The consumable cost calculated per tube is $25 / 250 tubes = $0.10. The total cost per patient is the total monthly reagent cost ($500 + $25) divided by the number of patients per month plus the analyst cost per patient (525 / 52 = $10.10 + $3 = $13.10).

This cost/test information is extremely useful for making pricing decisions for choosing between alternative contracts to offer in response to managed care or group purchasing organization bids and deciding about send-out testing.

Cost information is necessary to perform a break-even analysis. The **break-even analysis** is an analysis of dollars versus the volume of tests produced. The *break-even point* is where revenue matches the expense to produce the services. Figure 20–3 depicts a graph of total costs versus total revenues. The break-even point is identified by the intersection of the total cost line and the revenue line.

By utilizing this information and performing a **sensitivity analysis,** the projection of costs and revenues given a change in volume and pricing, the necessary information to make decisions on potential service contracts with insurers or managed care organizations is available (Table 20–2). Take a hypothetical situation in which a proposed contract will increase the current volume of testing to 1,500. In this example, the manager will project anticipated revenue ($1,000). A decision will then be made based on this information to

Table 20–1. Cost per Test Analysis

PTS/MO	RUNS/MO	PTS/RUN	REPS	CNTRS/RUN	REPS	CALS/RUN
52	4	13	2	3	2	7

REPS	PT DILNS/MO	FAILED RUNS/MO	RGT $/MO	TUBES/MO	RGT $/TUBE	RGT $/PT
2	8	1	500	250	2.00	9.62

ANALYST TIME /RUN	ANALYST TIME /PT	HOURLY SALARY AND BENEFITS	ANALYST $/PT	CONSUMABLE $/MO	CONSUMBALBE $/TUBE	CONSUMABLE $/PT
130 min	10 min	$18.00/hr	3.00	25	0.10	0.48

TOTAL $/PT
13.10

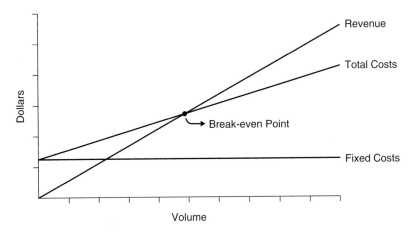

Figure 20–3. Break-even point between revenue and total costs.

determine if the proposed contract is a wise choice to make in keeping consistent with the goals of the organization.

Another method of cost allocation involves assigning **direct costs**—costs that can be traced directly to a specific procedure or process that generates revenue. Many procedures, processes, or departments can share **indirect costs,** such as utilities. **Unit costs** are the sum of both direct and indirect costs and are representative of the cost of producing one reportable test. Unit costs can be used to measure productivity and to demonstrate the laboratory's economy of scale.

C. FINANCIAL ACCOUNTING

Financial accounting is a system for providing financial information and serves as a means to document the financial history of an organization. The statement:

$$\text{Assets} = \text{Liabilities} + \text{Owners' Equity}$$

is the framework of all financial accounting. The primary unit of measurement for accounting information is dollars. *Assets* are the resources owned by the organization. Assets can be designated as tangible assets that have physical form and a value, such as a building, or intangible assets that have no physical form and are difficult to value, such as customer base. *Liabilities* are the obligations that an organization has to its owners. *Owners' equity* represents the value of the organization to its owners whether they are individuals, partners, or shareholders. This equity is termed *retained earnings* in the case of nonprofit organizations.

Table 20–2. Sensitivity Analysis

COMPONENT	CURRENT	PROJECTED
Volume	1,250	1,500
Revenue @ $10/test	$12,500	$15,000
Variable cost @ $6/test	$7,500	$9,000
Fixed cost	$5,000	$5,000
Net revenue	$0	$1,000

▶   **Financial statements** are a system to organize the financial information of the organization. The most frequently used statements are:

▶   • **Balance sheet**—the financial position of an organization at a particular point in time, usually at the end of an accounting period
▶   • *Income statement*—a document indicating what has transpired financially within the organization over a period of time, usually monthly or yearly
▶   • *Statement of cash flows*—an analysis that indicates the source of cash resources and their utilization

D. BUDGETING

Budgeting is a system for formalizing in writing a quantitative financial plan, supportive of the institution's financial mission, for a given time period. The budget planning process begins with establishing goals and objectives and prioritizing the use of financial resources.

▶ **1. Capital Budget.** The **capital budget** is a system for formalizing in writing a quantitative financial plan for a given period of time for those capital projects that require significant financial resources. Examples of capital items include building projects, new service offerings, new instrumentation, or additional equipment. The local governing boards and the institution may require a *certificate of need* (CON), an approval process for capital projects, for those projects with capital outlay above a specified level. The CON is prepared by the institution along with budget projections and provides the information for a governmental authority to approve or deny the proposed project.

▶ **2. Capital Project Funding.** **Capital acquisitions** require a plan for the significant cash outlay required for purchasing major equipment, proceeding with building projects, or expanding services. Managers provide a written justification to request the funds for a capital project; this justification will accompany the institution's CON documentation.

The leadership of the laboratory that approves capital projects must prioritize capital requests. Both the time frame of acquisition and the relationship to other projects are decision factors employed by the institution's financial officers in the process of prioritizing. *Opportunity costs,* the cost or value of what is given up to pursue an alternative project, is another component of the decision process.

Financial analyses provide the quantitative rationale for requesting funding for capital projects and are used to evaluate the profitability associated with these acquisitions. Analytical tools useful in determining the correct decisions concerning capital expenditures include:

• *Payback analysis,* which identifies the length of time it will take to recover the initial capital expenditure
• *Payback period* = P/I, where P is the purchase price, and I is the annual income generated
• *Rate of return,* the ratio of the annual income to the price of the project. The rate of return = I/P, where I is the annual income generated and P is the price of the project
• *Net present value,* a calculation used to indicate the current value of income generated in the future discounted at an esti-

mated percentage equal to the inflation rate over the years of income generation

Options to fund capital requests include purchase, lease, and reagent rental. The outright purchase option utilizes monies from allocations in the capital budget; thus, the current assets or cash of the institution are reduced by the purchase cost. Lease options are expensed from either the capital or operating budget, depending on how the lease contract was written, with the cost of the project being spread over the length of the lease agreement. Reagent rental methods allow manufacturers to provide equipment to institutions with the instrument cost being offset through the purchase of supplies for daily operation and testing. Reagent rental expenses are included in the operating budget as day-to-day expenses.

3. Operating Budget. The **operating budget** is a financial plan ◄ for allocating funds to cover the day-to-day expenditures, such as supplies, salaries, and overhead, involved in the operation of an organization.

Budget preparation is performed in one of two ways:

1. *Flexible budgeting*—budget forecasting based on a known variability of workload volume
2. *Zero-based budgeting*—budget forecasting based on a prioritization of goals and objectives

A successful budget blends past expenditure history with projections of expected expenses for a future given time period. Comparing the actual expenses with the projected budgeted expenses will provide important information to the laboratory manager and help him or her to direct the allocation of and control over the available financial resources for a given period of time.

Typically, personnel are the most expensive resource. Wage and salary administration accounts for approximately 50% of operating budget funds; an additional 25 to 30% of the salary or hourly wages is budgeted for employee benefits. Operating expenses account for the remainder of the budget and include items such as reagents, supplies, utilities, and instrument repair costs (Table 20–3).

E. MATERIALS MANAGEMENT

Materials management is the systematic process of managing and ◄ controlling the acquisition and use of supplies (inventory) to ensure both availability and cost effectiveness. Factors associated with inventory management include:

- *Annual usage*—where both historical and anticipated usage are analyzed to estimate future use
- *Daily usage*—the annual usage analysis divided by the days of operation
- *Cost of ordering*—an expense incurred with each purchase that is in addition to the item's price
- *Lead time*—the time between placing an order and the arrival of the items
- *Cost per unit*—a function of ordered volume and contract pricing
- *Inventory holding cost*—the space and personnel expense to maintain the inventory

Table 20–3. Budget Report (in Thousands)

	APRIL BUDGET	APRIL ACTUAL	VARIANCE	% OF TOTAL
Salaries and Wages				
Salaries (technical)				
Regular	25,000	26,686	1,686	
Wages (nontechnical)				
Regular	6,600	6,700	100	
Overtime	1,700	2,602	902	
Subtotal	33,300	35,988	2,688	
Benefits				
Retirement (9.6%)	3,197	3,455	258	
Health (18.4%)	6,127	6,622	495	
Total	9,324	10,077	753	
Subtotal SWB*	42,624	46,065	4,193	54
Operating Expenses				
Reagents	19,900	20,756	856	
Supplies	14,900	16,575	1,675	
Maintenance	800	725	–75	
Repairs	200	225	25	
Equipment lease	225	225	0	
Utilities	400	358	–42	
Total Operating Expenses	36,425	38,864	2,439	46
Total Expenses	79,049	84,929	6,632	100

* SWB = Salary, wages, and benefits

F. INVENTORY MANAGEMENT

▶ **Inventory management** is the process of continually checking, rotating, and ordering supplies to ensure adequate levels. Inventory is the second most costly expense in the laboratory and a thorough and organized inventory management process yields cost savings. An efficient process addresses these questions:

- When should the order be placed? (*Reorder point*)
- How much should be ordered? (*Ordering volume*)

The level of inventory is monitored by:

- *Periodic inventory review*—where supplies are ordered at the time inventory is checked. The laboratory or departmental staff performs weekly or monthly inventory checks.
- *Perpetual inventory review*—where items are removed from a checklist at the actual time of usage. This check is a duty of the stock room staff.
- *Random inventory review*—verifies the perpetual inventory. Random reviews are typically performed on a yearly basis to account financially for the inventory volume on hand.

The ordering volume and the reorder point can be determined by analyzing the annual usage, average daily usage, cost of ordering, cost of the items, lead time, and inventory cost. Reorder points are based on volume of in-stock items or time since last order. *Minimum/maximum,* a volume reorder technique, sets a minimum level of inventory that will generate an order and a maximum level of inventory that cannot be exceeded at any one time. The *just-in-time* inventory management technique commits a vendor or supplier to deliver items as they are needed. With the just-in-time system, inventory financial resources are kept to a minimum and costs associated with verifying and holding inventory are minimized. Cost efficiency is often realized by ordering as much as possible from a single vendor to obtain volume discounts.

VI. TECHNOLOGY MANAGEMENT

A. INFORMATION SYSTEMS

With an estimated 70% of diagnoses involving results of clinical laboratory diagnostic testing, the information flow in the laboratory is extensive. Correct patient information and testing results must be delivered to the correct location/clinician in a medically acceptable time frame. The laboratory information system (LIS) is an electronic means of organizing and incorporating every aspect of the laboratory's operations into a unified network for all users. Users will consist of both laboratory personnel and the customers utilizing laboratory services. The clinical LIS not only processes and reports patient information, but also monitors quality control and quality assurance surveillance tasks, provides management support functions, and performs mathematical calculations of patient data (see Chapter 18).

Most of the difficulties with an LIS stem from the lack of effort devoted to the initial design; planning for computerization is vital. The effectiveness of computerization begins with establishing goals based on the issues presented in Table 20–4.

The LIS should have interface capabilities with other systems in the institution, for instance, coordination with the hospital information system for compatibility of billing, patient management systems, and client servers. Software applications to consider include admissions, discharge, and transfer (ADT) functions; test order entry; work list building (e.g., phlebotomy draw lists, manual data entry, instrument load lists); online result entry and review; online result verification and reporting; specimen status reports; workload recording; quality control data collection and statistical analysis; patient and medical record result reports; and billing information.

Many LISs now provide PC-based applications that allow managers to manipulate and analyze LIS data and to produce a variety of

---

Software Applications for LIS

- Admissions, Discharge, and Transfer (ADT) functions
- Test order entry
- Work list building
- Online result entry and review
- Online result verification and reporting
- Specimen status reports
- Workload recording
- Quality control data collection and stastical analysis
- Patient and medical record result reports
- Billing information

Table 20–4. Planning for Computerization

1. Assess the needs of the laboratory and its clients. Collection of data and the analysis of existing needs and operations should be performed. Projection of future needs is important for outlining the specifications of the LIS and determining if it will be compatible with emerging technologies.
2. Define both hardware and software specifications and ensure they are compatible with one another to avoid system crashes.
3. Research LIS vendors to identify a match with the identified goals and to determine hardware and software options.
4. Prepare an extensive request for proposal (RFP) and submit to select vendors. The manufacturer's RFP response should include complete information on its LIS.
5. Require a presentation in the laboratory or institution for as many of the potential users as possible.

managerial reports. Word processing, quality assurance, utilization reports, budgeting, and decision-making supportive functions are all computer tools for the laboratory manager.

Implementation of a new LIS involves a logical organization of tasks, building files, planning for contingencies, defining critical paths, setting schedules, estimating resources, training users, and managing the start-up.

Clinical validation of the LIS addresses regulatory, contractual, and practical requirements. Documentation of the proper operation of the software satisfies regulatory agencies. The LIS validation process establishes accountability if the LIS does not meet agreed-upon specifications.

B. LABORATORY DESIGN

The physical design of laboratories is evolving to provide maximum efficiency. Centralized laboratories maintain high-volume instrumentation and provide economies of scale, while decentralized or point-of-care testing (POCT) laboratories provide more immediate testing at the patient site. In the design or redesign process, factors to consider are presented in Table 20–5.

C. ANALYTICAL TESTING

An effective laboratory manager is adept at not only managing financial and human resources, but also the technical/analytical aspects of testing. A thorough comprehension of each test provides the manager with the knowledge to serve as a credible resource for staff on matters of troubleshooting and for clients on matters of patient care. Additionally, the manager must keep abreast of the latest technological improvements to remain competitive and to provide continually improving quality service. A manager must review certain aspects related to testing, including quality control, method evaluations, and procedure manuals to verify appropriate documentation.

The clinical laboratory-testing field is subject to local, state, and federal regulations. A manager must be current with all applicable

Table 20–5. Factors for Consideration in the Design of Clinical Laboratory Testing Space

1. Production areas need sufficient space for personnel, equipment, and supplies to perform the testing.
2. Storage areas require adequate space and proximity to production areas.
3. Adequate space for staff is required as well as appropriate areas designated as clean for eating, lockers, and rest rooms.
4. Utilities, including special electric requirements, appropriate air handling, plumbing, and lighting, are addressed in laboratory designs.
5. Safety and ergonomic considerations are dictated by regulatory agencies. Appropriate safety equipment, hazardous material accommodations, and work areas designed for safe and efficient operation allow for appropriate and healthy access.
6. Traffic flow—the movement of equipment, supplies, specimens, patients, and personnel—must be optimized.
7. Interdepartmental relationships are considered in laboratory design for efficient interfacing between areas.

laws and ensure technological compliance (as discussed in Chapter 19).

A manager is responsible for assessing the technical competency of the testing personnel. Competency assessment is performed initially at the end of the training period, whenever procedures or policies change, and annually thereafter.

VII. HUMAN RESOURCE MANAGEMENT

Management has been defined as the process of getting things done through people. Staffing, the most expensive resource a manager has to allocate, greatly affects the organization's profitability. Ideally, managers should hire their own staff and thus build their own team. Realistically, however, managers must work with existing "inherited" staff. The manager's personality, temperament, and leadership style determine how well he or she interacts with employees. Analysis of the workload to determine the number and function of employees dictates the composition of the team. *Scheduling* is the process of integrating personnel with the expected workload to efficiently perform the duties required for testing. The manager must delegate responsibility and check the effectiveness of the delegation. Delegation of responsibilities is one of the most important and the most difficult skills a manager must master. When hiring, the manager matches the applicant's skills and competencies with the job requirements, the summary of all the factors necessary for an individual to perform successfully, including knowledge, technical abilities, judgement, attendance, and experience. Checking applicants' references and abiding by all applicable legal and regulatory requirements associated with the hiring process avoids potentially costly hiring mistakes.

A. *MOTIVATIONAL THEORIES*

The manager is responsible for motivating employees for optimal performance. Motivation must come from within each employee, but the manager's leadership style serves to stimulate employees. Since each person's motivators are unique, the manager must learn to individualize the provided stimulus. Motivational theories provide insight into individual performance. The most frequently cited motivational theories are Maslow's Hierarchy of Needs, Herzberg's Two-Factor Theory, and Vroom's Expectancy Theory.

Maslow's Hierarchy of Needs states there are five levels of individual needs:

1. Physiological needs, such as clothing and food
2. Safety needs, such as good health and protection from harm
3. Social needs of acceptance and giving
4. The need for esteem or recognition of self and a sense of worthiness
5. The need for self-actualization or self-fulfillment

Maslow's theory states that these needs are satisfied from lowest to highest order and that behavior is determined by the lowest unfulfilled need.

Herzberg's Two-Factor Theory, based on Maslow's ideas, stratifies motivational factors into two groups: satisfiers and dissatisfiers. Satisfiers or motivators are those factors that produce good feelings while the dissatisfiers are those factors that produce dissatisfaction with one's job. Where Maslow focused on the needs of the psychological person, Herzberg focuses on how the job conditions, usually those relating the Maslow's higher level needs, affect employee performance. Satisfaction and dissatisfaction are not mutually exclusive when used in the context of job satisfaction. The absence of motivational factors in one's job does not necessarily mean the employee is dissatisfied.

Vroom's Expectancy Theory, which further expands on the concepts of Maslow and Herzberg, defines motivation as the force or process in which a behavior facilitates the attainment of an award or outcome, and that an individual can make a choice between alternative behaviors.

Regardless of what motivational theory a manager subscribes to, its application in the workplace determines the success. The leadership and management of an organization are ultimately responsible for motivating employees. Those organizations that strive to make the actual job performance rewarding will achieve the greatest motivation from their employees.

B. LEADERSHIP STYLE

All leaders are managers, but not all managers are leaders. The laboratory manager plays a critical role in motivating employees and communicating to them the organization's mission. The effectiveness ▶ of a manager is determined by his or her **leadership style.** The Myers–Briggs and the Keirsey Temperament Sorter are often used to identify personality characteristics. Effective leaders use this information to choose the most effective management techniques to complement their personality and the group dynamics within their organization. The manager/leader is responsible for directing his or her employees to accomplish specific goals and objectives.

C. COMMUNICATION

The communication process, both oral and written, is an essential management tool. Organizational and interpersonal communication processes are means of exchanging information between individuals within the organization. Organizational communication processes are the formal communication systems established within the firm to exchange information and relay institutional policies to all employees. Interpersonal communication processes are more informal and occur between two individuals or within small groups. Both forms of communication are necessary for a manager to inform and manage employees. Without effective communication, employee performance will suffer.

▶ The **performance appraisal** is the formal process used by managers to communicate and provide feedback to employees about their job performance. Typically, the performance appraisal is completed at the end of the initial probationary training period and annually thereafter. Criteria for successful achievement are based on the job requirements stated in the employee's job description and a standard of acceptable performance. When expectations are not met, the issues affecting performance need to be addressed through counseling and corrective actions, including remedial training, to improve performance to acceptable standards.

Employee performance must equal or exceed the established performance standards to ensure quality results. Employees not achieving these standards receive disciplinary counseling by the manager. Disciplinary action is typically a progressive process (unless the offense is egregious) that is performed in coordination with human resources, the employee assistance program, and other available resources. During the counseling session, the manager clearly states the deficiencies, communicates expected performance, and outlines a remedial plan. Once the discipline process has begun, the manager must provide performance feedback to the employee. Employees repeatedly failing to achieve performance standards or not abiding by the organizational policies and procedures are dismissed. Successful termination involves the interplay between the manager and the human resources representative to ensure the rights of the employee and the organization are recognized and protected.

Employee education and training begin at the time of hiring and continue throughout employment. New employees receive orientation, where certain mandated training, such as safety, is presented and documented. Also, new employees must be trained concerning the technical aspects of their job with the effectiveness of the training being assessed and documented. Manuals serve as training tools and references. A **policy manual** is a collection of informative documents that outline the administrative and operational policies of the organization. A **technical procedure manual** is a collection of step-by-step instructions for a specific task or assay. These manuals are written according to the guidelines established by the National Committee on Clinical Laboratory Standards (NCCLS). Continuing education (CE) provides training to professional staff to ensure their competency. Continuing education can be provided internally by the institution or externally by laboratory industry vendors, professional organizations and societies, and colleges and universities.

VIII. CLIENT RELATIONS/MARKETING MANAGEMENT

Marketing is vital if laboratory management wishes to remain competitive within the laboratory market. Peter Drucker professes that "Marketing is so basic that it cannot be considered a separate function . . . it is the whole business seen from the point of view of its final result, that is, from the customer's point of view." Marketing has evolved from the mere selling and distribution of products and services to a comprehensive philosophy describing the dynamic relationship between customers and vendors. To be successful in the health care industry, the laboratory must be client focused and develop and implement services to meet the client's needs. The ultimate goal of the marketing plan is not only to meet the expectations of the client but to "dazzle" them. As with any business, a satisfied customer is a customer who will return. Ultimately, the laboratory must provide the client with the quality results and services desired. Quality is best assured if the laboratory can control as many service facets as possible, including specimen collection, specimen transportation, testing, and timely results reporting. The most efficient means today of sending results to the clinical practitioners is via electronic communication, including fax or modem transmissions. As Web security advances, information will be available to all authorized individuals. Timely resolution of complaints and responsiveness to requests also influences the clients' perception of laboratory quality.

Professional communication with clients, including phone etiquette, also enhances customer satisfaction.

IX. SUMMARY

The laboratory manager must be prepared to face the challenges presented by the health care industry. Economic forces, accreditations, and regulations have radically changed the process of clinical laboratory management. The manager must become adept at financial, human resource, and technological utilization for success in this rapidly changing industry.

SUGGESTED READINGS

Bickford GR. The vendor/laboratory manager relationship: Some practical negotiation tips. *Clin Lab Manage Rev* 7(4): 328–34, 1993.

Camp RR, Hermon MV. The employment interview: Avoiding the traps with effective strategies. *Clin Lab Manage Rev* 7(6): 500–51, 1993.

Cembrowski GS, Carey RN. *Laboratory quality management.* Chicago: ASCP Press, 1989.

Covey S. *Seven habits of highly effective people.* New York: Simon & Schuster, 1989.

Dadoum R. Impact on human resources: Core laboratory versus laboratory information system versus modular robotics. *Clin Lab Manage Rev* 12(4): 248–55, 1998.

DePree M. *Leadership is an art.* New York: Bantam-Doubleday-Dell, 1989.

Drucker P. Management: tasks, responsibilities, practices. New York, NY: Harper Business, 1985.

Finkler SA. *Finance and accounting for non-financial managers.* Englewood Cliffs, NJ: Prentice Hall, 1992.

Friedman BA. Integrating laboratory processes into clinical processes, Web-based laboratory reporting, and the emergence of the virtual clinical laboratory. *Clin Lab Manage Rev* 12(5): 333–38, 1998.

Ghorpade J, Chen MM. Appraising the performance of medical technologists in a clinical laboratory. *Clin Lab Manage Rev* 11(2): 132–41, 1997.

Goleman D. What makes a leader? *Harvard Business Review* November–December: 93–102, 1998.

Holland CA. Reengineering the mind, career, system, department. *Clin Lab Manage Rev* 12(3): 169–75, 1998.

Marchwinski J, Coggins F. Marketing skills for hospital-based laboratory managers in a managed care environment. *Clin Lab Manage Rev* 11(5): 296–300, 1997.

Mortland KK. Facility redesign for your future laboratory requirements. *Clin Lab Manage Rev* 11(3): 145–52, 1997.

National Committee for Clinical Laboratory Standards. *Cost accounting in the clinical laboratory: Tentative guideline.* Villanova, PA: NCCLS, 1994.

National Committee for Clinical Laboratory Standards. *Laboratory design: Proposed guidelines, document GP18-P.* Villanova, PA: NCCLS, 1994.

National Committee for Clinical Laboratory Standards. *Training verification for laboratory personnel, document GP-21P.* Villanova, PA: NCCLS, 1994.

Timm PR. From slogans to strategy: A workable approach to customer satisfaction and retention. *Clin Lab Manage Rev* 11(3): 153–58, 1997.

Travers EM, Wilkinson DS. Developing a budget for the laboratory. *Clin Lab Manage Rev* 11(1): 56–66, 1997.

Weber LJ, Bissell MG. Marketing testing to the public. *Clin Lab Manage Rev* 10(1): 49–51, 1996.

Westgard O. Westgard Quality Corporation. *www.westgard.com* (accessed January 1999).

Wing A. Architectural concerns of lab automation. *Adv Admin Lab* 7(8): 94–100, 1998.

Items for Further Consideration

1. Calculate the cost per test for "serum rhubarb," an esoteric test.

   Calibration frequency (*every run*)   Number of calibrators/run (*5*)

   Calibrator replicates (*duplicate*)   Control replicates (*duplicate*)

   Control frequency (*every run*)   Number of controls/run (*3*)

   Number of runs/month (*13*)   Patient replicates (*duplicate*)

   Frequency of failed runs/month (*1*)   Specimen dilutions (*2/run*)

   Number of specimens/month (*325*)   Reagent wastage/month (*10%*)

   Analyst salary and benefits (*$20.00/hr*)

   Reagent, calibrator, and control costs (*reagents: $250 for 100 test kit; calibrators included in kit; controls: $2/mL with 100 uL needed per tube*)

   Ancillary supply costs (test tubes, pipettes tips, etc.) (*disposables $5/run*)

   Analyst time for the analysis and reporting of one run (*1.5 hr/run*)

2. Prepare a budget for April of next year using the April budget information from Table 20–3 and the following assumptions:

   Salary and wage (↑ 5.3%)   Overtime (↓ 12%)

   Health benefits (17.5%)   Reagents (↑ 2%)

   Supplies (↓ 25%)   Maintenance (↓ 10%)

   Repairs (no change)   Equipment lease (↑ 50%)

   Utilities (↑ 3.9%)

3. Perform a sensitivity analysis (project costs and revenues) using the following information:

Volume (25,000 tests)   Revenue ($6.50/test)

Variable costs ($3.10/test)   Fixed cost ($150,000)

4. Design a laboratory facility addressing all issues listed in Table 20–5 and include the following functional areas in the design.

Patient service area (registration and   Administration, pathology, and
    phlebotomy)   manager offices

Employee facilities   Anatomic pathology

Core lab (chemistry, hematology,   Transfusion medicine
    immunology, and urinalysis)

Microbiology and TB   Molecular diagnostics

5. Write a memo to an employee who is habitually late. This memo should include an action plan, state consequences of noncompliance, provide documentation of the problem, and illustrate good communication techniques.

Answers to Pretest Questions

Part I: Classic Analytes

Chapter 1.	1. (c);	2. (d);	3. (a);	4. (b);	5. (a);	6. (b);	7. (d);	8. (d);
	9. (b);	10. (a);	11. (d);	12. (b);	13. (a);	14. (c);	15. (c);	
	16. (b);	17. (c);	18. (d);	19. (a);	20. (d)			
Chapter 2.	1. (b);	2. (d);	3. (c);	4. (a);	5. (c);	6. (b);	7. (d);	8. (c);
	9. (b);	10. (a);	11. (b);	12. (d);	13. (c);	14. (c);	15. (d);	16. (a);
	17. (c);	18. (b);	19. (c);	20. (a);	21. (c);	22. (d)		
Chapter 3.	1. (b);	2. (a);	3. (a);	4. (d);	5. (b);	6. (d);	7. (b);	8. (c);
	9. (b);	10. (d);	11. (c);	12. (b);	13. (a);	14. (c);	15. (c)	
Chapter 4.	1. (c);	2. (b);	3. (d);	4. (b);	5. (c);	6. (d);	7. (b);	8. (c);
	9. (d);	10. (c);	11. (b);	12. (d);	13. (a);	14. (c);	15. (c);	16. (d);
	17. (c);	18. (b);	19. (d);	20. (c);	21. (d);	22. (c);	23. (b);	24. (d)
Chapter 5.	1. (d);	2. (a);	3. (b);	4. (b);	5. (d);	6. (c);	7. (c);	8. (c);
	9. (c);	10. (c);	11. (a);	12. (b);	13. (b);	14. (c);	15. (d);	16. (b);
	17. (a);	18. (b);	19. (d);	20. (c);	21. (a);	22. (b);	23. (d);	24. (a);
	25. (c)							
Chapter 6.	1. (b);	2. (d);	3. (c);	4. (c);	5. (d);	6. (d);	7. (a);	8. (c);
	9. (c);	10. (c);	11. (d);	12. (a);	13. (d);	14. (b);	15. (c);	16. (a);
	17. (c);	18. (d)						
Chapter 7.	1. (c);	2. (c);	3. (b);	4. (a);	5. (d);	6. (b);	7. (c);	8. (d);
	9. (b);	10. (c);	11. (d);	12. (b);	13. (a);	14. (d);	15. (a);	16. (c);
	17. (b);	18. (b);	19. (c);	20. (b)				
Chapter 8.	1. (d);	2. (a);	3. (c);	4. (c);	5. (b);	6. (a);	7. (b);	8. (a);
	9. (b);	10. (b);	11. (b);	12. (c);	13. (c);	14. (c);	15. (d);	16. (a);
	17. (b);	18. (c);	19. (c);	20. (d)				
Chapter 9.	1. (b);	2. (d);	3. (a);	4. (b);	5. (b);	6. (b);	7. (d);	8. (c);
	9. (b);	10. (c);	11. (d);	12. (c);	13. (b);	14. (a);	15. (d);	16. (b);
	17. (b);	18. (c);	19. (c);	20. (b)				
Chapter 10A.	1. (c);	2. (b);	3. (d);	4. (a);	5. (b);	6. (a);	7. (c);	8. (b);
	9. (c);	10. (b);	11. (c)					
Chapter 10B.	1. (b);	2. (c);	3. (c);	4. (b);	5. (a);	6. (c);	7. (c);	8. (c)
Chapter 10C.	1. (c);	2. (a);	3. (b)					
Chapter 10D.	1. (b);	2. (b);	3. (d);	4. (b);	5. (a)			

Part II: Instrumentation and Analytical Techniques

Chapter 11.	1. (a);	2. (c);	3. (b);	4. (b);	5. (d);	6. (a);	7. (b);	8. (b);
	9. (b);	10. (b);	11. (a);	12. (d);	13. (c);	14. (b);	15. (b);	16. (d);
	17. (a);	18. (a);	19. (d);	20. (c);	21. (d);	22. (b);	23. (d);	24. (a);
	25. (d)							
Chapter 12.	1. (c);	2. (b);	3. (b);	4. (b);	5. (a);	6. (c);	7. (c);	8. (d);
	9. (c);	10. (c);	11. (d);	12. (c);	13. (b);	14. (a);	15. (b)	

Chapter 13. 1. (b);   2. (a);   3. (d);   4. (c);   5. (b);   6. (d);   7. (b);   8. (b);
             9. (a);   10. (c);   11. (c);   12. (d);   13. (c);   14. (c);   15. (d);   16. (d);
            17. (d);   18. (c);   19. (c);   20. (d);   21. (c);   22. (b);   23. (c);   24. (b)
Chapter 14. 1. (c);   2. (a);   3. (a);   4. (b);   5. (a);   6. (a);   7. (b);   8. (d);
             9. (a);   10. (b);   11. (c);   12. (b);   13. (a);   14. (d);   15. (a);   16. (b);
            17. (c);   18. (b);   19. (c);   20. (c)
Chapter 15. 1. (a);   2. (b);   3. (c);   4. (c);   5. (d);   6. (b);   7. (a);   8. (d);
             9. (a);   10. (c);   11. (a);   12. (b);   13. (d)

Part III: Laboratory Operations

Chapter 16. 1. (c);   2. (a);   3. (a);   4. (b);   5. (d);   6. (b);   7. (b);   8. (c);
             9. (b);   10. (c);   11. (b);   12. (b);   13. (d);   14. (c);   15. (a);   16. (b);
            17. (c);   18. (a);   19. (a);   20. (a)
Chapter 17. 1. (b);   2. (d);   3. (c);   4. (b);   5. (d);   6. (b);   7. (a);   8. (b);
             9. (c);   10. (c);   11. (c);   12. (d);   13. (d);   14. (a);   15. (b);   16. (b);
            17. (a);   18. (d);   19. (c);   20. (a)
Chapter 18. 1. (b);   2. (c);   3. (a);   4. (c);   5. (b);   6. (c);   7. (d);   8. (c);
             9. (a);   10. (c);   11. (c);   12. (d);   13. (c);   14. (b);   15. (a);   16. (d)
Chapter 19. 1. (b);   2. (a);   3. (d);   4. (b);   5. (d);   6. (b);   7. (c);   8. (b);
             9. (c);   10. (d);   11. (a);   12. (c);   13. (a);   14. (a)
Chapter 20. 1. (b);   2. (c);   3. (a);   4. (b);   5. (c);   6. (c);   7. (d);   8. (d);
             9. (d);   10. (a);   11. (b);   12. (d);   13. (c);   14. (a);   15. (d);   16. (b);
            17. (d);   18. (c);   19. (a);   20. (d)

Index